Lecture Notes in Computer Science 1272

Edited by G. Goos, J. Hartmanis and J. van Leeuwen

Advisory Board: W. Brauer   D. Gries   J. Stoer

Springer
Berlin
Heidelberg
New York
Barcelona
Budapest
Hong Kong
London
Milan
Paris
Santa Clara
Singapore
Tokyo

Frank Dehne   Andrew Rau-Chaplin
Jörg-Rüdiger Sack   Roberto Tamassia (Eds.)

Algorithms and Data Structures

5th International Workshop, WADS'97
Halifax, Nova Scotia, Canada
August 6-8, 1997
Proceedings

 Springer

Series Editors

Gerhard Goos, Karlsruhe University, Germany

Juris Hartmanis, Cornell University, NY, USA

Jan van Leeuwen, Utrecht University, The Netherlands

Volume Editors

Frank Dehne
Jörg-Rüdiger Sack
Carleton University, School of Computer Science
1125 Colonel By Drive, Ottawa, Canada K1S 5B6
E-mail: (dehne/sack)@scs.carleton.ca

Andrew Rau-Chaplin
Technical University of Nova Scotia, Department of Computer Science
P.O. Box 1000, Halifax, Nova Scotia, Canada B3J 2X4
E-mail: arc@tuns.ca

Roberto Tamassia
Brown University, Center for Geometric Computing
Providence, RI 02912-1910, USA
E-mail: rt@cs.brown.edu

Cataloging-in-Publication data applied for

Die Deutsche Bibliothek - CIP-Einheitsaufnahme

Algorithms and data structures : 5th international workshop ;
proceedings / WADS '97, Halifax, Nova Scotia, Canada, August 6 -
8, 1997. Frank Dehne ... (ed.). - Berlin ; Heidelberg ; New York ;
Barcelona ; Budapest ; Hong Kong ; London ; Milan ; Paris ; Santa
Clara ; Singapore ; Tokyo : Springer, 1997
  (Lecture notes in computer science ; Vol. 1272)
  ISBN 3-540-63307-3

CR Subject Classification (1991): F.1-2, E.1, G.2, I.3.5, H.3.3

ISSN 0302-9743
ISBN 3-540-63307-3 Springer-Verlag Berlin Heidelberg New York

© Springer-Verlag Berlin Heidelberg 1997
Printed in Germany

Typesetting: Camera-ready by author
SPIN 10548848   06/3142 – 5 4 3 2 1 0   Printed on acid-free paper

PREFACE

The papers in this volume were presented at the 1997 Workshop on Algorithms and Data Structures (WADS '97). The workshop took place from August 6 to 8, 1997, at TUNS/Dalhousie University in Halifax, Nova Scotia and was sponsored
Carleton University and TUNS/Dalhousie University.

In response to the program committee's call for papers, 8 1 papers where submitted. From these submissions, the program committee selected 37 papers for presentation at the workshop. In addition to these papers, the workshop included invited presentations Bernard Chazelle, David Dobkin, S. Rao Kosaraju, Ketan Mulmuley, and Christos Papadimitriou.

F. Dehne, A. Rau-Chaplin, J.-R. Sack and R. Tamassia
August, 1997

WADS Steering Committee
F. Dehne (Carleton), I. Munro (Waterloo), J.-R. Sack (Carleton), N. Santoro (Carleton), R. Tamassia (Brown)

WADS'97 Program Committee
Co-chairs: F. Dehne (Carleton), A. Rau-Chaplin (TUNS/Dal.), J.-R. Sack (Carleton), R. Tamassia (Brown).
A. Anderson (Lund), A. Apostolico (Purdue and Padova), G. Ausiello (Rome), C. Bajaj (Purdue), R. Cypher (Johns Hopkins), L. De Floriani (Genova), L. Devroye (McGill), D. Eppstein (Univ. of Calif. at Irvine), M. Farach (Rutgers), A. Ferreira (Ecole Norm. Sup. de Lyon), P. Fraignaud (Ecole Norm. Sup. de Lyon), G. Frederickson (Purdue), M. Goodrich (Johns Hopkins), H. Juergensen (Univ. of Western Ontario), E. Kranakis (Carleton), H.P. Kriegel (Muenchen), D.T. Lee (Northwestern), T. Lengauer (GMD), L. Pagli (Pisa), G. Plaxton (Univ. of Texas at Austin), J. Reif (Duke), N. Santoro (Carleton), P.G. Spirakis (Patras), H. Sudborough (Univ. of Texas at Dallas), P. Vitanyi (CWI), P. Widmayer (ETH), C.K. Wong (Chinese Univ. of Hong Kong).

WADS'97 Conference Chair
A. Rau-Chaplin (TUNS/Dal.)

TABLE OF CONTENTS

Discrepancy Theory and Computational Geometry

BERNARD CHAZELLE

Department of Computer Science
Princeton University
Princeton, NJ 08544, USA
chazelle@cs.princeton.edu

Abstract

The recent development of a theory of computational-geometric sampling has revolutionized the design of geometric algorithms, and led to the solution of some of the most outstanding problems in the field. Much of this development owes to the interplay between computational geometry and discrepancy theory. This talk will discuss some intriguing aspects of this development, including the use of data structuring ideas to prove theorems in discrepancy theory.

In recent years, a number of longstanding questions in computational geometry have been settled by derandomizing optimal probabilistic algorithms. While hardly unique to computational geometry, in no other field has derandomization played such a critical role. For example, the only known (worst-case) optimal solutions for computing convex hulls, Voronoi diagrams, smallest enclosing ellipsoids, etc, in deterministic fashion, have been obtained by starting from a probabilistic algorithm and then removing the use of random bits.

The key to this success has been the development of a sophisticated theory of sampling for (low-dimensional) geometric spaces. Its mathematical foundation is discrepancy theory: roughly speaking, this is the study of how one measure can be used to approximate another one. Typically, the measure to be approximated is complicated and perhaps not even computable, while the approximating measure is simple and computationally tractable. Several deep results from discrepancy theory were instrumental in creating the tools necessary for derandomizing geometric algorithms. The favor has been returned in a few cases, and theorems in discrepancy theory have been proven by using techniques specific to computational geometry. This talk will illustrate both aspects of the interplay between discrepancy theory and computational geometry. For books or surveys that tell fragments of this story, the reader should consult [1], [2], [3], [4], [5], [6].

Acknowledgments

This work was supported in part by NSF Grant CCR-93-01254, NSF Grant CCR-96-23768, and by the US Army Research Office under Grant DAAH04-96-1-0181.

References

[1] Chazelle, B. *Computational geometry: a retrospective*, in "Computing in Euclidean Geometry" (2nd Edition), Eds. D.-Z. Du and F. Hwang, World Scientific Press (1995), 22–46.

[2] Chazelle, B. *Discrepancy and Derandomization*, book in preparation, 1997.

[3] Clarkson, K.L. *Randomized geometric algorithms*, in *Computing in Euclidean Geometry*, D.-Z. Du and F.K. Kwang ed., Lecture Notes Series on Comput. 1 (1992), World Scientific, 117–162.

[4] Matoušek, J. *Derandomization in computational geometry*, J. Algorithms 20 (1996), 545–580.

[5] Matoušek, J. *Geometric Discrepancy*, book in preparation, 1997.

[6] Mulmuley, K. *Computational Geometry: An Introduction Through Randomized Algorithms*, Prentice-Hall, 1994.

Dynamic Motion Planning in Low Obstacle Density Environments [*]

Robert-Paul Berretty   Mark Overmars   A. Frank van der Stappen

Dept. of Computer Science, Utrecht Univ., P.O. Box 80.089, 3508 TB Utrecht, The Netherlands.

Abstract. A fundamental task for an autonomous robot is to plan its own motions. Exact approaches to the solution of this motion planning problem suffer from high worst-case running times. The weak and realistic low obstacle density (L.O.D.) assumption results in linear complexity in the number of obstacles of the free space [11]. In this paper we consider the dynamic version of the motion planning problem in which a robot moves among moving polygonal obstacles. The obstacles move along constant complexity polylines, but respect the low density property at any given time. We will show that in this situation a cell decomposition of the free space of size $O(n^2 \alpha(n) \log^2 n)$ can be computed in $O(n^2 \alpha(n) \log^2 n)$ time. The dynamic motion planning problem is then solved in $O(n^2 \alpha(n) \log^3 n)$ time. We also show that these results are close to optimal.

1   Introduction

Robot motion planning concerns the problem of finding a collision-free path for a robot $\mathcal{B}$ in a workspace W with obstacles $\mathcal{E}$ from an initial placement Z_0 to a final placement Z_1. The parameters required to specify a placement of the robot are referred to as the *degrees of freedom* of the robot. The motion planning problem is often studied as a problem in the configuration space C, which is the set of parametric representations of the placements of the robot $\mathcal{B}$. The free space FP is the sub-space of C of placements for which the robot does not intersect any obstacle in $\mathcal{E}$. A feasible motion for the robot corresponds to a curve from Z_0 to Z_1 in FP (or its closure).

Motion planning is a difficult problem. In general, many instances of the robot motion planning problem have appeared to be P-SPACE-complete, even if the obstacles are stationary [5]. For a constant-complexity robot moving amidst static obstacles polynomial time algorithms have been shown to exist. The running time complexity is exponential in the number of degrees of freedom of the robot [7]. For an f-DOF robot, the complexity of the free space, can be as high as $\Omega(n^f)$ and the motion planning problem will, therefore, in general have a worst case running time close to $\Omega(n^f)$. However, the actual bound is often smaller; the motion planning problem is often output sensitive, and the running time depends on the complexity of the free space FP. Most stationary motion planning algorithms tend to have very high running times, though.

We address the motion planning problem for a robot operating in an environment with moving obstacles. This problem is also referred to as the dynamic motion planning problem. In general, when the obstacles in the workspace are allowed to move, the motion planning problem becomes even more complicated. For example, Reif and

---

[*] Research is partially supported by the Dutch Organization for Scientific Research (N.W.O.).

Sharir [6] showed that, when obstacles in a 3-dimensional workspace are allowed to rotate, the motion planning problem is PSPACE-hard if the velocity modulus is bounded, and NP-hard otherwise. (A similar result was obtained by Sutner and Maass [8].) Canny and Reif [2] showed that dynamic motion planning for a point in the plane, with a bounded velocity modulus and an arbitrary number of convex polygonal obstacles, is NP-hard, even when the obstacles are convex and translate at constant linear velocities. They also showed that the 2-dimensional dynamic motion planning problem for a translating robot B with bounded velocity modulus, among polygonal obstacles $\mathcal{E}$ that translate at fixed linear velocity, can be solved using an algorithm that is polygonal in the total number of vertices of B and $\mathcal{E}$, if the number of obstacles is bounded. However, their algorithm takes exponential time in the number of moving obstacles.

Van der Stappen and Overmars [11] (see also [10]) showed that modelling robots in realistic workspaces has a profound influence on the complexity of solving the static motion planning problem, mainly independent of the number of degrees of freedom of the robot. They gave a description of environments with a so-called *low obstacle density* which leads to a surprising gain in efficiency for several instances of the motion planning problem. An environment has the low obstacle density property if any region in the workspace intersects a constant number of obstacles that are larger than the size of the region. (See below for a more precise definition.) Under the low obstacle density assumption, the exact motion planning problem for an f-DOF robot was efficiently solved, using the cell decomposition approach. An important property for an f-DOF robot B in a low obstacle density environment is the *linear* combinatorial complexity of the free space. For a robot B moving amidst n stationary obstacles in 2-space the cell decomposition of the free space is computable in $O(n \log n)$ time; in the case of obstacles in 3-space, the decomposition can be computed in $O(n \log^2 n)$ time [12]. Vleugels [12] extended these results to multiple robots simultaneously operating in the same workspace.

We demonstrate that the low obstacle density property can also be used to plan a motion for a robot B with f degrees of freedom moving in a 2-dimensional workspace with non-stationary obstacles. The obstacles are allowed to translate in the workspace along polyline trajectories, with a fixed speed per segment. The motion planning problem is then solved in $O(n^2 \alpha(n) \log^3 n)$ time, using $O(n^2 \alpha(n) \log^2 n)$-sized cell decomposition. Note that these bounds do not depend on f (assuming f is constant). We also show that this result is close to optimal, by giving an example where the robot has to perform $\Omega(n^2)$ simple motions to get from its start to its goal position.

2 Low Obstacle Density

In this section we recall some of the definitions and results from the paper by Van der Stappen and Overmars [11] on motion planning in low density environments. The authors focus in particular on the large class of motion planning problems with configuration spaces of the form $C = W \times D$, where W is the d-dimensional workspace and D is some $(f - d)$-dimensional rest space. Let us use the *reach* of a robot as a measure for its maximum size; the reach ρ_B of B is defined as the maximum radius that the minimal robot enclosing hypersphere, centered at its reference point, can ever have (in any placement of B). The reach of the robot is assumed to be comparable to the size of the smallest obstacle. The robot has constant complexity and moves in a workspace with constant-complexity obstacles. The workspace satisfies the static low obstacle density property which is defined as follows.

Property 1. Let $\mathbf{R}^d$ be a space with a set $\mathcal{E}$ of non intersecting obstacles. Then $\mathbf{R}^d$ is said to be a *static low (obstacle) density* space if there is a constant c, such that for any region $R \subset \mathbf{R}^d$ with minimal enclosing hyper-sphere radius $c \cdot \rho$, the number of obstacles $E \in \mathcal{E}$ with minimal enclosing hyper-sphere radius at least ρ intersecting R is bounded by a constant.

Van der Stappen and Overmars [11] showed that, under the circumstances outlined above, the complexity of the free space is linear in the number of obstacles.

A configuration space contains hyper-surfaces of the form $f_{\phi,\Phi}$, consisting of placements of the robot $\mathcal{B}$ in which a robot feature ϕ is in contact with an obstacle feature Φ. We shall denote the fact that ξ is a feature of some object or object set X by $\xi \in_f X$. The arrangement of all (constant-complexity) constraint hyper-surfaces $f_{\phi,\Phi}(\phi \in_f \mathcal{B}, \phi \in_f \mathcal{E})$ divides the higher-dimensional configuration space into free cells and forbidden cells. Van der Stappen and Overmars [11] considered so-called *cylindrifiable* configuration spaces $C = B \times D$ which have the property that the subspace B — referred to as the *base space* — can be partitioned into regions R satisfying

$$|\{f_{\phi,\Phi}|\phi \in_f B \wedge \Phi \in_f \mathcal{E} \wedge f_{\phi,\Phi} \cap (R \times D) \neq \emptyset\}| = O(1).$$

A partition that satisfies this constraint is called a *cylindrical partition*. In words, the lifting of the region R into configuration space is intersected by a constant number of constraint hyper-surfaces. These hyper-surfaces subdivide the cylinder $R \times D$ into $O(1)$ constant-complexity free and forbidden cells. The cylindrical partition of B therefore almost immediately gives us a cell decomposition of the free portion of C. Theorem 2 states that transformation from cylindrical partition to cell decomposition can be done efficiently.

Theorem 2. *[11] Let V be the set of regions of a cylindrical partition of a base space B and let E be the set of region adjacencies. Let the regions of B be of constant complexity. Then the cell decomposition of the free space calculated by lifting the regions R of the base partition into the configuration space consists of constant complexity subcells. Furthermore, the complexity of the decomposition and the time to compute it is $O(|V| + |E|)$.*

Note that the size of the cylindrical partition determines the size of the cell decomposition. The low obstacle density motion planning problem outlined above was shown to yield a cylindrifiable configuration space, in which the workspace W is a valid base space. Small and efficiently computable cylindrical partitions of W exist. This leads to efficient cell decompositions and thus efficient solutions to the motion planning problem (see [11] for details).

In this paper, we show that the configuration space of the dynamic version of the low obstacle density motion planning problem is cylindrifiable as well. We find a cylindrical partition of an appropriate base space that leads to an almost optimal size cell decomposition.

3 A Dynamic Base Space

3.1 Problem Statement

We now focus on the dynamic robot motion planning problem, subjected to low obstacle density. We show that the framework outlined in Section 2 can be used to plan a motion for a robot B with f degrees of freedom, moving in a 2-dimensional workspace with non-stationary obstacles. The obstacles translate in the workspace, and can only change speed or direction a constant number of times. We will use a cell-decomposition based on a cylindrical partition, similar to Section 2. Since dynamic motion planning is tedious to deal with, we split the problem into sub-problems. We first formally define the problem and state some useful properties of the base space for the dynamic motion planning problem. In Section 4, we construct a cylindrical decomposition, and in Section 5, we compute the actual path for the robot.

The *L.O.D. dynamic motion planning problem* is defined as follows. The workspace W of the robot B is the 2-dimensional Euclidean space $\mathbf{R}^2$ and contains a collection of n polygonal obstacles $E \in \mathcal{E}$, each moving along a polyline at constant speed per line segment. The robot B has constant complexity and its reach is bounded by $\rho_B \leq b \cdot \rho$, where $b \geq 0$ is a constant and ρ is a lower bound on the minimal enclosing hypersphere radii of all obstacles $E \in \mathcal{E}$. Each obstacle $E \in \mathcal{E}$ has constant complexity and is polygonal. Any hyper-surface in the configuration space corresponding to the set of robot placements in which a certain robot feature is in contact with a certain obstacle, is algebraic of bounded degree. The robot is placed at the initial placement Z_0, at time t_0 and has to move along a *time-monotone* path, to the goal placement Z_1, at time t_1. At any time between t_0 and t_1, the workspace with obstacles satisfies the static low obstacle density property.

A standard approach when dealing with moving obstacles, is to augment the stationary configuration space with an extra time dimension T. In this manner, we obtain the configuration-time space. When planning the motion of our robot through the configuration-time space, we have to make sure that the path is time-monotone—the robot is not allowed to move back in time. The first objective in solving the dynamic low obstacle density motion planning problem is to obtain a cylindrical partition that consists of constant complexity regions. An appropriate choice for a partitionable base space B is the Cartesian product of the 2-dimensional workspace and time. This way, the configuration time space is of the form $CT = W \times T \times D = \mathbf{R}^2 \times \mathbf{R} \times D \, (= \mathbf{R}^3 \times D)$, where D is some $(f - 2)$-dimensional rest space.

3.2 Characteristics of the Base Space

The base space $B = W \times T$ can be considered as a 3-dimensional Euclidean space. In our dynamic motion planning setting, we only consider the work-time space slice $\mathbf{R}^2 \times [t_0, t_1]$. We first look at the situation where the obstacles move along a line in the workspace. Later, we extend the result to the polyline case.

Definition 3. Let $S \subseteq W$ and let γ be a curve in $W \times T$. Then the column $col_\gamma(S)$ is defined by $col_\gamma(S) = \{(x, y, 0)|(x, y) \in S\} \oplus \gamma$, where $\oplus$ denotes the Minkowski sum operator.

The column $col_\gamma(S)$ is the volume swept by S as its reference point follows the curve γ. In our application, the curve γ describes the translational motion of an obstacle and is therefore time-monotone. A point (x, y, t) belongs to $col_\gamma(S)$ if and only if S covers the point (x, y) at time t. We use a partition of the base space to construct a cell decomposition of the free space. The base partition is based upon the arrangement in the work-time space of the moving obstacles grown with the reach of the robot. The regions of the base partition are sub-cells of 3-faces of the arrangement defined by the boundaries of the columns of grown obstacles. For cell decomposition purposes, we want the grown obstacles to be polyhedral, so we use the Minkowski sum of E and a square with side length $2\rho_B$ to grow the obstacles.

Definition 4. Let Q_{O,ρ_B} be a square centered at the origin, having side length $2\rho_B$. Then $H(E) = E \oplus Q_{O,\rho_B}$.

The Minkowski sum $H(E)$ encloses E. No point in $H(E)$ has a distance larger than $\sqrt{2} \cdot \rho_B$ to E. We denote the arrangement of the boundaries $\partial col_\gamma(H(E))$ of the grown obstacle columns by $\mathcal{A}(col \circ H)$. We will show that this arrangement is of $O(n^2)$ complexity. We say that an obstacle E is in the proximity of another obstacle E' if $H(E)$ and $H(E')$ intersect at some time, hence $col_\gamma(H(E))$ and $col_{\gamma'}(H(E'))$ intersect.

Theorem 5. *The complexity of the arrangement $\mathcal{A}(col \circ H)$ of the boundaries of the grown obstacle columns is $O(n^2)$.*

Proof. The complexity of the arrangement is determined by the number of vertices. A vertex results from an intersection of three columns. A necessary condition for three columns to intersect is that the corresponding obstacles are less than $2\sqrt{2}\rho_B$ apart at some moment in time. We show that the number of such triples is $O(n^2)$. We charge each such triple to a pair of obstacles. For this we choose the smallest obstacle E of the three and the one (of the remaining two) that last entered E's proximity. Assume that an obstacle E' enters the proximity of E. (Note that E' can enter E's proximity at most $O(1)$ times because E and E' have constant complexity and both move along line paths.) A third obstacle E'' involved in a triple (E, E', E'') must already be in the proximity of E at the time of arrival of E' in order to be charged to the pair E, E'. By Property 1, there are only $O(1)$ larger obstacles in E's proximity at any time, so E'' is chosen from a set of $O(1)$ size. As a result, only $O(1)$ triples are charged to each of the $O(n^2)$ pairs E, E'. Each of these $O(n^2)$ triples (E, E', E'') contribute a constant number of vertices to $\mathcal{A}(col \circ H)$ because the obstacles E, E', and E'' have constant complexity and move along line paths. Therefore, the complexity of $\mathcal{A}(col \circ H)$ is bounded by $O(n^2)$. $\square$

It is easy to see that a 2-face of a column in the final arrangement is divided into a number of parts, of which some are non-convex. The following theorem states that the 2-faces of the arrangement $\mathcal{A}(col \circ H)$ are polygons without holes. This property turns out to be important in the sequel.

Theorem 6. *$\mathcal{A}(col \circ H)$ has only polygonal faces without holes.*

Proof. The proof is not included in this version. It uses the fact that the grown obstacles columns are Minkowski sums of a square with a set of non-intersecting columns, and that all columns run from $t = t_0$ to $t = t_1$. A complete proof is provided in the full version of the paper. □

If we extend the setting to the case in which obstacles $E \in \mathcal{E}$ translate along polylines, the complexity of the arrangement $\mathcal{A}(col \circ H)$ does not increase asymptotically—in the proof of Theorem 5, the chargings to obstacle E'' are, in the worst case, multiplied by a constant factor. Unfortunately, the 2-faces of $\mathcal{A}(col \circ H)$ are no longer polygons without holes. We can resolve this by adding extra faces to the arrangement. For every time t_i at which one of the obstacles changes speed, we add a plane $t = t_i$. This way, the area between two successive planes is a work-time space slice where all obstacles move in a fixed direction with a fixed speed. The arrangements on the newly introduced planes are cross sections of the work-time space. They are arrangements of possibly intersecting grown obstacle boundaries in a static low obstacle density workspace and have linear complexity [11]. We compute a triangulation of these 2-dimensional arrangements to assure that their faces have no holes. Since we have $O(n)$ polyline vertices, the total added complexity is $O(n^2)$.

We will show that every cylinder $R \times D$, defined by any 3-cell R of the arrangement, is intersected by a constant number of constraint hyper-surfaces. We define the *coverage* of a region $R \subseteq B = W \times T$.

Definition 7. $Cov(R) = \{E \in \mathcal{E} | R \cap col_\gamma(H(E)) \neq \emptyset\}$.

In words, the coverage of a region is the set of obstacles whose columns intersect the region. The following result follows from the low density property and the observation that all points p in a single 3-cell of the arrangement of column boundaries lie in exactly the same collection of columns.

Lemma 8. *The regions R, defined by the cells of $\mathcal{A}(col \circ H)$ have $|Cov(R)| = O(1)$.*

Lemma 9 shows that the partition of the base space into regions R with $|Cov(R)| = O(1)$ is a cylindrical partition. The proof is very similar to the proof of Lemma 3.6 of Van der Stappen and Overmars [11].

Lemma 9. *Let $R \subseteq B$ be such that $|Cov(R)| = O(1)$. Then*

$$|\{f_{\phi,\Phi} | \phi \in_f B \wedge \Phi \in_f \mathcal{E} \wedge f_{\phi,\Phi} \cap (R \times D) \neq \emptyset\}| = O(1)$$

The only problem is that the complexity of the cells of $\mathcal{A}(col \circ H)$ is not necessarily constant. So, we must refine the partition to create constant complexity sub-cells. This is discussed in Section 4.

3.3 Complexity of the Free Space

In the previous subsection we showed that the work-time space of the robot can be partitioned into regions with total combinatorial complexity $O(n^2)$. Furthermore, by Lemmas 8 and 9, each region, when lifted into the configuration-time space is intersected by at most a constant number of constraint hyper-surfaces of bounded algebraic degree. Therefore, a decomposition of the configuration space into free and forbidden cells of combinatorial complexity $O(n^2)$ exists. Obviously, this $O(n^2)$ bound is an upper bound on the complexity of the free space for our dynamic motion planning setting.

Theorem 10. *The complexity of the free space of the low obstacle density dynamic motion planning problem is $O(n^2)$.*

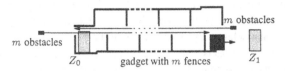

Fig. 1. The quadratic lowerbound construction.

We will now demonstrate that this bound is worst-case optimal, even in the situation where the robot is only allowed to translate and the obstacles move along lines. To this end, we give a problem instance with n obstacles, for which any path for the robot $\mathcal{B}$ has $\Omega(n^2)$ complexity. Consider the workspace in Figure 1. The grey rectangular robot must translate from position Z_0 to Z_1. The gadget in the middle forces the robot to make $\Omega(m)$ moves to move from left to right. It can easily be constructed from $O(m)$ stationary obstacles. The big black obstacle at the bottom right moves very slowly to the right. So it takes a long time before the robot can actually get out of the gadget to go to its goal. Now a small obstacle moves from the left to the right, through the gaps in the middle of the gadget. This forces the robot to go to the right as well. Only there it can move slightly further up to let the obstacle pass. But then a new obstacle comes from the right through the gaps, forcing the robot to move to the left of the gadget to let the obstacle pass above it. This is repeated m times after which the big obstacle is finally gone and the robot can move to its goal. The robot has to move $2m$ times through the gadget, each time making $\Omega(m)$ moves, leading to a total of $\Omega(m^2)$ moves. As $m = \Omega(n)$, the total number of moves is $\Omega(n^2)$. It is easily verified that at any moment the low obstacle density property is satified.

Theorem 11. *The complexity of the free space of the low obstacle density dynamic motion planning problem for a translating robot is $\Omega(n^2)$.*

Actually, the example shows a much stronger result. Not only does it give a bound on the complexity of the free space, but also on the complexity of a single cell in the free space and on the complexity of any dynamic motion planning algorithm.

Theorem 12. *The complexity of any algorithm for the low obstacle density dynamic motion planning problem (even for a translating robot) is lower bounded by $\Omega(n^2)$.*

4   Decomposing the Base Space

We still need to decompose the arrangement of columns $col_\gamma(H(E))$ $(E \in \mathcal{E})$ into constant complexity sub-cells. To this end, we construct a vertical decomposition of the arrangement. Since the vertical decomposition refines the cells of the arrangement, the sub-cells of the final decomposition still have constant coverage. The approach we use [3] requires that the columns in the work-time space, as described in Section 3.2, are in general position. This can be achieved by an appropriate perturbation of the vertices of the columns. Before we can calculate a vertical decomposition we have to triangulate the 2-faces of the columns. Triangulation does not increase the asymptotic complexity of the arrangement. After triangulation, the 2-faces of the arrangement might coincide, though. It is easily verified that the vertical decomposition algorithm still works with these introduced degeneracies. To bound the space we add two horizontal planes at time t_0 and t_1 (the start and goal time) and only consider the area in between. To bound the space in the x- and y-direction we also add a triangular prism far around the relevant region of the work-time space.

4.1 The Vertical Decomposition

Let $S = \{s_1, \ldots, s_n\}$ be a set of n possibly intersecting triangles in 3 space. The *vertical decomposition* of the arrangement $\mathcal{A}(S)$ decomposes each cell of $\mathcal{A}(S)$ into sub-cells, and is defined as follows (see [3]): from every point on an edge of $\mathcal{A}(S)$— this can be a part of a triangle edge or of the intersection of two triangles—we extend a vertical ray in positive and negative x_3-direction to the first triangle above and the first triangle below this point. This way we create a vertical wall for every edge, which we call a *primary wall*. We obtain a *multi-prismatic decomposition* of $\mathcal{A}(S)$ into cells, the *multi-prisms*, with a unique polygonal bottom and top face; the vertical projections of both faces are exactly the same. However, the number of vertical walls of a cylinder need not be constant and the cylinder may not be simply connected. We triangulate the bottom face as in the planar case. The added segments are extended upward vertically until they meet the top face. The walls thus erected are the *secondary walls*. Each sub-cell of the vertical decomposition is now a box with a triangular base and top, connected by vertical walls. (Note that, for navigation purposes, our notion of vertical decomposition is slightly different from other notions of vertical decomposition that construct secondary walls using a planar vertical decomposition of the projections of the top and bottom faces.)

Theorem 13. *The vertical decomposition of the arrangement $\mathcal{A}(col \circ H)$ in the work-time space consists of $O(n^2 \alpha(n) \log n)$ constant complexity sub-cells, and can be computed in time $O(n^2 \alpha(n) \log^2 n)$.*

Proof. Tagansky [9] proved that the vertical decomposition of the entire arrangement of a set of n triangles in $\mathbf{R}^3$ consists of $O(K + n^2 \alpha(n) \log n)$ cells where K is the complexity of the arrangement. Application of this result to the arrangement of grown obstacle column boundaries $\mathcal{A}(col \circ H)$, which satisfies $K = O(n^2)$, yields the complexity bound.

We can compute the vertical decomposition using an algorithm by De Berg, Guibas, and Halperin [3]. This algorithm runs in time $O(n^2 \log n + V \log n)$, where V is the combinatorial complexity of the vertical decomposition. As $V = O(n^2 \alpha(n) \log n)$, the bound follows. $\qquad\square$

To faciliate navigation, we want each subcell to have a constant number of neighbors. That property allows us to determine the navigation direction from a subcell in constant time. The common boundary of a subcell κ and one of its neighbors can be a secondary wall, a primary wall, or a 2-face of the arrangement $\mathcal{A}(col \circ H)$. It is easy to see that the number of neighbors sharing a primary or a secondary wall with κ is bounded by a constant. Let us now consider the maximum number of neighbors, sharing a part of a triangle of $\mathcal{A}(col \circ H)$ with κ. Unfortunately the arrangements of walls ending on the top and bottom side of the triangle can be very different, and can in general be as complex as the complexity of the full decomposition which is only upper-bounded by $O(n^2\alpha(n)\log n)$. Simply connecting the sub-cells at the top of the triangle to the sub-cells at the bottom of the triangle could result in a number of neighbors that is hard to bound by anything better than $O(n^2\alpha(n)\log^2 n)$ for each sub-cell κ. However, as we will show, we can connect the sub-cells at the top and bottom of a face by a symbolic, infinitely thin tetrahedralization. This tetrahedralization will increase the combinatorial complexity of the cell decomposition by a factor of at most $O(\log n)$, but assures that the number of neighbors per sub-cell is bounded by a constant. Since this method is quite complicated, we dedicate the following subsection to it. This will lead to the following result:

Theorem 14. *There exists a cylindrical decomposition of the base space B for the low obstacle density dynamic motion planning problem that consists of $O(n^2\alpha(n)\log^2 n)$ constant complexity sub-cells and a constant number of neighbors per sub-cell. This decomposition can be computed in $O(n^2\alpha(n)\log^2 n)$ time.*

4.2 Tetrahedralizing between Polygons

To reduce the number of neighbors of the sub-cells we will extend the vertical decomposition with a symbolic connecting structure, that increases the total combinatorial complexity of the vertical decomposition by a factor of $O(\log n)$. As a result, the number of neighbors per sub-cell of the cell decomposition with the connecting structure will be bounded by a constant. For each face of the arrangement $\mathcal{A}(col \circ H)$, this structure connects the sub-cells at the top side with the sub-cells at the bottom side. The structure we use is a symbolic, infinitely thin tetrahedralization. To simplify the discussion, we assume that the face for which we construct the connecting structure is horizontal. This is not a constraint, it is just a matter of definition. Throughout this section the vertical direction is parallel to the normal of the face.

Both top and bottom side of the face contain a triangulated 2-dimensional arrangement, say t_t and t_b, created by the other faces and the walls that end on it. Such triangulations with extra vertices in their interior are referred to as Steiner triangulations; the extra vertices are called Steiner points. The arrangements t_t and t_b are normally different; they do not share Steiner points. We separate the top and bottom of every face in the arrangement. Imagine that the top of the face is at height 1.0 and the bottom at height 0.0. We tetrahedralize the space between the top and bottom arrangement, by adding a number of Steiner points between the top and bottom face. (Remember that this is only done in a symbolic way. In reality, the top and bottom face lie in the same plane. The vertical distance is only used to define the adjacencies of the added (flat) sub-cells.)

Theorem 6 states that the faces of the arrangement $\mathcal{A}(col \circ H)$ are polygons without holes. We distinguish between the convex and the non-convex faces. We first show how to tetrahedralize the space between two different Steiner triangulations t_t and t_b of the same convex simple polygon P.

Our tetrahedralization has two layers joined at height 0.5 by a Steiner triangulation of P. This triangulation t_m has one Steiner point p: P is triangulated using a star of edges from p to all vertices of P. Both t_m and t_b are different Steiner triangulations of the same polygon P, therefore the vertical projections of the boundaries of t_b, t_m are equivalent. We tetrahedralize between t_m and t_b by adding a face from every edge of t_b to p. The result is a tetrahedralized pyramid where each tetrahedron corresponds to a triangle of t_b.

To triangulate the complement of this pyramid in the layer between t_b and t_m, we connect the boundaries of t_b and t_m by vertical faces between the boundary edges. For every face f_i introduced by connecting the boundaries, we add a Steiner point q_i in the middle of f_i. We connect q_i to all vertices on f_i and connect each resulting triangle to p (see Figure 2). These triangles complete the tetrahedralization of the space between t_b and t_m. The tetrahedralization between t_t and t_m is constructed in the same way. It is easy to see that the number of tetrahedra created is linear in the complexity of the triangulations t_b and t_t.

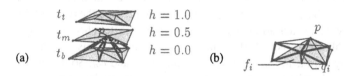

Fig. 2. (a) The construction of the pyramid of tetrahedra between p and the triangles of t_b. (The added edges are bold.) (b) The tetrahedra created in the second step between boundary face f_i and the pyramid.

Unfortunately, faces need not be convex. So we must also show how to tetrahedralize the space between two different Steiner triangulations of the same non-convex simple polygon P. (As indicated above we know that the polygon has no holes. This is crucial here.) We again add a Steiner triangulation t_m of P between t_t and t_b. In the non-convex case we have to use a more sophisticated Steiner triangulation. For this we use a triangulation by Hershberger and Suri [4] that was originally designed for ray shooting in simple polygons. This triangulation t_m has three important properties. Let k be the number of edges of P: (1) It introduces $O(k)$ Steiner points with each Steiner point directly connected to the boundary of P by at least one triangulation edge; (2) Every line segment that lies inside P intersects at most $O(\log k)$ triangles of t_m; (3) The triangulation can be computed in $O(k)$ time. We can derive the following lemma and corollary from the properties of t_m.

Lemma 15. *Let P be a polygon without holes, let k be the number of vertices of P, let t_m be the triangulation described in [4], let t be a triangle inside P, and let A_t be the arrangement of t_m inside triangle t. Then A_t has only $O(\log k)$ constant complexity faces.*

Proof. Let v be a vertex of t_m inside t. Because, by property (1), every vertex of the triangulation t_m is connected to the boundary of P by at least one edge, one outgoing edge from v must intersect the boundary of t. By property (2), the boundary of t intersects at most $O(\log k)$ edges, and, hence, there are also only $O(\log k)$ vertices inside t. Since each face of A_t is the intersection of a triangle from t_m and the triangle t, it has constant complexity. It follows that the whole arrangement $\mathcal{A}_t$ has complexity $O(\log k)$. □

Corollary 16. *Let P be a polygon without holes. Let t_m be the triangulation described in [4]. Let t_a be another triangulation of P with complexity m. The arrangement we obtain by overlaying t_a and t_m has $O(m \log m)$ constant complexity faces.*

Let $t_{b \times m}$ be the triangulation of P we obtain by overlaying t_b and t_m and triangulating the resulting faces. Let m_b denote the complexity of t_b. Corollary 16 shows that the complexity of $t_{b \times m}$ is $O(m_b \log m_b)$. To connect t_b to t_m (at height 0.5) we place $t_{b \times m}$ between t_b and t_m at height 0.25. First we tetrahedralize the layer between t_b and $t_{b \times m}$. We start by adding vertical faces from every edge of t_b to its corresponding edge in $t_{b \times m}$. This results in m_b prisms that have the triangles of t_b as their top and bottom faces. The top faces still contain a number of other edges, that are part of t_m. We tetrahedralize each prism by adding a vertex in the center and connecting it to top, bottom and sides, in the way described for the convex case. (Note that Steiner points might exist on the edges of triangles of $t_{b \times m}$. However, the triangles on the other side of these edges shares these Steiner points, because they are the result of the intersection of two fully connected arrangements. Therefore each tetrahedralized prism perfectly fits its neighboring prisms.) The number of tetrahedra in this layer is $O(m_b \log m_b)$. We similarly tetrahedralize the layer between $t_{b \times m}$ and t_m. So the total area between t_b and t_m can be filled with $O(m_b \log m_b)$ tetrahedra. In the same way we can fill the area between t_t and t_m using the triangulation $t_{t \times m}$. This tetrahedralization will have $O(m_t \log m_t)$ tetrahedra, where m_t is the complexity of t_t.

Summarizing, we can symbolically create a tetrahedralization between the top and the bottom side of the faces of the arrangement. As the original arrangement has combinatorial complexity $O(n^2 \alpha(n) \log n)$ (see Theorem 13), the extended arrangement has complexity $O(n^2 \alpha(n) \log^2 n)$. It can be computed in $O(n^2 \alpha(n) \log^2 n)$ time. The sub-cells in this arrangement each have constant complexity and a constant number of neighbors.

4.3 The Decomposition of the Free Space

We can use the same approach as in [11] to compute the complete cell decomposition of the free space (Theorem 2). This result is a graph CG. Each node of CG corresponds to a constant complexity cell in the free part of the configuration-time space. Each edge corresponds to an adjacency between two such cells. Here two cells are called adjacent if and only if they share an f-dimensional face. (This is important because we want to compute free paths, not semi-free paths.) The degree of the nodes is bounded by a constant. The complexity of CG is the same as the complexity of the base space, so it has $O(n^2 \alpha(n) \log^2 n)$ nodes and edges. The computation time of $O(n^2 \alpha(n) \log^2 n)$ for the base partition dominates the computation time for CG.

5 Finding a Path

In this section we show how to use the cell decomposition to compute a time-monotone path through the free space. Since the path must be time-monotone we cannot do an arbitrary search through the configuration-time space; we will use a space sweep algorithm in the time direction to keep track of the reachable space, while time passes. We sweep with a hyper-plane $\mathcal{P}$, orthogonal to the time direction, from $t = t_0$ to $t = t_1$. Slightly abusing the notion, we will from now on use CG to denote both the connectivity graph and the cell decomposition it represents.

5.1 Preprocessing the Cell Decomposition

To compute the areas of the configuration-time space reachable from the start configuration of the robot by time-monotone paths, we cannot use the cell decomposition directly. The space is partitioned into not necessarily convex constant complexity sub-cells. The sub-cells and their adjacencies are represented by the graph CG; each sub-cell in CG has a constant number of neighbors. A problem with CG is that the time-monotonicity restrictions are not incorporated in the graph. There can be a path between two configurations according to CG, while there exists no time-monotone path between those two configurations. Figure 3 shows a 2-dimensional example in which time increases in the vertical upward direction. Although the graph contains a path between the sub-cells κ_1 and κ_4 there exists no time-monotone path from (any configuration in) κ_1 to (any configuration) in κ_4. Note also that there exists a time-monotone path from κ_1 to only some of the configurations in κ_3.

Fig. 3. (a) A 2-dimensional space with four connected free sub-cells. (b) The connectivity graph CG. Although there is no time-monotone path from κ_1 to κ_4, this fact is not represented by CG.

Since the sub-cells in the cell decomposition are not necessarily convex, there can even exist a pair of configurations in the same sub-cell, that cannot be connected by a time-monotone path. In conclusion, the connectivity graph CG does not contain all necessary data to find a time-monotone path. It is possible, however, to decompose the $(f + 1)$-dimensional sub-cells of CG into smaller sub-cells for which there is a time-monotone path for every pair of configurations in the same sub-cell. If a hyper-plane $\mathcal{P}$, that is orthogonal to the time direction, intersects a sub-cell κ in a number of disconnected regions, then there might be configurations in κ that cannot be connected with a time-monotone path. We therefore decompose each such sub-cell κ into a constant number of smaller sub-cells, such that any cross-section of the hyper-plane $\mathcal{P}$ with a sub-cell consists of one connected region. If for some t, $\mathcal{P}$ is tangent to a feature of κ, then $\mathcal{P}$ decomposes κ into a constant number of constant complexity sub-cells. Since κ is of constant complexity, there are $O(1)$ of these tangencies. We decompose κ into $O(1)$ sub-cells generated by all possible tangencies of $\mathcal{P}$ with features of κ. To adapt CG, we replace the node of κ by nodes for the sub-cells of κ with the appropriate adjacencies. This extension does not increase the asymptotic combinatorial complexity of CG. Also the number of neighbors per sub-cell remains bounded by a (larger) constant.

It is easy to see that any pair of configurations in each resulting sub-cell can be connected by a time-monotone path. In Figure 3 for example, κ_5 is decomposed into three sub-cells by the dashed line.

5.2 Computing a Path

We take the refined connectivity graph CG and note that each sub-cell κ has the following property: if $(Z, \tau) \in \kappa$ is reachable by a time-monotone path, then all $\{(Z', \tau') \in \kappa | \tau' \geq \tau\}$ are reachable by a time-monotone path. In addition, all adjacent sub-cells κ' for which the intersection $\kappa \cap \kappa' \cap (t \geq \tau)$ is an f-dimensional face, are (at least partially) reachable through κ.

Our objective is to label each sub-cell κ with the earliest time τ at which it can be reached from (Z_0, t_0) and with a link to the sub-cell from which it can be reached at time τ. If the sub-cell $\kappa_1 \ni (Z_1, t_1)$ receives a label $\tau_1 < t_1$ then (Z_1, t_1) is reachable from (Z_0, t_0) by a time-monotone path. The sequence of sub-cells containing this path can be found by tracing back the links from each sub-cell, starting from κ_1.

To obtain the labelling outlined above, we perform a sweep-like search of CG starting from $\kappa_0 \ni (Z_0, t_0)$. An event occurs if a sub-cell κ' is reachable from another sub-cell κ at some time τ. We keep the events (κ, κ', τ) of the sweep in a priority queue Q—the event with the smallest τ is dequeued prior to every other event. A more detailed description of the sweep can be found in the full paper [1].

Summarizing, in order to solve a low obstacle density dynamic motion planning problem, we perform the following steps:

LOD-Dyn-Mot
1. COMPUTE vertical decomposition of $\mathcal{A}(col \circ H)$.
2. **for all** faces f of $\mathcal{A}(col \circ H))$ **do**
       COMPUTE flat tetrahedralization between the top and the bottom side of f.
3. TRANSFORM $\mathcal{A}(col \circ H)$ into a decomposition of the *configuration-time space*.
4. COMPUTE a strictly time-monotone *free* path.

It remains to analyse the complexity of the algorithm. Let $|CG|$ denote the size of the graph. Every sub-cell of the decomposition has a constant number of neighbors and, thus, every sub-cell creates a constant number of events. This sums up to $O(|CG|)$ events. So the size of Q is $O(|CG|)$ and enqueueing and dequeueing an event takes $O(\log |CG|)$ time. It is easy to see that handling an event, therefore, takes $O(\log |CG|)$ time. The time for the sweep is therefore upper bounded by $O(|CG| \log |CG|)$. Since $|CG| = O(n^2 \alpha(n) \log^2 n)$ by Theorem 14, the computation of a time-monotone path takes $O(n^2 \alpha(n) \log^3 n)$ time. The following theorem summarizes the result.

Theorem 17. *The low obstacle density motion planning problem for a robot among a set of n obstacles that move with constant speed along polylines, can be solved in $O(n^2 \alpha(n) \log^3 n)$ time.*

6 Conclusion

In this paper we addressed a dynamic extension of the robot motion planning problem in a low obstacle density environment setting. We developed an approach for the exact motion planning problem for a single robot in a low obstacle density environment with multiple moving objects whose motion is represented by a constant complexity polygonal line. We proved that the complexity of the free space of this motion planning problem is $\Theta(n^2)$. In this case our approach takes $O(n^2\alpha(n)\log^3 n)$ to compute a free, time-monotone path, for the robot. We are able to construct low obstacle density workspaces, with moving obstacles, for which the path of the robots is of combinatorial complexity $\Omega(n^2)$, so our result is close to optimal. It remains to be seen whether such a bound exists for a robot with bounded velocity modulus, or bounded acceleration.

It is an interesting question whether the results can be extended to a 3-dimensional workspace. We can again define columns in the (now 4-dimensional) work-time space. Theorem 5 can easily be extended, leading again to a bound of $O(n^2)$ on the complexity of the free space, like in the two-dimensional case. Also the lowerbound of $\Omega(n^2)$ easily carries over. The problem is to compute a constant-complexity partition of the work-time space that has constant coverage. It is easy to obtain an $O(n^4)$ partition by extending the faces of the columns to hyper-planes, building the complete arrangement of these hyper-planes, and subdividing the resulting cells into simplices. This leads to a close to $O(n^4)$ algorithm for motion planning in dynamic 3-dimensional low obstacle density environments. It is at the moment unclear how to improve this result.

References

1. R.-P. Berretty, A.F. van der Stappen, and M. Overmars. Dynamic motion planning in low obstacle density environments. Technical report, *in preparation*, Dept. of Computer Science, Utrecht University, 1997.
2. J. Canny and J. Reif. New lower bound techniques for robot motion planning problems. In *Proc. 28th IEEE Symp. on Foundations of Computer Science*, pages 49–60, Los Angeles, 1987.
3. M. de Berg, L. Guibas, and D. Halperin. Vertical decomposition for triangles in 3-space. *Discrete & Computational Geometry*, 15:35–61, 1996.
4. J. Hershberger and S. Suri. A pedestrian approach to ray shooting: Shoot a ray, take a walk. *J. Algorithms*, 18:403–431, 1995.
5. J. Reif. Complexity of the mover's problem and generalizations. In *Proc. of the 20th IEEE Symp. on Foundations of Computer Science*, pages 421–427, 1979.
6. J. Reif and M. Sharir. Motion planning in the presence of moving obstacles. In *Proc. 25th IEEE Symp. on Foundations of Computer Science*, pages 144–154, 1985.
7. J. Schwartz and M. Sharir. On the piano movers' problem: II. general techniques for computing topological properties of real algebraic manifolds. *Advances in Applied Mathematics*, 4:289–351, 1983.
8. K. Sutner and W. Maass. Motion planning among time dependent obstacles. *Acta Informatica*, 26:93–133, 1988.
9. B. Tagansky. A new technique for analyzing substructures in arrangements. In *Proc. 11th Annual ACM Symp. on Computational Geometry*, pages 200–210, 1995.
10. A. van der Stappen. *Motion planning amidst fat obstacles*. PhD thesis, Dept. of Computer Science, Utrecht University, 1994.
11. A. van der Stappen and M. Overmars. Motion planning in environments with low obstacle density. Technical report, UU-CS-1995-33, Dept. of Computer Science, Utrecht University, 1995.
12. J. Vleugels. On fatness and fitness – realistic input models for geometric algorithms. PhD-thesis, Dept. of Computer Science, Utrecht University, 1997.

Visibility-Based Pursuit-Evasion in a Polygonal Environment

Leonidas J. Guibas, Jean-Claude Latombe, Steven M. LaValle, David Lin and
Rajeev Motwani

Computer Science Department
Stanford University
Stanford, CA 94305 USA
{guibas,latombe,lavalle,dlin,rajeev}@cs.stanford.edu

Abstract. *This paper addresses the problem of planning the motion of
one or more pursuers in a polygonal environment to eventually "see" an
evader that is unpredictable, has unknown initial position, and is capable
of moving arbitrarily fast. This problem was first introduced by Suzuki and
Yamashita. Our study of this problem is motivated in part by robotics
applications, such as surveillance with a mobile robot equipped with a
camera that must find a moving target in a cluttered workspace.*
*A few bounds are introduced, and a complete algorithm is presented for
computing a successful motion strategy for a single pursuer. For simply-
connected free spaces, it is shown that the minimum number of pursuers required is $\Theta(\lg n)$. For multiply-connected free spaces, the bound
is $\Theta(\sqrt{h} + \lg n)$ pursuers for a polygon that has n edges and h holes. A
set of problems that are solvable by a single pursuer and require a linear
number of recontaminations is shown. The complete algorithm searches a
finite cell complex that is constructed on the basis of critical information
changes. It has been implemented and computed examples are shown.*

1 Introduction

The general problem addressed in this paper is an extension or combination of
problems that have been considered in several contexts. Interesting results have
been obtained for pursuit-evasion in a graph, in which the pursuers and evader
can move from vertex to vertex until eventually a pursuer and evader lie in the
same vertex [16, 19]. The *search number* of a graph refers to the minimum number of pursers needed to solve a pursuit-evasion problem, and has been closely
related to other graph properties such as cutwidth [15, 17]. Pursuit-evasion scenarios in continuous spaces have arisen in a variety of applications such as air
traffic control [1], military strategy [10], and trajectory tracking [9]. Although
interesting decision problems arise through the differential motion models, geometric free-space constraints are usually not considered in classical pursuit-
evasion games. Once these constraints are introduced, the problem inherits the
additional complications that arise in geometric motion planning.

A region of capture is often associated with a pursuit-evasion problem, and
the "capture" for our problem is defined as having the evader lie within a line-
of-sight view from a pursuer. A moving visibility polygon in a polygonal environment adds geometric information that must be utilized, and also leads to

connections with the static art gallery problems [18, 21]. In the limiting case, art gallery results serve as a loose upper bound on the number of pursuers by allowing a covering of the free space by static guards, guaranteeing that any evader will be immediately visible. Far fewer guards are needed when they are allowed to move and search for an evader; however, the required motion strategies can become quite complex. A closely related art gallery variant is the watchman tour problem [4]. In this case a minimum-length closed path is computed such that any point in the polygon is visible from some point along the path. In our case, however, the pursuers have the additional burden of ensuring that an evader cannot "sneak" to a portion of the environment that has already been explored. The problem that we consider and other variations have been considered previously in [5, 22]. It was stated in [21] that it remained an interesting challenge to determine if a polygon is searchable by a single pursuer.

Several applications can be envisioned for problems and motion strategies of this type. For example, suppose a building security system involves a few mobile robots with cameras or range sensors that can detect an intruder. A patrolling route can be automatically computed that guarantees that any mobile intruder will eventually be found. To optimize expenses, it would also be important to know the minimum number of robots that would be needed. Applications are not necessarily limited to adversarial targets. For example, the task might be to automatically locate another mobile robot, items in a warehouse or factory that might get moved during the search process, or possibly even people in a search/rescue effort. Such strategies could be used by automated systems or by human searchers.

2   Problem Definition

The pursuers and evader are modeled as points that translate in a polygonal free space, F. Let $e(t) \in F$ denote the position of the *evader* at time $t \geq 0$. It is assumed that $e : [0, \infty) \to F$ is a continuous function, and the evader is capable of moving arbitrarily fast. The initial position $e(0)$ and the path e are assumed unknown to pursuers. Any region in F that might contain the evader will be referred to as *contaminated*, otherwise it will be referred to as *cleared*. If a region is contaminated, becomes cleared, and then becomes contaminated again, it will be referred to as *recontaminated*.

Let $\gamma^i(t)$ denote the position of the ith *pursuer* at time $t \geq 0$. Let γ^i represent a continuous path of the ith pursuer of the form $\gamma^i : [0, \infty) \to F$. Let γ denote a *(motion) strategy*, which refers to the specification of a continuous path for every pursuer: $\gamma = \{\gamma^1, \ldots, \gamma^N\}$.

For any point, $q \in F$, let $V(q)$ denote the set of all points in F that are visible from q (i.e., the linear segment joining q and any point in $V(q)$ lies in F). A strategy, γ, is a *solution strategy* if for every continuous function $e : [0, \infty) \to F$ there exists a time $t \in [0, \infty)$ and an $i \in \{1, \ldots, N\}$ such that $e(t) \in V(\gamma^i(t))$. This implies that the evader will eventually be seen by one or more pursuers, regardless of its path. Let $H(F)$ represent the minimum number of pursuers for which there exists a solution strategy for F.

Section 3 presents some bounds on $H(F)$ for classes of free spaces, and also shows that some polygons for which $H(F) = 1$ only admit solutions that require a linear number of recontaminations. Section 4 addresses the problem of computing a solution strategy, γ, for a given F.

3   Worst-Case Bounds

Several new bounds are presented in this section. For a simply-connected free space, F, with n edges, it is shown that $H(F) = \Theta(\lg n)$. For a free space, F, with h holes, it is shown that $H(F) = \Omega(\sqrt{h}+\lg n)$ and $H(F) = O(h+\lg n)$. For the class of problems in which $H(F) = 1$, it is shown that the same region can require recontamination as many as $\Omega(n)$ times. This result is surprising because pursuit-evasion in a graph is known not to require any recontaminations [11].

Consider the problem of determining the minimum number of pursuers, $H(F)$, required to find an evader in a given free space F. This number will generally depend on both the topological and geometric complexity of F. In [22] a class of simple polygons is identified for which a single pursuer suffices (referred to as "hedgehogs"). For any F that has at least one hole, it is clear that at least two pursuers will be necessary; if a single pursuer is used, the evader could always move so that the hole is between the evader and pursuer. In some cases subtle changes in the geometry significantly affect $H(F)$.

Consider $H(F)$ for the class of simply-connected free spaces. Let n represent the number of edges in the free space, which is represented by a simple polygon in this case. A logarithmic worst-case bound can be established:

Theorem 1. *For any simply-connected free space F at worst $H(F) = O(\lg n)$.*

Proof: The proof is built on the following observation. Suppose that two vertices of F are connected by a linear segment, thus partitioning F into two simply-connected, polygonal components, F_1 and F_2. If $H(F_1) \leq k$ and $H(F_2) \leq k$ for some k, then $H(F) \leq k + 1$ because the same k pursuers can be used to clear both F_1 and F_2. This requires placing a static $(k + 1)^{th}$ pursuer at the edge common to F_1 and F_2 to keep F_1 cleared after the k pursuers move to F_2 (assuming arbitrarily that F_1 is cleared first).

In general, if two simply-connected polygonal regions share a common edge and can each be cleared by at most k pursuers, then the combined region can be cleared at most $k + 1$ pursuers. Recall that for any simple polygon, a pair of vertices can always be connected so that polygon is partitioned into two regions, each with at least one third of the edges of the original polygon [3]. This implies that F can be recursively partitioned until a triangulation is constructed, and each triangular region only requires $O(\lg n)$ recombinations before F is obtained (i.e., the recursion depth is logarithmic in n). Based on the previous observation and the fact that each triangular region can be trivially searched by a single pursuer, $H(F) = O(\lg n)$. $\square$

The remaining question for simply-connected free spaces is whether there actually exist problems that require a logarithmic number of pursuers. Some results from graph searching will first be described and utilized to construct difficult worst-case problem instances. Let *Parsons' problem* refer to the graph-searching problem presented in [16, 19]. The task is to specify the number of pursuers required to find an evader that can execute continuous motions along the edges of a graph. Instead of using visibility, capture is achieved when one of the pursuers "touches" the evader. Let G represent a graph, and $S(G)$ represent the number of needed pursuers, referred to as the search number of G.

The following lemma implies that a geometric realization of any planar graph instance can be constructed:

Lemma 2. *For every planar graph, G, there exists a polygonal free space F such that Parsons' problem on G is equivalent to the visibility-based pursuit evasion problem on F.*

Proof: Consider a planar representation of G in which linear segments correspond to the edges. Each linear segment can be replaced by a corridor of sufficient length as shown in Figure 1, with $\epsilon > 0$ chosen such that no pair or corridors intersect. Portions of the corridor edges can be removed at corridor junctions to prevent overlap.

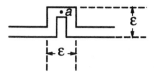

Fig. 1. Second-order visibility between the two entrances is not maintained.

It remains to establish that searching the environment constructed by the network of bent corridors is equivalent to searching a planar graph. If a solution strategy for Parsons' problem is specified as an ordered set of traversed edges for each pursuer, then traversing the corresponding corridors will clearly solve the geometric problem. Consider a given solution for the geometric problem. If the pursuers move from junction to junction without reversing within a corridor, then traversing the equivalent edges also solves the planar graph problem. For the geometric problem, a static pursuer can be placed in a corridor (as in position a in Figure 1), allowing two other pursuers to clear the corridor from each end without leaving the junctions unguarded. In this case and in similar cases, the pursuer that is fixed in the geometric problem can clear the corresponding edge by traveling between its endpoints in G; thus, the graph problem can still be solved using the same number of pursuers. $\square$

This theorem, from [19], is useful for proving Theorem 4:

Lemma 3. (Parsons) *Let G be a tree. Then $S(G) = N + 1$ if and only if there exists a vertex in G whose removal separates G into three components, G_1, G_2, and G_3, such that $S(G_i) \geq N$ for $i \in \{1, 2, 3\}$.*

Theorem 4. *There exist simply-connected free spaces F with n edges such that $H(F) = \Omega(\lg n)$.*

Proof: Using Lemma 3, a tree, G, can be constructed recursively that has a constant branching factor of three, height $N - 1$, and requires N pursuers (an example is given in [19]). By Lemma 2, an equivalent geometric instance can be constructed for any N. $\square$

Figure 2.a shows an example, which relies on Lemma 3. Theorem 1 and Theorem 4 together imply a tight logarithmic bound, $H(F) = \Theta(\lg n)$.

Next consider the class of problems for which F has h holes.

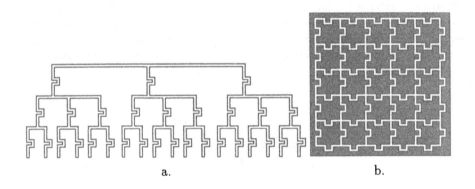

a. b.

Fig. 2. a) An instance from a sequence of simply-connected free spaces that require $\Omega(\lg n)$ pursuers. b) An instance from a sequence of problems that requires a number of pursuers that is at least proportional to the square root of the number of holes.

Theorem 5. *For any free space F with h holes at worst $H(F) = O(\sqrt{h} + \lg n)$.*

Proof: Divide the pursuers into two groups: $O(\sqrt{h})$ pursuers will be used to reduce the polygon to simply-connected components, and $O(\lg n)$ pursuers will be used to clear each component. Construct an arbitrary triangulation of F. Let a *trichromatic triangle* be defined as a triangle that touches three distinct connected components of the boundary of F. Using at most $O(h)$ trichromatic triangles, F can be partitioned into simply-connected components whose boundaries are comprised of the boundary of F and edges of trichromatic triangles. To establish this, form a planar graph by placing a vertex in each hole of F and one vertex outside of F. For every trichromatic triangle edge joining two boundaries, form an edge for this graph by joining the vertices corresponding to these boundaries by a path along the trichromatic edge, in the obvious way. From the planarity of this graph we can easily argue that the overall number of trichromatic edges, and therefore of trichromatic triangles, is $O(h)$.

Consider the dual graph of the triangulation, which has $O(n)$ edges and vertices. Take the subgraph induced by taking only the vertices that correspond to trichromatic triangles in the original triangulation. The planar graph separator theorem [14] implies that at most $O(\sqrt{h})$ edges can be chosen to partition the graph into two portions with at least one third of the edges on each side of the partition. Each edge in the induced subgraph corresponds to a simply-connected region of F that can be cleared with $O(\lg n)$ pursuers by Theorem 1. In fact, the same set of pursuers can be used for each simply connected component. The $O(\sqrt{h})$ pursuers then form a barrier that maintains the cleared areas obtained by other pursuers on either side, much in the same way as the partitioning edges in the proof of Theorem 1. The planar graph separator theorem can be applied recursively to the remaining portions of F on either side of a barrier, and the free space can be cleared using the same progression as for Theorem 1. At the ith level of recursion, at most $\frac{2}{3}$ as many pursuers will be needed to form a barrier in comparison to the $(i-1)^{th}$ level of recursion. The free space is reduced to simply-connected components that can be cleared using $O(\lg n)$ pursuers. The total number of pursuers needed to form barriers is $O(\sqrt{h} + \sqrt{\frac{2}{3}h} + \sqrt{\frac{4}{9}h} + \cdots) = O(\sqrt{h})$. Thus, F can be cleared using at most $O(\sqrt{h} + \lg n)$ pursuers. □

Figure 2.b shows an illustrative example of the following theorem:

Theorem 6. *There exist free spaces F with h holes such that $H(F) = \Omega(\sqrt{h} + \lg n)$.*

Proof: For any positive integer k, a planar graph of cutwidth k can be constructed using $O(k^2)$ vertices and edges. Recall that the cutwidth, $CW(G)$, is the minimum cutwidth taken over all possible linear layouts of G. A linear layout of G is a one-to-one function mapping the vertices of G to integers, and the cutwidth for a particular layout is the maximum over all i of the number of edges connecting vertices assigned to integers less than i to vertices assigned to integers as large as i. Define a sequence of planar graphs, $G_1, G_2, \ldots$. Let the vertices of G_k correspond to the set of all points with integer coordinates, (i, j), such that $0 \leq i, j \leq k$. Let the edges of G_k connect any two vertices for which one coordinate differs by one unit (i.e., a standard four-neighborhood). The cutwidth of G_k is k.

It is established in [15] that for all graphs G, the search number $S(G)$ is related to the cutwidth as $S(G) \leq CW(G) \leq \lfloor deg(G)/2 \rfloor \cdot S(G)$, in which $deg(G)$ is the maximum vertex degree of G. Since $deg(G_k) = 4$, $S(G_k) \leq k \leq 2S(G_k)$. Using Lemma 2, geometric instances of G_k can be constructed. Both G_k and each geometric instance require $\Omega(k)$ pursuers. There is a quadratic number of holes in each geometric instance; hence, $H(F) = \Omega(\sqrt{h})$. This corridor structure can be combined with the structure from Theorem 4 to yield an example that requires $\Omega(\sqrt{h} + \lg n)$ pursuers. $\square$

Theorem 6 and Theorem 5 together imply a tight bound, $H(F) = \Theta(\sqrt{h} + \lg n)$.

The final theorem of this section pertains to the class of free spaces that can be searched by a single pursuer. A similar result is also obtained in [5]. It states that there exist examples that require recontaminating some portion of the free space a linear number of times. This result is surprising because for Parsons' problem it was shown in [11] that no recontamination is necessary (a shorter proof of this appears in [2]). In [22] a free space was given that requires two recontaminations, which at least established that recontamination is generally necessary for visibility-based pursuit evasion. Theorem 7 establishes that a linear number of recontaminations can be needed, and it still remains open to determine whether the number of recontaminations can be bounded from above by a polynomial, which would imply that the problem of deciding whether $H(F) = 1$ lies in NP.

Theorem 7. *There exists a sequence of simply-connected free spaces with $H(F) = 1$ such that $\Omega(n)$ recontaminations are required for n edges.*

Proof: It will be shown that the example in Figure 3 requires $k - 2$ recontaminations by visiting the point $a \in F$ a total of $k - 1$ times to repeatedly clear the "peak." Without loss of generality, consider the set of strategies that can be specified by identifying the sequence of points, $a, b_1, \ldots, b_k, c_1, \ldots, c_k$, that are visited. Assume that the shortest-distance path is taken between any pair of points. Consider visiting b_i for some $1 < i < k$, followed by a visit to another "leg", say b_j (or c_j). If any legs between b_i and b_j are contaminated, then b_i will get contaminated, which undoes previous work. If all legs are initially contaminated, then they must be visited in one of two orders: $(b_1, c_1, b_2, c_2, \ldots, b_k, c_k)$ or $(b_k, c_k, b_{k-1}, c_{k-1}, \ldots, b_1, c_1)$. Because of symmetry, consider visiting the legs from left to right without loss of generality. Assume that the peak is initially contaminated. The points b_1 and c_1 can be visited to clear the leftmost set of legs; however, these will get contaminated when b_2 is visited. By traveling from c_1 to a to b_2, the leftmost set of legs remain cleared because the peak is cleared. When c_2 is visited, the leftmost three legs remain cleared; however, the peak becomes recontaminated. Thus, a will have to be visited again before clearing b_3. By induction on i for $1 < i \leq k$, the peak will have to be cleared by visiting a each time between visits to c_i and b_{i+1}. This implies that a will be visited $k - 1$ times, resulting in $k - 2$ recontaminations. $\square$

4 Computing a Solution Strategy

Section 4.1 defines a general information space (or space of knowledge states) for this problem, and provides a general method for partitioning the information space into equivalence classes, which can reduce the general problem to finite cell searching. NP-hardness is also established in Section 4.1. Section 4.2 presents a

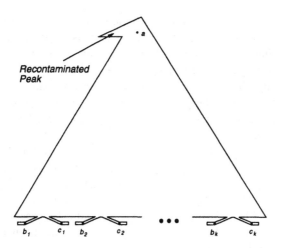

Fig. 3. A linear number of recontaminations is required. Although this polygon can be searched by a single pursuer, the peak must be visited $k - 1$ times.

complete algorithm for the case in which $H(F) = 1$ that computes a solution strategy by decomposing F into convex cells based on edge visibility. This algorithm is quite efficient in practice, and was used to compute the examples shown in Section 5.

4.1   General Concepts

A *complete* algorithm must compute a solution strategy for a given number of pursuers, if such a strategy exists. It is natural to compare the notion of completeness for this problem to completeness for the basic motion planning problem (i.e., the algorithm will find a collision-free path if such a path exists). One important difference, however, is that the *minimum* number of pursuers is crucial, but does not have a correspondence for the basic path planning problem. A variety of simple, heuristic algorithms can be developed that might use more pursuers than necessary.

The general problem is intractable if $P \neq NP$:

Theorem 8. *Computing $H(F)$ is NP-hard.*

Proof: It is shown in [17] that Parsons' problem for a planar graph with maximum vertex degree 3 is NP-complete (i.e., computing the search number, $S(G)$). By Lemma 2, equivalent geometric instances can be constructed, which implies that computing $H(F)$ is NP-hard. □

This significantly reduces hopes that an efficient algorithm can be determined for the general problem. A complete algorithm for $H(F) = 1$ is detailed in this paper, and the general techniques apply to the case in which $H(F) > 1$. In related work [13] we have developed a greedy algorithm that efficiently solves many multiple-pursuer problems.

Because the position of the evader is unknown, one does not have direct access to the state at a given time. This motivates the consideration of an information space that identifies all unique situations that can occur during the execution of a motion strategy. Let a *state space*, X, be spanned by the coordinates $x = (x^1, \ldots, x^N, x^e)$, in which x^i for $1 \leq i \leq N$ represents the position of the ith pursuer, and x^e represents the position of the evader. Since the positions of the pursuers are always known, let X^p denote the subspace of X that is spanned by the pursuer positions, $x^p = (x^1, \ldots, x^N)$.

It will be useful to analyze a strategy in terms of manipulating the set of possible positions of the evader. Let $S \subseteq F$ represent the set of all contaminated points in F. Let $\eta = (x^p, S)$ for which $x^p \in X^p$ and $S \subseteq F$ represent an *information state*. Let the *information space*, $\mathcal{I}$, represent the set of possible information states. The information space is a standard representational tool for problems that have imperfect state information, and has been useful for other motion planning problems [6, 12].

For a fixed strategy, γ, a path in the information space will be obtained by $\eta(t) = (\gamma^1, \ldots, \gamma^N, S(t))$ in which $S(t)$ can be determined from an initial $S(0)$ and the trajectories $\{\gamma^i(t') | t' \in [0, t]\}$ for each $i \in \{1, \ldots, N\}$. Let $\Psi(\eta, \gamma, t_0, t_1)$ represent the information state that will be obtained by starting from information state η, and applying the strategy γ from t_0 to t_1. The function Ψ can be thought of as a "black box" that produces the resulting information state when a portion of a given strategy is executed.

We next describe a general mechanism for defining critical information changes. This is inspired in part by a standard approach used in motion planning, which is to preserve completeness by using a decomposition of the configuration space that is constructed by analyzing critical events. For example, in [20] a cell decomposition is determined by analyzing the contact manifolds in a composite configuration space that is generated by the positions of several disks in the plane.

The next definition describes an information invariance property, which allows the information space, $\mathcal{I}$, to be partitioned into equivalence classes. A connected set $D \subseteq X^p$ is *conservative* if $\forall \eta \in \mathcal{I}$ such that $x^p \in D$, and $\forall \gamma : [t_0, t_1] \rightarrow D$ such that γ is continuous and $\gamma(t_0) = \gamma(t_1) = x^p$, then the same information state, $\eta = \Psi(\eta, \gamma, t_0, t_1)$, is obtained. This implies that the information state cannot be altered by moving along closed paths in D. Just as in the case of motions in a conservative field, the following holds:

Theorem 9. (Path invariance) *If D is conservative then for any two continuous trajectories, γ_1, γ_2, mapping into D such that $\gamma_1(t_0) = \gamma_2(t_0)$ and $\gamma_1(t_1) = \gamma_2(t_1)$ then $\Psi(\eta, \gamma_1, t_0, t_1) = \Psi(\eta, \gamma_2, t_0, t_1)$, for any η.*

Proof: Select any third continuous trajectory, $\gamma_3 : [t_0, t_1] \rightarrow D$, such that $\gamma_3(t_0) = \gamma_1(t_1)$ and $\gamma_3(t_1) = \gamma_1(t_0)$ (i.e., heading in the opposite direction). Form a new trajectory, γ_{132}, by concatenating the trajectories γ_1, γ_3, and γ_2. The resulting information state will be $\Psi(\eta, \gamma_1, t_0, t_1)$ because γ_3 followed by γ_2 forms a closed-loop path, and thus yields the same information state by conservativity of D. Note that γ_1 followed by γ_3 is also a closed-loop path, which implies that γ_2 must bring the information state from η to $\Psi(\eta, \gamma_1, t_0, t_1)$. Hence, $\Psi(\eta, \gamma_1, t_0, t_1) = \Psi(\eta, \gamma_2, t_0, t_1)$. $\square$

Thus, the information state from moving between $x_1^p \in D$ and $x_2^p \in D$ is invariant with respect to the chosen path. This partitions the information space into equivalence classes. Within is each class the particular chosen path is insignificant, which leads to a finite graph search problem.

4.2 A Complete Algorithm for a Single Pursuer

Since the general problem is NP-hard, it is worth focusing on the complete algorithm for the case of a single pursuer. The basic idea is to partition the free space into convex cells that maintain completeness, and perform a search on the resulting quotient information space. This algorithm has been implemented and tested on a variety of examples, two of which are shown in Section 5.

Suppose the pursuer is at a point $q \in F$. Consider the circular sequence of edges in the resulting visibility polygon. The edges generally alternate between bordering an obstacle and bordering free space. Let each edge that borders free space be referred to as a *gap edge*. Consider associating a binary label with each gap edge. If the portion of the free space that borders the gap edge is contaminated, then it is assigned a "1" label; otherwise, it is assigned a "0" label indicating that it is clear. Let $B(q)$ denote a binary sequence that corresponds to labelings that were assigned from $q \in F$. Note that the set of all contaminated points is bounded by a polygon that must contain either edges of F or gap edges from the visibility polygon of the pursuer. Thus, the specification of q and $B(q)$ uniquely characterizes the information state.

Consider representing the information state using q and $B(q)$, and let a pursuer move in a continuous, closed-loop path that does not cause gap edges to appear or disappear at any time. Each gap edge will continuously change during the motion of the pursuer; however, the corresponding gap edge label will not change. The information state cannot change unless gap edges appear or disappear. For example, consider the problem shown in Figure 4 which shows a single pursuer that is approaching the end of a corridor. If the closed-loop motion on the left is executed, the end of the corridor remains contaminated. This implies that although the information state changes during the motion, the original information state is obtained upon returning. During the closed-loop motion on the right, the gap edge disappears and reappears. In this case, the resulting information state is different. The gap label is changed from "1" to "0".

Hence, a cell decomposition that maintains the same corresponding gap edges will only contain conservative cells. The idea is to partition the free space into convex cells by identifying critical places at which edge visibility changes. A decomposition of this type has been used for robot localization in [8, 23], and generates $O(n^3)$ cells in the worst case for a simple polygon (which is always true if $H(F) = 1$). The free space can be sufficiently partitioned in our case by extending rays in the three general cases. Obstacle edges are extended in either direction, or both directions if possible. Pairs of vertices are extended outward

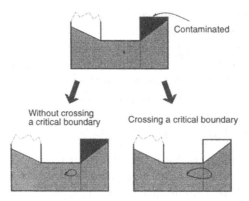

Fig. 4. A critical event in the information space can only occur when edge visibility changes.

only if both directions are free along the line drawn through the pair of points. This precludes the case in which one direction cannot be extended; although edge visibility actually changes for this case, it does not represent a critical change in information.

The next issue is searching the information space for a solution, which corresponds to specifying a sequence of adjacent cells. The solution strategy must take the form of a path that maps into F. This can be constructed by concatenating linear path segments, in which each segment connects the centroids of a consecutive pair of cells in the sequence.

The cells and their natural adjacency relationships define a finite, planar graph, G_c, referred to as the *cell graph*. Vertices in G_c are generally visited multiple times in a solution sequence because of the changing information states. For each vertex in G_c, a point, $q \in F$, in the corresponding cell can be identified, and the labels $B(q)$ can be distinct at each visit. Initially, the pursuer will be in some position at which all gap labels are "1". The goal is to find any sequence of cells in G_c that leave the pursuer at some position at which all gap labels are "0".

A directed *information state graph*, G_I, can be derived from G_c, for which each vertex is visited at most once during the execution of a solution strategy. For each vertex in G_c, a set of vertices are included in G_I for each possible labeling of the gap edges. For example, suppose a vertex in G_c represents some cell D, and there are 2 gap edges for $B(q)$ and any $q \in D$. Four vertices will be included in G_I that all correspond to the pursuer at cell D; however, each vertex represents a unique possibility for $B(q)$: "00", "01", "10", or "11". Let a vertex in G_I be identified by specifying the pair $(q, B(q))$.

To complete the construction of G_I, the set of edges must be defined. This requires determining the appropriate gap labels as the pursuer changes cells. Suppose the pursuer moves from $q_i \in D_i$ to $q_j \in D_j$. For the simple case shown in the lower right of Figure 4, assume that the gap edge on the left initially has a label of "0" and the gap edge on the right has a label of "1". Let the first bit denote the leftmost gap edge label. The first transition is from "01" to "0", and the second transition is from "0" to "00". The directed edges in G_I are $(q_i, \text{"01"})$ leads to $(q_j, \text{"0"})$, $(q_j, \text{"0"})$ leads to $(q_i, \text{"00"})$.

In the case of multiple gap edges, correspondences must be determined to correctly compute the gap labels. In general, if any n gap edges are merged, the corresponding gap edges will receive a "1" label if any of the original gap edges contain a "1" label. Once the gap edge correspondences have been determined, the information state graph can be searched using Dijkstra's algorithm with an edge cost that corresponds to the distance traveled in the free space by the pursuer. Unfortunately, the precise complexity of the complete algorithm cannot be determined. In the worst-case, examples can be constructed that yield an exponential number of information states, but it is not clear whether these information states necessarily have to be represented and searched to determine a solution (it is not even known if optimal-length solutions to the single-pursuer problem can be verified in polynomial time).

5 Computed Examples

The complete algorithm is implemented in C++ and executed on an SGI Indigo2 workstation with a 200 Mhz MIPS R4400 processor. Most problems we encountered were solved in a few seconds or less. The implementation uses the quad-edge structure from [7] to maintain the topological ordering of the conservative cells. The search strategy is Dijkstra's shortest path algorithm, in which the distance is measured from the adjacent cell centroids. Figure 5 shows two computed examples. We have also computed solutions for the example shown in Figure 3, the hookpin example described in [22] that requires two recontaminations, and several other problems.

6 Conclusions

We proved some new bounds and introduced a complete algorithm for the polygon searching problem. A logarithmic bound on the number of needed pursuers was shown for the case of simply-connected free spaces, and a square-root bound was expressed in terms of the number of holes for multiply-connected free spaces. It was also shown that there exist problems requiring a linear number of recontaminations. A few open problems remain, such as determining tight bounds on the number of pursuers for general polygons, and determining whether a polynomial-time algorithm exists to decide whether $H(F) = 1$. The complexity of our complete algorithm also remains open. It also remains an interesting pursuit to attempt to characterize the set of simple polygons such that $H(F) = 1$; interesting subsets have been characterized in [22], and our information space concepts might be useful in this endeavor.

Information space concepts were used to provide a natural characterization of the unique problem states. The visibility-based pursuit-evasion problem was

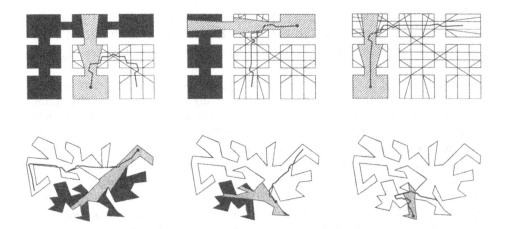

Fig. 5. Two computed examples are shown, each with three snapshots of the solution. The black area represents the contaminated region, and the white area represents the cleared region. The thick curve shows a portion of computed trajectory, which is continued in each frame. The shaded region indicates the visibility region at the final time step of the indicated portion of the trajectory. The thin lines in the cleared region indicate the cell boundaries.

established as NP-hard. The general concept of partitioning the information space on the basis of critical information changes was introduced to develop a complete algorithm. For the case in which $H(F) = 1$, the complete algorithm was implemented and tested on several examples.

References

1. T. Başar and G. J. Olsder. *Dynamic Noncooperative Game Theory*. Academic Press, London, 1982.
2. D. Bienstock and P. Seymour. Monotonicity in graph searching. *J. Algorithms*, 12:239–245, 1991.
3. B. Chazelle. A theorem on polygon cutting with applications. In *Proc. 23rd Annu. IEEE Sympos. Found. Comput. Sci.*, pages 339–349, 1982.
4. W.-P. Chin and S. Ntafos. Optimum watchman routes. *Information Processing Letters*, 28:39–44, 1988.
5. D. Crass, I. Suzuki, and M. Yamashita. Searching for a mobile intruder in a corridor – the open edge variant of the polygon search problem. *Int. J. Comput. Geom. & Appl.*, 5(4):397–412, 1995.
6. M. Erdmann. Randomization for robot tasks: Using dynamic programming in the space of knowledge states. *Algorithmica*, 10:248–291, 1993.
7. L. Guibas and J. Stolfe. Primitives for the manipulation of general subdivisions and the computation of Voronoi diagrams. *AMC Trans. Graphics*, 4(2):74–123, 1985.

8. L. J. Guibas, R. Motwani, and P. Raghavan. The robot localization problem. In K. Goldberg, D. Halperin, J.-C. Latombe, and R. Wilson, editors, *Proc. 1st Workshop on Algorithmic Foundations of Robotics*, pages 269–282. A.K. Peters, Wellesley, MA, 1995.

9. O. Hájek. *Pursuit Games*. Academic Press, New York, 1975.

10. R. Isaacs. *Differential Games*. Wiley, New York, NY, 1965.

11. A. S. Lapaugh. Recontamination does not help to search a graph. *J. ACM*, 40(2):224–245, April 1993.

12. S. M. LaValle. *A Game-Theoretic Framework for Robot Motion Planning*. PhD thesis, University of Illinois, Urbana, IL, July 1995.

13. S. M. LaValle, D. Lin, L. J. Guibas, J.-C. Latombe, and R. Motwani. Finding an unpredictable target in a workspace with obstacles. In *Prof. IEEE Int'l Conf. on Robotics and Automation*, 1997.

14. R. J. Lipton and R. E. Tarjan. A separator theorem for planar graphs. *SIAM Journal of Applied Mathematics*, 36:177–189, 1979.

15. F. Makedon and I. H. Sudborough. Minimizing width in linear layouts. In *Proc. 10th ICALP, Lecture Notes in Computer Science 154*, pages 478–490. Springer-Verlag, 1983.

16. N. Megiddo, S. L. Hakimi, M. R. Garey, D. S. Johnson, and C. H. Papadimitriou. The complexity of searching a graph. *J. ACM*, 35(1):18–44, January 1988.

17. B. Monien and I. H. Sudborough. Min cut is NP-complete for edge weighted graphs. *Theoretical Computer Science*, 58:209–229, 1988.

18. J. O'Rourke. *Art Gallery Theorems and Algorithms*. Oxford University Press, New York, NY, 1987.

19. T. D. Parsons. Pursuit-evasion in a graph. In Y. Alani and D. R. Lick, editors, *Theory and Applcation of Graphs*, pages 426–441. Springer-Verlag, Berlin, 1976.

20. J. T. Schwartz and M. Sharir. On the piano movers' problem: III. Coordinating the motion of several independent bodies. *Int. J. Robot. Res.*, 2(3):97–140, 1983.

21. T. Shermer. Recent results in art galleries. *Proc. IEEE*, 80(9):1384–1399, September 1992.

22. I. Suzuki and M. Yamashita. Searching for a mobile intruder in a polygonal region. *SIAM J. Comput.*, 21(5):863–888, October 1992.

23. R. Talluri and J. K. Aggarwal. Mobile robot self-location using model-image feature correspondence. *IEEE Trans. Robot. & Autom.*, 12(1):63–77, February 1996.

Acknowledgments

The research of Leonidas J. Guibas is supported by NSF grant CCR-9623851 and US Army MURI grant 5-23542-A. Rajeev Motwani's research is supported by an Alfred P. Sloan Research Fellowship, an IBM Faculty Partnership Award, an ARO MURI Grant DAAH04-96-1-007, and NSF Young Investigator Award CCR-9357849, with matching funds from IBM, Mitsubishi, Schlumberger Foundation, Shell Foundation, and Xerox Corporation. The remaining researchers are supported by ARO MURI grant DAAH04-96-1-007 and ONR grant N00014-94-1-0721. The authors thank Jian Bao, Julien Basch, Frédéric Cazals, Bruce Donald, Héctor González-Baños, Gary Kalmanovich, Jon Kleinberg, Suresh Venkatasubramanian, Li Zhang, and the anonymous reviewers, for their helpful suggestions.

Maintaining the Extent of a Moving Point Set

Pankaj K. Agarwal[1]*, Leonidas J. Guibas[2]**,
John Hershberger[3], and Eric Veach[2]

[1] Department of Computer Science, Duke University, Durham, NC 27708, USA
[2] Computer Science Department, Stanford University, Stanford, CA 94305, USA
[3] Mentor Graphics Corp., 1001 Ridder Park Drive, San Jose, CA 95131, USA

Abstract. Let S be a set of n moving points in the plane. We give new efficient and compact kinetic data structures for maintaining the diameter, width, and smallest area or perimeter bounding rectangle of the points. When the points in S move with pseudo-algebraic motions, these structures process $O(n^{2+\epsilon})$ events. We also give constructions showing that $\Omega(n^2)$ combinatorial changes are possible in these extent functions even when the points move on straight lines with constant velocities. We give a similar construction and upper bound for the convex hull, improving known results.

1 Introduction

Suppose S is a set of n moving points in the plane. In this paper we investigate how to maintain various descriptors of the *extent* of the point set, such as diameter, width, smallest enclosing rectangle, etc. These extent measures give an indication of how spread out the point set S is and are useful in various virtual reality applications such as clipping, collision checking, etc. As the points move continuously, the extent measure of interest (e.g., diameter) changes continuously as well, though its combinatorial realization (e.g., the pair of points defining the diameter) only changes at certain discrete times. Our approach is to focus on these discrete changes or events and track through time the combinatorial description of the extent measure of interest.

We do so within the framework of *kinetic data structures* (KDSs for short), as developed by Basch, Guibas, and Hershberger [3] and further elaborated in Section 2. There are two notable and novel aspects of that framework. Firstly, while extensive work has been done on dynamic data structures in computational geometry [4, 5], this is all focused on handling insertions/deletions of objects and not continuous change. Kinetic data structures by contrast gain

---

* Support was provided by National Science Foundation research grant CCR–93–01259, by Army Research Office MURI grant DAAH04–96–1–0013, by a Sloan fellowship, by a National Science Foundation NYI award and matching funds from Xerox Corp, and by a grant from the U.S.-Israeli Binational Science Foundation.
** Support was provided in part by National Science Foundation grant CCR–9623851 and by US Army MURI grant 5–23542–A.

their efficiency by exploiting the continuity or coherence in the way the system state changes. Secondly, unlike Atallah's dynamic computational geometry framework [2], which was introduced to estimate the maximum number of combinatorial changes in a geometric configuration under predetermined motions in a certain class, the KDS framework is fully *on-line* and allows each object to change its motion at will, due to interactions with other moving objects, the environment, etc.

Section 3 presents new kinetic algorithms for diameter, width, and smallest enclosing rectangle in both the area and perimeter senses. If we assume that the points of S follow pseudo-algebraic motions (defined below), then the number of events processed by each of our algorithms is $O(n^{2+\epsilon})$ (for all $\epsilon > 0$). In particular these bounds prove that none of the extent measures mentioned can change combinatorially more than $O(n^{2+\epsilon})$ times. A quadratic bound is natural for diameter, as it is defined by two points of the set S, but it is somewhat surprising for the other measures, as width is defined by three points, and the minimum bounding rectangles by four or five of the points. The data structures we give are efficient and compact in the KDS sense, though not local.

Section 4 is devoted to giving lower bound constructions for these extent measures under *linear* point motions: we show that diameter, width, and the two flavors of smallest bounding rectangle can all change $\Omega(n^2)$ times as the n points of S move on straight line trajectories with constant velocities (possibly different for each point). Such lower bound constructions are much easier if we allow quadratic or other higher degree motions—the fact that the same bounds hold with linear motions is quite interesting. Our constructions employ a key component consisting of cocircular (or nearly cocircular) points that move on straight lines while maintaining their (near-) cocircularity. Finally in Section 5 we give a similar construction showing that the convex hull of n points moving linearly in the plane can also change $\Omega(n^2)$ times. We also prove a tighter upper bound than was previously known for the number of combinatorial changes to the convex hull. This bound is $O(n\lambda_s(n))$, where $\lambda_s(n)$ is the length of a Davenport-Schinzel sequence [7], and the parameter s bounds the number of times three points can become collinear. The bound specializes to $O(n^2)$ for linearly moving points—which is therefore tight.

2 Kinetic data structure preliminaries

A kinetic data structure maintains a *configuration function* of continuously moving data (e.g., diameter, width, etc., of moving points). It does so by maintaining a set of *certificates* that jointly imply the correctness of the computed configuration function. Each certificate is a geometric predicate on a constant number of data elements, such as, for example, "points A and B are farther apart than points C and D." The certificates are typically derived from a static algorithm for computing the configuration function. For example, the certificates for maintaining the diameter might include a set of distance comparisons establishing a partial order on the relevant pairwise distances, with a single maximum element.

The certificates are stored in a priority queue, ordered by the next time at which a certificate will be violated. Each data element has a *flight plan* that gives full or partial information about the current motion of the element, and these flight plans are used to compute the next violation time for each certificate. When the next violation time is reached, the algorithm removes the violated certificate from the queue and computes certificates for the new data configuration. Some number of certificates may have to be removed from the queue, and some number of new certificates added.

When a data element changes its flight plan, all the certificates in the priority queue that depend on it must have their times of next violation recomputed, and their positions in the queue must be updated. A KDS is called *local* if the number of certificates that depend on a single data element is polylogarithmic in the total number of data elements.

The violation of a certificate is called an *event*. *External events* cause the configuration function to change. *Internal events* do not affect the configuration function, but must be processed for the integrity of the data structure. We evaluate a KDS by counting events under the assumption that the data motions are *pseudo-algebraic*, i.e., each certificate predicate changes sign a bounded number of times when applied to any fixed subset of data elements. A KDS is called *efficient* if the worst-case number of total events (internal plus external) is asymptotically the same as, or only slightly larger than, the worst-case number of external events, under the assumption of pseudo-algebraic motion.

A KDS is called *compact* if the number of certificates stored in the priority queue is roughly linear in the number of data elements.

Efficient, local, and compact kinetic data structures are known for maintaining the convex hull and closest pair of points moving in the plane, and for computing the maximum of points moving along a line [3]. The data structure for computing the maximum is known as a *kinetic tournament*.

3 Algorithms

In this section we present kinetic data structures for maintaining three different versions of the extent of a planar point set: diameter, width, and minimum enclosing box. Each of these data structures is based on the kinetic data structure of Basch, Guibas, and Hershberger for maintaining the convex hull of a point set in motion [3]. On top of that data structure we build a kinetization of the rotating calipers algorithm [6, 8], specialized to the desired version of the extent.

3.1 Diameter

The diameter of a point set is the maximum pairwise separation of two points in the set. It is realized by a pair of antipodal vertices of the convex hull. (By *antipodal* points we mean two points on opposite sides of the hull whose supporting lines are parallel.) A standard way to compute the diameter of a static point set is to compute the convex hull, find all pairs of antipodal vertices by a linear scan around the hull boundary, and then identify the pair with maximum separation [6]. In this section we show how to kinetize this algorithm.

Before we proceed, let us dualize the problem, because the computation of antipodal points is easier to describe in the dual setting. In the dual, each point (p, q) of the point set maps to the line $y = px + q$. The upper convex hull of the point set dualizes to the upper envelope of the set of dual lines. Likewise the lower convex hull dualizes to the lower envelope. Each convex hull vertex dualizes to a segment of the envelope, and the range of slopes of the vertex's supporting lines dualizes to the x-interval spanned by the segment.

If we consider the dual envelopes as x-ordered lists of intervals, such that each interval represents a convex hull vertex and its range of supporting slopes, then we can find all antipodal pairs simply by merging the lists for upper and lower dual envelopes in x-order. Any two convex hull vertices whose intervals overlap have a common supporting slope, i.e., they are antipodal.

To kinetize this static algorithm for computing the diameter, we use the merged x-order of the two envelope lists as certificates to guarantee the correctness of the current set of antipodal pairs. We store the merged list in a balanced binary tree so that updates to the list can be performed in time $O(\log n)$ plus time proportional to the number of antipodal pairs affected. When a certificate is violated, it means that two interval endpoints have exchanged places. A constant number of certificates involving those intervals need to be updated to restore the certificates to correctness. When the underlying convex hull changes combinatorially, intervals may be added to or deleted from one of the envelope lists, and we search the binary tree to find the location to modify in the merged list.

Theorem 1. *The data structure for maintaining the diameter is compact and efficient.*

Proof. The underlying convex hull data structure and the kinetic tournament are both compact and efficient. It is also easy to see that the merged list structure is compact: its size is $O(n)$.

To prove efficiency, we must bound the number of events in the merged list structure, under the assumption that the points move with pseudo-algebraic motion. The key quantity to bound is the number of pairs of points that become antipodal over the life of the algorithm, since the list changes only when antipodal pairs change.

We extend the 2-D upper envelope structure into 3-D by considering time as a static third dimension. Each line (dual to a point of the set) becomes a pseudo-algebraic surface when the third dimension is added. The upper envelope at any point in time is given by a 2-D slice through the 3-D upper envelope of surfaces. Because the convex hull of n points in pseudo-algebraic motion changes $O(n^{2+\epsilon})$ times, for any $\epsilon > 0$ [2], the upper envelope of surfaces has the same complexity.

At any instant in time, the antipodal pairs of the hull are determined by the overlay of two 2-D envelopes. Each pairwise overlap between the projections of two envelope edges, one from the upper envelope and one from the lower, corresponds to an antipodal pair of points. When we add time as a third dimension, we see that the total number of antipodal pairs created is equal to the number of envelope surface patches whose projections into the xt-plane overlap. This quantity is bounded by $O(n^{2+\epsilon})$ [1].

The separation of each antipodal pair is a pseudo-algebraic function of time, and hence the upper envelope of these functions has $O(n^{2+\epsilon})$ complexity; the kinetic tournament computes this upper envelope (the diameter) within essentially the same time bound. Theorem 4 of Section 4.1 shows that the total number of different diametral pairs is $\Omega(n^2)$, and hence our kinetic data structure is efficient. □

Note that the data structure is *not* local: one point may belong to $O(n)$ antipodal pairs. It may be possible to achieve locality by making the pairing relationship more sophisticated, but this would require some additional insight.

3.2   Width

The width of a point set is the minimum separation of two parallel lines that sandwich the point set between them. It is well known that one of the lines contains an edge of the convex hull and the other passes through a hull vertex.

In the dual, a convex hull edge maps to a vertex of the upper (or lower) envelope. An antipodal edge-vertex pair, therefore, corresponds to a vertex on the upper or lower envelope whose x-coordinate lies in the x-interval of a segment on the other envelope. To find all antipodal edge-vertex pairs, we merge the x-ordered intervals of the two envelopes, just as in the diameter algorithm, except we note overlapping interval-vertex pairs, instead of interval-interval pairs. To compute the width we simply find the minimum separation among all antipodal edge-vertex pairs.

The kinetic data structure for computing the width is almost identical to the one for the diameter. The only difference is that the basic antipodal pairs are edge-vertex pairs (both primally and dually), and the kinetic tournament computes the minimum separation, rather than the maximum.

Theorem 2. *The data structure for maintaining the width is compact and efficient.*

Proof. Because the width data structure is virtually identical to the diameter structure, the bounds are the same—the worst-case time spent processing events is $O(n^{2+\epsilon})$, for any $\epsilon > 0$. Theorem 6 of Section 4.2 shows that the number of combinatorial changes to the width is $\Omega(n^2)$ in the worst case, so the kinetic data structure is efficient. □

The similarity of the diameter and width data structures masks a rather surprising difference. We expect the number of combinatorial changes to the diameter to be $O(n^{2+\epsilon})$, because there are only $O(n^2)$ pairs of points. Because the width is determined by *triples* of points—two edge endpoints and an opposing vertex—the naïve bound on the number of combinatorial changes to the width is $O(n^{3+\epsilon})$. However, the points of the triples are not independent, and our algorithm shows that the actual number of changes is only $O(n^{2+\epsilon})$.

3.3 Extremal boxes

A common way of reducing the complexity of spatial algorithms is to approximate a complex geometric structure by a rectilinear box. Queries (e.g., intersection tests) are first performed on the box, then on the actual structure only if the approximate test shows it to be necessary. In this way many queries on the complex structure may be avoided. The box approximation is often chosen to be axis-aligned, but in situations in which a better approximation is desired, an arbitrarily oriented box may be computed.

Any of a number of criteria may be used to choose the approximating box. For a 2-D point set, we may wish to minimize the area or the perimeter of a rectangle enclosing the set. We may wish to find the smallest square enclosing the set. We may even wish to maximize the area or perimeter of a rectangle each of whose sides touches the point set.

We can maintain an optimal enclosing box for any of these criteria with a single technique. For concreteness, let us focus on maintaining the enclosing rectangle of minimum area. The basic idea is the same as in previous sections: we maintain the convex hull, extract antipodal points from it, and minimize the rectangle over all the antipodal sets of points. To maintain boxes, we need not just antipodal points, but sets of four points—two antipodal pairs with perpendicular supporting lines.

In the dual setting in which points map to lines, we compute four envelopes—upper, lower, left, and right. These are not all independent: if we split the convex hull at its topmost and bottommost vertices, the left and right envelopes are duals of the corresponding portions of the hull boundaries. We merge the four envelope lists into one list. An interval in the merged list corresponds to a slope range in which the four convex hull vertices supported by lines parallel and perpendicular to the slope are constant. For this slope range and set of four vertices, the minimum area rectangle is trivially computed. By minimizing over all intervals in the merged list, we find the global minimum-area rectangle.

The kinetic data structure for computing extremal boxes is essentially similar to those described above. The difference is that instead of maintaining the merge of two envelope lists, we maintain the merge of four lists. For each interval in the merged list we maintain the minimum-area rectangle enclosing the four extreme points, subject to the condition that the slope of one side of the rectangle must lie inside the range given by the interval. A kinetic tournament on the rectangles selects the one with global minimum area.

Theorem 3. *The data structure for extremal boxes is compact and efficient.*

Proof. Compactness follows as in Theorem 1. Efficiency follows from the lower bound of Theorem 9, which shows that there are $\Omega(n^2)$ combinatorial changes to the minimum-area box, and from the envelope overlay theorem [1]: the number of combinatorially different boxes examined is $O(n^{2+\epsilon})$, since that is the maximum complexity of the overlay of the projections of the four dual envelopes. □

The preceding theorem shows that the number of combinatorially different extremal boxes for points in pseudo-algebraic motion is $O(n^{2+\epsilon})$.

4 Lower bounds with linear motion

In this section we give a collection of lower bounds on the number of combinatorial changes to the extent of a point set when each point moves linearly. Each of our constructions uses cocircular points whose linear motion maintains cocircularity. Note, however, that the lower bounds hold even if we perturb the points slightly to place them in general position.

Let $\bar{c}(\theta) = (\cos\theta, \sin\theta)$ be the point on the unit circle at angle θ from the origin. Suppose that a point p moves linearly along a chord of the unit circle:

$$p(t) = (1-t)\,\bar{c}(\alpha) + t\,\bar{c}(\alpha + \phi).$$

The position of $p(t)$ can be expressed in polar coordinates, $p(t) = (r_p(t), \theta_p(t))$, in terms of $\theta(t) = \tan^{-1}((2t-1)\tan\frac{\phi}{2})$, which varies in the range $[-\phi/2, \phi/2]$ as t varies in $[0, 1]$.

$$r_p(t) = \frac{\cos\frac{\phi}{2}}{\cos\theta(t)}$$

$$\theta_p(t) = \alpha + \frac{\phi}{2} + \theta(t)$$

Note that the initial position $p(0)$ does not appear in the expression for $r_p(t)$. If multiple points start on the unit circle, then move at the same rate along chords of the same length, the points will remain cocircular through the whole motion. If all the motions are clockwise (or all counterclock-

Fig. 1. Cocircularity is preserved by linear motion along equal-length chords

wise), then the angular separation of each pair of points is constant: $\theta_p(t) - \theta_q(t)$ is just the difference of the initial angular positions of p and q. See Figure 1.

4.1 Diameter

In this section we give an $\Omega(n^2)$ lower bound on the number of distinct diametral pairs that can appear in a set of n points moving linearly. We first discuss diametral pairs for points lying on two concentric circles, then specify a particular set of linearly moving points, and finally argue that our set has $\Omega(n^2)$ diametral pairs over time.

Consider points on two concentric circles C_1 and C_2 with radii r_1 and r_2. Suppose that the points on C_i lie on an arc of length at most $\pi r_i/4$, for each $i \in \{1, 2\}$. If a point on C_1 and another on C_2 are collinear with the circles' common center and on opposite sides of it, then their separation is $r_1 + r_2$, and this is the diameter of the set. If there is only one such pair, it is the unique diametral pair.

We define a point set with m stationary points and m moving points, for a total of $n = 2m$ points. The set $P = \{p_0, \ldots, p_{m-1}\}$ of stationary points is defined by

$$p_i = \bar{c}\left(\frac{\pi}{8m^2}i\right).$$

The moving points are $Q = \{q_0, \ldots, q_{m-1}\}$. They move linearly along chords of the unit circle:

$$q_j(t) = (1-t)\bar{c}\left(\frac{7}{8}\pi + \frac{\pi}{8m}j\right) + t\bar{c}\left(\frac{9}{8}\pi + \frac{\pi}{8m}j\right).$$

Theorem 4. *The diameter of the set of n linearly moving points described above is defined by $\Omega(n^2)$ different pairs of points during the time interval $t \in [0, 1]$.*

Proof. As noted above, the points of Q lie on a common circle whose radius varies with time. The angular position of $q_j(t)$ is $\theta_j(t) = \pi + \frac{\pi}{8m}j + \theta(t)$, for $-\frac{\pi}{8} \le \theta(t) \le \frac{\pi}{8}$. Thus Q lies in a constant-size angular range $\theta_{m-1}(t) - \theta_0(t) = \frac{\pi(m-1)}{8m} < \frac{\pi}{8}$. The angular range of P is also less than $\frac{\pi}{8}$.

Points p_i, $q_j(t)$, and the origin are collinear iff

$$\frac{\pi}{8m^2}i + \pi = \theta_j(t) = \pi + \frac{\pi}{8m}j + \theta(t),$$

that is, iff

$$\theta(t) = \frac{\pi}{8m}\left(\frac{i}{m} - j\right).$$

Each (i, j) pair determines a unique value of $\theta(t)$ in the interval $[-\frac{\pi}{8}, \frac{\pi}{8}]$, which corresponds to a unique value of $t \in [0, 1]$; call this value t_{ij}. Thus there are $m^2 = \Theta(n^2)$ distinct values $t_{ij} \in [0, 1]$ such that $(p_i, q_j(t_{ij}))$ is the unique diametral pair of the point set at time t_{ij}. $\qquad\square$

4.2 Width

In this section we give an $\Omega(n^2)$ lower bound on the number of distinct vertex triples that determine the width of a set of n linearly moving points. The construction uses two sets of points, one stationary and one moving, each set cocircular.

Let us define the *slab* of a line segment s to be the set of all points that project perpendicularly onto s. The basis of our construction is the following observation:

Observation 5 *Suppose that the width of a point set is determined by a convex hull edge e and a hull vertex v. Then v lies in the slab of e.*

The stationary points of our set are

$$a = \bar{c}\left(\frac{\pi}{8}\right)$$

$$b = \bar{c}\left(-\frac{\pi}{8}\right)$$

$$p_i = \bar{c}\left(-\pi + \frac{\pi(i - m/2)}{64m^2}\right) + (2\cos\frac{\pi}{8}, 0), \quad \text{for } i \in \{0, \ldots, m\}.$$

The moving points are

$$q_j(t) = (1 - t)\bar{c}\left(-\frac{3\pi}{32} + \frac{\pi}{8m}j\right) + t\bar{c}\left(-\frac{\pi}{32} + \frac{\pi}{8m}j\right), \quad \text{for } j \in \{0, \ldots, m\}.$$

See Figure 2. The total number of points is $n = 2m + 4$. These points lie on two unit circles that intersect in arcs of length $\pi/4$. The intersection points are a and b; the p_i lie in a tight clump at the center of the left arc; the q_j lie in a $\pi/8$ sector of the right arc.

Theorem 6. *The width of the set of n linearly moving points described above is defined by $\Omega(n^2)$ different triples of points during the time interval $t \in [0, 1]$.*

Proof. As in the previous subsection, $q_j(t)$ can be expressed in polar form as

$$r_j(t) = \frac{\cos\frac{\pi}{32}}{\cos\theta(t)}$$

$$\theta_j(t) = -\frac{\pi}{16} + \frac{\pi}{8m}j + \theta(t)$$

for $\theta(t) \in [-\frac{\pi}{32}, \frac{\pi}{32}]$.

In their initial positions, all the points appear on the convex hull, in the counterclockwise order $a, p_0, \ldots, p_m, b, q_0, \ldots, q_m$. In fact, we show in the full paper that all the points appear on the convex hull for all $t \in [0, 1]$, in the same order.

For each j such that $\lceil m/4 \rceil \le j < \lfloor 3m/4 \rfloor$, the convex hull edge (q_j, q_{j+1}) intersects the line $y = 0$ during the angular interval

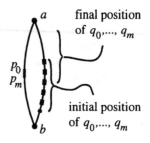

$$\frac{\pi}{16} - \frac{\pi}{8m}\left(j + \frac{3}{4}\right) \le \theta(t) \le \frac{\pi}{16} - \frac{\pi}{8m}\left(j + \frac{1}{4}\right).$$

Fig. 2. The width lower bound construction

(The restriction on j ensures that this is a valid interval of $\theta(t)$.) The full paper shows that during this $\theta(t)$ interval, only the edge (q_j, q_{j+1}) that intersects $y = 0$ satisfies Observation 5, and so only it can determine the width. As $\theta(t)$ varies in the interval, each point p_i becomes antipodal to (q_j, q_{j+1}) in turn, hence determining the width.

We have exhibited roughly $m/2$ convex hull edges, each of which in its turn determines the width with $m + 1$ different hull vertices. Thus the triple of hull vertices determining the width changes $\Omega(n^2)$ times. $\qquad\square$

4.3   Extremal boxes

This section exhibits a configuration of n points in linear motion such that the minimum-area (or minimum-perimeter) enclosing rectangle undergoes $\Omega(n^2)$ combinatorial changes.

The construction involves $n/2$ closely spaced points $p_1, \ldots, p_{n/2}$ that always lie on a circular arc of large radius, rotating counterclockwise around the origin. There are an additional $n/2$ points $q_1, \ldots, q_{n/2}$ near the origin, whose convex hull forms a sequence of squares. In particular, at integer times $t = j$ (for $j = 1, \ldots, n/8$), the convex hull of the q_i will be a square Q_j defined by the four points $q_{4j-3}, \ldots, q_{4j}$. All squares have a side length between 2 and

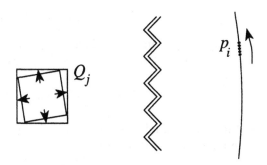

Fig. 3. The lower bound for minimum boxes

3, although each Q_j is slightly bigger than Q_{j-1}. The squares also have different orientations: the base of Q_j makes an angle of $j\theta$ with respect to the x-axis (where θ is a function of n). See Figure 3.

The idea is that at time $t = j$, the p_i will be just below the line L_j through the origin with angle $j\theta$. We will show that the bounding box B has sides parallel to Q_j, and thus its combinatorial description depends on which of the p_i is farthest from the origin in the direction of L_j. Each of the p_i will become the farthest point in turn, as the points rotate through L_j, thereby producing $n/2$ combinatorial changes to the bounding box. This is repeated at times $t = 1, \ldots, n/8$, yielding $\Theta(n^2)$ changes in total.

The following arguments make this construction more precise.

Lemma 7. *Let $p_1, \ldots, p_n$ be points on a circle with radius $r \gg 1$, centered at the origin. Assume that these points satisfy $p_{i,x} > 0$ and $p_{i,y} \in [-1,1]$ for all i, i.e., they lie on a short arc near the positive x-axis. Also let Q be the square whose vertices $q_1, \ldots, q_4$ are the points $(\pm 1, \pm 1)$. Then the minimum-area (or minimum-perimeter) bounding box that encloses the p_i and Q is given by $B = [-1, x_{\max}] \times [-1, 1]$, where $x_{\max} = \max_i p_{i,x}$.*

Proof. We will ignore all of the p_i except for the point p_{i_0} whose x-coordinate is largest (this can only shrink the optimal bounding box). For each angle $\alpha \in [-\pi/4, \pi/4]$, let $B(\alpha)$ denote the unique bounding box that just encloses p_{i_0} and Q, and whose base makes an angle of α with the x-axis. We must show that area (and perimeter) are minimized when $\alpha = 0$.

By symmetry, we can assume that $\alpha \geq 0$. We now consider two cases. First, suppose that $\alpha \leq \alpha_0$, where α_0 is the angle that the lower tangent to p_{i_0} and Q makes with the x-axis. In this case, the height $h(\alpha)$ of the bounding box is determined by the points $(-1, 1)$ and $(1, -1)$, while the width $w(\alpha)$ is determined by $(-1, -1)$ and p_{i_0}. A simple calculation reveals that

$$h(\alpha) = 2\sqrt{2}\sin(\alpha + \pi/4) \approx 2\left(1 + \alpha - \alpha^2/2\right)$$

$$w(\alpha) \geq (x_{\max} + 1)\cos\alpha \approx (x_{\max} + 1)(1 - \alpha^2/2)$$

where the approximations $\sin\alpha = \alpha$ and $\cos\alpha = 1 - \alpha^2/2$ are valid because $x_{\max} \gg 1$ and $\alpha \leq \alpha_0 < 2/(x_{\max} - 1)$. Thus,

$$h(\alpha)w(\alpha) \approx 2\left(x_{\max} + 1\right)\left(1 + \alpha - \alpha^2\right)$$

so that area is minimized when $\alpha = 0$. Similarly, to show that the perimeter is minimized for $\alpha = 0$, we must have

$$2(\alpha - \alpha^2/2) - (x_{\max} + 1)\alpha^2/2 > 0.$$

This is true when $\alpha < 4/(x_{\max} + 3)$, which holds because $\alpha_0 < 2/(x_{\max} - 1)$.

For the case when $\alpha > \alpha_0$, the height is now defined by the points $(-1, 1)$ and p_{i_0}, and is approximately

$$h(\alpha) \approx (x_{\max} + 1)\sin\alpha.$$

It is easy to verify that $h(\alpha)w(\alpha)$ and $h(\alpha) + w(\alpha)$ are monotonically increasing for $\alpha > \alpha_0$. □

The construction now proceeds as follows. For each $i = 1, \ldots, n/8$, the vertices of Q_i are obtained by taking the square Q whose vertices are $(\pm 1, \pm 1)$, and rotating it by an angle of $i\theta$ counterclockwise around the origin. The size of Q_i varies with time, according to the scale factor

$$s_i(t) = 1 + 4\theta(2it - i^2).$$

In other words, each vertex of Q_i is on a linear trajectory through the origin, such that its distance from the origin at time t is $\sqrt{2}s_i(t)$. We let $\theta = 8/n^2$, which ensures that all $s_i(t)$ lie in the range $[1/2, 3/2]$ for $0 \leq t \leq n/8$.

Note that Q_j is clearly the largest square at time $t = j$, since we can rewrite $s_i(t)$ as

$$s_i(t) = 1 + 4\theta[t^2 - (i - t)^2].$$

The following lemma is a slightly stronger version of this.

Lemma 8. *When $j - 1/4 \leq t \leq j + 1/4$, the square Q_j contains all other squares Q_i.*

Proof. Consider two squares Q and Q' whose orientations differ by θ, and such that Q barely contains Q'. It is easy to check that their side lengths satisfy

$$s = s' \sin \theta + s' \cos \theta,$$

which implies that $s/s' \approx 1 + \theta$. Thus for Q_j to contain Q_{j-1}, it is enough to require that

$$s_j \geq s_{j-1} + 2\theta, \tag{1}$$

since then from $s_{j-1} < 2$ we have $s_j \geq (1+\theta)s_{j-1}$. It is easy to check that (1) is satisfied whenever $t \geq j - 1/4$. Since this condition applies to all $i < j$, we have $Q_1 \subset \cdots Q_j$ whenever $t \geq j - 1/4$.

In a similar way, we can show that $Q_j \supset Q_{j+1} \supset \cdots Q_{n/8}$ whenever $t \leq j + 1/4$. □

Finally, the points $p_1, \ldots, p_n$ are placed on a circle C of radius $r = 4/\theta$, equally spaced along an arc of length $1/2$. At time $t = 0$, they all have y-coordinates in the range $[-1/4, 1/4]$. All points move counterclockwise along chords that subtend an angle of $\theta^* = (n/8)\theta$, such that they intersect C at times $t = 0$ and $t = n/8$. The approximate chord length is $r\theta^* = n/2$, so that the speed of each p_i is approximately 4, and the y-coordinates of the points lie in the range $[4t - 1/4, 4t + 1/4]$ (noting that the chords are all nearly vertical for large n).

Now, for each $j = 1, \ldots, n/8$, consider the points near time $t = j$. The bounding box is determined by Q_j and the point p_i that is farthest from the origin along the line L_j that makes an angle of $j\theta$ with the x-axis. Now, at time $t = j - 1/8$, all p_i have y-coordinates in the range $[4j - 3/4, 4j - 1/4]$, and lie below the intersection of L_j with C (at $y \approx 4j$). As each p_i crosses L_j, it is clearly the farthest point in direction L_j. By time $t = j + 1/8$, all the p_i have crossed, and we are done.

We have established the following theorem.

Theorem 9. *The combinatorial description of the minimum-area (or minimum-perimeter) bounding box of n points moving linearly in the plane can change $\Omega(n^2)$ times.*

5 Tight bounds for kinetic convex hulls

In this section we give tight bounds on the number of combinatorial changes that may occur in the convex hull of points moving linearly in the plane. The lower bound construction is an easy application of the linear-motion-on-circles technique of Section 4. The upper bound is an improvement on the known bounds for points in general pseudo-algebraic motion; when specialized to the case of linear motion, it shows that the convex hull may undergo $\Theta(n^2)$ combinatorial changes.

5.1 Lower bound

We exhibit a configuration of $2n$ points in linear motion for which the convex hull undergoes $\Omega(n^2)$ combinatorial changes. This improves the lower bound example given by Sharir and Agarwal [7], which uses quadratic motions.

We define two convoys of oppositely moving points. The points always lie on a common circle (which varies in size), so all are on the convex hull, but their order along the circle changes.

Let

$$p_i(t) = (1-t)\,\bar{c}\left(\frac{\pi}{4n}i\right) + t\,\bar{c}\left(\frac{\pi}{4} + \frac{\pi}{4n}i\right)$$
$$q_j(t) = (1-t)\,\bar{c}\left(\frac{\pi}{4} + \frac{\pi}{8n^2}j\right) + t\,\bar{c}\left(\frac{\pi}{8n^2}j\right),$$

for $i,j \in \{1,\dots,n\}$. At any time $t \in [0,1]$, all the p_i and q_j lie on a common circle with radius $r(t) = \cos\frac{\pi}{8}/\cos\theta(t)$, for $\theta(t) = \tan^{-1}((2t-1)\tan\frac{\pi}{8})$. The angular position of $p_i(t)$ is $\theta(p_i, t) = \pi i/4n + \pi/8 + \theta(t)$ and the angular position of $q_j(t)$ is $\theta(q_j, t) = \pi j/8n^2 + \pi/8 - \theta(t)$. Point p_i coincides with q_j at

$$\theta(t) = \frac{\pi}{8n}\left(\frac{j}{2n} - i\right).$$

Thus each (i,j) pair determines a unique $\theta(t) \in [-\frac{\pi}{8}, 0]$ at which p_i and q_j exchange on the convex hull. We have established the following theorem.

Theorem 10. *There is a set of n linearly moving points whose convex hull undergoes $\Omega(n^2)$ combinatorial changes as the points move.*

5.2 Upper bound

We bound the number of combinatorial changes to the convex hull in terms of the number of times any three points become collinear. It is well known that if the point trajectories are algebraic of degree k, then three points become collinear at most $s = 2k$ times. The theorem below shows that in this case there are $O(n\lambda_{2k}(n))$ changes to the convex hull. This improves the bound of $O(n\lambda_{2k+2}(n))$ given in [7]. In particular, it implies that for linear motion the number of changes is $O(n^2)$, matching the lower bound of the preceding section.

Theorem 11. *Given n points moving in the plane such that no three points become collinear more than s times, the combinatorial description of their convex hull changes at most $O(n\lambda_s(n))$ times.*

Proof. Let the points be identified by integers, $P = \{1,\dots,n\}$, and define the *left-neighbor* function $l_i(t)$ as follows. If i does not belong to the convex hull at time t, then $l_i(t) = \epsilon$. Otherwise, $l_i(t)$ is the point j on the convex hull that is adjacent to i in the counterclockwise direction.

For each i, let L_i be the sequence of values assumed by $l_i(t)$ as t ranges from $-\infty$ to ∞. We remove all occurrences of ϵ from L_i, and replace any strings of identical symbols by a single occurrence, to yield a reduced sequence L_i^*.

First, we show that $\sum |L_i^*|$ is an upper bound on the number of changes to the convex hull (where $|S|$ denotes the length of a sequence S). We do this by charging each change to a unique symbol in some L_i^*.

Observe that the convex hull can change in only two ways (assuming no degeneracies): either a current vertex is deleted, or a new vertex is inserted. Suppose that a current vertex i_1 is being deleted, and let i_0 and i_2 be its left and right neighbors just before the deletion. In this case, L_{i_2} will contain the substring $i_1 i_0$, and we charge the deletion to the symbol i_0. Similarly, if i_1 was just inserted, then L_{i_2} contains the substring $i_0 i_1$, and we charge the insertion to the symbol i_1. It is clear that no symbol is charged twice in this way, and that all charged symbols are present in the reduced sequences L_i^* (since each one is preceded by a different non-ϵ symbol).

Furthermore, each L_i^* is a $(n-1, s)$ Davenport-Schinzel sequence. To see this, observe that when symbol j appears in L_i^*, it means that ij is an edge of the convex hull, and thus all triangles ijk have a signed area that is positive. Similarly, if k appears in L_i^*, then all triangles ikj have positive area (and all triangles ijk have negative area). Thus, if the alternation $j \ldots k$ appears in L_i^*, then the signed area of triangle ijk is zero at some intermediate time, implying a collinearity of i, j, and k.

Given that any three points are collinear at most s times, there are at most s alternations between any two symbols j and k. Thus, each L_i^* is a $(n-1, s)$ Davenport-Schinzel sequence (recalling that L_i^* contains no repeated symbols by construction), and we have $\sum |L_i^*| \le n\lambda_s(n)$. $\qquad\square$

References

1. P. Agarwal, O. Schwarzkopf, and M. Sharir. The overlay of lower envelopes and its applications. *Discr. Comput. Geom.*, 15:1–13, 1996.

2. M. J. Atallah. Some dynamic computational geometry problems. *Comput. Math. Appl.*, 11:1171–1181, 1985.

3. J. Basch, L. Guibas, and J. Hershberger. Data structures for mobile data. In *Proceedings of the Eighth Annual ACM-SIAM Symposium on Discrete Algorithms*, pages 747–756, 1997.

4. Y.-J. Chiang and R. Tamassia. Dynamic algorithms in computational geometry. *Proc. IEEE*, 80(9):1412–1434, September 1992.

5. D. Eppstein. Average case analysis of dynamic geometric optimization. *Comp. Geom.: Theory and Appl.*, 6:45–68, 1996.

6. F. P. Preparata and M. I. Shamos. *Computational Geometry*. Springer-Verlag, New York, 1985.

7. M. Sharir and P. K. Agarwal. *Davenport-Schinzel Sequences and Their Geometric Applications*. Cambridge University Press, New York, 1995.

8. G. Toussaint. Solving geometric problems with the "rotating calipers". In *Proceedings of IEEE MELECON '83*, 1983.

Finding Cores of Limited Length

Stephen Alstrup[1]   Peter W. Lauridsen[1]
Peer Sommerlund[1]   Mikkel Thorup[1]

Department of Computer Science, University of Copenhagen, Universitetsparken 1,
DK-2100, Denmark (e-mail : (stephen,waern,peso,mthorup)@diku.dk)

Abstract. In this paper we consider the problem of finding a core of limited length in a tree. A core is a path, which minimizes the sum of the distances to all nodes in the tree. This problem has been examined under different constraints on the tree and on the set of paths, from which the core can be chosen. For all cases, we present linear or almost linear time algorithms, which improves the previous results due to Lo and Peng, J. Algorithms Vol. 20, 1996 and Minieka, Networks Vol. 15, 1985.

1 Introduction

In 1971 Goldman [3] gave a linear time algorithm for determining a node in a tree, called a *median*, minimizing the sum of the distances to all other nodes. Later, in 1982, Slater [14] proposed to determine the path which minimizes this sum, called a *core*. More specificly, a core is path for which the total sum of distances from all nodes to the path is minimum among all paths. In 1980 Morgan and Slater [12] gave a linear time algorithm for determining a core in a tree in which all edges have length one. This algorithm is easily extended to a tree in which edge lengths are arbitrary nonnegative. A simpler linear time algorithm for finding a core in a tree with arbitrary nonnegative edge lengths was given in 1993 by Peng, Stephens and Yesha [13].

In 1983 Minieka and Patel [11] proposed the problem of finding a path of length l that minimizes the distance sum over all paths of length l. Such a path is called a core of length l. Since we cannot be certain that a path of length l exist in a tree, partial edges are usually allowed in cores of length l. Minieka and Patel do not give an algorithm for determining a core of specified length, but list a number of problems in giving such an algorithm. For instance the core of length l can have distance sum larger than a core of length $< l$. In 1985, Minieka showed that a core of a specified length can be found in $O(n^3)$ time [10]. In 1996, Lo and Peng [8] gave an $O(n \log n)$ algorithm for finding a core of a specified length in the case where all edges have length one. Furthermore they claim that their algorithm is easily extended to cases where partial edges are allowed and the edges have arbitrary nonnegative length. This is true, but if the edge weights are arbitrary nonnegative integers the complexity increases to $O(nl \log n)$, which for constant l is still $O(n \log n)$.

In 1993 Hakimi, Labbé and Schmeichel [4] examined the problem of finding a core with length $\leq l$ using either full or partial edges. This is a natural extension of the core problem since in both cases paths with length $< l$ can exist which have cost less than any path of length $= l$ [11]. For the case allowing partial edges they show that the $O(n^3)$ algorithm [10] can be used. For full edges they show the existence of a polynomial time algorithm for the problem. However it is also shown that for locating the core in an arbitrary network the problem becomes NP-hard. Finding cores in trees using parallel algorithms have also been examined, see e.g. [7,8]. Finally we note that Minieka's $O(n^3)$ algorithm is easily modified to all cases mentioned above.

To summarize : Placing a core in a tree has been investigated for partial/full edges, core length $= l/\leq l$ and uniform/arbitrary edge weights. For the cases of uniform edge weights an $O(n \log n)$ algorithm has been given [8]. For this case we present an $O(n)$ algorithm. For arbitrary edge weights the algorithms given so far have complexities $O(nl \log n)$ and $O(n^3)$ [8,10]. In this paper we give two algorithms for all cases. The first determines the core in $O(nl)$ time and the second uses $O(n \log n \alpha(n, n))$ time. If only full edges are allowed the complexity of the second algorithm is $O(n \log n)$. The factor $\alpha(n, n)$ comes in because of a strong relation between the core problem and Davenport-Schinzel sequences [2]. A strong relation between algorithmic geometry and Davenport-Schinzel sequences has previously been established.

In [11], Minieka and Patel studied conditions under which there is a core containing a median. If their conditions are not satisfied, they write that "we do not know if a core of length l will contain [the median] m. Unfortunately, this situation remains unexplored and as this question remains open, the development of an efficient algorithm for locating a core of a specified length remains a difficult problem." We solve Minieka and Patel's question in [1, Figure 1], presenting a tree with 35 nodes in which the core does not contain the median.

2 Preliminaries

Let T be a tree with node set $V(T)$ and edge set $E(T)$ and let T be rooted at an arbitrary node. For a node $v \in V(T)$, T_v is the subtree in T rooted at v, hence T_v is the tree induced by v and descendants of v. With each edge $(v, w) \in E(T)$ a nonnegative integer length $length(v, w)$ is associated. The path from a node u to a node v in T is denoted $P_{u,v}$ and the length of a path P is denoted $|P|$.

For a path Q in T we use $dist(v, Q)$ to denote the distance from a node $v \in V(T)$ to Q, thus $dist(v, Q) = min_{u \in Q}|P_{v,u}|$. In the following we will allow endpoints of paths to be points on edges, thus the "u" in this definition should denote a point on Q. More precisely if $(v, w) \in E(T)$, $length(v, w) = x$ and P is a path of length $y < x$ starting at v towards w, then $dist(w, P) = x - y$. The case where points on edges are allowed is called *partial* and the case where only full edges are allowed is called *discrete*. Since we are considering distances between points we will also use the notation $dist(x, y)$ to denote the distance between the points x and y in T, hence $dist(x, y) = |P_{x,y}|$.

For a path P we define $cost_T(P) = \Sigma_{v \in V(T)} dist(v, P)$. Let $\mathcal{P}$ be a set of paths. We say $Q \in \mathcal{P}$ is a core with respect to $\mathcal{P}$, if $cost_T(Q) = min\{cost_T(P)|P \in \mathcal{P}\}$.

We will use the name $Core(\mathcal{P})$ to denote a core with respect to $\mathcal{P}$. The sets $\mathcal{P}$ considered as possible core candidates in this paper satisfies the following conditions:

(a) The paths have length $= l$ or length $\leq l$, where l is a nonnegative integer.
(b) The paths are partial or discrete.

The problems can furthermore be divided into whether edge lengths are uniform (i.e. have length one) or arbitrary nonnegative integers. We thus consider eight problem instances, however in the case of uniform edge lengths, the partial and discrete problems are the same. Analogous to other path problems the actual problem in finding a core is not so much finding a path, but finding the cost. We will thus concentrate on finding $cost_T(Core(\mathcal{P}))$ in this paper. The algorithms described are easily extended to finding a path attaining the cost of the core.

As mentioned in the introduction the core problem is a generalization of the problem of finding a median. In the median problem it is common that nodes have costs associated. Thus, in order to get the full generalization, this should also be the case for the core problem. More specifically this would mean that for a path P, $cost_T(P) = \Sigma_{v \in V(T)} dist(v, P) * cost(v)$. For reasons of clarity we will assume that $cost(v) = 1$ for all $v \in V$ unless anything else is stated. The formulas and lemmas deriven are easily extended to the case where costs of nodes are arbitrarily nonnegative.

3 An $O(nl)$ algorithm for all cases

In this section we give an $O(nl)$ algorithm for finding the core of a tree in all cases. To simplify the description of the algorithm we assume that partial edges are allowed and the length of the core should be $= l$. With minor changes the algorithm will, within the same complexity, also solve the problem in the other cases. These changes are briefly discussed at the end of the section.

For each subtree we will compute a core containing the root of the subtree. We define $MinCost(v)$ as the cost of a core containing v in T_v. By considering $MinCost(v)$ for all v we can compute the cost of the core of T. In order to compute $MinCost(v)$ we will need the values defined as follows.

Definition 1. We define $Size_{down}(v)$ to be the number of nodes in T_v. Furthermore $Sum_{down}(v)$ is defined as $\Sigma_{w \in V(T_v)} |P_{w,v}|$. Analogously $Size_{up}(v)$ is the number of nodes in $V(T) \backslash V(T_v)$ and $Sum_{up}(v) = \Sigma_{w \in V(T) \backslash V(T_v)} |P_{w,v}|$. Finally we define the extended Sum, $Sum^*_{down}(v)$ as $\Sigma_{w \in V(T_v)} |P_{w,parent(v)}|$, hence $Sum^*_{down}(v) = Sum_{down}(v) + length(parent(v), v) * Size_{down}(v)$.

Using recursion formulas the $Size$ and Sum values can be found in linear time by traversing T in bottom-up and top-down fashions [1]. $MinCost(v)$ can be found by combining the cost of two paths that start at v and propagates towards two different children of v. The cost of such paths will be denoted as $DownCost(v, k)$. More precisely we have

Definition 2. $DownCost(v, k)$ is the cost of the minimum cost path in T_v of length k, which starts at the root of T_v, hence $DownCost(v, k) = min\{cost_{T_v}(P_{v,x}) \mid x$ is a point in $T_v, |P_{v,x}| = k\}$. The extended $Downcost$, $DownCost^*(v, k)$, is the cost of the minimum cost path of length k, which starts at the root of $T_v \cup \{(parent(v), v)\}$.

To compute $MinCost(v)$ we will compute $DownCost^*(w, k)$, for $k = 0..l$ for all children w of v. Note that since the largest distance from a leaf to v could be $< l$, we can only compute $DownCost$ values for $k = 0..min\{l, max\{|P_{v,x}| \mid x \in T_v\}\}$. To make formulas more simple we will w.l.o.g. ignore this in the following. The lemma below shows how the $DownCost$ values can be computed bottom-up.

Lemma 3. *Let $v \in V(T)$ and let c denote the length of the edge from v to its parent in T. We have the following:*
$DownCost^*(v, k) = Sum_{down}(v) + (c - k) * Size_{down}(v)$, *if $c \geq k$*
$DownCost^*(v, k) = DownCost(v, k - c)$, *otherwise.*
$DownCost(v, k) = 0$, *if v is a leaf.*
$DownCost(v, k) = Sum_{down}(v) - max\{Sum_{down}^*(w) - DownCost^*(w, k) \mid w$ is a child of $v\}$, *otherwise.* □

If we traverse T in a bottom-up fashion and apply the formulas in lemma 3, we can compute $DownCost$ values for a node v with j children in $O(j * l)$ time. We can therefore compute $DownCost$ values for all nodes in $V(T)$ in $O(nl)$ time.

We will now show how to compute $MinCost(v)$. If v has only one child $MinCost(v) = DownCost(v, l)$. Assume that v has only two children, w_1 and w_2. We then have values $DownCost^*(w_1, k)$ and $DownCost^*(w_2, k)$ for $k = 0..l$. We can thus in $O(l)$ time compute $MinCost(v)$ using the formula :

$$MinCost(v) = min\{DownCost^*(w_1, k) + DownCost^*(w_2, l-k) \mid k = 0..l\} \quad (1)$$

In the general case where v has more than two children we do the following. Assume that $w_1, ..., w_j$ are the children of v. We cannot use formula 1 directly, since the best core involving nodes from two subtrees, say T_{w_1} and T_{w_2}, is $min\{DownCost^*(w_1, k) + DownCost^*(w_2, l-k) \mid k = 0..l\} + \Sigma_{3 \leq i \leq j} Sum_{down}^*(w_i)$. To get a simple formula we will instead compare how much is saved by using a path in any subtree T_{w_i}. To be more specific we could express formula 1 as

$$MinCost(v) = Sum_{down}(v) - max\{Save^*(w_1, k) + Save^*(w_2, l - k) \mid k = 0..l\} \quad (2)$$

where $Save^*(w, k) = Sum_{down}^*(w) - DownCost^*(w, k)$.

We now proceed as follows. First we compute a core candidate for the first two children by applying formula 2. We have thus computed a list of $Save(\cdot, k)$ values for both w_1 and w_2. These lists are now merged into one, by taking $min\{Save^*(w_1, k), Save^*(w_2, k)\}$ for each $k = 0..l$. By using the merged list of $Save$ values together with $Save^*(w_3, k)$ values in formula 2, we can compute $MinCost$ in the tree $T_{w_1} \cup T_{w_2} \cup T_{w_3} \cup \{v\}$. By continuing this process we find $MinCost(v)$.

Before we state the main theorem of this section we look at the case, in which a core should have length $\leq l$. In order to find $MinCost(v)$ in this case we only need to ensure that the *DownCost* values are correct. More precisely we should have $DownCost^*(v, k) = min\{DownCost^*(v, j) \mid 0 \le j \le k\}$ for any node v. By using the *DownCost* values computed using lemma 3 as a basis, this is however easily obtained in $O(l)$ time. Finally we can modify the algorithm to handle the discrete case, by changing the first part of the formula for $DownCost^*(v, k)$ in lemma 3 to: $DownCost^*(v, k) = \infty$, if $length(parent(v), v) \ge k$. Since the cost of a core in T is $MinCost(v) + Sum_{up}(v)$ for some v, we conclude:

Theorem 4. *Given a tree T and a nonnegative integer l, let $\mathcal{P}$ be a set of discrete or partial paths with length $= l$ or $\le l$. We can compute $cost(Core(\mathcal{P}))$ in $O(nl)$ time.* $\qquad \Box$

4 An almost linear algorithm for all cases

In this section we give an $O(n \log n \gamma(n))$ for finding a core in all cases listed in section 2. The function $\gamma(n)$ depends on whether paths are partial, in which case $\gamma(n) = O(\alpha(n, n))$, or discrete, in which case $\gamma(n) = O(1)$. We first present an algorithm with complexity $O(n * h * \gamma(n))$, where h is the height of the tree, measured in the number of edges. Secondly we show how to compress the tree so that $h = O(\log n)$, which establishes the promised complexity.

In this section we assume that T is binary, if this is not true a trivial binarization can be computed in linear time [1]. In the following subsections, we will restrict our attention to looking for a core should of length $= l$ allowing for partial edges. At the end of the section, we will outline how to deal with the other cases.

4.1 The structure of *DownCost*

In the following, we will discuss how to represent *DownCost* so as to facilitate an efficient computation of *MinCost*.

For a given node v, we can draw $DownCost(v,)$ in a coordinate system as follows. First, for every point a in T_v, where a may be on the middle of an edge, insert the point $(dist(v, a), cost_{T_v}(P_{v,a}))$. Clearly, the points on an edge form a straight line. $DownCost(v, \cdot)$ is now the lower envelope of the inserted points, i.e. $DownCost(v, k)$ is the minimum inserted y-value for $x = k$. Now

Lemma 5 ([6,15]). *The lower envelope of q straight line segments jumps at most $\Theta(q\alpha(q))$ times between the segments.* $\qquad \Box$

Thus $DownCost(v, \cdot)$ is a piecewise linear function, dividing into $m = O(|T_v|\alpha(|T_v|))$ pieces. Such a function could, say, be represented as a sequence of point pairs:
$$((x_0, y_1), (x_1, z_1)), ((x_1, y_2), (x_2, z_2)) \cdots, ((x_{m-1}, y_m), (x_m, z_m))$$
where (x_{i-1}, y_i) and (x_i, z_i) are the boundary points of the ith piece. Thus $DownCost(v, x_i) = min\{z_i, y_{i+1}\}$ and if $x_{i-1} < x < x_i$, $DownCost(v, x) = y_i + \frac{(z_i - y_i)}{(x_i - x_{i-1})}(x - x_{i-1})$. If $x < x_0$ or $x > x_m$, we define $DownCost(v, x) = \infty$. We refer to the x_i as *break points*. In the following we will think of any piecewise

linear function f, as being of the above form, and by $|f|$ we then denote the number of break points.

Lemma 6. *In the following, let f and g be piecewise linear functions and let $\delta, \Delta, a, b \in \Re$.*

- *Define $f_1 : x \mapsto f(x + \delta) + \Delta$. Then $|f_1| = |f|$ and f_1 can be constructed in time $O(|f|)$.*
- *Define $f_2 : x \mapsto \min\{f(x), g(x)\}$. Then $|f_2| \leq 2(|f| + |g|) - 1$, and f_2 can be constructed in time $O(|f| + |g|)$.*
- *Define $f_3 : x \mapsto f(x) + g(x)$. Then $|f_3| \leq |f| + |g|$, and f_3 can be constructed in time $O(|f| + |g|)$.*
- *$\mu = \min_{a \leq x \leq b} f(x)$ is found in time $O(|f|)$.*

Proof. Concerning f_1, note that we just need to subtract δ from all the x_i, and add Δ to all the y_i and z_i.

To prove $|f_2| \leq 2(|f| + |g|) - 1$, consider any break point p of f_2 which is neither a break point of f nor of g. Then p is the intersection of two straight-line segments of f and g, but then f_2 cannot break again until either f or g has broken. Similarly, the breakpoint before p in f_2 must be from f or g.

Now f_2 is constructed by a merge style procedure, where in each step, we either identify a new piece of f_2, or finish the processing of a piece from f or g. The time of this procedure is $O(|f_2| + |f| + |g|) = O(|f| + |g|)$.

To prove $|f_3| \leq |f| + |g| - 1$, we simply observe that any break point in f_3 must be a breakpoint in f or in g. The construction of f_3 is done in a merge style procedure in time $O(|f| + |g|)$. □

4.2 An $O(nh\alpha(n))$ algorithm

In this subsection, we are so far still restricting our attention to cores of a length $= l$, on which partial edges are allowed. In particular, this means that we are working with piecewise linear functions

Lemma 7. *Let v be a node with children w_1 and w_2. Given (representations of) $DownCost(w_1, \cdot)$ and $DownCost(w_1, \cdot)$, we can construct $DownCost^*(w_1, \cdot)$, $DownCost^*(w_2, \cdot)$, $DownCost(v, \cdot)$, and $MinCost(v)$ in time $O(|DownCost(w_1, \cdot)| + |DownCost(w_2, \cdot)|) = O(|T_v|\alpha(|T_v|))$.*

Proof. We find $DownCost^*(w_1, x)$ as the concatenation of $((0, Sum^*_{Down}(w_1)))$, $(length(v, w_1), Sum_{Down}(w_1))$ and $DownCost(w_1, x - length(v, w_1))$. $DownCost^*(w_2, \cdot)$ is constructed symmetrically. Now $DownCost(v, x)$ is found as $\min\{DownCost^*(w_1, x) + Sum^*_{Down}(w_2), DownCost^*(w_2, x) + Sum^*_{Down}(w_1)\}$. Finally $MinCost(v) = \min_{0 \leq k \leq l} DownCost^*(w_1, k) + DownCost^*(w_2, l - k)$. Thus, by lemma 6, all the desired values are found in time $O(|DownCost(w_1, \cdot)| + |DownCost(w_2, \cdot)|)$. By Lemma 5, $O(|DownCost(w_1, \cdot)| + |DownCost(w_2, \cdot)|) = O(|T_{w_1}|\alpha(|T_{w_1}|) + |T_{w_2}|\alpha(|T_{w_2}|)) = O(|T_v|\alpha(|T_v|))$ □

Theorem 8. *The core of a tree can be computed in $O(n\alpha(n)h)$, where h is the height of the tree.*

Proof. We apply Lemma 7 bottom-up on all vertices v. For a given vertex v, the computation time is $O(|T_v|\alpha(|T_v|))$. Since no vertex w participates in more than h trees T_v, the result follows. □

4.3 An almost linear algorithm

In the above algorithms, whenever we visit a node v, we look for an optimal core with v as the top-most node. As a result, a node is involved in a computation every time we visit one of its ancestors, and hence our complexity has the height h of the tree as a multiplicative factor.

In this section, we will replace the h-factor by a $(\log n)$-factor. When visiting a node v, we will still look for a core containing v, but the "subtree" it is restricted to, will no longer just be descending from v. Based on balancing techniques, we will assign subtrees to nodes, so that each node is involved in only $O(\log n)$ subtrees. The previous algorithms are then modified to work with these new subtrees.

Flattening a tree Consider a tree T with n nodes. We will now construct a *flattened* version $F(T)$ of T of height $O(\log n)$. The tree $F(T)$ for T will only be used to describe in which subtree we look for a core. Thus, cost, length etc. of additional nodes/edges in $F(T)$ have no meaning, since the additional nodes are only used to explain in which the subtree computation is done.

In order to obtain a tree with height limited to $O(\log n)$ we use heavy path division, as described by Harel and Tarjan [5]. First for each internal $v \in T$, let $heavy(v)$ denote the child of v with the maximum number of descending nodes. In case of a tie, we pick $heavy(v)$ as the left child. The other child of v is called *light*. We say that the edge from a light child to its parent is a light edge. Edges which are not light are called heavy. The heavy edges partition the nodes in T on heavy paths, such that each node belongs to exactly one heavy path. To compress the heavy path we will not follow the standard by Harel and Tarjan, but instead use the following lemma.

Lemma 9 ([9]). *Given a sequence $d_1 \cdots d_m$ of positive real weights, in time $O(m)$, we can construct an ordered binary tree, such that the depth of the ith leaf is $O(\lceil \log D - \log d_i \rceil)$ where $D = \sum_{i=1}^{m} d_i$.* □

$F(T)$ is constructed by taking each heavy path, and replace it by the weight balanced tree described in the above lemma. More specifically, let $P = v_1 \cdots v_m$ be a heavy path in T and let d_i denote the number of descendants of the light child of v_i, below denoted as $w(v_i)$. We then construct an ordered binary tree as described in the above lemma and let the i'th leaf be v_i.

Lemma 10. *$F(T)$ has height $O(\log n)$.*

Proof. Let $v_1, \ldots, v_k$ be the nodes from T that we meet on a path from a leaf v_1 to the root in $F(T)$. For each heavy path on this path we meet one node v_i, thus $k \le \log_2 n$. To get from v_i to v_{i+1} we generally traverse an edge from one balanced tree to another, and $O(\lceil \log w(v_{i+1}) - \log w(v_i) \rceil)$ balance edges, that is, $O(2 + \log w(v_{i+1}) - \log w(v_i))$ edges. Thus a total of

$$\sum_{i=1}^{k} O(2 + \log w(v_{i+1}) - \log w(v_i)) = O(\log n + \log w(v_k)) = O(\log n) \qquad \square$$

We note that $F(T)$ only is an abstraction we will use to explain the complexity of finding a core in T. In [1, Figure 2] the construction of $F(T)$ is illustrated.

Flattened core computation Consider the weight balanced tree t over some heavy path $P = v_1 \cdots v_m$ where v_1 is the node nearest the root. Let x be a node in t, and let $P(x) = v_i \cdots v_j$, $i \le j$, be the segment of P descending from x. Let $tree(x)$ be the subtree in T descending from v_i excluding the subtree rooted at the heavy child of v_j, that is, $tree(x)$ consists of $P(x)$ as well as all light subtrees of nodes on $P(x)$. For a node x let v_i and v_j be as above, $lower(x) = v_j$ and $upper(x) = v_i$. Let T^v be the subtree obtain by deleting the heavy child of v and its descendant from T. If v is a leaf $T^v = T$. Thus, $tree(x) = T^{lower(x)} \cap T_{upper(x)}$. Note that if $x \in T$, $P(x) = x$.

We say that a path Q in T *belongs* to x if it is contained in $tree(x)$ and there is no child y of x in the flattened tree such that Q is contained in $tree(y)$. Clearly, for any path Q in T we have a unique node x that Q belongs to. Thus, we identify the core, if for each node x in the flattened tree, we find the path belonging to it, minimizing the cost in all of T. In order to facilitate a bottom-up computation of cores in the flattened tree, we need some functions analogous to the function $DownCost$ from the previous section.

For any node x in F, define the restricted upcost, $RestrUpCost(x, k)$ as the minimum cost in $T^{lower(x)}$ of a path of length k in $tree(x)$ starting in $lower(x)$. Similarly, define the restricted DownCost, $RestrDownCost(x, k)$, as the minimum cost in $T_{upper(x)}$ of a path of length k in $tree(x)$ starting in $upper(x)$. The realization that both cost functions are needed is a main point in deriving an efficient algorithm. Observe that both $RestrUpCost$ and $RestrDownCost$ are the lower envelopes of a set of straight line segments, one for each edge in $tree(x)$. Hence both of them have $O(|tree(x)| \alpha(|tree(x)|))$ break points.

We are now ready to describe the bottom-up computation of optimal cores. The vertices x of $F(T)$ are visited in bottom-up order. For each x, we make a series of computations, each taking $O(|tree(x)| \alpha(|tree(x)|))$ time. In the computation, it is important whether x is a node in T. Note that the computation will determine $DownCost(v, \cdot)$ for each root of a heavy path in T.

- Let x be a leaf in T. For $k = 0$, $RestrDownCost(x, k) = 0$ and $RestrUpCost(x, k) = Sum_{up}(x)$. For $k > 0$, each of the above values become ∞.
- If x is in T but not a leaf and w is the light child of x, $RestrDownCost(x, k) = DownCost^*(w, k) + Sum^*_{down}(heavychild(x))$ and $RestUpCost(x, k) =$

$DownCost^*(w, k) + Sum_{up}(x)$. If x does not have a light child, replace $DownCost^*(w, k)$ by 0 in the above formulas.

- If x is the root of the balanced tree over some heavy path, $DownCost(upper(x), k) = RestrDownCost(x, k)$. $DownCost^*(upper(x), \cdot)$ is computed from $DownCost(upper(x), \cdot)$ as described in the previous section.

- If x is not in T and x has children y_1 and y_2 where $P(y_1)$ is above $P(y_2)$, $RestrDownCost(x, k) =$
min$\{RestrDownCost(y_1, k), RestrDownCost(y_2, k-\delta)+\Delta, f(x)\}$. Here $\delta = dist(upper(y_1), upper(y_2))$ and

$$\Delta = \sum \{Sum^*_{down}(w) | w \text{ is a light child of a node on } path(y_1)\}.$$

Finally f is the function corresponding to the edge $(lower(y_1), upper(y_2))$ which has boundary points $(dist(upper(y_1), lower(y_1)), \Delta + Sum^*_{Down}(lower(y_1)))$ and $(dist(upper(y_1), upper(y_2)), \Delta + Sum_{Down}(lower(y_1)))$.

The function $RestrDownCost(x, \cdot)$ is computed symmetrically.

Assuming that we are not in the (trivial) case, where the core is contained in the edge $(lower(y_1), upper(y_2))$, we can finally compute $MinCost(x)$ as
min$_k \{RestrDownCost(y_2, k)+$
$RestUpCost(y_1, l - k - length(upper(y_2), lower(y_1)))\}$.

The above computation takes time $|tree(x)|\alpha(|tree(x)|, |tree(x)|)$. Since each node participates in $O(\log n)$ $tree(x)$-values, we conclude

Theorem 11. *For a tree T with n nodes, lengths on the edges, and weights on the nodes, we can solve the discrete core problem in $O(n \log n)$ time and the partial core problem in $O(n \log n\alpha(n))$ time.* $\square$

We will now briefly consider the other cases of cores. For the case where cores should have length $\leq l$, for each leaf w in T_v, for all $k \geq dist(v, w)$, we have to insert the point (k, c), where c is the cost in T_v of the path from v to w. That is, each leaf gives rise to an extra horizontal line in our coordinate system. Asymptotically, this does not affect any of the above bounds, so again we get that $DownCost(v, \cdot)$ is a piecewise linear function, dividing into $O(|T_v|\alpha(|T_v|))$ pieces.

In the discrete case, we only need to store the at most $|T_v|$ points (k, c) where k is the distance from v to some node in T and c is the least cost of a path to such a node. Thus, we derive the same complexity for these cases.

5 A linear time algorithm for the uniform cases

In this section we assume that all edges have length one. This means that the discrete and partial cases are the same. We present a linear time algorithm for finding a core of length $= l$ or $\leq l$. Becasue of the lack of space will we only give the main ideas in the algorithm. A more detailed description can be found in [1]. First we speed up the algorithm from section 3 by processing simple paths in the tree differently. Essentially we do as follows. For a node with only one child we "copy" the *DownCost*-values, giving an algorithm with complexity $O(bl + n)$, where b is the number of leaves in the tree. Secondly we define a *leaf tree* as a subtree in T, in which all nodes has less than $(l/2) - 1$ descendants. Let R be the tree from which all leaf trees has been removed. The tree R has $O(n/l)$ leaves. Now we can compute $MinCost(v)$ for each node $v \in R$ in $O(n)$ time. The best core of length $\leq l$ in any leaf tree is computed by using the linear time algorithm for finding a core of unlimited length [12]. It thus only remains to find the core candidates, which contain nodes from both R and leaf trees. In order to do this, we compute in linear time $DownCost(r, \cdot)$ for any root, r, of a leaf tree, in a top-down fashion. The discussion above yields the following:

Theorem 12. *In the case where all edges have length one, we can compute* $cost(Core(\mathcal{P}))$ *in* $O(n)$ *time.* □

References

1. S. Alstrup, P.W. Lauridsen, P. Sommerlund, and M. Thorup. Finding cores of limited length. Technical report, Department of Computer Science, University of Copenhagen, 1997. See also http://www.diku.dk/ stephen/newpapers.html.
2. H. Davenport and A. Schinzel. A combinatorial problem connected with differential equations. *Amer. J. Math.*, 87:684–694, 1965.
3. A.J. Goldman. Optimal center location in simple networks. *Transportation Sci.*, 5:212–221, 1971.
4. S.L. Hakimi, M. Labbé, and E.F. Schmeichel. On locating path- or tree-shaped facilities on networks. *Networks*, 23:543–555, 1993.
5. D. Harel and R.E. Tarjan. Fast algorithms for finding nearest common ancestors. *Siam J. Comput*, 13(2):338–355, 1984.
6. S. Hart and M. Sharir. Nonlinearity of davenport-schinzel sequences and of general path compression schemes. *Combinatorica*, 6:151–177, 1986.
7. W. Lo and S. Peng. An optimal parallel algorithm for a core of a tree. In *International conference on Parallel processing*, pages 326–329, 1992.
8. W. Lo and S. Peng. Efficient algorithms for finding a core of a tree with a specified length. *J. Algorithms*, 20:445–458, 1996.
9. K. Mehlhorn. *Data Structures and Algorithms 1: Sorting and Searching*. EATCS. Springer, 1 edition, 1984.
10. E. Minieka. The optimal location of a path or tree in a tree network. *Networks*, 15:309–321, 1985.
11. E. Minieka and N.H. Patel. On finding the core of a tree with a specified length. *J. Algorithms*, 4:345–352, 1983.
12. C.A. Morgan and P.J. Slater. A linear algorithm for a core of a tree. *J. Algorithms*, 1:247–258, 1980.
13. S. Peng, A.B. Stephens, and Y. Yesha. Algorithms for a core and k-tree core of a tree. *J. Algorithms*, 15:143–159, 1993.
14. P.J. Slater. Locating central paths in a graph. *Transportation Sci.*, 16:1–18, 1982.
15. A. Wiernik. Planar realizations of nonlinear davenport-schinzel sequences by segments. In *Foundations of Computer Science*, pages 97–106, 1986.

On Bipartite Crossings, Largest Biplanar Subgraphs, and the Linear Arrangement Problem

Farhad Shahrokhi[1,*], Ondrej Sýkora[2,**], László A. Székely[3,***], Imrich Vrt'o[2]

[1] Department of Computer Science, University of North Texas
P.O.Box 13886, Denton, TX, USA
[2] Institute for Informatics, Slovak Academy of Sciences
P.O.Box 56, 840 00 Bratislava, Slovak Republic
[3] Department of Mathematics, University of South Carolina
Columbia, SC 29208, USA

Abstract. We study the bipartite crossing number problem. When the minimum degree and the maximum degree of the graph are close to each other, we derive two polynomial time approximation algorithms for solving this problem, with approximation factors, $O(\log^2 n)$, and $O(\log n \log \log n)$, from the optimal, respectively, where n is the number of vertices. This problem had been known to be NP-hard, but no approximation algorithm which could generate a provably good solution was known. An important aspect of our work has been relating this problem to the linear arrangement problem. Indeed using this relationship we also present an $O(n^{1.6})$ time algorithm for computing the bipartite crossing number of a tree.

We also settle down the problem of computing a largest weighted biplanar subgraph of an acyclic graph by providing a linear time algorithm to it. This problem was known to be NP-hard when graph is planar and very sparse, and all weights are 1.

1 Introduction

Throughout this paper $G = (V_0, V_1, E)$ denotes a bipartite graph, where V_0, V_1 are the two classes of independent vertices. A *bipartite drawing*, or *2-layer drawing* of G consists of placing vertices of V_0 and V_1 on two parallel lines and then drawing each edge using a straight line segment. Let $bcr(G)$ denote the *bipartite crossing number* of G, that is, $bcr(G)$ is the minimum number of edge crossings over all bipartite drawings of G. Bipartite crossing number was first studied in [10], [12], and [25]. In [12], it was observed that $bcr(G) = 0$ iff G is a caterpillar, and exact values $bcr(G)$ for certain graphs were obtained.

* The research of the first author was supported by NSF grant CCR-9528228.
** Research of the 2nd and the 4th author was partially supported by grant No. 95/5305/95 of Slovak Grant Agency and Alexander von Humboldt Foundation.
*** Research of the third author was supported in part by the Hungarian National Science Fund contracts T 016 358 and T 019 367.

Computing $bcr(G)$ is NP-hard [9], but can be solved in polynomial time for bipartite permutation graphs [22]. The problem of obtaining nice multiple layer drawings of graphs (i.e drawings with small number of crossings), has been extensively studied by the graph drawing, VLSI, and CAD communities [4, 5, 13, 18, 23, 24]. It is remarkable that very recently [13], the integer programming methods have been used inorder to compute $bcr(G)$ exactly, or to estimate it, nevertheless, no polynomial time approximation algorithm whose performance is guaranteed has been previously known for approximating $bcr(G)$. A remarkable result in this field is a polynomial algorithm of Eades and Wormald [5] which approximates the bipartite crossing number by a factor of 3, when the positions of vertices in V_0 are fixed. This restricted problem which is still NP-hard [5] and frequently appears in drawing hierarchical graphs [4].

In this paper we explore an important relationship between the bipartite drawings and the linear arrangement problem, which is another well-known problem in the theory of VLSI [1, 3, 14, 17, 21]. We then use this and derive a polynomial time approximation algorithm for computing $bcr(G)$ whose output is $O(\log n \log \log n)$ times the optimal value, for a large class of graphs. Moreover, we derive a lower bound of $\Omega(\delta n B(G))$, for $bcr(G)$, where $B(G)$ and δ are the size of the minimal bisection and minimum degree of G, respectively. This significantly improves the well-known lower bound of $\Omega(B^2(G))$ [15] which was derived for the planar crossing number. This improved lower bound, then, is used to design a second polynomial time algorithm for approximation of $bcr(G)$ with an approximation factor of $O(\log^2 n)$ from the optimal, based on recursively bisecting the graph. Moreover, using the similarities between the linear arrangement problem and bipartite drawings, we give an $O(n^{1.6})$ time algorithm for computing $bcr(G)$ when G is a tree.

We also study biplanar graphs. $G = (V_0, V_1, E)$ is called a *biplanar graph*, if it has a bipartite drawing in which no two edges cross each other. Assume that a *weight* w_{ij} is assigned to any $ij \in E$ and let $w(H)$ denote the sum of the weights on the edges of any subgraph H of G. We refer to $w(H)$ as the weight of H. A *largest biplanar subgraph* (first defined in [19] for the weighted case) (LBSG) of G has maximum weight among all biplanar subgraphs of G. We denote by $b(G)$ the weight of a LBSG in G. Eades and Whitesides [6] have shown that computing $b(G)$ is NP-hard even when G is planar, vertices in V_0 and V_1 have degrees at most 3 and 2, respectively, and all weights are 1. This raised the question of whether or not computing a LBSG can be done in polynomial time when G is acyclic. In this paper we present a linear time dynamic programming algorithm for computing a LBSG in an acyclic graph.

2   Linear Arrangement and Bipartite Crossings

Let $G = (V_0, V_1, E)$. We denote by d_v the degree of a vertex $v \in V_0 \cup V_1$, and denote by d_v^1 denote the number its neighbors of degree 1. We denote by δ_G and Δ_G the minimum and maximum degrees of G, respectively, where when the context is clear we write δ and Δ. Let D_G be a bipartite drawing of G. Whenever, the context is clear, we omit the subscript G and write D. For any $e \in E$ let $bcr_D(e)$ denote the number of crossings on the edge e. Let $bcr(D)$ denote the total number of crossings in D. We may assume that at most 2 edges cross at a point in D, for otherwise we can re-arrange the position of vertices without changing their order, and therefore without changing the number of crossings, so that our assumption holds. Clearly, $bcr(G) = \min_D bcr(D)$.

We assume throughout this paper that the vertices of V_0 are placed on the line y_0 which is taken to be the x-axis, and vertices of V_1 are placed on the line y_1 which is taken to be the line $y = 1$. For a vertex $v \in V_0 \cup V_1$ let $x_D(v)$ denote its x-coordinate in the drawing D. We call the function $x_D : V \to \mathbb{R}$ the coordinate function of D. Throughout this paper, we often omit the y coordinates.

Given an arbitrary graph $G = (V, E)$, and a real function $f : V \to \mathbb{R}$, define the *length* of f, as

$$L_f = \sum_{uv \in E} |f(u) - f(v)|.$$

The *linear arrangement problem* is to find a bijection $f : V \to \{1, 2, 3, ..., |V|\}$, of minimum length. This minimum value is denoted by $\hat{L}(G)$.

In this section we derive a relation between the bipartite crossing number and the linear arrangement problem. This result immediately implies an approximation algorithm for computing $bcr(G)$.

Let D be a bipartite drawing of G. As the number of crossings depends only on the order of vertices V_0 and V_1, we will assume throughout this paper that the vertices of V_0 are placed into the points

$$(1, 0), (2, 0), ..., (|V_0|, 0).$$

For $v \in V_1$, let $u_1, u_2, ..., u_{d_v}$ be its neighbors satisfying $x_D(u_1) < x_D(u_2) < ... < x_D(u_{d_v})$. Define the *median vertex* of v, $med(v) = u_{\lfloor \frac{d_v}{2} \rfloor}$ [5]. We say that D has the *median property* if the vertices of G have distinct x-coordinates and the x-coordinate of any vertex v in V_1 is larger but very close to $x_D(med(v))$, with the restriction that if a vertex of odd degree and a vertex of even degree have the same median vertex, then the odd degree vertex has a smaller x-coordinate.

When the bipartite drawing D does not have the median property, one can always convert it to a drawing which has the property, by placing each $v \in V_1$ on a proper position so that the median property holds. Such a construction is called the *median construction* and was utilized by Eades and Wormald [5] to obtain the following remarkable result.

Theorem 2.1 [5] *Let* $G = (V_0, V_1, E)$, *and* D *be a bipartite drawing of* G. *Assume* D' *is obtained using the median construction from* D, *then*

$$bcr(D') \leq 3bcr(D).$$

$\square$

Let $G = (V_0, V_1, E)$ and D be bipartite drawing of G. Consider an edge $e = ab \in E$, and $u \in V_0 \cup V_1$. We say e *covers* u in D, if the line parallel to the y axis passing through u intersects e at a point different from a or b. Thus the end points of any edge are not covered by that edge. Let $N_D(e)$ denote the number of those vertices in V_1 which are covered by e in D. We will use the following lemma later.

Lemma 2.1 *For $G = (V_0, V_1, E), |E| = m$, let D be a bipartite drawing of G which has the median property. Recall that x_D is the coordinate function of D. Then,*

(i) Let $e \in E, e = ab, a \in V_0, b \in V_1$, then

$$N_D(e) \leq \begin{cases} \frac{2bcr_D(e)}{\delta - 1}, & \text{for } \delta \geq 2, \\ bcr_D(e) + d_a^1, & \text{for } \delta = 1. \end{cases}$$

(ii) There is a bijection $f^* : V_0 \cup V_1 \to \{1, 2, ..., |V_0| + |V_1|\}$ so that

$$L_{f^*} \leq \begin{cases} \frac{4bcr(D)}{\delta - 1} + L_{x_D}, & \text{for } \delta \geq 2, \\ 2bcr(D) + \sum_{a \in V_0} d_a d_a^1 + L_{x_D}, & \text{for } \delta = 1. \end{cases}$$

Proof. For (i), assume that $x(a) < x(b)$, the opposite case is similar. Let v be any vertex in V_1 covered by e in D. Since D has the median property, at least $\lfloor d_v/2 \rfloor$ of vertices adjacent to v are separated from v in D by the straight line segment e. This means that each vertex v covered by e generates at least $\lfloor \delta/2 \rfloor \geq (\delta - 1)/2$ crossings on e. In particular, for $\delta = 1$, observe that among the vertices covered by e only vertices of degree 1 adjacent to a do not produce crossings on e. Consequently, $bcr_D(e) \geq N_D(e)(\delta - 1)/2$ if $\delta \geq 2$ and $bcr_D(e) \geq N_D(e) - d_a^1$ if $\delta = 1$, respectively, proving (i), To prove (ii), we construct f^* by moving all vertices in V_1 to integer locations. Formally, let $w_1, w_2, ..., w_n$ be the order of vertices of $V_0 \cup V_1$ such that $x_D(w_1) < x_D(w_2) < ... < x_D(w_n)$. For i, $1 \leq i \leq |V_0| + |V_1|$, define $f^*(w_i) = i$. It is easy to see that then for any $e = ab \in E$, we have

$$|f^*(a) - f^*(b)| \leq N_D(e) + |x_D(a) - x_D(b)|.$$

Now using (i) and summing over all $e \in E$, we finish the proof. $\square$

Lemma 2.2 *Let $G = (V_0, V_1, E), |E| = m, |V_0 \cup V_1| = n$ and D be any bipartite drawing of G. Then the following hold.*

$$L_{x_D} \geq \sum_{v \in V_1} \left\lfloor \frac{d_v}{2} \right\rfloor \left\lceil \frac{d_v - 2}{2} \right\rceil + \sum_{v \in V_1} \left\lfloor \frac{d_v}{2} \right\rfloor$$

Proof. For $v \in V_1$, let $u_1, u_2, ..., u_{d_v}$ be its neighbors with $x_D(u_1) < ... < x_D(u_{d_v})$. Observe that for $i = 1, 2, ..., \lfloor \frac{d_v}{2} \rfloor$ and the edge pair vu_i and $vu_{d_v - i + 1}$

$$|x_D(v) - x_D(u_i)| + |x_D(u_{d_v - i + 1}) - x_D(v)|$$
$$\geq |x_D(v) - x_D(u_i) + x_D(u_{d_v - i + 1}) - x_D(v)|$$
$$= |x_D(u_{d_v - i + 1}) - x_D(u_i)| = x_D(u_{d_v - i + 1}) - x_D(u_i) \geq d_v - 2i + 1$$

It follows from the above inequality that

$$\sum_{i=1}^{d_v} |x_D(v) - x_D(u_i)| \geq \sum_{i=1}^{\lfloor \frac{d_v}{2} \rfloor} |x_D(v) - x_D(u_i)| + |x_D(v) - x_D(u_{d_v - i + 1})|$$

$$\geq \sum_{i=1}^{\lfloor \frac{d_v}{2} \rfloor} d_v - 2i + 1 \tag{1}$$

But it is easy to show that

$$\sum_{i=1}^{\lfloor \frac{d_v}{2} \rfloor} d_v - 2i = \sum_{v \in V_1} \left\lfloor \frac{d_v}{2} \right\rfloor \left\lceil \frac{d_v - 2}{2} \right\rceil$$

and the claim follows from summing (1) for all $v \in V_1$. □

Remark. Note that in the above Lemma we did not require D to have the median property.

Theorem 2.2 *Let* $G = (V_0, V_1, E)$, $|E| = m$, $|V_0 \cup V_1| = n$. *Then, for* $\delta \geq 2$

$$\frac{\delta}{60} \hat{L}(G) \leq bcr(G) \leq \Delta \hat{L}(G)$$

and for $\delta = 1$

$$\frac{1}{24} \hat{L}(G) - \frac{1}{12} \sum_{v \in V_1} \left\lfloor \frac{d_v}{2} \right\rfloor \left\lceil \frac{d_v - 2}{2} \right\rceil - \frac{1}{12} \sum_{a \in V_0} d_a d_a^1 \leq bcr(G) \leq \Delta \hat{L}(G).$$

Proof. First we prove the lower bound on $bcr(G)$. For D be a bipartite drawing of G, we will construct an appropriate bijection $f^* : V_0 \cup V_1 \to \{1, 2, ...n\}$. Let D' be a drawing which is obtained by applying the median construction to D. Let vertex $v \in V_1$. Let $u_1, u_2, ..., u_{d_v}$ be its neighbors with $x_{D'}(u_1) < x_{D'}(u_2) < ... < x_{D'}(u_{d_v})$. For $i = 1, 2, ..., \lfloor d_v/2 \rfloor$, and any j, $i < j < d_v - i + 1$, observe that u_j generates d_{v_j} crossings if it is adjacent to v, and $d_{v_j} - 1$ crossings if it is not adjacent to v, respectively, on the edges $u_i v$ and $u_{d_v - i + 1} v$. Thus

$$bcr_{D'}(u_i v) + bcr_{D'}(u_{d_v - i + 1} v) \geq (x_{D'}(u_{d_v - i + 1}) - x_{D'}(u_i) - 1)\delta - (d_v - 2i)$$

$$= (x_{D'}(u_{d_v - i + 1}) - x_{D'}(v) + x_{D'}(v) - x_{D'}(u_i) - 1)\delta - (d_v - 2i) \tag{2}$$

Note that D' has the median property, thus for $i = 1, 2, ..\lfloor d_v/2 \rfloor$,

$$x_{D'}(u_i) < x_{D'}(v) < x_{D'}(u_{d_v - i + 1})$$

and hence (2) implies

$$bcr_{D'}(u_i v) + bcr_{D'}(u_{d_v - i + 1} v) \geq (|x_{D'}(v) - x_{D'}(u_{d_v - i + 1})| + |x_{D'}(v)$$

$$- x_{D'}(u_i)| - 1)\delta - (d_v - 2i) \tag{3}$$

Moreover when d_v is odd, we have

$$bcr_{D'}(u_{\lfloor \frac{d_v}{2} \rfloor}v) + bcr_{D'}(u_{\lceil \frac{d_v}{2} \rceil}v) \geq (x_{D'}(u_{\lceil \frac{d_v}{2} \rceil}) - x_{D'}(u_{\lfloor \frac{d_v}{2} \rfloor}))\delta$$

since no vertex adjacent to v is between $u_{\lceil \frac{d_v}{2} \rceil}$ and $u_{\lfloor \frac{d_v}{2} \rfloor}$. Consequently, for odd d_v, we obtain,

$$bcr_{D'}(u_{\lfloor \frac{d_v}{2} \rfloor}v) + bcr_{D'}(u_{\lceil \frac{d_v}{2} \rceil}v) \geq \delta(|x_{D'}(v) - x_{D'}(u_{\lceil \frac{d_v}{2} \rceil})| + |x_{D'}(v) - x_{D'}(u_{\lfloor \frac{d_v}{2} \rfloor})| \tag{4}$$

Using (3) observe that the contribution of crossings on edges incident to v is

$$\sum_{i=1}^{d_v} bcr_{D'}(u_i v) \geq \delta \sum_{i=1}^{\lfloor \frac{d_v}{2} \rfloor} (|x_{D'}(v) - x'_D(u_i)| + |x_{D'}(v) - x_{D'}(u_{d_v - i+1})|)$$

$$- \delta \left\lfloor \frac{d_v}{2} \right\rfloor - \sum_{i=1}^{\lfloor \frac{d_v}{2} \rfloor} (d_v - 2i)$$

$$= \delta \sum_{i=1}^{\lfloor \frac{d_v}{2} \rfloor} (|x_{D'}(v) - x'_D(u_i)| + |x_{D'}(v)$$

$$- x_{D'}(u_{d_v - i+1})|) - \delta \left\lfloor \frac{d_v}{2} \right\rfloor - \left\lfloor \frac{d_v}{2} \right\rfloor \left\lceil \frac{d_v - 2}{2} \right\rceil$$

Combining this and (4) we obtain

$$2 \sum_{i=1}^{d_v} bcr_{D'}(u_i v) \geq \delta \sum_{i=1}^{d_v} |x_{D'}(v) - x_{D'}(u_i)| - \delta \left\lfloor \frac{d_v}{2} \right\rfloor - \left\lfloor \frac{d_v}{2} \right\rfloor \left\lceil \frac{d_v - 2}{2} \right\rceil$$

Summing up the last expression over all $v \in V_1$ we get

$$bcr(D') \geq \delta \sum_{uv \in E} |x_{D'}(v) - x_{D'}(u)| - \delta \sum_{v \in V_1} \left\lfloor \frac{d_v}{2} \right\rfloor - \sum_{v \in V_1} \left\lfloor \frac{d_v}{2} \right\rfloor \left\lceil \frac{d_v - 2}{2} \right\rceil$$

$$= \delta L_{x_{D'}} - \delta \sum_{v \in V_1} \left\lfloor \frac{d_v}{2} \right\rfloor - \sum_{v \in V_1} \left\lfloor \frac{d_v}{2} \right\rfloor \left\lceil \frac{d_v - 2}{2} \right\rceil - \frac{\delta |V_1|}{2n^2}. \tag{5}$$

We consider 2 cases:
Case I . Assume $\delta = 1$. Consider the bijection f^* in Part (ii) of Lemma 2.1, case $\delta = 1$. Then

$$L_{x_{D'}} \geq L_{f^*} - 2bcr(D') - \sum_{a \in V_0} d_a d_a^1.$$

Now (5) implies that

$$4bcr(D') \geq L_{f^*} - \frac{m}{2} - \sum_{v \in V_1} \left\lfloor \frac{d_v}{2} \right\rfloor \left\lceil \frac{d_v - 2}{2} \right\rceil - \sum_{a \in V_0} d_a d_a^1$$

Observe that $L(G) \geq m$ and therefore, $\frac{L_{f^*}}{2} \geq \frac{m}{2}$, to obtain

$$bcr(D') \geq \frac{1}{8} L_{f^*} - \frac{1}{4} \sum_{v \in V_1} \left\lfloor \frac{d_v}{2} \right\rfloor \left\lceil \frac{d_v - 2}{2} \right\rceil - \frac{1}{4} \sum_{a \in V_0} d_a d_a^l.$$

To finish the proof observe that $bcr(D') \leq 3bcr(D)$.

Case II. Assume $\delta \geq 2$. Use the lower bound of Lemma 2.2 in (5) to obtain

$$2bcr(D') \geq (\delta - 1)L_{x_{D'}} - (\delta - 1) \sum_{v \in V_1} \left\lfloor \frac{d_v}{2} \right\rfloor \tag{6}$$

Let f^* be the bijection in Part (ii) Lemma 2.1, case $\delta \geq 2$. Then

$$L_{x_{D'}} \geq L_{f^*} - \frac{4bcr(D')}{\delta - 1}.$$

Hence

$$bcr(D') \left(2 + \frac{4\delta}{\delta - 1} \right) \geq \delta L_{f^*} - \frac{\delta m}{2} \tag{7}$$

Since $L_{f^*} \geq m$, we have

$$bcr(D') \left(1 + \frac{4\delta}{\delta - 1} \right) \geq \frac{\delta}{2} L_{f^*}. \tag{8}$$

and using the above inequality, observing that $\frac{\delta}{\delta-1} \leq 2$, and the result follows, since $L_{f^*} \geq L(G)$ and $bcr(D') \leq 3bcr(D)$. To prove the upper bound, consider a solution (not necessarily optimal) of the optimal linear arrangement of G, realized by a bijection $f^* : V_0 \cup V_1 \rightarrow \{1, 2, ..., |V_0| + |V_1|\}$. The mapping f^* induces an ordering of vertices of $V_0 \cup V_1$ in y_o. Project vertices of V_1 into y_1 to obtain a bipartite drawing D. Note that

$$L_{x_D} = \sum_{uv \in E} |x_D(u) - x_D(v)| = L_{f^*}, \tag{9}$$

for this drawing. Now consider a new drawing D' obtained by the median construction with ε being half of the minimum distance between any vertex in V_1 and its median vertex in V_0. It is easy to verify that $\sum_{uv \in E} |x_{D'}(u) - x_{D'}(v)| \leq \sum_{uv \in E} |x_D(u) - x_D(v)|$ or

$$L_{x_{D'}} \leq L_{x_D} = L_{f^*}. \tag{10}$$

Using a similar method used to derive the lower bound, we obtain

$$bcr(D') < \Delta \sum_{uv \in E} |x_{D'}(u) - x_{D'}(v)| = \Delta L_{x_{D'}}. \tag{11}$$

Combining (10) and (11) we get

$$bcr(D') \leq \Delta L(f^*).$$

To complete the proof we take f^* to be the optimal solution to the linear arrangement problem, that is, $L_{f^*} = L(G)$. □

Even et al. [7] derived a polynomial algorithm for the linear arrangement problem with the approximation factor $O(\log n \log \log n)$ from the optimal. Combining their result with the Theorem 3.1 we obtain the following.

Corollary 2.1 *Let G be a connected bipartite graph which is not a tree with $\delta = \Theta(\Delta)$. Then $bcr(G)$ can be computed within a factor of $O(\log n \log \log n)$ from the optimal in polynomial time, provided that $\delta \geq 2$. Moreover, if $\delta = 1$, then $bcr(G)$ can be computed within a factor of $O(\log n \log \log n)$ from the optimal in polynomial time, provided that $m \geq n(1 + \epsilon)$, for some $\epsilon > 0$.*

Proof. Consider the drawing D' in Theorem 2.2 obtained form an approximate solution to the linear arrangement problem provided in [7].
When $\delta_G \geq 2$, Theorem 2.2 implies that $bcr(D') = O(\delta \log n \log \log n \hat{L}(G))$, and the claim follows because $bcr(D') \geq bcr(G) = \Omega(\delta \hat{L}(G))$.
When $\delta = 1$, distinguish 2 cases:
If

$$\hat{L}(G) \geq 3 \sum_{v \in V_1} \left\lfloor \frac{d_v}{2} \right\rfloor \left\lceil \frac{d_v - 2}{2} \right\rceil + 3 \sum_{a \in V_0} d_a d_a^1$$

then by Theorem 2.2 we get $bcr(G) = \Omega(L(G))$ and the rest is similar as in the case $\delta \geq 2$.
If

$$\hat{L}(G) < 3 \sum_{v \in V_1} \left\lfloor \frac{d_v}{2} \right\rfloor \left\lceil \frac{d_v - 2}{2} \right\rceil + 3 \sum_{a \in V_0} d_a d_a^1$$

then Theorem 2.2 implies $bcr(G) = O(\log n \log \log n L(G)) = O(n \log n \log \log n)$. On the other hand, May and Szkatula [18] proved that $bcr(G) \geq m - n + 1 = \Omega(n)$, for $m \geq (1 + \varepsilon)n$, which finishes the proof. □

Let $B(G)$ denote the size of the minimal (1/3-2/3) bisection of G. Leighton [15] proved that planar crossing number of any degree bounded graph H is $\Omega(B^2(H))$. Another very interesting consequence of Theorem 3.1 is providing a stronger version of Leighton's result, for $bcr(G)$.

Corollary 2.2 *For any n vertex bipartite graph with m edges and $\delta \geq 2$,*

$$bcr(G) = \Omega(n \delta B(G)).$$

Proof. The claim follows from the lower bound in Theorem 3.1, and the well known observation that $L(G) = \Omega(n B(G))$. (See for instance [11].) □

An algorithmic consequence of Corollary 2.2 is a bisection based divide an conquer algorithm which approximates $bcr(G)$ to within a factor of $O(\log^2 n)$ from the optimal. Our algorithm uses an approximation algorithm for bisecting a graph such as those in [8, 16]. These algorithms have a performance guarantee of $O(\log n)$ from the optimal [8, 16].

Theorem 2.3 *Let A be a polynomial time algorithm to approximate the bisecting a graph with performance guarantee $O(\log n)$. Consider a divide and conquer algorithm which recursively bisects G using A, obtains the two 2-layer drawings, and then inserts the edges of the bisection into these two drawing. Let G be bipartite, with n vertices, m edges and $\Delta = O(\delta)$. Then in polynomial time we generate a 2-layer drawing D so that $bcr(D)$ is within $O(\log^2 n)$ from the optimal, where in case $\delta = 1$ we assume $m \geq (1 + \varepsilon)n$, for any $\varepsilon > 0$.*

Proof. Assume that using A, we partition the graph G to 2 vertex disjoint subgraphs G_1 and G_2 recursively. Let $\bar{B}(G)$ denote the number of those edges having one endpoint in the vertex set of G_1, and the other in the vertex set of G_2. Let D_{G_1}, and D_{G_2} be the 2-layer drawings already obtained by the algorithm for G_1 and G_2, respectively. Let D denote the drawing obtained for G. We have

$$bcr(D) \leq bcr(D_{G_1}) + bcr(D_{G_2}) + \bar{B}^2(G) + \bar{B}(G)m$$

But since we use, the approximation algorithm A for $B(G)$, we have $\bar{B}(G) = O(\log n B(G))$. Now observing that $m \leq \Delta n$, and $nB(G) = O(L(G))$, we obtain,

$$bcr(D) \leq bcr(D_{G_1}) + bcr(D_{G_2}) + O(\Delta \hat{L}(G) \log n)$$

which implies

$$bcr(D) = O(\Delta \hat{L}(G) \log^2 n)).$$

Assume $\delta \geq 2$, then by Theorem 2.2 $bcr(G) = \Omega(\delta L(G))$, and the claim follows. Assume $\delta = 1$. Then the results follows by the same analysis as in the proof of Corollary 2.1, case $\delta = 1$. $\square$

Remarks. The previously known lower bound of $\Omega(B^2(G))$ can not be used to show the quality of the approximation algorithm, since it does not bound from above the error term in our recurrence relation. Thus our lower bound of $\Omega(n \delta B(G))$ is crucial to show the suboptimality of the solution.

The above approximation algorithm although in theory weaker than the first approximation algorithm, it is easier to implement and in fact may be superior to the first algorithm in practice. This is due to the fact that the method of Even et.al requires the usage of specific interior linear programming algorithms which could be hard to code, whereas, many existing VLSI and CAD tools are divide and conquer heuristics and can be used to approximately bisect a graph and to implement our second algorithm.

Now we prove a stronger relation of the bipartite crossing number to the linear arrangement problem in the case of trees.

Theorem 2.4 *For any n-vertex tree T*

$$bcr(T) = \hat{L}(T) - n + 1 - \sum_{v \in T} \left\lfloor \frac{d_v}{2} \right\rfloor \left\lceil \frac{d_v - 2}{2} \right\rceil.$$

Moreover, bipartite drawing D_T can be constructed in $O(n^{1.6})$ time.

Proof. We show the equality by induction on n. The claim is true for $n = 1, 2$. Let $n \geq 3$. Assume that the claim is true for all m-vertex trees, $m < n$. Consider an optimal linear arrangement f of the tree T. According to Seidvasser [20], $f^{-1}(1)$ and $f^{-1}(n)$ are leaves. Let $P = v_0 v_1 ... v_k$ be the path between $v_0 = f^{-1}(1)$ and $v_k = f^{-1}(n)$ in T. Deleting the edges of P we get trees T_i, rooted in $v_i, i = 1, 2, ..., k - 1$. It holds [20]:

$$\hat{L}(T) = \sum_{i=1}^{k-1} \hat{L}(T_i) + n - 1.$$

By the inductive hypothesis we have

$$\hat{L}(T) = \sum_{i=1}^{k-1} (bcr(T_i) + n_i - 1 + \sum_{v \in T_i} \left\lfloor \frac{d_v}{2} \right\rfloor \left\lceil \frac{d_v - 2}{2} \right\rceil) + n - 1 \qquad (12)$$

Consider the optimal bipartite drawings of T_i, $i = 1, 2, ..., k - 1$, and place them consecutively such that T_i does not cross T_j, for $i \neq j$. Then draw the path P without self crossings such that v_0 (v_k) is placed to the left (right) of the drawing of $T_1 (T_{k-1})$. We get a drawing D_T which satisfies

$$bcr(D_T) = \sum_{i=1}^{k-1} bcr(T_i) + n - 1 - k - \sum_{i=1}^{k-1} (d_{v_i} - 2). \qquad (13)$$

Note that $n - 1 - k - \sum_{i=1}^{k-1} (d_{v_i} - 2)$ denotes the number of crossings on P which equals the number of edges in T not incident to P. Substituting (13) into (12) we get

$$\hat{L}(T) = bcr(D_T) + n - 1 + \sum_{i=1}^{k-1} (\sum_{v \in T_i} \left\lfloor \frac{d_v}{2} \right\rfloor \left\lceil \frac{d_v - 2}{2} \right\rceil + d_{v_i} - 2)$$

$$\geq bcr(T) + n - 1 + \sum_{v \in T} \left\lfloor \frac{d_v}{2} \right\rfloor \left\lceil \frac{d_v - 2}{2} \right\rceil$$

$$- \sum_{i=1}^{k-1} (\left\lfloor \frac{d_{v_i}}{2} \right\rfloor \left\lceil \frac{d_{v_i} - 2}{2} \right\rceil - \left\lfloor \frac{d_{v_i} - 2}{2} \right\rfloor \left\lceil \frac{d_{v_i} - 4}{2} \right\rceil - d_{v_i} + 2)$$

$$= bcr(T) + n - 1 + \sum_{v \in T} \left\lfloor \frac{d_v}{2} \right\rfloor \left\lceil \frac{d_v - 2}{2} \right\rceil.$$

We omit the proof of the the opposite inequality.

Our proof immediately implies that any optimal linear arrangement of a tree T induces an optimal bipartite drawing of the tree T by projecting the vertices of V_1 on y_1. Since the optimal linear arrangement of an n-vertex tree can be found in $O(n^{1.6})$ time [1], computing D_T can also be done in $O(n^{1.6})$ time. $\square$

3 A Linear time algorithm for computing largest biplanar subgraphs in acyclic graphs

A graph is a *caterpillar*, if it is a path to which some vertices of degree 1 are attached. The path is called the *backbone* of the caterpillar. The two endpoints of the path are called the endvertices and the other vertices of degree 1 are called leaves. We note that given a caterpillar, the choice for the backbone and the endvertices may not be unique. For example a star on n vertices is a caterpillar; we can select any 2 vertices of degree 1 to be the endvertices of the backbone.

It is well known and easy to show that a graph is biplanar iff it is a collection of vertex disjoint caterpillars. This is equivalent to saying that a graph is biplanar iff it does not contain a double claw. We will frequently use these facts in this section, sometimes without explicitly referring to them.

Let $T = (V_T, E_T)$ be an unrooted tree and $r \in V_T$. Then, we view r as the root of T and obtain a plane embedding of T with no edge crossing in which all edges are oriented in the direction of root to leaves. Then any vertex $x \in V_T$ will have a unique parent and a set of children denoted by N_x. For any $x \in V_T$ we denote by T_x, the largest subtree of T which is rooted at x. Let A_x be a biplanar subgraph of T_x. We will treat A_x as an edge set. We say A_x spans a vertex a, if there is an edge $ab \in A_x$. Let $x \in V_T$ and B_x be a biplanar subgraph in T_x which spans a vertex b in T_x. Assume that b is located on the backbone of a caterpillar in B_x. Then, we say b is good vertex in B_x, provided that, the distance of b (in B_x) is at most one from one end point of the caterpillar, otherwise, we say b is a bad vertex in B_x. Now assume that b is a leaf in B_x. Then we call b a good leaf in B_x, if the other end point of the (unique) edge incident to b (in B_x) is of distance one from one end point of the backbone of the caterpillar (in B_x) containing it, otherwise, b is called a is a bad leaf in B_x.

Let $T = (V_T, E_T)$, $x \in V_T$, and w_{ij} be a weight assigned to each edge $ij \in E_T$. Let B_x be a biplanar subgraph of T_x. B_x is of type 0, if it has the largest total weight among all biplanar subgraphs of T_x which do not span x. B_x is of type 1, if it has the largest total weight among all biplanar subgraphs of T_x which span x as a good vertex, and is of type 2, if it has the largest total weight among all biplanar subgraphs which span x as a bad vertex. Finally, B_x is of type 3 if it has the largest total weight among all subgraphs of T_x which span x as a bad leaf.

Let $x \in V_T$, and define $b^i(T_x)$ to be the weight of a biplanar subgraph of G of type i, $0 \le i \le 3$ in T_x. Note that $b^i(T_x) \le b(T_x)$, $0 \le i \le 3$.

Lemma 3.1 Let $T = (V_T, E_T)$ $x \in V_T$, and B_x be any biplanar subgraph of T_x. Let $\mathbf{B_x} = \{B_x^0, B_x^1, B_x^2, B_x^3\}$, where B_x^i, $0 \le i \le 3$, is a biplanar subgraph of T_x of type i. Then the following hold.

(i)- If B_x spans x as good leaf in T_x, then it spans x as a good vertex.
(ii)- The subgraph in $\mathbf{B_x}$ which has largest weight is a LBSG in T_x.

Proof. For (i), let l be backbone of the caterpillar in B_x which spans x. and let a be the vertex adjacent to x in B_x. Since x is a good leaf, a must be adjacent to one end point of l, denoted by b. We can now view $l \cup \{ax\} - \{ab\}$ as the backbone of a caterpillar in B_x, where x is taken to be an end point of the backbone. This means B_x spans x as a good vertex. Proof of (ii), follows from (i). □

Remark. Part (ii) in Lemma 3.1, means that, if we can compute the subgraphs of type 0, 1, 2, and 3 in T_x, then we can compute $b(T_x)$. Thus, throughout the rest of this section we will focus on computing subgraphs of type $i = 0, 1, 2, 3$. In fact the idea behind the algorithm is to compute the subgraphs of i, $0 \leq i \leq 3$ in subtrees first, and then at the tree which contains these subtrees.

Let $x \in V_T$ and $z \in N_x$, define $\epsilon_z = max\{w_{zx} + b^0(T_z), b(T_z)\}$. Moreover, define

$$\epsilon'_z = w_{zx} + max\{b^0(T_z), b^1(T_z)\} - \epsilon_z,$$

$$\epsilon^*_z = w_{zx} - b(T_z) + b^2(T_z),$$

and

$$\bar{\epsilon}_z = w_{zx} + b^1(T_z) - \epsilon_z.$$

Theorem 3.1 Let $T = (V_T, E_T)$, $x \in V_T$, and let w_{ij} be the weight on any $ij \in E_T$. For any $y \in N_x$, let B_y be a LBSG in T_y and let B^i_y, be a biplanar subgraph of type $i = 0, 2$ in T_y. Then, the following hold
   (i)- Let

$$\mathbf{B}^0_x = \cup_{y \in N_x} B_y,$$

then, $\mathbf{B}^0_x$ is a biplanar subgraph of type 0 in T_x.
   (ii)- Let $t \in N_x$ so that $\epsilon^*_t = max_{y \in N_x} \epsilon^*_y$. Also let

$$\mathbf{B}^3_x = \cup_{y \in N_x - \{t\}} B_y \cup B^2_t \cup \{xt\},$$

then $\mathbf{B}^3_x$ is a biplanar subgraph of type 3 in T_x.

Proof. Proof is omitted. □
   We now state a lemma which will be used later.

Lemma 3.2 Let $T = (V_T, E_T)$, $x \in V_T$, and let w_{ij} be the weight on any $ij \in E_T$. For any $y \in N_x$, let B_y be a LBSG in T_y and let B^i_y, be a biplanar subgraph of type i, $i = 0, 1$ in T_y. Moreover, for any $y \in N_x$, define subgraphs of T_x,

$$A_y = \begin{cases} B^0_y \cup \{yx\} & \text{if } w_{yx} + b^0(T_y) > b(T_y) \\ B_y & \text{otherwise,} \end{cases}$$

$$A'_y = \begin{cases} B^0_y \cup \{yx\} & \text{if } w_{yx} + b^0(T_y) > b^1(T_y) \\ B^1_y \cup \{yx\} & \text{otherwise,} \end{cases}$$

and

$$\bar{A}_y = B_y^1 \cup \{yx\}.$$

Then for any $y \in N_x$, A_y, A'_y, and $\bar{A}_y$ are biplanar and have $w(A_y) = \epsilon_y$, $w(A'_y) = \epsilon'_y - \epsilon_y$ and $w(\bar{A}_y) = \bar{\epsilon}_y - \epsilon_y$.

$\square$

Theorem 3.2 Let $T = (V_T, E_T)$, $x \in V_T$, and let w_{ij} be the weight on any $ij \in E_T$. For any $y \in N_x$, let B_y be a LBSG in T_y and let B_y^i, be a biplanar subgraph of type i, $i = 0, 1$ in T_y. Moreover, for any $y \in N_x$, let A_y, A'_y and $\bar{A}_y$ be defined as in Lemma 3.2. Then the following hold.
    (i)- Let $s \in N_x$ so that $\epsilon'_s = max_{y \in N_x} \epsilon'_y$. Then the subgraph

$$\mathbf{B}_x^1 = A'_s \cup_{y \in N(x) - \{s\}} A_y$$

is of type 1 in B_x.
    (ii)-Let $t, r \in N_x$ so that $\bar{\epsilon}_t = max_{y \in N_x} \bar{\epsilon}_y$, and $\bar{\epsilon}_r = max_{y \in N_x - \{t\}} \bar{\epsilon}_y$. Then the subgraph

$$\mathbf{B}_x^2 = \bar{A}_t \cup \bar{A}_r \cup_{y \in N(x) - \{r, t\}} A_y$$

is of type 2 in B_x.

Proof. Proof uses the previous lemma; details are omitted. $\square$

Theorem 3.3 For any forest $T = (V_T, E_T)$ a LBSG of weight $b(T)$ can be computed in $O(|V_T|)$ time.

Proof. Proof follows from Theorems 3.2 and 3.1. $\square$

Acknowledgement. The research of the second and fourth author was done while they were visiting Department of Mathematics and Informatics of University in Passau. They thank Prof. F.-J. Brandenburg for perfect work conditions and hospitality.

References

1. Chung, F. R. K., On optimal linear arrangements of trees, *Computers and Mathematics with Applications* **10**, (1984), 43–60.
2. Chung, F. R. K., A conjectured minimum valuation tree, *SIAM Review* **20** (1978), 601–604.
3. Díaz, J., Graph layout problems, in: Proc. *International Symposium on Mathematical Foundations of Computer Sciences*, Lecture Notes in Computer Scince 629, Springer Verlag, Berlin, 1992, 14–21.
4. Di Battista, J., Eades, P., Tamassia, R., Tollis, I. G., Algorithms for drawing graphs: an annotated bibliography, *Computational Geometry* **4** (1994), 235–282.
5. Eades, P., Wormald, N., Edge crossings in drawings of bipartite graphs, *Algorithmica* **11** (1994), 379–403.

6. Eades, P., Whitesides, S., Drawing graphs in 2 layers, *Theoretical Computer Sceince* 131, 1994, 361-374.

7. Even, G., Naor, J. S., Rao, S., Scieber, B., Divide-and-Conquer approximation algorithms via spreading metrices, in Proc. *36th Annual IEEE Symposium on Foundation of Computer Science*, IEEE Computer Society Press, 1995, 62-71.

8. Even, G., Naor, J. S., Rao, S., Scieber, B., Fast Approximate Graph Partition Algorithms, 8th Annual ACM-SIAM Symposium on Disc. Algo., 1997, 639-648.

9. Garey, M. R., Johnson, D. S., Crossing number is NP-complete, *SIAM J. Algebraic and Discrete Methods* **4** (1983), 312-316.

10. Harary, F., Determinants, permanents and bipartite graphs, *Mathematical Magazine* **42** (1969), 146-148.

11. Hansen, M., Approximate algorithms for geometric embeddings in the plane with applications to parallel processing problems, *30th FOCS*, 1989, 604-609.

12. Harary, F., Schwenk, A., A new crossing number for bipartite graphs, *Utilitas Mathematica* **1** (1972), 203-209.

13. Jünger, M., Mutzel, P., Exact and heuristic algorithm for 2-layer straightline crossing number, in: *Proc. Graph Drawing'95*, Lecture Notes in Computer Science 1027, Springer Verlag, Berlin, 1996, 337-348.

14. Juvan, M., Mohar, B., Optimal linear labelings and eigenvalues of graphs, *Discrete Mathematics* **36** (1992), 153-168.

15. Leighton, F.T., Complexity issues in VLSI, MIT Press, 1983.

16. Leighton F. T., Rao, S., An approximate max flow min cut theorem for multicommodity flow problem with applications to approximation algorithm, *29th Foundation of Computer Science*, IEEE Computer Society Press, 1988, 422-431.

17. Lengauer, T., Combinatorial algorithms for integrated circuit layouts, *Wiley and Sons*, Chichester, UK, 1990.

18. May, M., Szkatula, K., On the bipartite crossing number, *Control and Cybernetics* **17** (1988), 85-98.

19. Mutzel, P., An alrenative method to crossing minimization on hierarchical graphs, *Proceeding of Graph Drawing 96*, Lecture Notes in Computer Science, Springer Verlag, Berin, 1997.

20. Seidvasser, M. A., The optimal number of the vertices of a tree, *Diskretnii Analiz* **19** (1970), 56-74.

21. Shiloach, Y., A minimum linear arrangement algorithm for undirected trees, *SIAM J. Computing* **8** (1979), 15-32.

22. Spinrad, J., Brandstädt, A., Stewart, L., Bipartite permutation graphs, *Discrete Applied Mathematics* **19**, 1987, 279-292.

23. Sugiyama, K., Tagawa, S., Toda, M., Methods for visual understanding of hierarchical systems structures, *IEEE Transactions on Systems, Man and Cybernetics* **11** (1981), 109-125.

24. Warfield, J., Crossing theory and hierarchy mapping, *IEEE Transactions on Systems, Man and Cybernetics* **7** (1977), 502-523.

25. Watkins, M.E., A special crossing number for bipartite graphs: a research problem, *Annals of New York Academy Sciences* **175** (1970), 405-410.

Approximation Algorithms for a Genetic Diagnostics Problem

S. Rao Kosaraju[*1] and Alejandro A. Schäffer[2] and Leslie G. Biesecker[3]

[1] Department of Computer Science,
Johns Hopkins University, Baltimore
[2] National Center for Human Genome Research,
National Institutes of Health, Bethesda
[3] National Center for Human Genome Research,
National Institutes of Health, Bethesda

Abstract. We define and study a combinatorial problem called WEIGHTED DIAGNOSTIC COVER (WDC) that models the use of a laboratory technique called *genotyping* in the diagnosis of a important class of chromosomal aberrations. An optimal solution to WDC would enable us to define a genetic assay that maximizes the diagnostic power for a specified cost of laboratory work. We develop approximation algorithms for WDC by making use of the well-known problem SET COVER for which the *greedy* heuristic has been extensively studied. We prove worst-case performance bounds on the *greedy* heuristic for WDC and for another heuristic we call *directional greedy*. We implemented both heuristics. We also implemented a local search heuristic that takes the solutions obtained by *greedy* and *dir-greedy* and applies swaps until they are locally optimal. We report their performance on a real data set that is representative of the options that a clinical geneticist faces for the real diagnostic problem. Many open problems related to WDC remain, both of theoretical interest and practical importance.

1 Introduction

We apply tools from combinatorial algorithms to improve the estimated detection power of a recently proposed diagnostic method for a class of genetic abnormalities. These abnormalities are called *segmental aneusomies*, and they are cases in which a small piece of a chromosome has been inherited in a medically problematic way. We begin by defining (a model of) the diagnostic problem in mostly combinatorial terms and tying the problem to previous work on combinatorial algorithms. In the next section we provide the relevant genetics terminology and motivation from the medical genetics literature.

Segmental aneusomies affect a contiguous interval along one arm of a chromosome. We model a chromosome arm as an interval $[l, r]$ on the real line. We are interested in detecting an abnormality of length b that lies somewhere in the

* Supported by NSF under grant CCR-9508545 and ARO under grant DAAH 04-96-1-0013

interval $[l, r]$. Very little is known about the distribution of real-life aneusomies, so we have little choice but to assume that the possible aberrations are uniformly distributed. We are given a large collection I of *markers*, also called *probes*, that correspond to points in $[l, r]$. Each marker m has an associated detection probability p_m, such that if m is contained in the aberrant interval of length b a genetic test using m can detect the aberration with probability p_m.

In the combinatorial representation, we reverse the roles of point and interval for markers and aneusomies. Thus a marker is represented as a subinterval $[\max(l, m - b), m]$ and an aberration of length b is represented by its left endpoint, which is somewhere in $[l, r - b]$. An aberration is detected with probability p_m if it lies in the marker interval. All marker intervals are of length b, except those that have l as the left endpoint. Since p_m is always < 1, there may be false negatives, but if the experiments are carried out properly, there will be no false positives.

Our objective is to select k markers out of the set I so as to maximize the overall detection probability. If two markers c and d both contain the same aneusomy, we assume that they act independently, so that the detection probability is:

$$p_c + p_d - (p_c \times p_d).$$

The cost of each marker test per patient depends very little on the choice of the marker, provided one ignores the initial overhead of setting up the laboratory and the same set of markers is used to test all patients. Finding a best marker set of fixed size models the clinician's problem of *maximizing the diagnostic power for a pre-specified cost of the test*. Formally we define:

WEIGHTED DIAGNOSTIC COVER(WDC). Inputs: a base interval $B = [l, r]$, a collection of marker subintervals I, a detection probability p_i for each $i \in I$, a desired number of markers k, and an overall target detection probability d. Let C be a subcollection of I. The detection probability of C at the point x is

$$D(C, x) = 1 - \prod_{i \in C} [1 - (p_i \times \chi_{x \in i})],$$

where $\chi_{x \in i}$ is the indicator function that takes the value 1 if $x \in i$ and the value 0 if $x \notin i$. The diagnostic power of C is the integral

$$D(C) = \int_l^r D(C, x) dx.$$

Is there a subcollection $C \subset I$ of at most k markers, such that $D(C) \geq d$?

Giving each point x uniform weight in the integral reflects the assumption that the possible aberrations are uniformly distributed. For any C the integral $D(C)$ can be replaced by a finite sum over at most $2k + 1$ terms because there are at most $2k + 1$ subintervals with distinct values of $D(C, x)$.

In the diagnostic application, we are also interested in the special case where all the intervals in I that do not intersect the endpoints l, r have the same length, b. The intervals that intersect the endpoints have length $\leq b$ because they are truncated at the endpoint. Like the uniform distribution assumption, the equal length assumption reflects a lack of genetic knowledge about the sizes and placement of real aneusomies. We call this problem, EQUAL-LENGTH WEIGHTED DIAGNOSTIC COVER (EWDC).

If we think of B and I as sets, ignoring the geometry, WEIGHTED DIAGNOSTIC COVER is similar to the well-known NP-complete [14] problem:

WEIGHTED SET COVER. Inputs: a base set B, a collection of subsets I, a weight p_i for each $i \in I$, and a target weight d. Let C be a collection of sets in I. Then the weight $D(C) = \sum_{i \in C} p_i$. Is there a collection $C \subset I$ of such that $\cup_{i \in C} i = B$ and $D(C) \leq d$?

The non-linear weight function in WEIGHTED DIAGNOSTIC COVER is hard to deal with. However, the geometric nature of the sets may make the problem easier; we prove that WEIGHTED SET COVER can be solved in polynomial time when the sets are intervals on the line.

Let H be any heuristic algorithm for WDC. Then for any instance of WDC, we define the performance ratio of H:

$$r_H(k) = \frac{\text{diagnostic power of } k \text{ markers selected by } H}{\text{maximum diagnostic power achievable with } k \text{ markers}}.$$

This ratio is different from the ratio of full solution sizes usually used for SET COVER [13, 16, 4, 11, 22, 7, 21].

By analogy to the widely-studied greedy heuristic for WSC, we define a *greedy* heuristic for WDC: repeatedly choose the marker that increases the diagnostic power D by the largest amount. We also define a different heuristic called *dir-greedy* that takes advantage of the geometric nature of the problem and the assumption that all the markers not at the ends of B are of the same length. We prove that $r_{greedy}(k) \geq 0.318$ for WEIGHTED DIAGNOSTIC COVER. We prove that $r_{dir-greedy}(k) \geq 0.5$ for EQUAL-LENGTH WEIGHTED DIAGNOSTIC COVER.

In human genetics, the patient genomes are comprised of 23 chromosome pairs and 41 pairs of chromosome arms, and we wish to distribute markers across all the arms. Therefore, we wish to solve many instances of WDC using 41 disjoint intervals and sets of markers (one set per arm), with distinct choices of k for each arm. If the clinician wants to use K markers overall, dynamic programming can be used to optimally distribute the K markers among the 41 arms so as to maximize the overall diagnostic power.

We implemented both *greedy* and *dir-greedy* and tried them on a real data set for the full human genome. On this data set *dir-greedy* is comparable in power to a heuristic studied previously [2]. *Greeedy* gives a consistent 2–5% improvement in diagnostic power, depending on the number of markers K and the size of defect b. We also implemented a local search heuristic that takes the solutions obtained by *greedy* and *dir-greedy* and applies marker swaps iteratively until the solutions cannot be improved by further swaps. The performance of *greedy* + local search and *greedy* + local search are almost indistinguishable. Based on our estimates of number of patients and cases in the next section, a 2–5% improvement in diagnostic power could be important in the clinical setting. The anomaly that *greedy* performs better than *dir-greedy* on real data despite the weaker performance guarantee calls for a better theoretical understanding of the *greedy* heuristic for WDC.

The theoretical performance bounds on the approximation algorithms and some other theoretical results are described in Section 3. The implementation is described and evaluated in Section 4. We conclude with a long list of open problems in Section 5.

2 Genetics Background

Recent genetic studies suggest that a substantial number of cases of unexplained mental retardation and other serious phenotypes are caused by small chromosomal aberrations that disrupt multiple nearby genes (see [15, 8] and references therein). The most common aberrations are *monosomy* and *trisomy* in which a child inherits only one or three copies of a chromosomal region, instead of the usual two copies. For example, Miller-Dieker syndrome, a common cause of mental retardation, is caused by a segmental monosomy on chromosome 17, usually involving a small piece of the chromosome.

Aneusomies that have been recognized in a variety of patients to be similar in both the region affected and the phenotype are called *contiguous gene syndromes*(CGS) [15]. Besides Miller-Dieker syndrome, some other named CGS's include: DiGeorge syndrome (monosomy of part of chromosome 22), Smith-Magenis syndrome (monosomy of a part of chromosome 17, different from the Miller-Dieker region), and Beckwith-Weidemann syndrome (trisomy of part of chromosome 11). A survey of the intensive research on CGS's can be found in [15]. Most of the research focuses on identifying the critical region and genes affected in a particular CGS and devising new diagnostics for a particular CGS. The genome-wide study of the frequency and distribution of aneusomies is just beginning [5]. It is expected that there are many CGS's as yet uncharacterized genetically or phenotypically.

Genome-wide study of aneusomies is essential because in large clinical genetic centers patients with a broad range of phenotypes arrive for genetic testing, seeking a genetic explanation of their malady. If there are chromosomal aberrations, they may be anywhere along the genome. Clinical genetics centers currently use a set of techniques called *karyotyping* to look for visible aberrations in any of the chromosomes. In karyotyping chromosomes are stained with a dye that makes different parts of a chromosome exhibit light or dark *bands*. The pattern of bands from the patient's chromosomes is compared under the microscope against a "normal" banding pattern to look for aberrations. An indication of the importance of detecting chromosomal aberrations is that in 1992, the large regional United States genetic testing centers performed over 15,000 peripheral blood cell karyotypes and over 160,000 prenatal karyotypes at a cost of hundreds of dollars per test [6]. Karyotyping suffers from limited and variable resolution, which makes it difficult to detect small defects. The survey [15] describes some laboratory methods that may eventually replace or supplement karyotyping.

One possible diagnostic method is *genotyping*. The basic principle of genotyping in this context is to sample the genome of the father, mother, and child at *markers* that are variable across the population. If the father and mother have sufficiently different DNA sequences at a marker, it is possible to detect that a child inherited only one or three copies of that marker by simple application of the logical rules of Mendelian inheritance. More definitions and some examples are given below.

Mansfield [17] and Celi et al. [3] were among the first to propose using genotyping to detect monosomies and trisomies and to demonstrate that the method works in practice. Wilkie [23] considered some of the methodological issues and Flint et al. [8] conducted the first substantial study. In the study, 99 patients with mental retardation that could not be explained by karyotyping were genotyped using markers only at the chromosome ends because the ends are thought to be hot spots for aneusomies. The study found 6 patients with detectable aberrations on at least one chromosome end. Biesecker et al. [1] have recently completed a pilot study using genome-wide genotyping on 12 patients with serious phenotypes that could not be explained by karyotyping. Biesecker and Schäffer [2] defined the theoretical model considered here and gave statistical evidence that genome-wide genotyping is a cost-effective alternative or supplement to karyotyping. *In this paper we pursue the algorithmic question of how one should optimize a genome-wide genotyping assay for monosomy and trisomy.*

For simplicity we restrict our most of our attention to monosomies, since they are thought to be more common [9]. The detection probabilities that a marker has for monosomy and trisomy are highly correlated; some experiments we did suggested that optimizing for monosomy or optimizing for trisomy yield almost exactly the same assay.

Humans have 23 chromosome pairs; except for X, the chromosomes are numbered $1, 2, 3, \ldots, 22$ in approximately decreasing order of length. All the chromosomes have a *centromere* that plays a pivotal role in chromosome stability and duplication. All the chromosomes except 13,14,15,21, and 22 have the centromere somewhere in the middle of the genetic material. The sides of the centromere are called *chromosome arms*; the shorter arm is p and the longer arm is q. Because of the pivotal role of the centromere, we assume that a monosomy/trisomy cannot span the centromere and be viable. Therefore, we separately consider 41 chromosome arms that we wish to test. For simplicity, we optimize and report results for female children since they have two copies of all chromosomes. Our software also correctly computes the detection probability (and could instead optimize for) male children that have only one copy of X.

Genotyping uses location-specific *markers* to trace inheritance. Markers used for genotyping exhibit significant variation or *polymorphism* across the population, but can be localized by unique flanking DNA strings that are not very polymorphic and can be used to set up a polymerase-chain-reaction(PCR) duplication. Most markers used in genotyping today are *short tandem repeat polymorphisms* (STRP), which means that a a small DNA string, such as CA, repeats multiple times. The number of repeats varies from person to person, and is called an *allele*. Each person has homologous chromosome pairs (except for males and the X chromosome), so each person has two alleles at each marker. Even though the alleles vary, they are almost always inherited according to Mendelian rules. A child has two alleles, one is identical to one allele of the father and the other is identical to one allele of the mother. The pair of alleles at a marker is called the *genotype* at that marker. If the two alleles are identical, the genotype is *homozygous*; the two alleles differ, the genotype is *heterozygous*. Current genotyping technology shows which allele(s) are present, but not how many copies. Thus we assume that it is not possible to directly distinguish a case of normal homozygosity at a marker (two copies of the same allele) from a case of abnormal monosomy (one copy of one allele, and no other alleles) without using extra information. The extra information we will use are the parental genotypes.

STRP markers typically lie outside genes, but they serve as excellent landmarks to locate nearby genes. Hundreds of genes that control single-gene diseases have been localized by correlating the inheritance of the disease-causing allele(s) with alleles of nearby STRP markers. Substantial early effort in the Human Genome Project has gone into identifying thousands of STRP markers and delineating the different alleles that occur (see for example the map in [10]). The French research group CEPH (Centre d'Etude du Polymorphisme Humain) has collected a panel of reference families and genotyped those families at many STRP markers. The genotypes are used to find the sorted order of the markers along each chromosome.

Another result of the CEPH effort is a table of estimated allele frequencies for many STRP markers. These allele frequencies are used in disease studies to average over the possible genotypes for family members who cannot be genotyped. In aneusomy testing, we want to theoretically plan an assay for thousands of nuclear families (father, mother, child) before we know their genotypes. Therefore, we will assume that the father and mother have their alleles chosen independently at random from the allele frequencies available (in our experiments we use the CEPH-tabulated frequencies). This assumption is commonplace in medical genetics and goes by the name *Hardy-Weinberg equilibrium*.

We now give a couple examples of how monosomy and trisomy of a child can be detected by genotyping the mother, father, and child. Suppose the father has heterozygous genotype AB and the mother has heterozygous genotype CD; i.e., all four parental alleles are distinct. Then if the child has only a single allele, the child must have a monosomy. Similarly, the child can have a trisomy if and only if the child has three distinct alleles.

A more interesting case occurs when the parents have genotypes AB and BC. Then a monosomy can be detected if the child has single allele A or single allele C. However, if the child has single allele B, we cannot distinguish between monosomy and homozygosity. When there is a monosomy, we assume that each of the 4 parental alleles is inherited with equal probability because there is little data to show which parent of origin is more common. Therefore, if we condition on the events that the parents have genotypes AB and BC and that a monosomy is present, the probability that a monosomy is detected is 2/4.

Assuming Hardy-Weinberg equilibrium and known allele frequencies, Wilkie [23] worked out a table of the probability that a marker detects a monosomy/trisomy. The probability is conditioned on the aberration being present and averaged over all possible parental genotypes. Let $a_1, a_2, \ldots$ be the allele frequencies. For $i = 1, 2, 3$, let f_i be the sum of the allele frequencies to the ith power. That is,

$$f_i = \sum_j a_j^i.$$

Then the probability of detecting monosomy is:

$$1 - 2f_2 + f_3$$

and the probability of detecting trisomy is:

$$1 - 3f_2 + 2f_3.$$

Jarnik et al. [12] have proposed a more general measure of detection probability, which they called *overall informativity*, intended as a measure of the usefulness of a marker for detecting a variety of chromsomal aberrations including monosomy and trisomy.

3   Two Heuristics

In this section we present our bounds for WEIGHTED DIAGNOSTIC COVER. As we noted in the Introduction, WEIGHTED DIAGNOSTIC COVER (WDC) has some similarity to WEIGHTED SET COVER (WSC). In this section we implicitly convert the WDC and WSC problems from the decision version stated in the introduction to the optimization version.

We first point out that when WSC is restricted to intervals (such as the sets in WDC), then WSC can be solved optimally in $O(n \log n)$ time. Consider an instance of WSC in which the base set B is the interval $[l, r]$ and the sets that may be used in the cover are subintervals $s_1, \ldots, s_n$. We assume that all the interval endpoints are fixed-precision decimal numbers, so they can be compared.

Theorem 1. *The WSC problem can be solved optimally in time $O(n \log n)$ when the n input sets are intervals.*

Proof. The following dynamic programming algorithm solves the problem in time $O(n^2)$.

Sort and renumber the subintervals in increasing order of left-endpoint.
Let $s_i = [l_i, r_i]$.
Let $\Delta(j) =$ minimum weight needed to cover $[l, r_j]$ making use of s_j
    and a subset of the first $j - 1$ intervals. $\Delta(j) = \infty$ if there is no cover.
$\Delta(1) = p_1$.
For $j = 1, \ldots n - 1$:
    $\Delta(j + 1) = p_{j+1} + \min_{k \leq j} \{\Delta(k) | s_k \text{ overlaps with } s_j \text{ and } r_k \leq r_{j+1}\}$.

The optimal weight is the minimum $\Delta(i)$ such that $r_i = r$.

By backtracking over the indices where Δ changes, we get the actual cover.

If there is a cover, the computation of $\Delta(i)$ finds the minimum cost of a cover because of: **Observation:** Let M be any subset of markers that covers $[x_0, x_n]$. Then for every j, there exists a $k \leq j$ such that $\{s_1, \ldots, s_k\} \cap S$ covers $[l, r_j]$. The backtracking step is the standard way to convert from a 1-dimensional dynamic program that computes an optimal cost to a solution achieving the optimal cost. Each step runs in $O(n)$ time and the total time for a naive implementation is $O(n^2)$. The time-consuming step is the computation of Δ.

The running time can be improved to $O(n \log n)$ by a better implementation of the step computing Δ, using a geometric approach. Define a Δ-pair to be a pair $[\Delta(j), r_j]$ such that $\Delta(j)$ is the cost of an optimal solution covering $[l, r_j]$. Two Δ-pairs $[\Delta(j), r_j]$ and $[\Delta(i), r_i]$ can be compared by asking:

1. Is $\Delta(j) \leq \Delta(i)$?
2. Is $r_j \geq r_i$?

and we say that the j pair *dominates* the i pair, if both answers are "yes". This is not the usual geometric domination because the two questions have the inequalities in opposite directions. Nevertheless, a standard solution to the problem of finding geometrically dominant points in 2-dimensions [20, p. 184] can be applied.

The $\Delta(k)$ used on the minimization to compute the new $\Delta(j+1)$ must come from an undominated pair. Therefore, we store all the undominated pairs; we use a balanced binary search tree for this purpose. Sorting the tree by increasing Δ is equivalent to sorting by increasing r. Therefore, the k we want in the minimization is the rightmost r_k such that the new left endpoint l_{j+1} is to the left of r_k. This can be found by one search in the tree.

We use the tree to:

1. Find the best r_k to compute $\Delta(j+1)$.
2. Insert the new pair $[\Delta(j+1), r_{j+1}]$.
3. Delete all the pairs in the tree that are dominated by the new pair.

The first two steps take $O(\log n)$ time because there are most n insertions overall. The elements to be deleted occur consecutively in the sorted order, so they can be deleted in $O(\log n)$ time each. Whether any pair needs to be deleted can be determined by doing one search in the tree to find the pair whose Δ is smallest among all those that are greater that $\Delta(j+1)$. If the new pair does not dominate this pair, then no pairs need to be deleted. There are at most n deletions overall, so the total running time is $O(n \log n)$.

3.1 Performance of *greedy*

Next we prove a lower bound guarantee on the performance of *greedy* for WDC. The lower bound proof explicitly divides the markers into two types depending on whether their detection probability is above or below some threshold T. Some steps require that $T \leq 0.5$, but we will defer optimizing T, within the interval $[0.0.5]$ until the end of the proof. Our proof does not make use of the fact that the internal markers all have the same length. Let the *weight* of an interval be its (average) detection probability multiplied by its length. We start with two basic facts about weight.

Fact 2. Suppose point x has detection probability p due to a set of intervals S and detection probability p' due to a subset $S' \subseteq S$. Then x has probability $\geq p' - p$ due to the complementary subset $S \setminus S'$.

Fact 3. Let a set of r intervals achieve a weight of M over a set of points. Then one of the r intervals achieves a weight of at least M/r by itself.

Lemma 4. *Let the markers chosen after some iteration of* greedy *result in a detection probability of $q \leq T$ for some point x. Let the next marker chosen have probability p and suppose the new marker contains x. Then the detection probability at x increases by least $p(1 - T)$.*

Proof. The new probability at x is

$$q + p - (qp)$$

$$\geq q + p - (Tp)$$

$$= q + (1 - T)p.$$

Let the maximum weight achievable by k markers be W. Then

Lemma 5. *For any $j \geq 0$, the* greedy *algorithm realizes a total weight of at least $WT(1 - (1 - \frac{1-T}{Tk})^j)$ when it chooses j markers.*

Proof. The proof is by induction on j. When $j = 0$ the weight is 0, and hence the basis holds. For the inductive hypothesis, let the *greedy* algorithm realize a weight of

$$\alpha \geq WT(1 - (1 - \frac{1 - T}{Tk})^j)$$

with the first j markers.

Define a point to be *heavy* if its probability with the first j greedy markers is at least T times its probability in the optimal solution. Out of α let weight α' come from integrating over the heavy points. If we remove their contribution to W, the remaining weight is at least $W - \alpha'/T$. Next remove the contribution of the points that are non-heavy and in the greedy solution. This weight is $\alpha - \alpha'$ in the greedy solution. By Fact 2, the remaining weight is at least:

$$W - \frac{\alpha'}{T} - (\alpha - \alpha').$$

Then by Fact 3, among the k markers in the optimal solution (for k markers), there is one that is not among the the j chosen greedily, such that it has weight of at least

$$\frac{1}{k}(W - \frac{\alpha'}{T} - (\alpha - \alpha')) = U,$$

from the non-heavy points.

The weight achieved by greedy after choosing $j + 1$ markers satisfies the following (in)equalities:

$$\geq \alpha + (1 - T)U$$

$$= \alpha + \frac{1-T}{k}(W - \alpha'\frac{1-T}{T} - \alpha)$$

$$= \alpha(1 - \frac{1-T}{k}) + W\frac{1-T}{k} - \alpha'\frac{(1-T)^2}{Tk}$$

$$\geq \alpha(1 - \frac{1-T}{Tk}) + W\frac{1-T}{k}$$

$$\geq W\frac{1-T}{k} + (WT - WT(1 - \frac{1-T}{Tk})^j)(1 - \frac{1-T}{Tk})$$

$$\geq WT(1 - (1 - \frac{1-T}{Tk})^{j+1}).$$

This completes the induction step and the lemma holds.

Theorem 6. $r_{\text{greedy}} \geq 0.318$ *for WDC.*

Proof. Using the previous lemma for choosing k markers we get a performance guarantee of

$$WT - WT(1 - \frac{1-T}{Tk})^k \geq WT(1 - e^{1-1/T}).$$

This function is maximized when T is approximately 0.466 and the resulting weight is approximately $0.318W$

The next two theorems show that *greedy* can be provably far from optimal, but there is still a big gap between the preceding lower bound and the succeeding upper bounds.

Theorem 7. *There exist instances of EWDC such that* $r_{\text{greedy}}(k) \leq 0.829$.

Proof. We construct a class of such instances. In each instance there are *heavy* markers of probability $1 - \epsilon$ and *light* markers of probability $1 - 2\epsilon$. There are k light markers, ideally placed, so that the optimal solution is to choose all the light markers. The *greedy* algorithm will choose all the heavy markers and then some light markers to get k markers in all. In the limit, we let ϵ decrease to 0, so that the probability advantage of the heavy markers is negated.

For simplicity we set $b = 1$; all the instances can be scaled up to larger integral values of b. Let $k = r - l - b$, the arm length. Put the light markers with left ends at $l, l+1, l+2, \ldots, r-1$. There is a fractional parameter $a < 1$, used to place the heavy makers; we will optimize the choice of a at the end of the proof. Put heavy markers with their left ends at $l, l+1+a, l+2(1+a), \ldots$. To make the calculation of the pessimal a precise, assume that the arm length, $r - l - 1 = k$, is an integer multiple of, $1+a$, the heavy marker spacing.

There is a cover that chooses all k light markers. This cover has power $1 - 2\epsilon$, which approaches 1 as ϵ approaches 0. The *greedy* algorithm will first choose all the heavy markers. There are $\frac{k}{1+a}$ of these. This leaves $k - \frac{k}{1+a} = \frac{ka}{1+a}$ markers to choose. The *greedy* algorithm can therefore choose $\frac{ka}{1+a}$ of the light markers. There are gaps of length a between each pair of adjacent heavy markers. In the *greedy* cover each light marker provides weight at most a for the uncovered region and essentially 0 for the region where it overlaps with 1 or 2 heavy markers, as ϵ goes to 0.

The *greedy* cover has probability approximately 1 wherever it places a marker, but has some uncovered regions of probability 0. The amount of the arm covered is $\frac{k}{1+a} + \frac{ka^2}{1+a}$, where the first term comes from the light markers and the second from the heavy markers. Since the total arm length is k, there is a length of $\frac{ka}{1+a} - \frac{ka^2}{1+a}$ uncovered.

We wish to choose a to maximize the uncovered fraction. We can factor out the k, and differentiate $\frac{a - a^2}{1+a}$ with respect to a to find a maximum. The derivative is $-a^2 - 2a + 1$, which has a 0, corresponding to the maximum of the function, at $\sqrt{2} - 1$. Therefore, the uncovered fraction is $\frac{\sqrt{2}-1}{1+\sqrt{2}} > 0.171$. By choosing a nearby rational value of a, we can make k large enough so that the arm length is a multiple of $1 + a$. Then $r_{greedy}(k) \leq 0.829$.

Theorem 8. *There exists an instance of WDC, such that*

$$r_{\text{greedy}}(k) \leq (1 - \prod_{i \geq 1}(1 - \frac{1}{2^i})) + \frac{\log_2 k}{k}$$

Proof. For simplicity, assume initially k is a power of 2. The length of the arm is k. Our instance has $2k - 1$ markers based on a complete binary tree with k leaves. For each leaf we specify an interval of length 1 with a probability close to $1 - \epsilon_0$. The marker corresponding to each internal node in the binary tree is the interval that is the union of the intervals of its children. Assume the leaves are at level 0, their parents at level 1, and so on. Then the detection probability for the markers at level ℓ is $2^{-\ell} - \epsilon_\ell$. The optimal choice is the k leaves, and in the limit as all the ϵ go to 0, it has a probability 1. To confuse *greedy*, we require

$$\epsilon_i \geq 2 \times \epsilon_{i+1}.$$

Then *greedy* chooses first the root, then its children, and so on— leaving out all but one leaf.

To find the diagnostic power for the *greedy* solution in the limit, observe that 1 level above the leaves it fails to diagnose 1/2 the defects, 2 levels above the leaves it fails to diagnose 3/4 of the defects, and so on. Therefore, the diagnostic probability is bounded above by the difference:

$$(1 - \prod_{i \geq 1}(1 - \frac{1}{2^i})) + 1.$$

The +1 term comes from the extra leaf.

Now consider any k, not necessarily a power of 2. Again, the instance will have $\leq 2k - 1$ markers organized in a binary tree, bottom up. One level above the leaves, every node has two children, but the rightmost leaf may be without a parent, if k is odd. In general on each, level at most one node is left out and we will assume that it is always rightmost. Every node/marker gets power $2^{-\ell}$ as above. The *greedy* heuristic will choose all the internal nodes again. There will be at most $\frac{\log_2 k}{k}$ markers left to choose, because at most 1 marker is left over per level of the tree. The remaining markers will be all leaves. The diagnostic probability is:

$$r_{greedy}(k) \leq (1 - \prod_{i \geq 1}(1 - \frac{1}{2^i})) + \frac{\log_2 k}{k}.$$

Corollary 9. *There exists an instance of WDC, such that $r_{\text{greedy}}(k) \leq 0.722$, for $k \geq 1024$.*

Proof. Apply the bound of the previous Lemma. The infinite product

$$P = \prod_{i \geq 1}(1 - \frac{1}{2^i})$$

was studied by Euler in the context of the theory of integer partitions(see [18, pp. 220-223]). He showed that the product P is equal to the following infinite sum:

$$1 + \sum_{n \geq 1}(-1)^n[2^{-n(3n-1)/2} + 2^{-n(3n+1)/2}].$$

The sum has the virtue that it is quadratically convergent. To show that $P \geq 0.288$ and know that the error is $\leq O(10^{-4})$ we need to evaluate only the first 3 terms of the sum [18, p.223].

The term $\frac{\log_2 k}{k} < 0.01$ if $k \geq 1024$. Add the bounds for the two terms together to obtain the overall bound 0.722.

3.2 Performance of *directional greedy*

We now describe a different heuristic that we call *directional greedy* or *dir-greedy*. It is particularly intended for the equal-length special case. Recall that b is the size of monosomy we are trying to detect, so in the instance of EWDC, every marker that does not hit a endpoint is of length b. We partition the arm interval $[l, r - b]$ into subintervals that are all of size b, except possibly the subinterval on one end. The partition is called P. We call each subinterval a *slice*. Each marker is divided into two pieces by one slice, s, that intersects it. The left piece extends from s to the left end of the original marker; the right piece extends from s to the right end of the original marker.

We can now define two instances of WDC, induced by P, in which the markers are of varying lengths. The *left* instance L contains as markers all the left pieces from the original instance. The *right* instance R contains as markers all the right pieces from the original instance. As described below, we find the optimal solutions L_O and R_O to L and R. Then we extend each chosen marker to its original size and define the corresponding (not necessarily optimal) solutions L_{tot} and R_{tot} to the original instance. We choose whichever of L_{tot} and R_{tot} has higher power.

Theorem 10. $r_{\text{dir-greedy}} \geq 0.5$ *for EWDC.*

Proof. Let Z be an optimal solution to the original EWDC instance. Apply the partition P to Z and the instance. This defines left and right instances of WDC (without equal lengths) and also yields, based on Z, (not necessarily optimal) solutions Z_ℓ and Z_r to those instances. Since $Z_\ell \cup Z_r = Z$, we have $D(Z_\ell) + D(Z_r) > D(Z)$. The solutions, Z_ℓ and Z_r need not be optimal, but our algorithm does find the optimal solutions L_O and R_O. Therefore, $D(L_O) \geq D(Z_\ell)$ and $D(R_O) \geq D(Z_r)$. Combining inequalities, we get $D(L_{tot}) + D(R_{tot}) \geq D(Z)$. Thus one of the two of L_{tot} and R_{tot} has at least $1/2$ the power of an optimal solution to the original instance. Our algorithm chooses a solution that has power at least as high as the better of L_{tot} and R_{tot}

Since the arm interval length, $r - l - b$ is not necessarily divisible by b, one could define the slices either starting at the left end l or at the right end r. This will yield two different pairs of sliced instances. In the implementation we try both pairs, find the two optimal solutions for each pair, extend them to the original instance, and choose the best of the four solutions.

To complete the explanation of *dir-greedy* we need to show how to solve the instances of WDC derived from the initial instance of EWDC. The WDC instances have the special property that on one side the endpoints of all the markers are identical. We will prove that the *greedy* algorithm is optimal for such instances of WDC. The proof is nontrivial. For the ensuing text, we assume that we are solving instances where all markers have the same left endpoint, 0. We assume that the *greedy* algorithm includes a tie-breaking rule, so that the solution is uniquely defined.

We need a simple inequality about how the probability changes when the same new marker is added to different covers.

Fact 11. Let a_1, a_2, c be probabilities. If $a_1 \geq a_2$, then $(a_1 + c - a_1 * c) \geq (a_2 + c - a_2 * c)$.

The proof that *greedy* is optimal for the common endpoint case of WDC is by contradiction. We will need to compare different candidate covers that differ in one element. The following technical lemmas cover three different cases that arise in the proof of the theorem later.

Lemma 12. *Let T be an instance of WDC in which all markers have the same left endpoint 0. Suppose that A and C are covers and $\{b_1\}$ and $\{b_2\}$ are markers such that*

- *$A, \{b_1\}, \{b_2\}, C$ are pairwise disjoint;*
- *$D(A \cup \{b_1\}) \geq D(A \cup \{b_2\})$;*
- *b_1 is shorter than b_2;*
- *every marker in C is longer than b_1 and b_2*

Then $D(A \cup \{b_1\} \cup C) \geq D(A \cup \{b_2\} \cup C)$.

Proof. Let the right endpoints of b_1 and b_2 be r_1 and r_2 respectively. Suppose A has average diagnostic probability a_1 in the interval $[0, r_1]$, a_2 in the interval $[0, r_2]$, a_3 in the interval $[r_1, r_2]$, and a_4 in the interval $[r_2, r]$. Then:

$$D(A) = [r_1 * a_1] + [(r_2 - r_1) * a_3] + [(r - r_2) * a_4]$$

$$D(A \cup \{b_1\} = [r_1 * (a_1 + b_1 - a_1 * b_1)] + [(r_2 - r_1) * a_3] + [(r - r_2) * a_4]$$

$$D(A \cup \{b_2\}) = [r_1 * (a_1 + b_2 - a_1 * b_2)] + [(r_2 - r) * (a_3 + b_2 - a_3 * b_2)] + [(r - r_2)] * a_4$$

Let d_1 be the average diagnostic power of $A \cup \{b_1\}$ in the interval $[0, r_2]$ and let d_2 be the average diagnostic power of $A \cup \{b_2\}$ in the interval $[0, r_2]$. Since the third term in all the three-term sums above is the same, it follows from the second assumption that $d_1 \geq d_2$.

Now consider the effect of adding C to each of the two covers. The covers are the same to the right of r_2, so C makes no difference there. To the left of r_2 all the markers of C participate, so they cumulatively contribute the same diagnostic probability c at every point of $[0, r_2]$. So, in the interval $[0, r_2]$, the average diagnostic probability d_1 increases to $d_1 + c - d_1 * c$ and the average diagnostic probability d_2 increases to $d_2 + c -- d_2 * c$. By Fact 11 $d_1 + c - d_1 * c \geq d_2 + c - d_2 * c$, and the Lemma follows.

Lemma 13. *Let T be an instance of WDC in which all the markers have the same left endpoint 0. Let A be a cover for T. Let b_1 and b_2 be two markers with probabilities p_1, p_2 such that:*

- *$A, \{b_1\}, \{b_2\}$, are pairwise disjoint;*
- *b_1 is longer than b_2;*
- *$p_1 \geq p_2$.*

Then $D(A \cup \{b_1\}) \geq D(A \cup \{b_2\})$.

Proof. Subdivide b_1 into two markers b_3 and b_4 both with probability p_1. Let b_3 have the same enpoints as b_2 and let b_4 go from the right endpoint of b_2 to the right endpoint of b_1. Then $D(A \cup \{b_1\}) = D(A \cup \{b_3, b_4\})$. Since b_3 has the same enpoints as b_2 and higher probability, $D(A \cup \{b_3\}) \geq D(A \cup \{b_2\})$. Since b_4 contributes nonnegative diagnostic probability, $D(A \cup \{b_3, b_4\}) \geq D(A \cup \{b_3\})$. The Lemma follows by transitivity of $\geq$.

Lemma 14. *Let T be an instance of WDC in which all markers have the same left endpoint 0. Suppose that A and C are covers and $\{b_1\}$ and $\{b_2\}$ are markers with probabilities p_1, p_2 such that:*

- *$A, \{b_1\}, \{b_2\}, C$ are pairwise disjoint;*
- *$D(A \cup \{b_1\}) \geq D(A \cup \{b_2\})$;*
- *b_1 is longer than b_2;*
- *$p_1 \leq p_2$;*
- *every marker in C is shorter than b_1 and b_2.*

Then $D(A \cup \{b_1\} \cup C) \geq D(A \cup \{b_2\} \cup C)$.

Proof. Let the right endpoints of b_1 and b_2 be r_1 and r_2 respectively. Let the furthest right endpoint in C be r_0. Suppose $A \cup \{b_1\}$ has average diagnostic probability a_0 in the interval $[0, r_0]$, a_1 in the interval $[r_0, r_2]$, a_2 in the interval $[r_2, r_1]$, and a_3 in the interval $[r_1, r]$. Suppose that $A \cup \{b_2\}$ has average diagnostic probability d_0 in the interval $[0, r_0]$, d_1 in the interval $[r_0, r_2]$, d_2 in the interval $[r_2, r_1]$, and a_3 in the interval $[r_3, r]$. Then.

$$D(A \cup \{b_1\}) = [r_0 * a_0] + [(r_2 - r_0) * a_1] + [(r_1 - r_2) * a_2] + [(r_3 - r_1) * a_3]$$

$$D(A \cup \{b_2\}) = [r_0 * d_0] + [(r_2 - r_0) * d_1] + [(r_1 - r_2) * d_2] + [(r_3 - r_1) * a_3]$$

Since $p_1 \leq p_2$ and both b_1 and b_2 span $[0, r_0]$, it follows that $a_0 \leq d_0$. Moreover, at any point x in $[0, r_0]$ the diagnostic probability a_x of the first cover is less than the diagnostic probability d_x of the second cover. Now consider the effect of adding C and suppose it has diagnostic probability c_x at x. Then the increase in the first cover's probability at x is is $c_x - c_x * a_x$ and the increase in the second cover is $c_x - c_x * d_x$. Since $a_x \leq d_x$ the first cover improves by a larger amount at x than the second cover by virtue of adding C. Since the first cover was already better, it remains better after adding C.

Theorem 15. *The* greedy *algorithm is optimal for instances of WDC that all have the same (left) endpoint (0).*

Proof. The proof is by contradiction. Let such a common-endpoint instance of WDC be given. Let the number of markers desired be m. Let the greedy solution be G. We need some notation to compare G to a non-*greedy* solution H. Let G be constructed by selecting the markers $g_1, g_2, \ldots, g_m$ in this order. Define j such that $g_1, \ldots, g_{j-1} \in H$, but g_j is not in H. Then we say that the set $\{g_1, \ldots, g_{j-1}\}$ is the greedy *prefix* of H, g_j is the *ultragreedy* element, and the set of markers in H which are not in the greedy prefix is the non-greedy suffix of H. It is possible that some markers in the non-greedy suffix of H are also in G. Because the *greedy* algorithm always chooses for immediate benefit, the non-greedy suffix of an optimal H must be non-empty.

Seeking to establish a contradiction, let H be an optimal solution with maximal greedy prefix. Let the greedy prefix of H be P and the non-greedy suffix of H be S. We will find an element $h \in S$ such that if we replace h by g_j to obtain

$H' = P \cup \{g_j\} \cup S \setminus \{h\}$, and $D(H') \geq D(H)$. Since we included the ultragreedy g_j in H', it contradicts the assumption that H has maximal greedy prefix among optimal solutions.

Sort the elements of S in increasing order of length, regardless of their probability, as $s_1, s_2, \ldots, s_n$. There are now several cases for choosing h. Whenever we apply either Lemma 12 or Lemma 14 below, the second assumption (that compares diagnostic power of two covers) in the statement of the Lemmas holds because the *greedy* algorithm prefers the marker we designate as b_1 to the marker we designate as b_2.

Case 1. g_j is longer than s_n and has higher probability. Let $h = s_n$ and apply Lemma 13 with $A = P \cup S \setminus \{h\}$, $b_1 = g_j$ and $b_2 = h$, to conclude that $D(H') \geq D(H)$.

Case 2. g_j is longer than s_n but has smaller probability. Let $h = s_n$ and apply Lemma 14 with $A = P$, $b_1 = g_j$, $b_2 = h$, $C = S \setminus \{h\}$, to conclude that $D(H') \geq D(H)$.

Case 3. g_j is shorter than s_{i+1}, but longer than any shorter elements of S; note that i is 0 if g_j is shorter than all the elements of s. Choose $g_j = s_i$. In this case, the argument that the replacement is better proceeds in two stages, unless $i = 0$. First we observe that $D(P \cup \{g_j, s_{i+1}, \ldots s_n\}) \geq D(P \cup \{s_i, s_{i+1}, \ldots, s_n\})$ by applying Lemma 12 with $A = P$, $b_1 = g_j$, $b_2 = h$, $C = \{s_{i+1}, \ldots, s_n\}$.

In the second stage, needed if $i > 0$, we show that adding in the shorter elements of S does not affect the conclusion. If g_j has higher probability than s_i (similar to Case 1), apply Lemma 13 with $A = P \cup S \setminus \{h\}$, $b_1 = g_j$, $b_2 = h$. to conclude that $D(H') \geq D(H)$. Otherwise, g_j has lower probability that s_i (similar to Case 2), and apply Lemma 14 $A = P \cup \{s_{i+1}, \ldots s_n\}$, $b_1 = g_j$, $b_2 = h = s_i$, and $C = \{s_1, \ldots, s_{i-1}\}$ to conclude that $D(H') \geq D(H)$.

Theorem 16. *The performance guarantee $r_{\mathrm{dir-greedy}} \geq 0.5$ is best possible for EWDC.*

Proof. We construct a set of extreme instances that make $r_{dir-greedy}$ perform badly. Let the interval $B = [0, nb]$ for any integer $n > 1$. The set of possible markers consists of two types, which we call *light-grid* markers and *heavy-off-grid* markers. The light-grid markers are $[0, b], [b, 2b], \ldots, [(n-1)b, nb]$ and all have detection probability $1/2$. The heavy-off-grid markers are $[b/2, 3b/2], [3b/2, 5b/2], \ldots, [nb-3b/2, nb-b/2]$ and all have detection probability $1 - \epsilon$. Suppose we want to choose $k < n$ markers.

The *dir-greedy* algorithm will always choose light-grid markers because they have weight $1/2$ in each of the two sliced instances, while the heavy-off-grid markers have weight $(1 - \epsilon)/2$ in each of the two sliced instances. The optimal solution is to choose k heavy-off-grid markers. The *dir-greedy* cover has detection power $k/2n$. The optimal cover has detection power $(k(1-\epsilon))/n$. Taking the limit as ϵ goes to 0 completes the proof.

4 Implementation

We modified a program called DECIDE (DEtecting Chromsomal Insertions and DEletions) for EWDC [2]. The previous implementation used a heuristic that seemed plausible, but is not justified by any performance guarantee. DECIDE implements a dynamic programming algorithm to combine solutions to the 41 arm instances of WDC for each choice of K into an overall genome-wide set of markers. The dynamic programming code is almost independent of what heuristic is used to solve the WDC instances. However, for *greedy* one can use an overall greedy strategy choosing one marker at a time from the overall pool of markers, without resorting to dynamic programming.

Let the arms be $a_1, \ldots, a_A$. Let the endpoints of arm a_i be l_i and r_i. The size of the arm, which is the interval where the left end of an aneusomy can be, is $s_i = r_i - l_i - b$. Suppose we have a collection of solutions to WDC for each arm and each choice of K. Let the detection power for arm a_i and K markers be power(i, K). Since the arms are of different sizes, denote the weighted power by wpower$(i, K) = $ power$(i, K) \times s_i$. The overall power for using j markers on the first h arms can be maximized by maximizing the objective function

$$F_h(j) = \max[\text{wpower}(1, K_1) + \cdots + \text{wpower}(h, K_h)]$$

subject to the constraint that

$$K_1 + \cdots + K_h = j.$$

The overall detection power for j markers is given by $F_A(j)/S$ where $S = s_1 + \cdots + s_{41}$ is the sum of the arm sizes. Given $F_{h-1}(j)$ for all j, DECIDE computes

$$F_h(j') = \max_{j \leq j'}(F_{h-1}(j) + \text{wpower}(h, j' - j)).$$

The previously implemented heuristic solves EWDC by placing K points uniformly across the arm. Then it finds for each point c the actual marker of highest probability within $b/2$ units of the point c. Reference [2] describes how to combine resources from three WWW sites to construct a genome-wide map of 2,501 real STRP markers in such a way that we have a physical location for each marker, we know its allele frequencies, and we have some assurance that the marker can be successfully genotyped quickly and cost-effectively. We implemented the *greedy* and *dir-greedy* heuristics in modified versions of DECIDE. Here we compare the detection power of the old heuristic, *greedy*, and *dir-greedy* for monosomy. We used a target monosomy size, b, ranging from 5Mbases to 14Mbases. We asked DECIDE to select anywhere from 200 to 1000 markers. The sizes are roughly based on the current resolution of karyotyping. The marker numbers are based on the current cost of genotyping and projected decreases due to improving and spreading technology. See [2] for a discussion of these choices. Some sample results are shown in Table 1.

Size(b)	No. of Markers	Old	Greedy	Dir-Greedy
5	200	0.227	0.254	0.238
5	300	0.324	0.364	0.340
5	400	0.414	0.458	0.429
5	500	0.488	0.531	0.498
5	600	0.545	0.586	0.546
5	700	0.587	0.630	0.584
5	800	0.618	0.665	0.619
5	900	0.642	0.693	0.650
5	1000	0.661	0.715	0.676
8	200	0.389	0.413	0.390
8	300	0.528	0.566	0.539
8	400	0.626	0.666	0.624
8	500	0.694	0.732	0.683
8	600	0.743	0.779	0.735
8	700	0.775	0.814	0.770
8	800	0.797	0.840	0.797
8	900	0.815	0.859	0.820
8	1000	0.829	0.872	0.838
11	200	0.529	0.562	0.540
11	300	0.675	0.713	0.669
11	400	0.761	0.799	0.749
11	500	0.817	0.852	0.808
11	600	0.855	0.888	0.846
11	700	0.878	0.911	0.874
11	800	0.894	0.926	0.895
11	900	0.906	0.936	0.911
11	1000	0.914	0.943	0.922
14	200	0.644	0.682	0.658
14	300	0.774	0.810	0.767
14	400	0.846	0.881	0.840
14	500	0.890	0.921	0.883
14	600	0.917	0.943	0.912
14	700	0.932	0.957	0.931
14	800	0.943	0.965	0.944
14	900	0.951	0.970	0.953
14	1000	0.956	0.973	0.960

Table 1. Monosomy detection power for 3 heuristics, size in Mbases

4.1 Speeding up the Dynamic Programming Algorithm

The dynamic programming algorithm has a running time of $O(AK^2)$. We prove that under the following condition a faster algorithm can be used to combine the solutions from each arm.

Monotonicity: wpower$(a, i+1)/(i+1) <=$ wpower$(a, i)/(i)$ for all arms a and all numbers of markers i. Since wpower is obtained from D by multiplying by a number that depends only on the arm length, this is equivalent to: $D(S_{i+1})/(i+1)) \leq D(S_i)/i$, where S_i and S_{i+1} are the solutions for i and $i+1$ markers respectively.

  We prove that if the candidate solutions satisfy the monotonicity property, only $K + A$ candidate solutions need to be computed, and they can be combined optimally in time $O(K \log A)$. We also prove that both the optimal single arm solutions and the greedy single arm solutions satisfy the monotonicity property. DECIDE could be modified to identify cases where the single arm solutions from the other two heuristics fail to satisfy monotonicity.

  A solution to an A arm problem can be summarized by a vector of A nonengative integers, such that:

  – the numbers sum to K
  – the ith number is the number of markers to use on arm i.

Let extra(a, i) be the extra weighted power obtained using i markers instead of $i - 1$ markers for arm a. That is, extra$(a, i) =$ wpower$(a, i) -$ wpower$(a, i - 1)$. Our algorithm fills in values of the function extra as needed, and we assume implicitly that the requisite values of wpower are known.

> Compute extra$(a, 1)$ for each arm.
> Insert the A values from the previous step into a decreasing priority queue.
> $m \leftarrow 0$. (m will count the number of markers)
> For each a, $m_a \leftarrow 0$. (m_a counts the number of markers on arm a)
> While $m < K$
> > Find an remove the maximum element extra$(b, m_b + 1)$ from the queue.
> > $m_b \leftarrow m_b + 1$.
> > $m \leftarrow m + 1$.
> > Compute extra$(b, m_b + 1)$ and insert it in the queue.
> The vector $[m_1, m_2, \ldots, m_A]$ describes the optimal multi-arm combination

  Ignoring the time to compute extra, the running time is $O(K \log A)$ because there are $O(K)$ priority queue operations, and the priority queue contains one item per arm. The number of times extra and wpower need to be evaluated are each at most $K + A$, K for the distribution of markers used in the output solution plus one additional value per arm.

Theorem 17. *The above algorithm combines the single arm solutions to get the best overall power possible using those solutions.*

Proof. The proof is by contradiction. Let B be the vector produced by the algorithm and let C be the optimal vector that has minimal total discrepancy from B. Since $B \neq C$, but they each sum to K, there exist two indices i, j, such that $B_i > C_i$ and $B_j < C_j$. We have the inequality extra$(i, B_i - 1) \geq$ extra$(j, B_j + 1)$ because the algorithm preferred to use another marker on arm i rather than arm j. We have the inequalities, extra$(i, C_i) \geq$ extra$(i, B_i - 1)$ and extra$(j, B_j + 1) \geq$ extra$(j, C_j - 1)$ by monotonicity. Combining the three inequalities, we get the inequality, extra$(i, C_i) \geq$ extra$(j, C_j - 1)$. Thus, if we modify C by adding a marker to arm i and taking away a marker from arm j, we get a solution that is at least as good and closer to B, a contradiction.

Both theorems that monotonicity is satisfied for certain solutions rely on the following lemma.

Lemma 18. *Let M be any marker set and S a subset of M. Let t be any marker not in S. Then $D(M \cup \{t\}) - D(M) \leq D(S \cup \{t\}) - D(S)$. In words, the benefit gained from adding marker t to the subset S is at least as large as the benefit from adding t to M.*

Proof. Let the detection probability for M, S, and t at point x, be $m(x)$ and $s(x)$ and $t(x)$ respectively. Let $m'(x)$ and $s'(x)$ be the detection probabilities for $M \cup \{t\}$ and $S \cup \{t\}$ respectively. Since S is a subset of M, $s(x) \leq m(x)$. Adding t to S boosts $s(x)$ by $t(x) - (t(x) \cdot s(x))$, while adding t to M boosts $m(x)$ by the *smaller* amount $t(x) - (t(x) \cdot m(x))$. That is, $m'(x) - m(x) \leq s'(x) - s(x)$.

Since the power D is obtained by integrating the detection probability over a fixed interval, the lemma follows.

Theorem 19. *The optimal single arm solutions satisfy monotonicity.*

Proof. The proof is by contradiction. Let M_j, M_{j+1} be optimal solutions with $j, j + 1$ markers having weighted powers W_j, W_{j+1}. Assume seeking a contradiction that $W_{j+1}/(j + 1) > W_j/j$.

For any marker t, let $N(t) = M_{j+1} \setminus \{t\}$. The set $N(t)$ may or may not be optimal for j markers, but its average power must be at most W_j/j. Therefore, the change in power by adding each element t is at least $W_{j+1}/(j + 1)$. The previous Lemma shows that if we add t to a smaller set, the increase in power is at least $W_{j+1}/(j + 1)$. Therefore, if we take the union of any j elements of W_{j+1}, their total power is at least $j \cdot W_{j+1}/(j + 1)$. But this means that the average power is at least $W_{j+1}/(j + 1)$, a contradiction.

Theorem 20. *The greedy single arm solutions satisfy monotonicity.*

Proof. Let a be an arbitrary arm and j be an arbitrary number of markers. Let G_{j-1}, G_j, G_{j+1} be the greedy solutions using $j - 1, j, j + 1$ markers respectively. Let s_{j+1} be the marker added to go from G to G_{j+1}. By Lemma 18, $D(G_{j+1}) - D(G_j) \leq D(G_{(j-1)} \cup \{s_{j+1}\}) - D(G_{j-1})$. By definition of the greedy algorithm, $D(G_{j-1} \cup \{s_{j+1}\}) - D(G_{j-1}) \leq D(G_j) - D(G_{j-1})$ Combining inequalities, we get $D(G_{j+1}) - D(G_j) \leq D(G_j) - D(G_{j-1})$.

4.2   Local Search

Local search is a technique applied to improve non-optimal solutions for a variety of difficult combinatorial optimization problems [19]. Local search iteratively moves from one solution to a better *neighbor* solution that is a small perturbation. For WDC (and many other problems) a natural neighborhood is the *swap* neighborhood: Solutions A and B are neighbors if B can be obtained by removing one marker from A and replacing it with another marker. We implemented a local search addendum to the *greedy* and *dir-greedy* heuristics using the swap neighborhood. Our addendum runs after the dynamic programming step, so swaps between arms are allowed.

In our implementation of local search we distinguished two types of swaps:

1. Swaps of markers whose detection intervals *do not* overlap.
2. Swaps of markers whose detection intervals *do* overlap.

The program always prefers the non-overlapping swap that gives the best power increase, if one is available. Otherwise, it takes the best overlapping swap. The program stops when there are no more swaps available that improve the overall detection power. The performance of the local search addendum is shown in Table 2. Local search helps negligibly for *greedy* and by a few percentage points for *dir-greedy*. The detection power obtained by *greedy* + local search and *dir-greedy* + local search are almost indistinguishable.

5   Conclusions and Open Problems

WEIGHTED DIAGNOSTIC COVER (WDC) models the clinical genetics problem of setting up a genome-wide genotyping assay for segmental aneusomy. We studied the performance of two heuristics, *greedy* and *dir-greedy* from both theoretical and practical points of view. We proved lower bounds on the quality of solutions they produce and showed that the solutions may be useful in practice. Our work leaves many open problems.

Problem 1: Is WDC NP-complete?

Problem 2: Can the lower bound of $0.318 \times$ OPT for *greedy* be improved?

Problem 3: How can one use the geometry to analyze *greedy*?

Problem 4: Are there better heuristics? For example, can the following approach lead to a better approximation factor? Partition the markers into different classes depending on their detection probability The cumulative probability of low-probability sets is almost linear. For high-probability sets the addition of a second set at a point does not boost the detection probability at that point very much.

Problem 5: Can one prove a better performance gurantee for *greedy* on EWDC by eploiting the fact that all markers are of the same length?

Problem 6: Can we design a polynomial time optimal algorithm for the special case of WDC in which all the marker intervals have a common point? We have shown that if the common point is the left end then the greedy algorithm is

Size(b)	No. of Markers	Greedy	Greedy + Loc.	Dir-Greedy	Dir-Greedy + Loc.
5	200	0.254	0.254	0.238	0.245
5	300	0.364	0.364	0.340	0.363
5	400	0.458	0.459	0.429	0.459
5	500	0.531	0.533	0.498	0.532
5	600	0.586	0.588	0.546	0.588
5	700	0.630	0.631	0.584	0.631
5	800	0.665	0.665	0.619	0.665
5	900	0.693	0.693	0.650	0.693
5	1000	0.715	0.716	0.676	0.715
8	200	0.413	0.413	0.390	0.410
8	300	0.566	0.568	0.539	0.567
8	400	0.666	0.668	0.624	0.668
8	500	0.732	0.733	0.683	0.733
8	600	0.779	0.780	0.735	0.779
8	700	0.814	0.815	0.770	0.815
8	800	0.840	0.841	0.797	0.841
8	900	0.859	0.859	0.820	0.859
8	1000	0.872	0.873	0.838	0.873
11	200	0.562	0.563	0.540	0.561
11	300	0.713	0.716	0.669	0.716
11	400	0.799	0.800	0.749	0.799
11	500	0.852	0.853	0.808	0.853
11	600	0.888	0.888	0.846	0.888
11	700	0.911	0.911	0.874	0.911
11	800	0.926	0.926	0.895	0.926
11	900	0.936	0.936	0.911	0.936
11	1000	0.943	0.943	0.922	0.943
14	200	0.682	0.686	0.658	0.681
14	300	0.810	0.811	0.767	0.811
14	400	0.881	0.882	0.840	0.882
14	500	0.921	0.921	0.883	0.921
14	600	0.943	0.944	0.912	0.944
14	700	0.957	0.957	0.931	0.957
14	800	0.965	0.965	0.944	0.965
14	900	0.970	0.970	0.953	0.970
14	1000	0.973	0.973	0.960	0.973

Table 2. Monosomy detection power before and after local search, size in Mbases

optimal. The *greedy* heuristic is not optimal for for this case as given by the following example. The markers are $[1, 3], [2, 4], [3, 5]$ with probabilities $1/2, 1/2 + \epsilon, 1/2$ respectively, and $k = 2$. The optimal solution consists of the markers $[1, 3], [3, 5]$, while the greedy solution chooses $[2, 4]$ first.

Problem 7: Consider the following local improvement heuristic: Start with any solution, and repeatedly replace a marker in the solution by a marker not in the solution when such a replacement increases the total realized weight. How does this strategy perform for EWDC and for the special case discussed in Problem 6?

Problem 8: Is there an order of swaps in local search for WDC that is guaranteed to converge to a local maximum after a polynomial number of swaps?

Acknowledgments

The research of S.R.K. is supported in part by NSF grant CCR-9508545 and ARO grant DAAH4-96-1-0013. A. A. S. thanks the Computer Science Department at Johns Hopkins University for its hospitality.

References

1. L. G. Biesecker, M. Rosenberg, Y. Ning, D. Vaske, J. Weber, and the UMCAS research group. Detection of inapparent chromosomal aberrations by whole genome STRP scanning: Results of a pilot project. *American Journal of Human Genetics*, 59:A50, 1996. Meeting Abstract.
2. L. G. Biesecker and A. A. Schäffer. Automated selection of STRP markers for whole genome screening for segmental aneusomy. *Human Heredity*, 1996. To appear.
3. F. S. Celi, M. M. Cohen, S. E. Antonarakis, E. Wertheimer, J. Roth, and A. R. Shuldiner. Determination of gene dosage by a quantitative adaptation of the polymerase chain reaction (gd-PCR): Rapid detection of deletions and duplications of gene sequences. *Genomics*, 21:304–310, 1994.
4. V. Chvátal. A greedy heuristic for the set-covering problem. *Mathematics of Operations Research*, 4:233–235, 1979.
5. O. Cohen, C. Cans, M. Cuillel, J. L. Gilardi, H. Roth, M.-A. Mermet, P. Jalbert, and J. Demongeot. Cartographic study: Breakpoints in 1574 families carrying human reciprocal translocations. *Human Genetics*, 97:659–667, 1996.
6. Council of Regional Networks for Genetic Services Data and Evaluation Committee. *The CORN Minimum Data Set Report: 1992*. CORN, 1994.
7. U. Feige. A threshold of ln n for approximating set cover. In *Proceedings of the 28th Annual ACM Symposium on Theory of Computing*, pages 314–318, 1996.
8. J. Flint, A. O. M. Wilkie, V. J. Buckle, R. M. Winter, A. J. Holland, and H. E. McDermid. The detection of subtelomeric chromosomal rearrangements in idiopathic mental retardation. *Nature Genetics*, 9:132–139, 1995.
9. R. J. M. Gardner and G. R. Sutherland. *Chromosome Abnormalities and Genetic Counseling*. Oxford University Press, 1989. Oxford Monographs on Medcial Genetics, Volume 17.

10. G. Gyapay, J. Morissette, A. Vignal, C. Dib, C. Fizames, P. Millasseau, S. Marc, G. Bernardi, M. Lathrop, and J. Weissenbach. The 1993–94 Généthon human genetic linkage map. *Nature Genetics*, 7:246–339, 1994.

11. M. Halldórsson. Approximating discrete collections via local improvements. In *Proceedings of the 6th Annual ACM-SIAM Symposium on Discrete Algorithms*, pages 160–169, 1995.

12. M. Jarnik, J.-Q. Tang, M. Korab-Lazowska, E. Zietkiewicz, G. Cardinal, I. Gorska-Flipot, D. Sinnett, and D. Labuda. Overall informativity, oi, in dna polymorphisms revealed by inter-ALU PCR: Detection of genomic rearrangements. *Genomics*, 36:388–398, 1996.

13. D. S. Johnson. Approximation algorithms for combinatorial problems. *Journal of Computer and System Sciences*, 9:256–278, 1974.

14. R. M. Karp. Reducibility among combinatorial problems. In R. E. Miller and J. W. Thatcher, editors, *Complexity of Computer Computations*, pages 85–103. Plenum Press, 1972.

15. D. H. Ledbetter and A. Ballabio. Molecular cytogenetics of contiguous gene syndromes: Mechanisms and consequences of gene dosage imbalances. In C. R. Scriver, A. L. Beaudet, W. S. Sly, and D. Valle, editors, *The Metabolic and Molecular Bases of Inherited Disease*, chapter 20, pages 118–132. McGraw-Hill, 1995.

16. L. Lovász. On the ratio of otimal integral and fractional covers. *Discrete Mathematics*, 13:383–390, 1975.

17. E. S. Mansfield. Diagnosis of down syndrome and other aneuploidies using quantitative polymerase chain reaction and small tandem repeat polymorphisms. *Human Molecular Genetics*, 2:43–50, 1993.

18. Z. A. Melzak. *Companion to Concrete Mathematics: Mathematical Technique and Various Applications*. John Willey & Sons, 1973.

19. C. H. Papadimitriou and K. Steiglitz. *Combinatorial Optimization: Algorithms and Complexity*. Prentice-Hall, 1982.

20. F. P. Preparata and M. I. Shamos. *Computational Geometry: An Introduction*. Springer-Verlag, 1985.

21. P. Slavík. A tight analysis of the greedy algorithm for set cover. In *Proceedings of the 28th Annual ACM Symposium on Theory of Computing*, pages 435–439, 1996.

22. A. Srinivasan. Improved approximation of packing and covering prblems. In *Proceedings of the 27th Annual ACM Symposium on Theory of Computing*, pages 268–276, 1995.

23. A. O. M. Wilkie. Detection of cryptic chromosomal abnormalities in unexplained mental retardation: A general strategy using hypervariable subtelomeric DNA polymorphisms. *American Journal of Human Genetics*, 53:688–701, 1993.

Cartographic Line Simplication and Polygon CSG Formulae in $O(n \log^* n)$ Time

John Hershberger
Mentor Graphics

Jack Snoeyink[*]
UBC Dept. of Comp. Sci.

Abstract

The cartographers' favorite line simplification algorithm recursively selects from a list of data points those to be used to represent a linear feature, such as a coastline, on a map. A constructive solid geometry (CSG) conversion for a polygon takes a list of vertices and produces a formula representing the polygon as an intersection and union of primitive halfspaces. By using a data structure that supports splitting convex hulls and finding extreme points, both were known to have $O(n \log n)$ time solutions in the worst-case. This paper shows that both are easier than sorting by presenting an $O(n \log^* n)$ algorithm for maintaining convex hulls under splitting at extreme points. It opens the question of whether there is a practical, linear-time solution.

1 Introduction: Line simplification and CSG formula computation

To draw a coastline or other linear feature on a simple and readable map, one may need to perform *line simplification* to reduce the detailed data available in a database. Cartographers have identified the recursive algorithm detailed in Douglas and Peucker [2] as best in mathematical [8] and perceptual [15] studies. This algorithm first approximates a polygonal line $p_1, p_2, \ldots, p_n$ with the segment $\overline{p_1 p_n}$. If the vertex $p_{\max}$ at maximum distance from the line $\overleftrightarrow{p_1 p_n}$ is within toler-

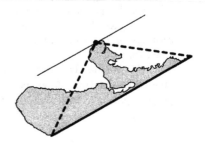

Fig. 1: Line simplification

ance, this approximation is accepted, otherwise, the two polygonal lines from p_1 to $p_{\max}$ and from $p_{\max}$ to p_n are approximated recursively, as in Figure 1. This algorithm has been called Ramer's algorithm [13] in vision and the sandwich algorithm [14] in computational geometry.

---

[*]Supported in part by an NSERC Research Grant and IRIS project IC–5.

If implemented in a straightforward fashion, this algorithm has a worst-case running time of $\Theta(n^2)$ (and best-case of $\Theta(n \log n)$ to compute a tree encoding all approximations). Because the vertices at maximum distance will be found on the convex hull, this same simplification can be computed in $O(n \log n)$ worst-case time [5, 6] using a convex hull data structure that supports splitting at extreme vertices. (Here and throughout this paper, we assume that the polygonal line $p_1, p_2, \ldots, p_n$ is *simple*: the only intersections between its line segments occur where adjacent segments share a common endpoint. It should be noted that the algorithm does not guarantee simplicity of the output, however.)

A plane polygon can be represented either by the sequence of vertices and edges around its boundary or by a boolean combination of "primitive" regions such as halfspaces. Solid modeling systems may convert from the former to the latter; this is a 2-dimensional example of a conversion from a *boundary representation* to a *constructive solid geometry (CSG) representation*.

Peterson [11] showed that a simple polygon always has a CSG formula using one primitive halfplane for each edge of the polygon. In fact, one can write down the formula by starting at the leftmost vertex and listing the halfspaces in the order that their edges appear around the polygon, inserting an "AND" for every convex corner and an "OR" for every reflex corner. The interesting part is to add parentheses appropriately.

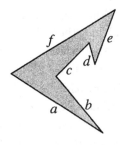

Fig. 2: CSG formula is $a(b + ((c + d)e))f$

In the dart at right, for example, the three terms a, $(b + ((c + d)e))$, and f can be joined by "AND"s since extensions of segments a and f, and of the polyline $bcde$ by extending segments b and e, do not return to intersect the polygon. Segment e cannot appear at the top level because its extension intersects edges a and b. Dobkin et al. [1] gave an $O(n \log n)$ algorithm to recursively add parentheses. They maintain convex hulls of fragments of the polygon and split at hull vertices that are extreme in a direction determined by the directions of the first and last edges of each fragment.

Thus, the problems of line simplification and CSG formula computation can both be solved by finding on-line a sequence of extreme vertices at which to split a polygonal line—the criterion for which extreme vertex is a valid splitting vertex depends on the current fragment. It is natural to ask whether these problems are easier than sorting. This paper gives an affirmative answer by presenting an $O(n \log^* n)$ algorithm for maintaining convex hulls of a polyline under splits. In Section 2 we review the "path hull" of Dobkin et al. [1] and other data structures for dynamic convex hulls. Section 3 describes our new data structure, which builds an augmented path hull data structure on "beads"—convex hulls of polylog-size fragments of the polygonal line. Section 4 analyzes the operations of finding extreme vertices and splitting for this structure.

Note: All logarithms in this paper are taken base 2. The iterated logarithm, $\log^* n$, is the number of times the logarithm function must be applied to reduce the argument to less than 2. It is a slowly growing function: $\log^* 16 = 3$, $\log^* 2^{16} = 4$, and the first n with $\log^* n \geq 5$ has about 20,000 decimal digits.

2 A brief review of 2-d dynamic convex hulls

As mentioned in the previous section, we wish to support splitting and finding extreme vertices of fragments of a polyline in the plane. At the cost of additional logarithmic factors, one could use general convex hull algorithms for points in the plane [12]. Overmars and van Leeuwen [10] showed that the divide-and-conquer algorithm can support finding extreme points in $O(\log n)$ time. Deletions and insertions of points take amortized $O(\log^2 n)$ time per operation; no algorithm is known to achieve amortized $O(\log n)$ time.

When the points lie on a simple polygonal line then the convex hull has additional structure, as many have observed. (See Figure 3.)

Observation 2.1 *If the vertices are numbered along the polyline, then the sequence of vertex numbers around the hull, when read counter-clockwise from the maximum, decreases to the minimum number and then increases to the maximum.*

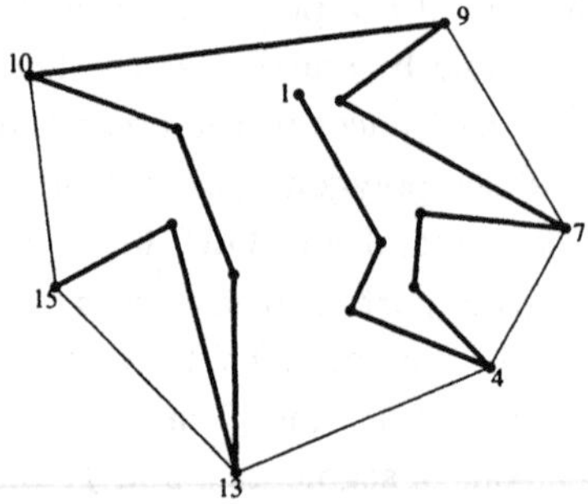

Fig. 3: Vertex numbers on the hull

Guibas et al. [3] used this observation to build an $O(n)$-size data structure for "subpath hull queries," which include the ability to find an extreme point of any contiguous fragment of the polyline in $O(\log n \log \log n)$ time after $O(n)$ preprocessing. This level of generality is not needed in the applications considered in this paper.

We do need one consequence of Observation 2.1 that Guibas et al. [3] applied to their data structure.

Lemma 2.2 *Given array representations of the convex hulls of two consecutive fragments of a polygonal line with m vertices, one can compute the at most two tangents between them in $O(\log^2 m)$ time.*

Proof: Because the polyline fragments are consecutive, if, say, the first hull is contained inside the second, then it is contained entirely inside one of the two bays determined by the hull edges incident to the lowest numbered vertex.

For a specific example with hull A inside hull B, consider Figure 4. Vertex b on hull B has the lowest index, and is incident to hull edges e and e'; the bay defined by e contains the hull A. To detect this, one can test, in $O(\log m)$ time, whether the extreme vertices of A in the directions normal to e and e' are left of the lines through e and e'. If so, then A is inside.

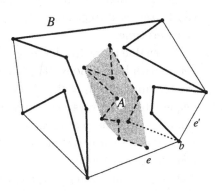

Fig. 4: Hull A (shaded) in B

If neither one is inside the other, then we will have found a pair of "helper points"—a point on each hull that is not contained in the other. Guibas et al. [3] show how to use helper points to reduce to the problem of finding a common tangent of two intersecting, upper convex hulls. This can be solved by nested binary search: choose a median vertex on one hull and use binary search to determine the tangent to the other hull, if it exists. ∎

Melkman's convex hull algorithm [9] uses Observation 2.1 in a different way to compute an array that stores the convex hull of a polyline with m vertices in $O(m)$ time. It actually computes convex hulls of prefixes of the polyline; Dobkin et al. [1] added a history stack so that these prefix convex hulls can be recovered as vertices are deleted from the end of the polyline.

Lemma 2.3 *Given a polygonal line $p_1, p_2, \ldots, p_m$, where p_1 is the start or anchor vertex, one can build a convex hull data structure in $O(m)$ time that supports deletion from the (non-anchor) end in amortized $O(1)$ time, and search for the extreme in a given direction finds the vertex p_d in $O(\log \min\{d, m - d + 1\})$ time.*

Proof: We sketch the data structure—see Dobkin et al. [1] for greater detail—and then note that the search can also be carried out.

Melkman's algorithm builds the convex hull incrementally, by adding vertices $p_1, p_2, \ldots, p_n$ in order. The algorithm maintains the current convex hull in a doubly-ended queue (a deque), which can be stored in an array of size $2m$. Start with p_2, p_1, p_2 in the middle of the array. To add p_i, check if p_i appears on the convex hull; as noted in the proof of Lemma 2.2, it sufficient to look at the edges incident on the highest numbered vertex on the current hull, which are found at the ends of the deque. If p_i appears on the hull, it may hide some vertices off the previous hull; pop these hidden vertices off the top and/or bottom of the deque. Pushing p_i onto both top and bottom produces the convex hull of $p_1 \ldots p_i$.

By recording the history of the changes to the deque, these operations can be reversed to delete vertices from the end of the polyline in amortized constant time.

To enable the search for an extreme vertex, maintain a pointer to the hull vertex with lowest index. The hull vertex with highest index appears at both the top and bottom of the deque. These lowest and highest indices split the deque into two arrays; one containing vertices with increasing numbers and the other with decreasing numbers. A constant-time computation on these vertices and their neighbors on the hull is sufficient to determine which array contains the extreme vertex. That array can be searched in $O(\log d)$ time by starting two increasing-increment searches in parallel from the ends: Check the first, second, fourth, eighth, ..., from the end until the extreme vertex lies in the inverval between the current vertex and the end, then finish with binary search on that interval. ∎

We could say that Melkman-plus-history supports "one-sided" splits—splitting at a vertex produces a valid convex hull data structure for the first part of the polygonal line by simply deleting vertices from the second part. Of course, no structure is produced for the second part, so splits that occur near the beginning of the polyline waste most of the computation of Melkman's algorithm.

Dobkin et al. [1] used this observation to define a "path hull" data structure that supports two-sided splits in amortized $O(\log n)$ time. They choose an anchor in the middle of a polyline and use Melkman-plus-history to build two convex hull structures outward from the anchor. Thus, the splits that waste computation are those near the middle; since these splits now break the problem into two equal-sized subproblems, a credit scheme shows that the total splitting time is $O(n \log n)$. Extreme vertices are found in $O(\log n)$ time apiece by finding the two candidate extreme vertices on the two convex hulls that make up the path hull and returning the true extreme.

3   The bead hull data structure and its construction

To improve on the path hull structure sketched at the end of the previous section, we need to reuse more computation—we cannot afford to have all previous computation wasted by a single split. Less evident, but equally important, is the fact that the cost of finding an extreme vertex must be related to the size of the smaller polyline that will be created by splitting there.

We break a polyline of length m at vertices to form m/k fragments of length k, where k can be chosen to be $\log^2 m$. (We justify the choice of k in Section 4.) Adjacent fragments share a common endpoint. For each fragment, we build a convex hull using Melkman-plus-history as described in Lemma 2.3. We call such a fragment-with-hull a *bead*. We actually build the convex hull twice, once from each end. As a corollary to Lemma 2.3,

Corollary 3.1 *Beads of size k for a polyline of length m can be built in $O(m)$ time and space.*

When a split occurs in a bead, two *broken beads* are produced. These are fragments of the polyline, with length at most k, that have Melkman-plus-history representations of their convex hulls built from the original endpoints out towards the split. Further splitting a broken bead produces a smaller broken bead and an unstructured fragment between the two splits.

A *string of beads* is a two-level convex hull structure that represents the hull of a sequence of consecutive whole beads, possibly ending with one broken bead. The lower level consists of the convex hull arrays for the beads. The upper level is an array of tangents between beads, with pointers into the corresponding bead hull arrays.

Lemma 3.2 *Given m beads with size at most k, a string of beads can be built using a modification of Melkman's algorithm in $O(m \log^2 k)$ time.*

Proof: As in Melkman's algorithm for points, we can maintain the current list of tangents between beads in a deque with the tangents to the most recently added bead at the top and bottom.

To add the ith bead, we pop tangents from the deque whose supporting lines cut the bead—these will no longer be edges of the convex hull. Then we use the nested binary search of Lemma 2.2 to compute at most two tangents between the ith bead and the beads remaining at the top and bottom of the deque and add these tangents to the deque.

Because there are at most $2m$ tangents added, there are at most $2m$ tangents popped at a total cost of $O(m \log k)$. Adding new tangents costs $O(m \log^2 k)$ time. ∎

A *bead hull* consists of the following parts, which are depicted schematically in figure 5.

- An *anchor vertex* that is the common endpoint of two beads (broken or whole). An anchor is initially chosen in the middle of a sequence of beads.
- Two, possibly empty, strings of beads constructed to the left and right of the anchor.
- Two, possibly empty, broken beads at the ends of the strings.
- Tangents (at most four) from the broken beads to their adjacent strings.
- Tangents (zero or two) between the structures to the right and left of the anchor.

Lemma 3.3 *Given m > 1 beads with size at most k, a bead hull can be built in $O(m \log^2 k)$ time.*

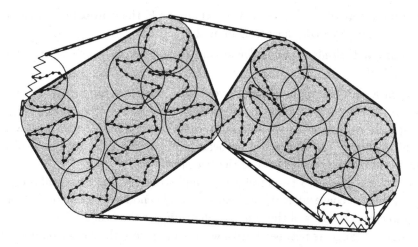

Fig. 5: Schematic depiction of a bead hull with anchor in the middle

Proof: The anchor can be chosen and two strings of beads constructed in this amount of time. By Lemma 2.2, the two tangents between the strings of beads can be found by nested binary search in $O((\log m + \log k)^2) \leq O(m \log^2 k)$ time. ∎

4 Analysis of bead hull operations

In a bead hull, finding an extreme vertex in a chosen direction is relatively easy because all hull edges are represented.

Lemma 4.1 *Given a bead hull representing $p_1, p_2, \ldots, p_n$, the extreme vertex p_d in a particular direction can be found in $O(\log \min\{d, n-d+1\})$ time.*

Proof: We can use increasing-increment searches in parallel from both ends, as in Lemma 2.3. The two-level structure increases the programming complexity, but not the asymptotic running time. ∎

To handle split operations, we must account for time spent at several levels: computing tangents for the bead hull, computing bead hulls, computing beads, and recursively handling fragments within beads. (Recall that, when a partial bead is split, a bead hull data structure must be built recursively for the polyline fragment that is contained entirely within the bead.) In the next lemma we analyze the cost of splitting a bead hull structure so that all the remaining fragments are contained within beads. We defer the recursive cost of handling fragments within beads until Theorem 4.3.

Lemma 4.2 *Suppose that we are given a polyline with n vertices and an on-line sequence of splitting vertices. Then, in $O(n + (n/k) \log n(\log n + \log^2 k))$ total*

time, we can build beads of size k and maintain bead hulls under splits for all fragments that are not strictly contained within the original beads.

Proof: Corollary 3.1 and Lemma 3.3 say that Melkman's algorithm can be used to build the initial beads and initial bead hull in the desired time. We record the history of these computations.

To prove this lemma, we will give bead hulls two types of *credits* with which to pay for all construction and tangent computation after the initialization. We maintain the following invariants: If a bead hull has l (whole) beads to the left of the anchor and r beads to the right, then it has $(l + r) \log(\max\{l, r\})$ *bead credits*. Each vertex also has up to three *vertex credits*: one if it is inside the convex hull, one if is inside the convex hull of its string, and one if it is inside the convex hull of its bead. We establish the invariants by giving $(n/k) \log(n/2k)$ bead credits to the initial bead hull, plus at most $3n$ vertex credits to polyline vertices.

Each split produces one or two new bead hulls (only one, if one of the fragments is completely contained in an original bead). We assign credits to the new bead hulls according to the invariants. The *credit budget* is the difference between the total number of credits before the split and after. We will see that the credit budget is non-negative. We charge $O(\log^2 k)$ computation to each bead credit and $O(1)$ to each vertex credit in the budget; we also charge $O(\log^2 n)$ to each bead when it is first broken. Together, these charges establish the lemma.

There are two cases to consider when splitting: either a whole bead is split for the first time or the splitting vertex is contained in one of the broken beads.

Case 1: When a whole bead is split, we charge tangent computation to the bead and spend bead credits to rebuild strings of beads. Assume that s whole beads are split off before the anchor; the analysis for splitting after the anchor is symmetric.

The bead hull for the fragment containing the anchor can be obtained in three steps. First, play back the history of Melkman's algorithm to give the string of whole beads between the anchor and splitting vertex. Next, break the bead containing the splitting vertex. Finally, compute tangents from the broken bead to the string, and between the strings before and after the anchor.

This bead hull must be given

$$(l - s - 1 + r) \log \max\{l - s - 1, r\} \le (l - s - 1 + r) \log \max\{l, r\}$$

bead credits, but no credits need be spent for its computation. The first two steps run hull construction algorithms backwards, so we can charge their computation to the initial build. By Lemma 2.2, the third step can be performed in $O(\log^2 n)$ time, which can be charged to the bead for a total of $O((n/k)\log^2 n)$ overall.

The bead hull for the fragment not containing the anchor must be built in $O(s \log^2 k)$ time (Lemma 3.3), which consumes s bead credits. Since an anchor is chosen in the middle,

$$s \log \lfloor s/2 \rfloor \leq s(\log s - 1) \leq s \log \max\{l, r\} - s$$

bead credits must be given to this fragment.

In case 1, the $(l+r)\log(\max\{l,r\})$ bead credits available are sufficient to pay for the build and satisfy the invariants for the resulting bead hulls. The vertex credits are also sufficient, since splitting can only increase the total number of hull vertices.

Case 2: When a broken bead is split, we spend vertex credits on updating tangents for the bead hull. No bead credits are spent on computation, as all are needed for the bead hull invariant. The new broken bead is formed by playing back the history of the bead's construction, which is charged to the initial construction. Notice that the vertices removed from the broken bead form a fragment that is entirely contained within the original bead. In this lemma, no computation is required for such fragments.

The bead hull has four tangents that may need to be updated. As a representative example, consider the tangent that goes counter-clockwise (ccw) from the string to the broken bead in Figure 6. We shrink the broken bead, playing back the history of Melkman's algorithm, until we reach the splitting vertex. This may cause new vertices to appear on the hull of the broken bead.

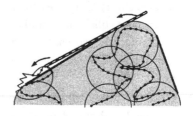

Fig. 6: Updating a tangent while shrinking a broken bead

We need to update the tangent if and only if the tangent endpoint is removed. The candidates for the tangent endpoint on the broken bead are the new vertices and those adjacent to them. Candidates on the string are at or ccw of the old tangent endpoint. If we begin by joining the clockwise-most candidates on both bead and string, then by testing incident edges we can determine whether we have found the tangent or which candidate endpoint should move ccw. We can advance until we find a tangent or determine that the broken bead is contained in the hull of the string; we charge the search time to vertex credits taken from vertices that now join the bead and string convex hulls.

When the splitting vertex is common to two beads, then two beads are affected. When the splitting vertex is the anchor, both strings are also affected. These reduce to combinations of cases 1 and 2, however, depending on whether the affected beads were previously whole or broken. ∎

Theorem 4.3 *For a simple polygonal line with n vertices, bead hulls of all fragments can be constructed and maintained under the operations of finding and splitting at extreme vertices in $O(n \log^* n)$ time and $O(n)$ space.*

Proof: By Lemma 4.1, each search for a splitting vertex can be performed in time logarithmic in the size of the smaller fragment. It is known that recursion trees with this behavior take linear time in total [4, 7].

For a polyline of n vertices, let us choose beads of size $k = \log^2 n$. (We need a small $k \geq \log^2 n + \log n \log^2 k$.) Then Lemma 4.2 says that one level of bead hull computation produces, in time

$$O(n + (n/\log^2 n) \log n (\log n + (\log(\log^2 n))^2)) = O(n),$$

a set of fragments of sizes $n_1, n_2, \ldots, n_f$, with each $n_i < k$.

Let $T(m)$ be the total time to handle a fragment of size m recursively: building beads and maintaining bead hulls until the fragments are of size less than $\log^2 m$. Since $T(m)$ is a concave function that is at least linear, we can bound $T(n)$ with a recurrence that is maximized when all n_i are as large as possible:

$$T(n) \leq O(n) + \sum_{1 \leq i \leq f} T(n_i) \leq O(n) + (n/\log^2 n) T(\log^2 n).$$

Unrolling once gives the inequality

$$T(n) \leq O(n) + (n/\log n) T(\log n) = O(n \log^* n).$$

Since only one bead hull is needed at any moment, it is easy to implement the algorithm using linear space. ∎

Corollary 4.4 *Douglas-Peucker line simplification and CSG formulæ for simple polygons can be computed in $O(n \log^* n)$ time and $O(n)$ space.*

5 Conclusion

We have given an $O(n \log^* n)$ algorithm for maintaining a convex hull under splits; this gives a theoretical improvement to the running time of algorithms for Douglas-Peucker line simplification and for building CSG formulæ for planar polygons—showing that both problems are easier than sorting. This opens the question whether these problems have a practical, linear-time solution.

References

[1] D. Dobkin, L. Guibas, J. Hershberger, and J. Snoeyink. An efficient algorithm for finding the CSG representation of a simple polygon. *Algorithmica*, 10:1–23, 1993.

[2] D. H. Douglas and T. K. Peucker. Algorithms for the reduction of the number of points required to represent a line or its caricature. *The Canadian Cartographer*, 10(2):112–122, 1973.

[3] L. Guibas, J. Hershberger, and J. Snoeyink. Compact interval trees: A data structure for convex hulls. *International Journal of Computational Geometry & Applications*, 1(1):1–22, 1991.

[4] L. J. Guibas, J. Hershberger, D. Leven, M. Sharir, and R. Tarjan. Linear time algorithms for visibility and shortest path problems inside triangulated simple polygons. *Algorithmica*, 2:209–233, 1987.

[5] J. Hershberger and J. Snoeyink. Speeding up the Douglas-Peucker line simplification algorithm. In *Proceedings of the 5th International Symposium on Spatial Data Handling*, pages 134–143. IGU Commission on GIS, 1992.

[6] J. Hershberger and J. Snoeyink. An $O(n \log n)$ implementation of the Douglas-Peucker line simplification algorithm. In *Proceedings of the Tenth Annual ACM Symposium on Computational Geometry*, pages 383–384, 1994. Video Review of Computational Geometry. 4:45 animation.

[7] K. Hoffmann, K. Mehlhorn, P. Rosenstiehl, and R. E. Tarjan. Sorting Jordan sequences in linear time. *Information and Control*, 68:170–184, 1986.

[8] R. B. McMaster. A statistical analysis of mathematical measures for linear simplification. *The American Cartographer*, 13:103–116, 1986.

[9] A. A. Melkman. On-line construction of the convex hull of a simple polyline. *Information Processing Letters*, 25:11–12, 1987.

[10] M. Overmars and J. van Leeuwen. Maintenance of configurations in the plane. *Journal of Computer and System Sciences*, 23:166–204, 1981.

[11] D. P. Peterson. Halfspace representation of extrusions, solids of revolution, and pyramids. SANDIA Report SAND84-0572, Sandia National Laboratories, 1984.

[12] F. P. Preparata and M. I. Shamos. *Computational Geometry—An Introduction*. Springer-Verlag, New York, 1985.

[13] U. Ramer. An iterative procedure for the polygonal approximation of plane curves. *Computer Vision, Graphics, and Image Processing*, 1:244–256, 1972.

[14] G. Rote. Quadratic convergence of the sandwich algorithm for approximating convex functions and convex figures in the plane. In *Proceedings of the Second Canadian Conference on Computational Geometry*, pages 120–124, Ottawa, Ontario, 1990.

[15] E. R. White. Assessment of line-generalization algorithms using characteristic points. *The American Cartographer*, 12(1):17–27, 1985.

Constrained TSP and Low-Power Computing

Moses Charikar[*1], Rajeev Motwani[1**1], Prabhakar Raghavan[***2], Craig Silverstein[†1]

[1] Department of Computer Science, Stanford University, Stanford, CA-94305, USA
[2] IBM Almaden Research Center, 650 Harry Road, San Jose, CA 95120-6099, USA

Abstract. In the *precedence-constrained traveling salesman problem (PTSP)* we are given a partial order on n nodes, each of which is labeled by one of k points in a metric space. We are to find a visit order consistent with the precedence constraints that minimizes the total cost of the corresponding path in the metric space. We give negative results on approximability by relating the problem to the Shortest Common Supersequence problem, helping to explain why there has been very little success in approximation algorithms for this problem. We also give approximation algorithms for a number of special cases, included cases appropriate for a problem in low-power computing; in the process, we show that algorithms for the k-server problem and the traveling salesman problem can be used to derive approximation algorithms for the PTSP. We give tight bounds on the approximation ratios achieved by natural classes of algorithms for this optimization problem (which include algorithms proposed and used in empirical studies of this problem). We briefly summarize results of experiments with several algorithms on a standard set of compiler benchmarks, comparing several known and new algorithms.

1 Overview

In the *precedence-constrained traveling salesman problem (PTSP)*, each node of a given directed, acyclic graph (DAG) is labeled by a point of a metric space $\mathcal{M}$. We wish to serialize the DAG so that the induced tour through its vertices is of the shortest possible length, where the distance between successive nodes on the tour is the distance between the corresponding points in $\mathcal{M}$. This classic generalization of the TSP has been extensively studied for its applications to vehicle routing [3, 4, 8, 12], with essentially no theoretical approximation guarantees. In this paper we explain this lack of success through

* E-mail: moses@cs.stanford.edu. Supported by Stanford School of Engineering Groswith Fellowship, an ARO MURI Grant DAAH04-96-1-0007 and NSF Award CCR-9357849, with matching funds from IBM, Schlumberger Foundation, Shell Foundation, and Xerox Corporation.

** E-mail: rajeev@cs.stanford.edu. Supported by an Alfred P. Sloan Research Fellowship, an IBM Faculty Partnership Award, an ARO MURI Grant DAAH04-96-1-0007, and NSF Young Investigator Award CCR-9357849, with matching funds from IBM, Mitsubishi, Schlumberger Foundation, Shell Foundation, and Xerox Corporation.

*** E-mail: pragh@almaden.ibm.com.

† E-mail: csilvers@cs.stanford.edu. Supported by the Department of Defense, with partial support from ARO MURI Grant DAAH04-96-1-0007 and NSF Award CCR-9357849, with matching funds from IBM, Schlumberger Foundation, Shell Foundation, and Xerox Corporation.

a negative result showing that the PTSP is hard to approximate even on simple metric spaces such as the line and the hypercube, showing that it is not possible to exploit geometry to obtain good approximation algorithms for PTSP. (Previous hardness results, due to Bhatia, Khuller and Naor [2], do not apply to these simple metric spaces.) The PTSP on the hypercube is important because of its connection to an application to low-power computing, which we describe below. Because of the difficulty of approximation, we attack the PTSP problem by developing heuristics that work well for special cases. We prove upper bounds for several heuristics, some previously proposed in the context of low-power computing and some newly motivated by theoretical connections between the PTSP and the k-server problem. For this latter class of heuristics we can show our upper bound is tight. One of our results is related to the *dial-a-ride* or *taxicab problem*.

Section 3 below proves our main negative result. This result uses a connection of the PTSP to the Shortest Common Supersequence problem (SCS), thereby explaining the lack of prior success in finding approximation algorithms for the PTSP. We give our positive results in Section 4. We conclude in Section 5 with an experimental summary.

1.1 Low-Power Computing

The demand for ubiquitous mobile computing [23] has greatly outstripped the improvement in the performance of lightweight batteries. All notebook computers on the market employ clever power management techniques, and researchers are converging to CPU power minimization as key issue in mobile systems [6, 9, 13, 19, 20, 21]. Recently, Yao *et al.* [24] have applied competitive analysis to an online model for a scheduling problem related to power minimization. Their approach has its limitations, partly due to processor technology and more importantly due to the fact that operating system scheduling is inapplicable to mobile embedded systems. For example, it does not help in embedded DSP (digital signal processing) chips designed for dedicated applications [13, 20].

The major source of power consumption in a CMOS processor is in energy dissipation due to state changes in the logic circuits. Thus, the main avenue for reducing CPU power consumption is in reducing the switching activity [1, 11]. Several papers in the low-power literature have made proposals on reducing switching activity during a clock cycle [16, 17, 22]. Su, Tsui, and Despain [19] model power consumption at the architecture level by the switching activity at the input/output of the processor modules/chips. Specifically, they model the power consumption by the number of bit switches at all positions of the instruction word as the CPU executes a stream of instructions.

Su *et al.* [19] perform instruction scheduling within basic blocks (that is, branch-free segments of code) to minimize the number of bit switches. Instructions within basic blocks can be freely reordered subject to data dependencies that are represented as a DAG. Su *et al.* [19] adopt the approach of scheduling basic blocks so as to minimize the total amount of bit switches rather than the common approach of trying to minimize running time. In experiments with a simple greedy scheduler, they discovered that it is possible to reduce the number of bit switches by as much as 30% over the schedule geared towards optimizing the running time, without degrading the running time performance by more than 5%.

This basic block scheduling problem is an instance of PTSP, where the metric space is a hypercube since the distance between instructions is the Hamming distance. Though the lower bounds in Section 3 apply to this metric space, experimental data in Section 5 show that many scheduling algorithms for PTSP work well on standard benchmarks.

2 Preliminaries

Let $\mathcal{M}$ be a metric space of size k. For points $i, j \in \mathcal{M}$, let $d(i, j)$ be the distance between i and j in $\mathcal{M}$. Let $X = x_1, x_2, \ldots, x_m$ be a sequence of points in $\mathcal{M}$. We define the *cost* of a sequence X as follows:

$$\text{cost}(X) = \sum_{i=1}^{m-1} d(x_i, x_{i+1}). \tag{1}$$

Definition 1 *A labeled directed acyclic graph (LDAG) is a DAG $G(V, E)$ in which each vertex $v \in V$ has a label $l(v) \in \mathcal{M}$, where $\mathcal{M}$ is a metric space.*

Definition 2 *A repetition-free LDAG is an LDAG where each vertex has a distinct label.*

A *schedule* for an LDAG $G(V, E)$ is an ordering of V. The cost of a schedule $S = s_1, \ldots, s_n$ is the cost of its label sequence, namely $\text{cost}(l(s_1), l(s_2), \ldots, l(s_n))$. A schedule S for an LDAG G is said to be *legal* for G if it respects the precedence constraints of G (i.e., for all $v_i, v_j \in V$, if v_i is a predecessor of v_j, then v_i occurs before v_j in S.)

Definition 3 *The Precedence-Constrained TSP (PTSP) problem is the following: given an LDAG $G(V, E)$, find a legal schedule of minimum cost.*

In applications such as low-power computing, the precedence constraints could necessitate revisiting vertices, making the solution differ from that of standard TSP. To verify that our hardness results do not depend on this property, we study the following variant:

Definition 4 *The PTSP without repetitions is defined as follows: given a repetition-free LDAG $G(V, E)$, find a legal schedule of minimum cost.*

3 Lower Bounds

In this section, we develop a connection between PTSP and SCS, the Shortest Common Supersequence problem. The SCS problem is as follows: Given a set of strings $X = s_1, s_2, \ldots s_m$, construct a string S, of minimum length, such that each s_i is a (possibly non-contiguous) subsequence of S. The SCS problem is known to be difficult to solve exactly [7] or even approximately [2]. We first review the hardness results for SCS from Bhatia, Khuller, and Naor [2].

Theorem 1. *For the SCS problem with a fixed alphabet size k, there is no polynomial time k^α-approximation algorithm for some $\alpha > 0$, unless $P = NP$.*

Theorem 2. *For the SCS problem of size n, and for any constant δ, there is no polynomial time $(\log n)^\delta$-approximation algorithm unless $NP \subseteq DTIME(n^{O(\log \log n)})$.*

These hardness results for SCS apply directly to PTSP on the uniform metric space. However, they do not rule out the possibility of finding better approximations for PTSP on simple geometric spaces such as the line, the plane and — the case of most interest in low-power computing — the hypercube. By proving hardness of approximation results similar to Theorems 1 and 2 for the line and the hypercube, we show that, in fact, we cannot hope to exploit the structure of these simple spaces to obtain better approximations for PTSP.

3.1 Hardness on the Line and Hypercube

By converting instances of SCS to "large" instances of PTSP, we can prove hardness results even when $\mathcal{M}$ is a simple metric space. We first consider when $\mathcal{M}$ is the line.

Consider an instance X of the SCS problem over a fixed alphabet $\{c_1, \ldots, c_k\}$. The instance X is a set of strings over the alphabet. Given X, we construct an instance $P(X)$ of PTSP, where $P(X)$ consists of a collection of chains, one for each string in X. The length of each chain is much longer than the length of its corresponding string, because we associate with each letter c_i an entire chain C_i to be described below. The chain corresponding to the string $c_{i_1}, c_{i_2}, \ldots, c_{i_r}$ is $C_{i_1} \prec C_{i_2} \prec \cdots \prec C_{i_r}$.

C_i (for $1 \leq i \leq k$) is defined to be the $m = 2n^2$-length chain $(a_i \prec b_i)^{m/2}$, where a_i and b_i are points with coordinates $2i-1$ and $2i$ respectively. D_i is the corresponding schedule $(a_i b_i)^{m/2}$. Clearly D_i is a legal schedule for C_i. A *regular* solution of a PTSP instance is a solution of the form $D_{i_1} D_{i_2} \ldots D_{i_r}$. If $S = c_{i_1} \ldots c_{i_r}$ solves an instance X of SCS then $R = D_{i_1} \ldots D_{i_r}$ solves the corresponding instance $P(X)$ of PTSP, and vice versa. S has cost r while $(m-1)r + r - 1 \leq \text{cost}(R) \leq (m-1)r + 2n(r-1)$. This leads to the following result.

Lemma 3. *Given a solution S of X, we can construct a (regular) solution R of $P(X)$ of cost at most $(m + 2n)\text{cost}(S)$. Conversely, given any regular solution R of $P(X)$ we can obtain a solution S for X with cost at most $\text{cost}(R)/m$.*

A solution U of $P(X)$ might schedule parts of different C_i's interspersed with each other. However, we can construct a regular solution which schedules the C_i's in the order in which they are completed by U. This gives us the following result.

Lemma 4. *Given any solution U of $P(X)$, we can obtain a regular solution of $P(X)$ with cost at most $(1 + 1/n)\text{cost}(U)$.*

Using our reduction of SCS to PTSP and the results of the previous lemmas, we can show that a c-approximation algorithm for PTSP on the line can be used to construct a $c(1 + 1/n)^2$-approximation algorithm for SCS. Thus the hardness results of Theorem 1 extend to PTSP on the line.

Theorem 5. *For the PTSP problem on the line with k points, there is no polynomial time k^α-approximation algorithm for some $\alpha > 0$, unless $P = NP$.*

We can modify the above construction to work for the hypercube as well. In this case the points a_i and b_i are chosen such that $d(a_i, b_i) = 1$. The construction and proof proceed as before.

Theorem 6. *For the PTSP problem on the hypercube, there is no polynomial time k^α-approximation algorithm for some $\alpha > 0$, unless $P = NP$.*

3.2 Hardness for PTSP Without Repetitions

The construction in the previous section used the fact that the precedence constraints could force multiple visits to the same point in the metric space. If we replace each point from the previous reduction by a closely packed cluster of points, we can extend the hardness results of PTSP to cover PTSP without repetitions.

Let m be as before. We replace a_i and b_i by sets A_i and B_i, where (1) the size of each set is at least $nm = 2n^3$, (2) the distance between two points in distinct sets is at least d_1, (3) the distance between points in A_i and B_i is at most d_2, (4) the maximum distance between any two points is at most d_3, and (5) the points in any set A_i (or B_i) have a TSP tour $T(A_i)$ (or $T(B_i)$) of length at most d_4.

Since, in the reductions of the last section, each a_i and b_i was visited at most nm times, we can replace each occurrence of a_i (or b_i) with a unique element from A_i (or B_i). The version of $P(X)$ formed this way, without repetition, is denoted $P_d(X)$ (d is for "distinct"). A *regular* solution generalizes in the obvious way: $D_i = a_i^1 b_i^1 a_i^2 b_i^2 \ldots$, where each a_i^j is a unique element of A_i. The analogue of Lemma 3 is as follows.

Lemma 7. *Given a solution S of X, we can construct a regular solution R of $P_d(X)$ of cost at most $(d_2 m + d_3)\mathrm{cost}(S)$. Conversely, given a regular solution R of $P_d(X)$ we can obtain a solution S for X with cost at most $\frac{1}{md_1}\mathrm{cost}(R)$.*

Again, we can convert an arbitrary solution to a regular solution. The regular solution is built from sequences of the form $(T(A_i)T(B_i))^{m/2}$.

Lemma 8. *Given a solution U of $P_d(X)$ we can obtain a regular solution of $P_d(X)$ with cost at most $\frac{m(d_4+d_2)+d_3}{md_1}\mathrm{cost}(U)$.*

For the line, we choose $A_i = [(3i-1)n^4 - 2n^3, (3i-1)n^4]$ and $B_i = [(3i)n^4, (3i)n^4 + 2n^3]$, for $1 \le i \le n$. For this choice of A_i, B_i, we have $d_1 = n^4$, $d_2 = n^4 + 4n^3$, $d_3 = 3n^5$, and $d_4 = 4n^3$. Thus, given an instance X of SCS of size n, our reduction constructs an instance of PTSP without repetitions on a discrete line of length $O(n^5)$. The previous lemmas imply that a c-approximation algorithm for PTSP without repetitions in the discrete line can be used to construct a $c(1+10/n)^2$-approximation algorithm for SCS. The hardness result of Theorem 2 now extends to PTSP without repetitions.

Theorem 9. *For PTSP without repetitions on the discrete line of length k, and for any $\delta > 0$, there is no polynomial time $(\log k)^\delta$-approximation algorithm, unless $NP \subseteq DTIME(n^{\log \log n})$.*

Observe that we can embed a discrete line of length l in the l-dimensional hypercube. We obtain the same approximation results as before, but now the size of the metric space is $2^{O(n^5)}$, where n is the size of the original SCS problem. This gives us the following hardness result.

Theorem 10. *For PTSP without repetitions on the hypercube of size k, and for any $\delta > 0$, there does not exist a polynomial time $(\log \log k)^\delta$-approximation algorithm, unless $NP \subseteq DTIME(n^{\log \log n})$.*

4 Upper Bounds

Before proving upper bounds, we define a connection between schedules for G and sequences on $\mathcal{M}$. There is a natural connection due to the fact that each vertex in G has an associated point in $\mathcal{M}$.

Recall that a *schedule* is a list of vertices in G. We define a *sequence* as a (possibly infinite) list of points in $\mathcal{M}$. Every schedule $v_1, \ldots, v_n$ has a sequence *corresponding to* it, namely $l(v_1), \ldots, l(v_n)$ with duplicates removed — that is, we remove $l(v_i)$ if $l(v_i) = l(v_{i-1})$. We say a sequence X is *legal* for G if some (possibly non-contiguous) subsequence of X is a sequence corresponding to a legal schedule. (Legal sequences are thus sequences associated with legal schedules, possibly augmented by dummy nodes.)

For a chain $C = v_1 \prec \cdots \prec v_m \in G$, let $l(C)$ be the sequence $l(v_1), \ldots, l(v_n)$ with duplicates removed. The following lemma characterizes legal sequences for G.

Lemma 11. *X is legal for G if and only if X is a supersequence of $l(C)$ for every maximal chain C in G.*

Note that the cost of a schedule, as defined in Section 2, is exactly the cost of its corresponding sequence, since removing duplicates does not affect the cost.

To define the flip side of this correspondence, we must introduce the concept of *eager schedules*. A *ready vertex* is a vertex all of whose predecessors have been scheduled. An *eager scheduling algorithm* proceeds as follows: initially, pick a point $p \in \mathcal{M}$. While there is still a ready vertex with label p, schedule it. Repeat this until no ready vertex with label p remains. Repeat this scheduling process with a new point, continuing until all vertices in G have been scheduled. An *eager schedule* is a schedule produced by an eager scheduling algorithm. Eager schedules differ from arbitrary schedules in that they do not move in the metric space unless they have to. Thus, an eager schedule is uniquely defined by the sequence of points picked by the algorithm. The schedule *corresponding to* a sequence X is the unique eager schedule produced by an eager scheduling algorithm which selects X as the sequence of points to be scheduled.

The cost of a sequence X is an upper bound on the cost of its corresponding schedule S. The sequence could be more expensive if the eager scheduling algorithm were to choose a point p for which there was no ready vertex. Then X has to pay the cost for switching to p while S does not. We can remove a point x_i from X if, while constructing an eager schedule based on X, there is no ready vertex with label x_i when we encounter x_i. This act of removing unused points is called *short-circuiting*. If we restrict our attention to short-circuited sequences and eager schedules, the correspondence mapping preserves cost. The following lemma justifies restricting ourselves to eager schedules.

Lemma 12. (a) *Given any schedule for G with cost c, there is an eager schedule for G with cost at most c.* (b) *Given any sequence legal for G with cost c, there is a short-circuited sequence legal for G with cost at most c.*

A special kind of short-circuiting occurs when X is much longer than is needed by any eager scheduling algorithm (for instance, when X is infinite). In this case, we wish to restrict our attention to the shortest prefix of X which is legal for G, denoted $\mathrm{pre}(X, G)$. We also define $f(X, X')$ to be the shortest prefix of X that is a supersequence of X'.

From now on, we will refer to our solution as a sequence X (on $\mathcal{M}$) instead of a schedule (on V). The schedule implied is the eager schedule corresponding to $\mathrm{pre}(X, G)$. Thus, the cost of scheduling X for G is at most $\mathrm{cost}(\mathrm{pre}(X, G))$, though it could be lower due to short-circuiting.

In the following sections, we give good approximation algorithms for several special cases of the PTSP. We should note that an exact solution can be easily obtained, using dynamic programming, in time n^w, where w is the size of the largest anti-chain [5]. In general, however, we will be searching for polynomial time algorithms.

We start with an upper bound for a special case of PTSP on the line. This case is exactly the *dial-a-ride problem* on the line [12].

Theorem 13. *There is a polynomial time exact algorithm for PTSP (without repetitions) on the line when the precedence constraints are a collection of chains of length 2.*

4.1 $k - 1$ Approximation via Universal Sequences

Definition 5 *A* universal sequence algorithm *for PTSP is an algorithm that produces a schedule corresponding to a fixed sequence U, where U depends only on the metric space $\mathcal{M}$.*

That is, a universal sequence algorithm ignores the structure of the LDAG. We refer to U as the *universal sequence* of the algorithm. For a finite sequence X, X^* denotes the sequence X repeated infinitely often. If a universal sequence U equals X^* for some finite sequence X, then the corresponding algorithm is called a *repeating sequence algorithm*.

For a universal sequence U on $\mathcal{M}$, we say that U is an (r, β)-*sequence* if for all finite sequences X without duplicates, $\mathrm{cost}(f(U, X)) \leq r \cdot \mathrm{cost}(X) + \beta$.

Lemma 14. *If a universal sequence is an (r, β)-sequence, then its corresponding universal sequence algorithm is an r-approximation algorithm for PTSP, with additive error at most β.*

There is an interesting connection between universal sequences and algorithms for the celebrated k-server problem. Consider a k point metric space $\mathcal{M}$. Let A be a $(k-1)$-competitive algorithm for the $(k - 1)$-server problem on $\mathcal{M}$ (see [14]). We use A to obtain a $(k-1, \beta)$-sequence, where β is a constant. By Lemma 14, this gives a $(k-1)$-approximation algorithm for PTSP with a constant additive error.

To define the PTSP sequence, note that the configuration of the $(k - 1)$ servers is completely specified by the single point in $\mathcal{M}$ not covered by any server. Define the *state* of A after i requests, s_i, to be this uncovered point. We generate a request sequence $s_0, s_1, s_2, \ldots$, at each step requesting the point left uncovered after the previous step. The universal sequence we derive from A is the request sequence $U = s_0, s_1, \ldots$.

Lemma 15. *U is a $(k - 1, \beta)$-sequence, where β is a constant depending only on $\mathcal{M}$.*

Besides being applicable to PTSP, our universal sequence construction gives a $(\rho - 1)$-approximation (with at most constant additive error) for the Loading Time Scheduling Problem (LTSP), improving the ρ approximation of [2]. This is not an insignificant improvement since ρ, the number of machines, is typically very small (about 4 or 5). The distance function for LTSP is not a metric, since $d(i, j)$ depends only on j. However, $d'(i, j) = (d(i, j) + d(j, i))/2$ defines a metric and has only an additive difference from $d(i, j)$. Thus, the algorithms described above apply to d'.

As the following lemma shows, the bound in Lemma 15 is tight.

Lemma 16. *No universal sequence U can be an (r, β)-sequence for any $r < k - 1$ and constant β.*

Despite Lemma 16, it is conceivable that a universal sequence algorithm might have an approximation ratio better than $k - 1$ due to short-circuiting. For repeating sequence algorithms, however, we can prove that this is not the case.

Theorem 17. *No repeating sequence algorithm for PTSP can have an approximation ratio better than $k - 1 - \epsilon$ for any $\epsilon > 0$.*

4.2 Approximation via TSP

A natural type of repeating sequence algorithm for a k-point metric space repeats sequences of length k; that is, it cycles through $\mathcal{M}$ in a predetermined order. The optimal algorithm of this type uses a TSP cycle. Let T be an optimal TSP cycle, and let T' be its reverse. Let $U_T = T^*$ and $U'_T = T'^*$. Let d_{min} be the minimum distance between two points in $\mathcal{M}$. If the length of the longest chain is ℓ, then $\text{cost}(U_T) + \text{cost}(U'_T) \leq \ell \cdot \text{cost}(T) + \beta$, while the optimal schedule has cost $\geq \ell \cdot d_{min}$. We have the following.

Theorem 18. *The algorithm that schedules instructions using the lower-cost sequence U_T or U'_T is an r-approximation algorithm for PTSP (with at most constant additive error), where $r = \text{cost}(T)/2d_{min}$.*

Note that the TSP-based algorithm is not a repeating sequence algorithm, but rather one that takes the best of two repeating sequence algorithms.

If $\mathcal{M}$ is uniform, or if $\mathcal{M}$ is the hypercube, $d_{min} = 1$ while $\text{cost}(T) = k$ (for the hypercube, take the points in Grey code order). In each case we get a $k/2$ approximation, beating the lower bound for repeating sequence algorithms.

4.3 Greedy Algorithms

There are several natural heuristic algorithms for PTSP. All of these produce eager schedules. They produce the schedule incrementally, choosing the next point of $\mathcal{M}$ at each stage by looking at the vertices remaining to be scheduled.

Let $p \in \mathcal{M}$ be the last point scheduled. Let $p' \in \mathcal{M}$ be a point that labels some ready vertex, and $n(p')$ be the number of ready vertices that have label $l(p')$. The COST GREEDY heuristic picks the point p' that minimizes $d(p, p')$; this is essentially the heuristic of Su *et al.* [19]. The NODE GREEDY heuristic picks the point p' minimizing $n(p')$. The WEIGHTED GREEDY heuristic picks the p' minimizing $d(p, p')/n(p')$.

A *critical path* in an LDAG is a chain p such that $\text{cost}(l(p))$ is maximized over all chains p. The CRITICAL PATH heuristic works as follows: at each step, it selects a ready vertex j which belongs to a critical path; the next point scheduled is the label of j.

As Figure 1 shows, all of these algorithms have unbounded competitive ratios, even when $|\mathcal{M}|$ is as small as 4.

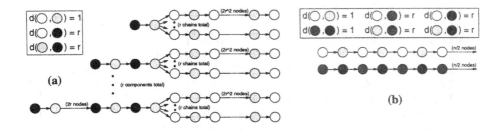

Fig. 1. (a) A lower-bound example for the greedy algorithms. The optimal algorithm schedules the points in the order of the bottom component, incurring cost $2r^2 - r + 2r^2 = 4r^2 - r$. All three greedy algorithms, however, will schedule the components from top to bottom, incurring cost $2r^2 - r + 2r^3$. Since r is arbitrary, this is an unbounded ratio. (b) A lower-bound example for CRITICAL PATH. An optimal algorithm incurs cost $n + r$ by scheduling all the points in the first chain and then all the points in the second. CRITICAL PATH will alternate chains, however, incurring cost at least $rn/2 + n$. This gives competitive ratio $r/4$ (for $n \geq r$), which is unbounded.

5   Experimental Results

We tested our heuristics on basic blocks generated from the standard SPEC92 benchmark suite [18] that is commonly used in compiler research. We used the experimental setup of Motwani, Palem, Reyen, and Sarkar [15], in that we used the input data generated by them (for an unrelated experimental study) as the input for our empirical testing. They compiled the SPEC92 benchmark programs for the SuperSPARC processor, with full optimization turned on in gcc, the Gnu C compiler. They added a small phase to gcc that, for each basic block, extracts the corresponding problem instance.

For each of the 19 benchmarks we tested, we broke each function of the benchmark into basic blocks. We threw out basic blocks with less than 10 instructions in gcc's internal representation. To test a benchmark, we ran the scheduling algorithm for each basic block in the benchmark and calculated the cost of following the schedule devised by the algorithm. We took the average cost per basic block to be the cost of the benchmark for that algorithm.

Each instruction is 32 bits long, and the cost to change from one instruction to the other is the number of bits by which the instructions differ. Thus, $\mathcal{M}$ is a 32-dimensional hypercube, and each instruction is a point in $\mathcal{M}$. In reality, not every 32-bit sequence corresponds to a legal instruction, and not all basic blocks use all instructions, so for a given basic block $|\mathcal{M}|$ may be significantly less than 2^{32}. Each vertex of the precedence graph is labeled by an instruction, so a *ready instruction* is one that labels a ready vertex.

We tested the following eager scheduling algorithms.

NGreedy – NODE GREEDY picks the most frequently occurring ready instruction.

CGreedy – COST GREEDY picks the instruction among the ready instructions that is closest (in terms of cost) to the current instruction. This is the algorithm of [19].

WGreedy – WEIGHTED GREEDY combines the other greedy schemes, picking the instruction among the ready instructions that minimizes cost/node count.

SPEC92 Benchmark	Number of basic blocks	Range of bb sizes (# instructions)	SPEC92 Benchmark	Number of basic blocks	Range of bb sizes (# instructions)
008.espresso	489	10 – 74	052.alvinn	11	10 – 17
013.spice2g6	1746	10 – 405	056.ear	113	10 – 74
015.doduc	575	10 – 903	072.sc	201	10 – 39
022.li	96	10 – 201	077.mdljsp2	108	10 – 303
023.eqntott	34	10 – 25	078.swm256	48	10 – 105
026.compress	18	10 – 29	089.su2cor	269	10 – 178
034.mdljdp2	108	10 – 308	090.hydro2d	205	10 – 281
039.wave5	739	10 – 274	093.nasa7	145	10 – 196
047.tomcatv	26	10 – 158	094.fpppp	124	10 – 4182
048.ora	19	10 – 100	**All 19 tests**	**5063**	**10 – 4182**

Table 1. Characteristics of basic blocks used in performance measurements.

Benchmark	NGreedy	CGreedy	WGreedy	TSP	KSrv	KSrv-NoS	CGreedy+KSrv
008.espresso	102%	100%	100%	100%	101%	267%	100%
013.spice2g6	108%	100%	102%	103%	103%	296%	100%
015.doduc	114%	100%	102%	106%	103%	294%	100%
022.li	102%	100%	100%	100%	102%	240%	100%
023.eqntott	101%	100%	100%	99%	100%	248%	100%
026.compress	102%	100%	100%	100%	102%	250%	100%
034.mdljdp2	112%	100%	104%	103%	105%	270%	101%
039.wave5	109%	100%	101%	103%	103%	289%	100%
047.tomcatv	110%	100%	105%	105%	107%	283%	100%
048.ora	108%	100%	100%	103%	102%	299%	100%
052.alvinn	104%	100%	100%	100%	101%	264%	100%
056.ear	104%	100%	100%	101%	101%	265%	100%
072.sc	102%	100%	100%	100%	101%	244%	100%
077.mdljsp2	108%	100%	103%	102%	103%	266%	101%
078.swm256	110%	100%	101%	103%	105%	314%	100%
089.su2cor	108%	100%	102%	103%	105%	306%	100%
090.hydro2d	106%	100%	102%	101%	104%	273%	100%
093.nasa7	109%	100%	102%	104%	105%	329%	100%
094.fpppp	117%	100%	105%	109%	104%	284%	100%

Table 2. The average cost of a basic block, taken as a percent of the average cost for CGREEDY.

Benchmark	NGreedy	CGreedy	WGreedy	TSP	KSrv	KSrv-NoS	CGreedy+KSrv
013.spice2g6	108%	100%	98%	102%	101%	380%	100%
015.doduc	117%	100%	104%	113%	104%	420%	101%
022.li	106%	100%	103%	101%	109%	234%	105%
034.mdljdp2	118%	100%	114%	105%	116%	286%	108%
039.wave5	141%	100%	109%	115%	106%	357%	100%
077.mdljsp2	115%	100%	111%	105%	114%	290%	107%
090.hydro2d	109%	100%	109%	100%	105%	164%	100%
094.fpppp	188%	100%	138%	154%	106%	266%	100%

Table 3. The average cost of large basic blocks (≥ 200 nodes) as a percent of CGREEDY's cost.

TSP – TSP orders the instructions in TSP order. (Actually, we use a 2-approximation algorithm to obtain an approximate TSP ordering.) It picks its next instruction to be the next instruction in TSP order, short-cutting non-ready instructions. It then tries ordering instructions in reverse TSP order and takes the better of the two runs.

KSrv – KSERVER uses the $k-1$-sever algorithm of [14] to obtain a universal sequence. It picks its next instruction to be the next instruction in this universal sequence, short-cutting non-ready instructions.

KSrv-NoS – KSERVER WITHOUT SHORTCUT is the same as KSERVER, but it does not short-cut, and therefore pays for instructions even if they are not used.

CGreedy+KSrv – COST GREEDY PLUS KSERVER combines the two algorithms: it runs COST GREEDY for 100 distinct instructions, adding weights to the $k-1$ servers as if the instructions changes were being made by KSERVER (that is, when changing from node i to node j, it updates server weights as if a server were moving from j to i). After 100 instructions, it switches to KSERVER, using the server weights from COST GREEDY as an initial configuration for the KSERVER algorithm.

The results of these experiments are shown in Figures 2 and 3. Figure 2 indicates that none of the theoretically motivated algorithms — TSP, KSRV, and their variants — outperformed CGREEDY, the best algorithm in our tests, and Figure 3 shows that the difference in performance often grew as the size of the basic block increased. Several of the theoretically motivated algorithms, however, were competitive with CGREEDY and all have superior theoretical properties. It is known that a greedy algorithm performs well on the average for the Shortest Common Supersequence problem [10]. The similarity between that problem and the PTSP could explain why greedy algorithms for the PTSP perform better than predicted by worst case analysis.

Acknowledgements

We are grateful to Sharad Malik for helpful discussions and pointers to the relevant literature. We are deeply indebted to the authors of [15] for sharing their experimental set-up, and in particular to Salem Reyen for his help with the experiments.

References

1. A. Aggarwal, A.K. Chandra and P. Raghavan. Energy Consumption in VLSI Circuits. In *Proceedings of the ACM Symposium on Theory of Computing*, 1988, pp. 205–216.
2. R. Bhatia, S. Khuller, and J. Naor. The Loading Time Scheduling Problem. In *Proceedings of the 36th Annual IEEE Symposium on Foundations of Computer Science*, 1995, pp. 72–81.
3. L. Bianco, A. Mingozzi, S. Riccardelli, and M. Spadoni. Exact and Heuristic Procedures for the Traveling Salesman Problem with Precedence Constraints, Based on Dynamic Programming. *INFOR*, 32(1):19–32, 1994.
4. N. Christofides. Vehicle Routing. In *The Traveling Salesman Problem: A Guided Tour of Combinatorial Optimization* (Ed: E.L. Lawler, J.K. Lenstra, A.H.G. Rinnooy Kan, and D.B. Shmoys), John Wiley & Sons (1985), pp. 431–448.
5. C.J. Colbourn and W.R. Pulleyblank. Minimizing Setups in Ordered Sets of Fixed Width. *Order*, 1:225–229 (1985).

6. S. Devadas and S. Malik. A Survey of Optimization Techniques Targeting Low-Power VLSI Circuits. Preprint, 1995.

7. M. Garey and D. Johnson. *Computers and Intractability — a Guide to the Theory of NP-completeness*. W.H. Freeman, San Francisco, 1979.

8. B.L. Golden and A.A. Assad, Eds. *Vehicle Routing: Methods and Studies*. North-Holland, Amsterdam (1988).

9. K. Govil, E. Chan, and H. Wasserman. Comparing Algorithms for Dynamic Speed-Setting of a Low-Power CPU. Preprint, 1995.

10. T. Jiang and M. Li. On the Approximation of Shortest Common Supersequences and Longest Common Subsequences. In *Proceedings of 21st International Colloquium on Automata, Languages and Programming*, 1994, pp. 191–202.

11. G. Kissin. Energy Consumption in VLSI Circuits: a Foundation. *ACM Symposium on Theory of Computing*, 1982.

12. M. Kubo and H. Kagusai. Heuristic Algorithms for the Single-Vehicle Dial-a-ride Problem. *Journal of the Operations Research Society of Japan*, 30(4):354–365, 1990.

13. M. T. Lee, V. Tiwari, S. Malik, and M. Fujita. Power Analysis and Low-Power Scheduling Techniques for Embedded DSP Software. Technical Report FLA-CAD-95-01, Fujitsu Labs of America, March 1996.

14. M. Manasse, L. McGeoch, and D. Sleator. Competitive Algorithms for Server Problems. *Journal of Algorithms*, 11:208–230 (1990).

15. R. Motwani, K. Palem, S. Reyen, and V. Sarkar. Combining Register Allocation and Instruction Scheduling. Submitted for publication, 1997. Preliminary version: *Technical Report STAN-CS-TN-95-22*, Department of Computer Science, Stanford University, 1996.

16. S. Prasad and K. Roy. Circuit Activity Driven Multilevel Logic Optimization for Low Power Reliable Operation. *Proceedings of the European Conference on Design Autoamtion (EDAC)*, 1993, pp. 368–372.

17. K. Roy and S. Prasad. SYSLOP: Synthesis of CMOS Logic for Low-Power Applications. In *Proceedings of the International Conference on Computer Design*, 1992.

18. SPEC Consortium. The SPECint95 and SPECfp95 Benchmarks. World-Wide Web URL http://www.specbench.org, 1995.

19. C-L. Su, C-Y. Tsui, and A.M. Despain. Low-Power Architecture Design and Compilation Techniques for High-Performance Processors. In *Proceedings of the IEEE COMPCON*, 1994, pp. 489–498.

20. V. Tiwari, S. Malik, and A. Wolfe. Power analysis of embedded software: A first step towards software power minimization. *IEEE Transaction on VLSI Systems*, 2:437–445 (1994).

21. V. Tiwari, S. Malik, and A. Wolfe. Compilation Techniques for Low Energy: An Overview. In *Proceedings of the 1994 Symposium on Low-Power Electronics*, 1994, pp. 38-39.

22. C.Y. Tsui, M. Pedram, and A.M. Despain. Technology Decomposition and Mapping Targeting Low-Power Dissipation. In *Proceedings of the 30th Design Automation Conference*, 1993, pp. 68-73.

23. M. Weiser. Some Computer Science Issues in Ubiquitous Computing. *Communications of the ACM*, 36:74–83 (1993).

24. F. Yao, A. Demers, and S. Shenker. A Scheduling Model for Reduced CPU Energy. In *Proceedings of the 36th Annual IEEE Symposium on Foundations of Computer Science*, 1995, pp. 374–382.

On-line Load Balancing for Related Machines

Piotr Berman[1], Moses Charikar[2], Marek Karpinski[3]

[1] The Pennsylvania State University, University Park, PA16802, USA
email: berman@cse.psu.edu
[2] Stanford University, Stanford, CA 94305-9045
email: moses@cs.stanford.edu
[3] University of Bonn, 53117 Bonn, and
International Computer Science Institute, Berkeley
email: marek@cs.uni-bonn.de

Abstract. We consider the problem of scheduling permanent jobs on related machines in an on-line fashion. We design a new algorithm that achieves the competitive ratio of $3 + \sqrt{8} \approx 5.828$ for the deterministic version, and $3.31/\ln 2.155 \approx 4.311$ for its randomized variant, improving the previous competitive ratios of 8 and $2e \approx 5.436$. We also prove lower bounds of 2.4380 on the competitive ratio of deterministic algorithms and 1.8372 on the competitive ratio of randomized algorithms for this problem.

1 Introduction

The problem of on-line load balancing was studied extensively over the years (cf., e.g., [11], [3], [4], and [2]). In this paper we study the on-line load balancing problem for related machines (cf. [2]). We are given a set of machines that differ in speed but are related in the following sense: a job of size p requires time p/v on a machine with speed v. While we cannot compare structurally different machines using a single speed parameter, it is a reasonable approach when the machines are similar; in other cases it may be a good approximation.

Our task is to allocate a sequence of jobs to the machines in an on-line fashion, while minimizing the maximum load of the machines. This problem was solved with a competitive ratio 8 by Aspnes *et al.* [2]. Later, it was noticed by Indyk [8] that by randomizing properly the key parameter of the original algorithm the expected competitive ratio can be reduced to $2e$. A similar randomization idea has been used earlier by several authors in different contexts (cf. [5, 9, 12, 10, 6]).

For the version of the problem where the speeds of all the machines are the same, Albers [1] proved a lower bound of 1.852 on the competitive ratio of deterministic algorithms and Chen *et al.* [7] proved a lower bound of 1.5819 on the competitive ratio of randomized algorithms.

Adapting the notation of Aspnes *et al.*, we have n machines with speeds $v_1, \ldots, v_n$ and a stream of m jobs with sizes $p_1, \ldots, p_m$. A schedule s assigns to each job j the machine $s(j)$ that will execute it. We define $load(s, i)$, the load of a machine i in schedule s and $Load(s)$, the load of entire schedule s as follows:

$$load(s,i) = \frac{1}{v_i} \sum_{s(j)=i} p_j, \quad Load(s) = \max_i load(s,i)$$

It is easy to observe that finding an optimum schedule s^* is NP-hard off-line, and impossible on-line. We want to minimize the competitive ratio of our algorithm, i.e. the ratio $Load(s)/Load(s^*)$ where s is the schedule resulting from our on-line algorithm, and s^* is an optimum schedule.

In Section 2, we describe an on-line scheduling algorithm with competitive ratio $3 + \sqrt{8} \approx 5.828$ for the deterministic version, and $3.31/\ln 2.155 \approx 4.311$ for its randomized variant. In Section 3, we prove lower bounds of 2.4380 on the competitive ratio of deterministic algorithms and 1.8372 on the competitive ratio of randomized algorithms for on-line scheduling on related machines.

2 Algorithm

2.1 Preliminaries

The idea behind the algorithm of Aspnes *et al.* [2] is the following. There exists a simple algorithm that achieves competitive ratio 2 if we know exactly the optimum load Λ: we simply assign each job to the slowest machine that would not increase its load above 2Λ. Because we do not know Λ, we make a safely small initial guess and later double it whenever we cannot schedule a job within the current load threshold.

Our innovation is to double (or rather, increase by a fixed factor r) the guess as soon as we can prove that it is too small, without waiting for the time when we cannot schedule the subsequent job. Intuitively, we want to avoid wasting the precious capacity of the fast machines with puny jobs that could be well served by the slow machines. Therefore we start with describing our method of estimating the necessary load, i.e. computing lower bounds on the optimal load for the sequence of jobs seen so far.

Let $V = \{0, v_1, \ldots, v_n\}$ (for later convenience, we assume that the sequence of speeds is nondecreasing). For $v \in V$ we define $Cap(v)$ as the sum of speeds of these machines that have speed larger than v. (Cap stands for capacity, note that $Cap(0)$ is the sum of speeds of all the machines and $Cap(v_n) = 0$.) For a set of jobs J and a load threshold Λ we define $OnlyFor(v, \Lambda, J)$ as the sum of sizes of these jobs that have $p_j/v > \Lambda$. ($OnlyFor$ stands for the work that can be performed only by the machines with speed larger than v if the load cannot exceed Λ.) The following lemma is immediate:

Lemma 1. *For a set of jobs J, there exists a schedule s with $Load(s) \leq \Lambda$ only if $OnlyFor(v, \Lambda, J) \leq \Lambda Cap(v)$ for every $v \in V$.*

Before we formulate and analyze our algorithm, we will show how to use the notions of Cap and $OnlyFor$ to analyze the already mentioned algorithm that keeps the load under 2Λ if load Λ is possible off-line. We reformulate it to make it more similar to the new algorithm. Machine i has *capacity* $c_i = \Lambda v_i$ equal to the amount of work it can perform under Λ load, and the *safety margin* m_i to assure that we will be able to accomodate the jobs in the on-line fashion. In this algorithm the capacity and the safety margin are given the same value, in the new one they will be different.

```
(* initialize *)
for i ← 1 to n do
    m_i ← c_i ← Λv_i
j ← 0
(* online processing *)
repeat
    read(p)
    j ← j + 1
    s(j) ← min{i| c_i + m_i > p}
    c_s(j) ← c_s(j) − p
forever
```

This algorithm shares the following property with the new one: the jobs are offered first to the machine 1 (the slowest), then to machine 2 etc., so each time the first possible machine accepts the new job. Given a stream of jobs J, we can define J_i as the stream of jobs that are passed over by machine i or that reach machine $i+1$ (for $1 \leq i < n$ these two conditions are equivalent, for $i = 0$ only the latter and for $i = n$ only the former applies). The correctness of the algorithm is equivalent to the fact that the stream J_n is empty—it consists of the jobs passed over by all the machines. From the correctness the load guarantee follows easily, because the sum of sizes of jobs assigned to machine i is less than the initial capacity plus the safety margin, i.e. $\Lambda v_i + \Lambda v_i$, and so the load is less than $2\Lambda v_i / v_i = 2\Lambda$.

For the inductive reasoning we define $V_i = \{0, v_i, \ldots, v_n\}$ and $Cap_i(v)$ which is the sum of speeds from V_i that exceed v.

Lemma 2. *If there exists a schedule s^* with $Load(s^*) = \Lambda$, then for every $i = 0, \ldots, n$ and every $v \in V_i$*

$$OnlyFor(v, \Lambda, J_{i-1}) \leq \Lambda Cap_i(v).$$

Proof. By induction on i. For $i = 0$ the claim is equivalent to Observation 1. For the inductive step, after assuming the claim for i, we have to show that

$$OnlyFor(v, \Lambda, J_i) \leq \Lambda Cap_{i+1}(v) \text{ for } v \in V_{i+1}.$$

For $v \neq 0$, the inequality follows from the inductive hypothesis directly: the left hand side can only become smaller (because J_i is a subsequence of J_{i-1}), while the right hand side remains unchanged. Thus it suffices to show that $OnlyFor(0, \Lambda, J_i) \leq \Lambda Cap_{i+1}(0)$.

We consider two cases according to the final value of c_i in the execution of the algorithm.

Case 1: $c_i > 0$.

In this case machine i accepted all jobs with size at most $m_i = \Lambda v_i$ from the stream J_{i-1}, hence $OnlyFor(0, \Lambda, J_i)$, which is the sum of job sizes in J_i, is at most $OnlyFor(v_i, \Lambda, J_{i-1})$, which in turn is less or equal to $\Lambda Cap_i(v_i)$. Because $Cap_i(v_i) \leq \sum_{j=i+1}^{n} v_j = Cap_{i+1}(0)$, the claim follows.

Case 2: $c_i \leq 0$.

In this case, the total size of the jobs accepted by machine i is at least Λv_i, the initial value of c_i, hence $OnlyFor(0, \Lambda, J_i) \leq OnlyFor(0, \Lambda, J_{i-1}) - \Lambda v_i$, while

$$Cap_{i+1}(0) = \sum_{v \in V_{i+1}} v = \left(\sum_{v \in V_i} v \right) - v_i = Cap_i(0) - v_i.$$

Because one of our assumption is $OnlyFor(0, \Lambda, J_{i-1}) \leq \Lambda Cap_i(0)$, the claim follows. □

Lemma 2 allows us to conclude that if a schedule with load Λ exists, then $OnlyFor(0, \Lambda, J_n) \leq \Lambda Cap_{n+1}(0) = 0$. Thus the stream J_n of unscheduled jobs is empty, which means that the algorithm is correct.

2.2   The New Algorithm

The next algorithm is similar, but it proceeds in phases, each phase having a different value of Λ. While it is correct for any value of the parameter $r > 1$, we will later find the optimum r's (they are different in the deterministic and randomized versions).

```
(* initialize *)
Λ ← something very small
for i ← 1 to n do
      m_i ← c_i ← 0
j ← 0, J ← empty string
(* online processing *)
repeat
      read(p)
      j ← j + 1, p_j ← p, append J with p_j
      (* start a new phase if needed *)
      while OnlyFor(v, Λ, J) > ΛCap(v) for some v ∈ V do
            Λ ← rΛ,    m_i ← Λv_i,    c_i ← c_i + m_i
      (* schedule p_j *)
      s(j) ← min{i| c_i + m_i > p_j}
      c_s(j) ← c_s(j) − p_j
forever
```

We will say that executing "$s(j) \leftarrow \min\{i|\ c_i + m_i > p_j\}$" schedules p_j (even though, for the sake of argument, we admit the case that the set of machines with sufficient capacity is empty). We need to prove that the algorithm is correct, i.e. that we never apply min to an empty set; in other words, for every job we can find a machine with sufficient remaining capacity.

Let Λ_0 be the value of Λ when the first job was scheduled. We view the execution as consisting of phases numbered from 0 to k, where l-th phase schedules jobs with $\Lambda = \Lambda_l = \Lambda_0 r^l$. Let J^l be the stream of jobs scheduled in phase l. Using the same convention as in the analysis of the previous algorithm, we define J_{i-1}^l to be the stream of jobs that in phase l machine i received or machine $i-1$ passed over. Now the correctness will mean that the stream J_n^l is empty for every phase l.

Because the initial estimate for Λ may be too low, machines may receive more work than in the previous algorithm. This is due to the fact that in the initial phases the machines from the beginning of the sequence needlessly refuse to pick jobs that they would gladly accept later, thus increasing the load at the end of the sequence. Nevertheless, as we shall show, this increase is limited.

As a preliminary, we need to analyze the consequences of the test that triggers a new phase as soon as Λ is not appropriate for the stream of jobs received so far. First of all, this implies that every Λ_l is appropriate for the stream $J^1 \cdots J^l$, and in particular, for the substream J^l. Therefore

$$OnlyFor(v, \Lambda_l, J^l) \leq \Lambda_l\, Cap(v) \quad \text{for every phase } l \text{ and every } v \in V. \quad (\#)$$

This allows to prove the following lemma by induction:

Lemma 3. *For every $i = 0, \dots, n$ and every phase l*

$$\sum_{t=0}^{l} OnlyFor(0, \Lambda_t, J_i^t) \leq \left(\sum_{t=0}^{l} \Lambda_t\right)\left(\sum_{j=i+1}^{n} v_j\right).$$

Proof. For $i = 0$ this follows simply from the fact that for every phase $t \leq l$

$$OnlyFor(0, \Lambda_t, J_0^t) = OnlyFor(0, \Lambda_t, J^t) \leq \Lambda_t\, Cap(0) = \Lambda_t\left(\sum_{j=1}^{n} v_j\right).$$

For $l = 0$ this follows from Lemma 2, as the phase 0 is identical to the first algorithm with $\Lambda = \Lambda_0$.

Therefore we may assume that the claim is true for $(i, l-1)$ and $(i-1, l)$. We will prove the claim for (i, l). We consider two cases, according to the value of c_i at the end of phase l.

Case 1: $c_i > 0$.

Subtract formally from both sides of the claim for i and l the respective sides of the claim for i and $l-1$; this way we see that it suffices to show that

$$OnlyFor(0, \Lambda_l, J_i^l) \leq \Lambda_l\left(\sum_{j=i+1}^{n} v_j\right)$$

Because the final value of c_i is positive, in phase l machine i accepted all jobs from the stream J_{i-1}^l that had size bounded by $\Lambda_l v_i$, and therefore the stream J_i^l consists only of the jobs that must be executed on machines faster than v_i. Thus the sum of sizes of all jobs in this stream, $OnlyFor(0, \Lambda_l, J_i^l)$, equals to $OnlyFor(v_i, \Lambda_l, J^l)$, which by (#) is at most $\Lambda_l Cap(v_i)$. Lastly, $Cap(v_i) \leq \sum_{j=i+1}^n v_j$.

Case 2: $c_i \leq 0$.

Suppose the final value of c_i in phase l equals some $c \leq 0$. This time subtract from both sides of the claim the respective sides of the claim for $i-1$ and l, this way we can see that it suffices to show that

$$\sum_{t=0}^l (OnlyFor(0, \Lambda_t, J_i^t) - OnlyFor(0, \Lambda_t, J_{i-1}^t)) \leq -\left(\sum_{t=0}^l \Lambda_t\right) v_i.$$

Equivalently,

$$\sum_{t=0}^l (OnlyFor(0, \Lambda_t, J_{i-1}^t) - OnlyFor(0, \Lambda_t, J_i^t)) \geq \left(\sum_{t=0}^l \Lambda_t\right) v_i. \quad (\#\#)$$

On the left hand side this inequality has the difference between the sum of jobs sizes that reach machine i and the sum of the job sizes that are passed over by machine i to the subsequent machines (during the phases from 0 to l). In other words, this is the sum of sizes of the jobs accepted by machine i during these phases. This sum, say s, is related in the following manner to c:

$$0 \geq c = \left(\sum_{t=0}^l \Lambda_t v_i\right) - s \text{ which implies } s \geq \left(\sum_{t=0}^l \Lambda_t\right) v_i \equiv (\#\#). \quad \square$$

Lemma 3 implies that

$$\sum_{t=0}^k OnlyFor(0, \Lambda_t, J_n^t) \leq 0$$

This means that, for every phase $t \leq k$,

$$OnlyFor(0, \Lambda_t, J_n^t) = 0$$

Observe that $OnlyFor(0, \Lambda_t, J_n^t)$ is simply the sum of the sizes of all jobs in J_n^t. Thus J_n^t is empty for every phase t, implying the correctness of the algorithm.

To analyze the competitive ratio, we may assume that $Load(s^*) = 1$. Then the penultimate value of Λ must be smaller than 1 and the final one smaller than r. Consider a machine with speed 1. The work accepted by a machine is smaller than the sum of all Λ's up to that time (additions to the capacity plus the last Λ given for the safety margin. Together it is $(r + 1 + r^{-1} + \ldots) + r = r(1/(1 - r^{-1}) + 1) = r(2r - 1)/(r - 1)$. To find the best value of r, we find zeros of the derivative of this expression, namely of $(2r^2 - 4r + 1)/(r - 1)^2$, and solve the resulting quadratic equation. The solution is $r = 1 + \sqrt{1/2}$ and the resulting competitive ratio is $3 + \sqrt{8} \approx 5.8284$.

One can observe that the worst case occurs when our penultimate value of Λ is very close to 1 (i.e. to the perfect load factor). We will choose the initial value of Λ to be of the form r^{-N+x} where N is a suitably large integer and x is chosen, uniformly at random, from some interval $< -y, 1 - y >$ (we shifted the interval $< 0, 1 >$ to compensate for the scaling that made $Load(s^*) = 1$). Therefore we can replace the factor r with the average value of the last Λ. For negative x this value is r^{x+1}, for positive it is r^x. The average is

$$\int_{-y}^{0} r^{1+x} dx + \int_{0}^{1-y} r^x dx = \int_{1-y}^{1} r^x dx + \int_{0}^{1-y} r^x dx = \int_{0}^{1} r^x dx = \frac{r-1}{\ln r}$$

Therefore the expected competitive ratio is

$$\frac{r-1}{\ln r} \frac{2r-1}{r-1} = \frac{2r-1}{\ln r}$$

The equation for the minimum value of this expression does not have a closed form solution, but nevertheless we can approximate it numerically. The minimum is achieved for r close to 2.155, and approximately equals 4.311.

3 Lower Bounds

In this section, we will prove deterministic and randomized lower bounds for load balancing on related machines.

As in the previous section, we will use very similar constructions for the deterministic and the randomized case. The set of machines and the sequences of jobs being considered are defined in terms of a parameter $\alpha > 1$; again, different values of the parameter will be used in these two cases.

We will consider a set of machines $\{0, 1, \ldots, n\}$, where i-th machine has speed

$$v_i = \begin{cases} \alpha^{-i} & \text{if } i < n \\ \frac{\alpha}{\alpha-1} \alpha^{-n} & \text{if } i = n \end{cases} \tag{1}$$

Note that $v_n = \sum_{i=n}^{\infty} \alpha^{-i}$, consequently $\sum_{i=0}^{n} v_i = \sum_{i=0}^{\infty} \alpha^{-i}$. In the deterministic case the only requirement on n will be that $v_n \le v_3 = \alpha^{-3}$. In the randomized case we will consider $n = 6$. In both settings, a lower bound for some particular value of n will implies the same result for all higher values.

The sequences of job sizes that we will consider will have the form $J_i = (j_0, \ldots, j_i)$, where $j_k = \alpha^k$. These sequences have two salient properties. Because all of them are prefixes of the same infinite sequences, after processing k initial jobs, all sequences of length at least k remain equally possible. Moreover, j_i, the size of the last job of J_i, is also the optimum load: on one hand, the last job alone requires load j_i, as the largest speed equals 1; on the other, we can achieve this load by assigning j_k to machine $\min(i - k, n)$.

3.1 Deterministic Lower Bound

The idea of the deterministic lower bound is to formulate a necessary condition for the existence of a scheduling algorithm that has a competitive ratio below α^3; finally arriving at a condition that can be effectively tested by a computer program. Then to show a lower bound of α^3 it will be sufficient to check that α fails this test. We start from the following lemma.

Lemma 4. *Suppose that a scheduling algorithm A achieves a competitive ratio lower than α^3. Then for every sequence of jobs sizes of the form J_i all jobs are scheduled on machines 0, 1, and 2.*

Proof. Suppose an algorithm A schedules $i - th$ job on machine $m > 2$ for some i. Then for the sequence J_i the algorithm will lead on machine m to the load $j_i/v_m \geq j_i/\alpha^{-3} = \alpha^3 j_i$ (remember that $v_m \leq \alpha^{-3}$). On the other hand the optimum load for J_i is j_i, hence the competitive ratio is at least α^3, a contradiction. □

The second step in developing our necessary condition is forming a representation of the configuration achieved by an algorithm after processing a sequence J_i. Obviously, we can represent this configuration by recording for each machine the sum of sizes of jobs scheduled on this machine. By Lemma 4, we may assume that jobs are scheduled only on machines 0, 1, and 2, and so it suffices to record the sums for these machines only, say that they form a vector $S^i = (S_0^i, S_1^i, S_2^i)$. By dividing the coefficients of this vector by the optimum load we obtain the vector of *relative loads* $R^i = \alpha^{-i} S^i$. This way the necessary condition for the competitive ratio to be below α^3 is that $R^i \prec U$ for every $i > 0$. Here the upper bound vector U is $(\alpha^3, \alpha^2, \alpha)$ and $X \prec Y$ means that for every coordinate the entry in X is lower than the entry in Y.

Let $E_0 = (1, 0, 0)$, $E_1 = (0, 1, 0)$ and $E_2 = (0, 0, 1)$. Suppose that the $(i+1)$st job is scheduled on machine m. Then $R^{i+1} = \alpha^{-1} R^i + E_m$. Obviously, $R^0 = (0, 0, 0)$. Consider the following infinite graph G_0: the set of vertices is $\mathbf{V} = \{V \in \Re^3 | V \prec U\}$ and the edges are of the form $(V, \alpha^{-1} V + E_m)$. Then if there exist a deterministic scheduling algorithm with competitive ratio below α^3, we have an infinite path in the graph G_0 that starts at the node $(0, 0, 0)$.

To reduce the graph to a finite size we *discretize* the relative load vector. More precisely, we define discretization as operation ', where V' is a vector obtained from V by replacing each coordinate with the largest multiple of δ that does not exceed this coordinate. Here $\delta > 0$ is a parameter of discretization. We define a new graph G_1 with the set of nodes $\mathbf{V}'$ and edges of the form $(V, (\alpha^{-1} V + E_m)')$. Obviously, if there exists an infinite path starting at $(0,0,0)$ in G_0, we also have such a path in G_1.

To simplify the problem further, we define $G(\alpha, \delta)$ to be a subgraph of G_1 consisting of nodes reachable from $(0,0,0)$. Note that the vertices of $G(\alpha, \delta)$ are of the form $\delta(c_0, c_1, c_2)$ where c_0, c_1, c_2 are nonnegative integers. Thus the size of $G(\alpha, \delta)$ is finite. Now we can phrase our necessary condition: for every $\delta > 0$

the graph $G(\alpha, \delta)$ contains a cycle. Consequently, to show a lower bound of α^3 it suffices to compute $G(\alpha, \delta)$ for some $\delta > 0$ and check that it is acyclic.

We have verified exactly that for $\alpha = 1.3459$ and $\delta = 0.0008$, thus obtaining a lower bound of $\alpha^3 > 2.4380$

We mention that it is possible to give a purely analytic proof of a weaker lower bound of 2.25. The idea is to fix $\alpha = 1.5$ and consider the loads on the fastest 2 machines over the last 3 jobs of the job sequence J_i as $i \to \infty$. We omit the details.

3.2 Randomized Lower Bound

Fix a constant m. We consider the distribution over the m job sequences $J_1, \ldots J_m$ where the sequence J_i, $1 \le i \le m$ is given with probability $\frac{1}{m}$. In other words, we give the job sequence J_m and stop the job sequence after i jobs where i is chosen uniformly and at random from the set $\{1, 2, \ldots m\}$. As noted before, the optimal load for J_i is α^i, hence the expected value of the optimal load is $\frac{1}{m} \cdot \sum_{i=1}^{m} \alpha^i$.

Consider the schedule produced by a deterministic algorithm for the job sequence J_m. Note that any schedule for J_m induces a schedule for J_i, $1 \le i \le m$. From this, we can compute the expected load incurred by A for the chosen distribution of job sequences.

We compute all possible schedules for J_m and for each schedule, compute the expected load for the distribution of job sequences. From this, we obtain the minimum expected load for any deterministic algorithm. By Yao's principle (cf. [13]), the ratio of the minimum expected load to the expected value of the optimal load gives us a lower bound for randomized algorithms versus oblivious adversaries.

A computer program tested all possible schedules for $n = 6$, $m = 14$ and $\alpha = 1.6$ and computed a lower bound of 1.8372. Note that this implies a randomized lower bound of 1.8372 for any $n \ge 6$ machines. For $n > 6$, we consider n machines with speeds v_i chosen as before. For the purpose of analysis, we group the slowest $n - 5$ machines into a single machine, i.e. pretend that any job scheduled on the $n - 5$ slowest machines is scheduled on a single machine of speed $\sum_{i=5}^{n-1} v_i = \frac{\alpha}{\alpha - 1} \cdot \frac{1}{\alpha^5}$. The load on this single machine is a lower bound for the maximum load on the slowest $n - 5$ machines. Observe that this gives us 6 machines whose speeds are the same as the speeds of the machines we use for the case $n = 6$. Hence the analysis for $n = 6$ applies and so does the lower bound of 1.8372.

4 Acknowledgements

The work of Moses Charikar was supported by Stanford School of Engineering Groswith Fellowship, an ARO MURI Grant DAAH04-96-1-0007 and NSF Award CCR-9357849, with matching funds from IBM, Schlumberger Foundation, Shell Foundation, and Xerox Corporation. The work of Marek Karpinski was partially supported by the DFG Grant KA 673/4-1, by the ESPRIT BR Grants 7097 and EC-US 030.

We would like to thank Yossi Azar, Amos Fiat, Piotr Indyk and Rajeev Motwani for valuable discussions and encouragement, and Susanne Albers for letting us read her paper before its publication.

References

1. S. Albers, *Better bounds for online scheduling*, Proc. 29th ACM STOC (1997), pp. 130-139.
2. J. Aspnes, Y. Azar, A. Fiat, S. Plotkin and O. Waarts, *On-line load balancing with applications to machine scheduling and virtual circuit routing*, Proc. 25th ACM STOC(1993), pp. 623-631.
3. Y. Azar, A. Broder, A. Karlin, *On-line load balancing*, Proc. 33rd IEEE FOCS (1992), pp. 218-225.
4. Y. Azar, J. Naor, R. Rom, *The competitiveness of on-line assignment*, Proc. 3rd ACM-SIAM SODA (1992), pp. 203-210.
5. A. Beck and D. Newman, *Yet more on the linear search problem*, Israel Journal of Math., 8:419–429, 1970.
6. S. Chakrabarti, C. Phillips, A. Schulz, D.B. Shmoys, C. Stein, and J. Wein, *Improved scheduling algorithms for minsum criteria*, Proc. 23rd ICALP, Springer, 1996.
7. B. Chen, A. van Vliet and G. J. Woeginger, *A lower bound for randomized on-line scheduling algorithms*, Information Processing Letters, vol.51, no.5, pp. 219-22.
8. P. Indyk, personal communication.
9. S. Gal, *Search Games*, Academic Press, 1980.
10. M. Goemans and J. Kleinberg, *An improved approximation ratio for the minimum latency problem*, Proc. 7th ACM-SIAM SODA, (1996), pp. 152–157.
11. R. L. Graham, *Bounds for certain multiprocessing anomalies*, Bell System Technical Journal 45 (1966), pp. 1563-1581.
12. R. Motwani, S. Phillips and E. Torng, *Non-clairvoyant scheduling*, Proc. 4th ACM-SIAM SODA (1993), pp. 422–431, see also: Theoretical Computer Science, 130 (1994), pp. 17–47.
13. A.C. Yao, *Probabilistic computations: Towards a unified measure of complexity*, Proc. 17th IEEE FOCS (1977), pp. 222–227.

A Linear-Time Algorithm for the 1-Mismatch Problem

Nikola Stojanovic[1], Piotr Berman[1], Deborah Gumucio[2], Ross Hardison[1] and
Webb Miller[1]

[1] Penn State, University Park PA 16802, USA
[2] University of Michigan Medical School, Ann Arbor MI 48109-0616, USA

Abstract. For sequence alignments (which can be viewed simply as
rectangular arrays of characters), a frequent need is to identify regions,
each consisting of a run of consecutive columns, that have some particular
property. The *1-mismatch problem* is to locate all maximal regions in a
given alignment for which there exists a (not necessarily unique) "center"
sequence such that inside the region alignment rows are within Hamming
distance 1 from the center. We first describe some properties of these
regions and their centers, and then use these properties to construct an
algorithm that for a *dxn* alignment runs in time $\Theta(nd)$ and extra space
$\Theta(d)$ (beyond that needed for the storage of the alignment itself).

1 Introduction

One effective approach to detecting potentially important regions within a DNA
sequence is to compare it with the corresponding sequences from species at a
moderate evolutionary distance from one another, say, the human sequence with
sequences from a number of other mammals. The result is an *alignment*, i.e., a
rectangular matrix in which rows are the respective DNA sequences "padded"
with gap characters. Given such an alignment, we look for stretches of columns
where the sequences differ "very little". As a rule, regions performing some
important function will be more highly conserved than "junk DNA", which constitutes the vast majority of human DNA.

There are number of useful ways in which this process, sometimes called
"phylogenetic footprinting" [2], can be made precise. For several years we have
operated an e-mail server [3] and World-Wide Web site that permit searching an
alignment for regions that are conserved according to a user-specified criterion.
No single definition is best for all cases [6, 4], but variants of the new approach
described here are proving effective [5] for locating potential *transcription factor
binding sites*, which is one of the most important uses of these (genomic DNA)
alignments.

Transcription factors are proteins that bind to the DNA strand and thereby
affect expression of a gene (i.e., production of the protein specified by that DNA
blueprint). Each of these proteins recognizes and binds to a particular string
of consecutive DNA letters, allowing for only small amount of variation. For instance, it might be that the protein attaches to any string of length 6 that differs

from the string AGATAG in at most one position. Recognition sites are known for a few hundred different transcription factors [7], and well-known algorithmic techniques [1] could be used to search for imprecise matches to any particular string. On the other hand, there are thousands of transcription factors whose recognition site is unknown.

Here we describe an algorithm for locating all regions of an alignment where the sequence conservation is sufficiently strong that a single unknown transcription factor might bind in all the represented species, assuming the model that the factor binds at just those strings that are within one mismatch of the factor's preferred recognition site. As a result, these regions satisfy the definition of 1-regions, provided below.

2 Definitions

For the purposes of this paper, an *alignment* is simply a rectangular array of characters. For alignments of DNA sequences, the alphabet of available characters consists of the four nucleotides, A, C, G and T, plus the gap character, dash (–).

Definition 1. A *1-region* is an interval of column positions within an alignment for which there exists a (not necessarilly unique) string, called the *center sequence*, such that each of the alignment rows differs from this string by at most one character within that interval. A *maximal 1-region* is a 1-region that it is not properly contained in any other 1-region.

Example 1. Consider the following alignment:

```
ATAAA
AATAA
TAATT
TTAAT
```

Columns 2, 3 and 4 form a 1-region with center AAA. Indeed AAA is the only center for those columns, since picking T as center character for column 2 forces a second mismatch in column 3 with either row 2 or row 3, and given an A for column 2 requires A's in columns 3 and 4 to avoid a second mismatch with row 1. Moreover, columns 2–4 are a maximal 1-region since any attempt to extend the 1-region creates a second mismatch with some row.

For this paper, we treat the gap character in the same way as any other character. In practice, one may want to treat them specially, e.g., not allowing them in the center sequence or considering only runs of columns that don't contain any gap characters. A variety of such alternative treatments are straightforward to develop.

Discussion of properties of 1-regions is complicated in inessential ways by *trivial columns*, i.e., columns that contain the same character in every row. We shall show later that finding maximal 1-regions in an arbitrary alignment easily reduces to finding them in the alignment obtained by skipping the trivial columns, so the following class of alignments is quite natural.

Definition 2. A *diverse alignment* contains only non-trivial columns, i.e., each column contains occurrences of at least two distinct characters.

The following conventions are adopted to make subsequent arguments more concise:

- The length of the alignment is n, its number of sequences (rows) is d, and the size of the alphabet underlying the aligned sequences is s.
- The character at row k and column p of the alignment is a_p^k.
- The run of columns starting at column p and ending at column q is called region $[p, q]$.
- For a non-trivial column, p, $h(p) = \min\{i \leq d \mid a_p^i \neq a_p^1\}$, i.e., the position of the first character in column p that differs from its top character.

3 Properties of Diverse Alignments

In this section, all alignments are assumed to be diverse alignments. Many of the statements made herein are false if trivial columns are allowed.

Maximal 1-regions may overlap and may also have more than one center sequence. Example 2 shows that two overlapping maximal 1-regions may have unique center sequences, yet with different character in a shared column.

Example 2. Consider the alignment:

```
ATA
TTA
TAT
TAA
```

It is easy to verify that the alignment has two overlapping maximal 1-regions, namely $[1, 2]$ and $[2, 3]$. Both regions have unique center sequences, TT and AA respectively, with different center character for position 2.

The illustrated situation may happen only for regions that overlap on one column. It will be shown shortly that if two maximal 1-regions overlap in two or more columns, their center sequences must be identical in the area of the overlap.

A center sequence for a region $[p, q]$ that has a mismatch with every row in a diverse alignment cannot be extended any further, and no 1-region can be longer than d (the number of aligned sequences). Moreover, once a character of a candidate center is selected, at some position $r \in [p, q]$, the rest of the center is uniquely determined. To see this, note that since the alignment is diverse, there is some sequence, S_i with $i \in [1, d]$, that differs from the center at position r.

In order for the sequence to remain the center of $[p, q]$, it must elsewhere have the same characters as S_i, which already has a mismatch. Observations of this kind lead to the following lemma, which divides adjacent column pairs into three cases.

Lemma 3. *Assume that $c_1 c_2$ is a center for $[p, p+1]$.*

1. *If $h(p) = h(p+1)$, then $c_1 c_2 = a_p^{h(p)} a_{p+1}^1$ or $c_1 c_2 = a_p^1 a_{p+1}^{h(p+1)}$;*
2. *if $h(p) < h(p+1)$, then $c_1 c_2 = a_p^{h(p+1)} a_{p+1}^1$;*
3. *if $h(p) > h(p+1)$, then $c_1 c_2 = a_p^1 a_{p+1}^{h(p)}$.*

Proof. 1. Assume that $h(p) = h(p+1)$. Any center sequence $c_1 c_2$ must satisfy either $c_1 = a_p^1$ or $c_2 = a_{p+1}^1$, since otherwise it has two mismatches with row 1. If $c_1 = a_p^1$, then $c_2 = a_{p+1}^{h(p)}$, since otherwise $c_1 c_2$ has two mismatches with row $h(p)$. Similarly, if $c_2 = a_{p+1}^1$, then $c_1 = a_p^{h(p)}$.
2. Assume that $h(p) < h(p+1)$. First suppose that $c_2 \neq a_{p+1}^1$. Then $c_1 = a_p^1$ causes $c_1 c_2$ to have two mismatches with row $h(p)$, while $c_1 \neq a_p^1$ leads to two mismatches with row 1. Therefore, $c_2 = a_{p+1}^1$. Then, $c_1 = a_p^{h(p+1)}$, since otherwise there are two mismatches with row $h(p+1)$.
3. The proof for this statement is symmetric to that of case 2. □

A two-column region $[p, p+1]$ can be:

- *ambiguous*, if $[p, p+1]$ is a 1-region with more than one center;
- *unambiguous*, if $[p, p+1]$ is a 1-region with unique center;
- *exclusive*, if $[p, p+1]$ is not a 1-region.

Lemma 3 shows that if a two-column region $[p, p+1]$ is ambiguous, then $h(p) = h(p+1)$ and $[p, p+1]$ has exactly two centers. Our algorithm relies also on the following property:

Lemma 4. *Every ambiguous two-column region $[p, p+1]$ is a maximal 1-region.*

Proof. If $[p, p+1]$ is ambiguous then, by Lemma 3, any row having no mismatch with one of the two centers automatically has two with the other, a contradiction. Hence every row has exactly one mismatch with both possible centers, and consequently neither center can be extended to include another (non-trivial) column. □

Lemmas 3 and 4 have the following consequences, which are used below to construct an efficient algorithm for locating all maximal 1-regions in an alignment. First, a 1-region of length 3 or more must have a unique center sequence. Moreover, if two different 1-regions overlap on at least two columns, their center sequences must agree within the overlap.

Before we describe our algorithm for a general alignment, we sketch briefly how the ideas would work in a diverse alignment. Assume that maximal 1-regions form a sequence $[p_1, q_1], [p_2, q_2], \ldots$ where the p_i's and q_i's are both increasing and $p_{i+1} \leq q_i + 1$ (the latter holds because every column belongs to a 1-region), and that we have information on $[p_i, q_i]$. Thus we know that p_{i+1} is the first colum $p > p_i$ such that $[p, q_i + 1]$ is a 1-region. Our observations show that the only case when we need to test multiple candidates for p_{i+1} is when $[p_i, q_i]$ has a unique center $c_{p_i} \ldots c_{q_i}$ and $[q_i, q_i + 1]$ has a unique center $c_{q_i} c_{q_i+1}$. In this case we can easily test each candidate in time $\Theta(d)$. The same idea applies to testing candidates for q_{i+1} once p_{i+1} is established. In both cases the multiple candidates are tested by updating the number of mismatches that each row has with the unique center. The details of this approach are described in the next section.

4 Algorithm

Given an alignment $\mathbf{A}$ with columns $A_1, \ldots, A_n$, we can delete trivial ones to obtain a diverse alignment $\mathbf{D} = A_{i_1}, \ldots, A_{i_l}$, for $l \leq n$. To simplify boundary cases, we augment $\mathbf{D}$ with (artificial) delimiting columns $i_0 = 0$ and $i_{l+1} = n+1$. There is a one-to-one correspondence between the maximal 1-regions of $\mathbf{A}$ and $\mathbf{D}$: $[p, q]$ is a maximal 1-region of $\mathbf{D}$ if and only if $[i_{p-1}+1, i_{q+1}-1]$ is a maximal 1-region of $\mathbf{A}$. Our algorithm works by maintaining a queue of non-trivial columns.

As we scan the alignment, the columns of the queue, except for the one at the head, form a 1-region that can not be extended to the left. Assume that $[p, q]$ is the next maximal 1-region of $\mathbf{D}$, and $h = Q.head.position = i_{p-1}$ and $t = Q.tail.position \leq i_q$; suppose also that yet next maximal 1-region of $\mathbf{D}$ is $[\bar{p}, \bar{q}]$. The new stage of the scan starts with enqueuing a new column number, say u. Next we check the status of $[t, u]$, using the technique of Lemma 3. If EXCLUSIVE, then obviously $t = i_q$ and $u = i_{q+1} = i_{\bar{p}}$, so we can report $[h+1, u-1]$ and reduce the queue content to $[t, u]$ (lines 8–10 of the pseudo-code of Figure 1). If AMBIGUOUS, then $[A_t, A_u]$ is a maximal 1-region of $\mathbf{D}$ (Lemma 4). Therefore it suffices to leave the last three columns in the queue (remember that the first one is excluded from the next region), after checking if a region should be reported (lines 11–14). If UNAMBIGUOUS, and in order to extend the enqueued region, first a center character for A_t computed in the status check must agree with the one stored in the queue so far (if any — thus the region can be extended only if we do not change already computed center characters); second, the number of mismatches between the extended center sequence and each row cannot exceed 1 (for these conditions, see lines 15–26 of Figure 1).

The meaning of the reporting $[p, q]$ depends on the immediate application. In the simplest form we can just output this pair. However, we may also check if the region satisfies some other criteria (e. g. requested minimal length) and then list all columns of this region, or its center sequence. We abbreviate this operation as "report the queue contents".

Lemma 5. *Every time the execution of the procedure* CENTERS *starts a new iteration (i.e., just prior to finding the next non-trivial column), the following invariants are valid:*

1. *The number of columns in the queue is between 2 and $d + 1$, inclusive;*
2. *There exists a 1-region containing all columns in the queue except the one at its head position;*
3. *There exists no 1-region containing all columns in the queue;*
4. *Only maximal 1-regions that end to the left of the column stored at the queue tail have been reported;*
5. *Every maximal 1-region that ends to the left of the column stored at the queue tail have been reported exactly once.*

Proof. We prove the invariants of Lemma 3 using induction on the number of iterations of the main loop of the procedure CENTERS.

Induction base. We view the initialization as iteration 1, as it enqueues the first non-trivial column, A_{i_1}. All invariants are trivially true; in particular note that the fictious column 0 does not belong to any 1-region.

Induction hypothesis. Assuming that the invariants hold at the beginning of the kth iteration, we show that they are valid at the beginning of iteration $k + 1$.

Induction step. Iteration $k + 1$ starts with adding column $A_{i_{k+1}}$ at the queue tail, which may lead to the violation of invariants 1, 2 and 5. Assume that A_{i_j} is the column represented by the queue head.

If the status of $\{A_{i_k}, A_{i_{k+1}}\}$ is EXCLUSIVE, then $[j + 1, k]$ is a maximal 1-region of $\mathbf{D}$, and the next such region must start at $A_{i_{k+1}}$, therefore executing line 9 restores invariant 5: we report maximal 1-region that is to the left of $A_{i_{k+1}}$, but not entirely to the left of A_{i_k}. This preserves invariant 4. However, invariant 2 is still false — in order to restore it we dequeue all columns that cannot be in the same 1-region with $A_{i_{k+1}}$ except one; this is accomplished by executing line 10. Because column A_{i_k} remains on the queue, invariant 3 remains true, as well as invariant 1.

Similarly, if the status of $\{A_{i_k}, A_{i_{k+1}}\}$ is AMBIGUOUS, then $[k, k + 1]$ is a maximal 1-region of $\mathbf{D}$, and if $j + 1 \neq k$, so is $[j + 1, k]$. Therefore executing CLEAR-QUEUE restores the invariants in a similar manner as lines 9–10 in the EXCLUSIVE case: to restore invariants 2 and 3, we must have exactly three columns on the queue, $A_{i_{k-1}}$, A_{i_k} and $A_{i_{k+1}}$. If this requires dequeuing, we report the 1-region of $\mathbf{A}$ corresponding to $[j + 1, k]$. Note that setting the mismatch counts to 1 assures that in the next iteration $[i_{k-1} + 1, i_{k+2} - 1]$ will be reported, regardless of the status of $\{A_{i_{k+1}}, A_{i_{k+2}}\}$.

If the status of $\{A_{i_k}, A_{i_{k+1}}\}$ is UNAMBIGUOUS, we need to distinguish three subcases.

If the queue contains only i_{k-1}, i_k and i_{k+1}, maximal 1-regions ending to the left of $A_{i_{k+1}}$ end to the left of A_{i_k}, thus invariants 1–5 remain true; we only need to initialize the mismatch counts to handle other cases properly in the future iterations.

If the queue contains more than three columns, the center character for A_{i_k} was already computed before. If it differs from the character computed in the status check, we know that $[k-1, k+1]$ is not a 1-region of $\mathbf{D}$, and consequently, by invariants 2 and 3, $[j+1, k]$ is a maximal 1-region of $\mathbf{D}$. Therefore executing CLEAR-QUEUE restores the invariants, in the same manner as it did in the AMBIGUOUS case.

In the remaining subcase we extend the already computed center sequence for $[j+1, k]$ with the character for $A_{i_{k+1}}$. This is the only subcase that requires the mismatch counts to be handled efficiently. We know the only possible center sequence for $A_{i_{j+1}}, \ldots, A_{i_{k+1}}$, thus the invariants 2 and 5 became false only if this sequence has 2 or more mismatches with one of the rows. Since we did not alter this sequence but merely appended with the character for $A_{i_{k+1}}$, the mismatch counts can be updated by reading the tail column only. Because of invariant 2, before the update all mismatch counts are either 0 or 1. If, after the update, one of the counts increases to 2, we know that $A_{i_{j+1}}$ and $A_{i_{k+1}}$ cannot belong to the same 1-region, so $[j+1, k]$ is a maximal 1-region of $\mathbf{D}$. We restore invariant 5 by executing line 23, and invariant 4 remains true. In order to restore invariant 2 while keeping 3 true we need to dequeue all columns that cannot be in the same 1-region as $A_{i_{k+1}}$, except one. Say that i_h should be the new head, so the queue should represent $[h+1, k+1]$ as the new 1-region of $\mathbf{D}$. If $h+1 < k$, then $[j+1, k]$ and $[h+1, k+1]$ overlap on at least two columns, so their center sequences must agree. If $h+1 = k$, then $[h+1, k+1] = [k, k+1]$, and in this subcase the only possible center sequence for $[k, k+1]$ is already stored in the queue. Therefore, in every case, the center sequence for $[h+1, k+1]$ is stored in the queue. As a result, we can use the mismatch counts to dequeue exactly the required number of columns: after dequeuing a column, the new head does not belong anymore to the 1-region represented by the queue, and the mismatch counts can be updated by reading the new head column only. Obviously, we should stop this process once all mismatch counts are below 2, and this is exactly what lines 24–26 do. As we dequeue at least one column in this case, and leave no less than three columns in the queue, invariant 1 also holds. □

Theorem 6. *The algorithm of Figure 1 determines precisely the maximal 1-regions of a given alignment.*

Proof. Since the invariants of Lemma 5 are valid at the beginning of any iteration of the procedure CENTERS, they are also valid at the beginning of the last one i.e., when the fictious column $n+1$ is at the queue tail. But that means that all maximal 1-regions ending to the left of it, and only these, are already reported only once each, which proves the correctness of the algorithm. □

5 Time and Space Performance of the Algorithm

The total time for all executions of line 5, i.e., finding the successive non-trivial columns, is $\Theta(nd)$. Suppose that to report a 1-region means to indicate its location and the determined center sequence. Then reporting a maximal 1-region can take time of $\Theta(n)$, and there may be up to n of them. However, there can never be more than $d+2$ columns in the queue at the same time (invariant 1 of Lemma 5, plus the new column). A new report never takes place before a new column gets enqueued at the tail, and the head of the queue always moves at least one position to the right after reporting. In consequence, no column of the alignment can be reported more than $d + 1$ times. There are exactly n columns, so reporting maximal 1-regions takes $O(nd)$ time cumulatively for the entire algorithm. It can be shown that this bound is tight, i.e. $\Theta(nd)$.

Basic queue operations (initializing, inserting, deleting, recording center characters, etc.) each take constant time. The main loop of the procedure CENTERS executes up to n times, in each iteration examining a pair of columns, in time $\Theta(d)$, and then branching depending on the status of the examination. The only other non-constant time actions are the repeated dequeuing of columns and updates of mismatch counts. At any given iteration, at most $d - 1$ columns are dequeued each updating the mismatch vector at the cost proportional to d. However, no column can lead to more than three updates of the mismatch vector, which yields the amortized cost of d per iteration for these updates, over n iterations. We thus conclude that the actions done by the procedure CENTERS are of $\Theta(nd)$ total time.

If it is only the location of maximal 1-regions we are interested in, then the above bounds on total time add up to $\Theta(nd)$. If we are interested in reporting the center sequence(s), too, then centers of length two or more do not add any complexity — there are always at most two such candidates. However, if a maximal 1-region contains only one non-trivial column, then there may be up to $\min\{s, d\}$ center sequences for it, and if we wish to select the "best" one (e.g. containing the majority character for the non-trivial column as its center) that may require extra time to do the selection.

Except for storing the alignment, or its part under examination, the algorithm needs space for the variables and the vector of mismatch counts, all of space $\Theta(d)$ or less, plus the structures needed to implement the queue. Since it has already been shown that the queue can never have more than $d + 2$ entries at the same time, it follows that it needs only $\Theta(d)$ space for its implementation. Therefore the whole algorithm needs $\Theta(d)$ extra space. However, if we need to determine the best center sequences, as mentioned above, additional space may be needed in order to select the majority character for a single non-trivial column center.

6 Discussion

It has already been noted that the treatment of gaps in different ways than as any other character leads to variations of the algorithm, many of them simple, involving additional conditions and restrictions. Different ways of treating gaps are motivated by distinct reasons for their existence. The approach of this paper is justified if a gap represents an actual difference between the aligned DNA sequences, however in many instances gaps represent either an artefact created by the alignment-generating software or the fragmentary nature of the data. The latter kind of a gap can be modelled by introducing a character that matches every other (a "wildcard"), but cannot be elected as a part of the center. Another useful variant is that when the minimal length for desired maximal 1-regions is set, so that only these longer then the threshold value get reported.

A natural extension of the 1-mismatch problem is the *k-mismatch problem*: finding all regions of an alignment such that there exists a (previously unknown) center sequence that does not differ in more than k positions from any of the alignment rows within the region, where k is a fixed non-negative number. While it is relatively easy to design a general algorithm for locating such regions, it is still an open question if there exists a polynomial-time solution. Some variants of this problem, e.g., the "wildcard" one (as above) in which the special symbol can occur anywhere in the alignment, have been proven to be NP-complete (unpublished work), but the nature of the general problem is yet unknown.

7 Acknowledgement

This work was supported by grant R01 LM05110 from the National Library of Medicine to W.M. and R.H., and by NIH grants HL33940 and HL48802 to D.G.

References

1. Galil, G., and R. Giancarlo (1988) Data structures and algorithms for approximate string matching. *J. Complexity* 4, 33-72.
2. Gumucio, D., D. Shelton, W. Zhu, D. Millinoff, T. Gray, J. Bock, J. Slightom and M. Goodman (1996) Evolutionary strategies for the elucidation of *cis* and *trans* factors that regulate the developmental switching programs of the β-like globin genes. *Mol. Phyol. and Evol.* 5, 18–32.
3. Hardison, R., K.-M. Chao, S. Schwartz, N. Stojanovic, M. Ganetsky and W. Miller (1994) Globin Gene Server: a prototype E-mail database server featuring extensive multiple alignments and data compilation for electronic genetic analysis. *Genomics* 21, 344–353.
4. Hardison, R., J. Slightom, D. Gumucio, M. Goodman, N. Stojanovic and W. Miller (1997) Phylogenetic and functional analyses of locus control regions of mammalian β-globin gene clusters. Submitted.
5. Shelton, D., L. Stegman, R. Hardison, W. Miller, J. Slightom, M. Goodman and D. Gumucio (1997) Phylogenetic footprinting of hypersensitive site 3 of the β-globin locus control region. To appear in *Blood*.
6. Slightom, J., J. Bock, D. Tagle, D. Gumucio, M. Goodman, N. Stojanovic, J. Jackson, W. Miller and R. Hardison (1997) The complete sequences of galago and rabbit β-globin locus control regions: extended sequence and functional conservation outside the cores of DNase hypersensitive sites. *Genomics* 39, 90–94.
7. Wingender, E., A. Kel, O. Kel, H. Karas, T. Heinemeyer, P. Dietze, R. Knüppel, A. Romaschenko and N. Kolchanov (1997) TRANSFAC, TRRD and COMPEL: towards a federated database system on transcriptional regulation. *Nucleic Acids Res* 25, 265–268.

```
    CENTERS ()
1.  if all alignment columns are trivial then
2.    Report that the entire alignment is a 1-region and exit
3.  Initialize the queue to column 0 and the first non-trivial column
4.  while the column at the tail of the queue is not n+1 do
5.    Enqueue the next non-trivial column
6.    if the new column is not n+1 then
7.      Determine the status of the last column pair
8.    if the status is EXCLUSIVE or the last column is n+1 then
9.      Report the queue contents
10.     Dequeue all but the last two queue entries
11.   else if the status is AMBIGUOUS then
12.     CLEAR-QUEUE
13.     Select one of the two centers for the last two columns
14.     Set the mismatch count to 1, for each row
   { Otherwise the status is UNAMBIGUOUS, so we examine the sub-cases }
15.   else if the queue contains 3 entries then
16.     Set the mismatch counts according to the last two columns in the queue
17.   else if next-to-last queue entry's center characters are inconsistent then
18.     CLEAR-QUEUE
19.     Set center characters and mismatch counts for the last 2 queue entries
20.   else
21.     Update the center character and mismatch counts for the new column
22.     if some row has two mismatches with the center then
23.       Report the queue contents
24.       while some row has two mismatches do
25.         Dequeue the column at the queue head
26.         Update the mismatch vector to disregard the new queue head

    CLEAR-QUEUE ()
1.  if the queue contains more than 3 columns then
2.    Report the queue contents
3.    Dequeue all but the last three entries
```

Fig. 1. Pseudo-code for the procedure CENTERS for reporting all maximal 1-regions in a given alignment.

On Some Geometric Optimization Problems in Layered Manufacturing*

Jayanth Majhi[1]   Ravi Janardan[1]   Michiel Smid[2]   Prosenjit Gupta[3]

[1] Dept. of Computer Science, Univ. of Minnesota, Minneapolis, MN 55455, U.S.A.
{majhi,janardan}@cs.umn.edu
[2] Fakultät für Informatik, Otto-von-Guericke-Universität Magdeburg, D-39106
Magdeburg, Germany.
michiel@isg.cs.uni-magdeburg.de
[3] Bell Laboratories, 600 Mountain Avenue, Murray Hill NJ 07974, U.S.A.
prosenjit@lucent.com

Abstract. Efficient geometric algorithms are given for optimization problems arising in layered manufacturing, where a 3D object is built by slicing its CAD model into layers and manufacturing the layers successively. The problems considered include minimizing the degree of stair-stepping on the surfaces of the manufactured object, minimizing the volume of the so-called support structures used, and minimizing the contact area between the supports and the manufactured object—all of which are factors that affect the speed and accuracy of the process. The stair-step minimization algorithm is valid for any polyhedron, while the support minimization algorithms are applicable to convex polyhedra only. Algorithms are also given for optimizing supports for non-convex, simple polygons. The techniques used include construction and searching of certain arrangements on the sphere, 3D convex hulls, halfplane range searching, ray-shooting, visibility, and constrained optimization.

1 Introduction

This paper describes efficient algorithms for certain geometric optimization problems arising in layered manufacturing. In *layered manufacturing*, a physical prototype of a 3D object is built from a (virtual) CAD model by orienting and slicing the model with parallel planes and then manufacturing the slices one by one, each on top of the previous one. Layered manufacturing is the basis of an emerging technology called *Rapid Prototyping and Manufacturing* (RP&M). This technology, which is used extensively in the automotive, aerospace, and medical industries, accelerates dramatically the time it takes to bring a product to the market because it allows the designer to create rapidly a physical version of the CAD model (literally on the desktop) and to "feel and touch" it, thereby detecting and correcting flaws in the model early on in the design cycle [13].

---

* JM's and RJ's work supported, in part, by a Grant–in–Aid of Research from the Graduate School of the Univ. of Minnesota. MS's work done, in part, while at the Dept. of Computer Science, King's College London, UK. PG's work done, in part, while at MPI-Informatik, Saarbrücken, Germany and at the Univ. of Minnesota.

Although there are many types of layered manufacturing processes, the basic principle underlying them all is as outlined above. Therefore, for concreteness, we will focus here on just one such method, called *StereoLithography*, which dominates the RP&M market [13].[4]

1.1 StereoLithography

The input to StereoLithography is a surface triangulation of the CAD model in a format called STL. The triangulated model is oriented suitably, sliced by xy-parallel planes, and then built slice by slice in the positive z direction, as follows:

In essence, the StereoLithography Apparatus (SLA) consists of a vat of photocurable liquid resin, a platform, and a laser. Initially, the platform is below the surface of the resin at a depth equal to the slice thickness. The laser traces out the contour of the first slice on the surface and then hatches the interior, which hardens to a depth equal to the slice thickness. In this way, the first slice is created and it rests on the platform. (See Figure 1.)

Next, the platform is lowered by the slice thickness and the just-vacated region is re-coated with resin. The second slice is then built in the same way. Ideally, each slice after the first one should rest in its entirety on the previous one. In general, however, portions of a slice can overhang the previous slice and so additional structures, called *supports*, are needed to hold up the overhangs. Supports are generated automatically during the process itself. For this the CAD model is analyzed beforehand and a description of the supports is generated and merged into the STL file. Supports come in shapes such as wedges, cylinders, and rectangular blocks.

Once the solid has been made, it is postprocessed to remove the supports. Additional postprocessing is often necessary to improve the finish, which has a stair-stepped appearance on certain surfaces due to the non-zero slice thickness used. (See Figure 2, which shows the cross-section of a long polyhedral "cylinder": the facets corresponding to the edges $\overline{12}, \overline{23}, \ldots \overline{41}$ are normal to the paper, so the cross-section is uniform.)

1.2 Issues

A key step in layered manufacturing is choosing an orientation for the model, i.e., the *build direction* [13]. Among other things, the build direction affects the quantity of supports used and the surface finish—factors which impact the speed and accuracy of the process. As a simple example, consider building the object in Figure 2: If it is built in the direction **d** shown, then the manufactured solid will have a stair-stepped finish and will require supports. However, if it is built in a direction normal to the paper, no supports are needed and there is no stair-stepping. In current systems, the build direction is often chosen by the human

---

[4] The recent report of the Computational Geometry Task Force identifies this process as one where geometric techniques could play a significant role [7, page 31].

operator, based on experience, so that the amount of supports used is "small" and the surface finish is "good". We seek to design computer algorithms that optimize these criteria automatically and lessen the need for human intervention.[5]

Let us define more formally the parameters of our problem. Throughout the paper, we denote by $\mathcal{P}$ the polyhedral object that we wish to build and by n the number of vertices in $\mathcal{P}$. (Note that there is no loss of generality in assuming a polyhedral model since the input—the STL representation—is polyhedral, even if the original part is not.) We let $\mathbf{d}$ denote the build direction and, for convenience, imagine it to be vertical so that notions such as "above" and "below" have their usual meaning. *Our problem is to find a* $\mathbf{d}$ *which minimizes the following three parameters, considered independently (i.e., in isolation from one another):*

Degree of stair-stepping: Due to the non-zero slice thickness, the manufactured part will have a stair-stepped finish on any facet f that is not parallel to $\mathbf{d}$. (See Figure 2. Notice that there is no stair-stepping on the facet corresponding to edge $\overline{34}$ since it is parallel to $\mathbf{d}$.) The degree of stair-stepping on a facet f depends on the angle, $\theta_f(\mathbf{d})$, between the facet normal and $\mathbf{d}$, and it can be mitigated by a suitable choice of $\mathbf{d}$. In [4], the notion of an *error-triangle* for a facet is introduced as a way of quantifying stair-stepping (Figure 2). We define the degree of stair-stepping as the maximum height of the error triangle, taken over all facets.

Volume of supports: The quantity of supports used affects both the building time and the cost. If $\mathcal{P}$ is convex, then the support volume is the volume of the region lying between $\mathcal{P}$ and the platform, i.e., the vertical polyhedral "cylinder" which is bounded below by the platform and above by the facets of $\mathcal{P}$ whose outward normals point downward. If $\mathcal{P}$ is non-convex, then the problem is more complex, as the supports for some facets may actually be attached to other facets instead of to the platform. (Figure 3 shows this—in 2D, for convenience.)

Contact area of supports: The amount of $\mathcal{P}$'s surface that is in contact with the supports affects the postprocessing time, since the supports that "stick" to $\mathcal{P}$ must be removed. If $\mathcal{P}$ is convex, then this is the total area of the downward-facing facets. If $\mathcal{P}$ is non-convex, then this is the total facet area that is in contact with the supports (it includes the areas of all downward facing facets and portions of certain upward-facing facets.)

1.3 Results

Our main results include:

(1) A simple and practical algorithm for minimizing the maximum height of the error-triangles of the facets. The algorithm runs in time $O(n \log n)$, with the most involved step being merely the computation of the convex hull of a point-set in 3D. The algorithm works for any polyhedron—even one which is non-convex and has holes.

---

[5] There is a commercial software package called Bridgeworks for generating supports. The algorithms it uses are proprietary and it is unclear whether they optimize the supports in any way.

(**2**) An $O(n^2)$-time algorithm for computing a build direction **d** which minimizes the volume of supports needed by a convex polyhedron $\mathcal{P}$. The algorithm consists of constructing a certain arrangement on the unit-sphere, $\mathbf{S}^2$, whose regions represent directions for which the combinatorial structure of the supports is invariant. We then show how to write the volume formula for a region in a way that allows us to quickly update it in an incremental fashion as we move from region to region. Within each region, we find the best direction using the method of Lagrangian Multipliers.

(**3**) An $O(n^2)$-time algorithm for minimizing the surface area of a convex polyhedron $\mathcal{P}$ that is in contact with supports. This algorithm involves computing a certain arrangement on $\mathbf{S}^2$, with weighted faces and finding the lightest face. We transform the problem to the plane and solve it using topological sweep [9]. We also give a faster algorithm for the case where $\mathcal{P}$ is built so that an entire facet is in contact with the platform (this is usually the case in practice because it provides more stability to the part). This algorithm runs in roughly $O(n^{4/3})$ time (ignoring polylog factors), and is based on transforming the supports problem to a halfplane range counting problem on weighted points in 2D.

(**4**) Efficient support optimization algorithms for 2D non-convex polygons. For an n-vertex polygon, we show that a build direction which minimizes the contact length (resp. the area) of the supports can be found in time $O(n \log n + np(n))$ (resp. $O(n^2 + nq(n))$), where $p(n)$ and $q(n)$ are the times it takes to compute the roots of certain polynomials of degree $\Theta(n)$ over the real line. ("Area" and "contact length", as used here, are the 2D counterparts of "volume" and "contact area".) If the polygon's sides have only a constant number of different orientations, then the time bounds improve to $O(n \log n)$ and $O(n^2)$, respectively. These results are based on visibility and ray-shooting techniques and are quite intricate. We view them as a first step towards solving the corresponding problems in 3D, which appear to be considerably more difficult.

(**5**) Evidence that the quadratic algorithms in (2) and (3) above for support optimization may not be improvable. We show that a closely related problem belongs to the class of *3SUM-hard problems* [11] for which no subquadratic algorithms have been found despite much effort. Specifically, we show that, for a convex polyhedron, it is 3SUM-hard to find a build direction **d** with a positive z-component which minimizes the number of facets needing support.

1.4 Prior work

To our knowledge, the only formal algorithmic work in layered manufacturing is by Asberg *et al.* [3] (see also [5]), where linear-time algorithms are given to decide if a given polygonal or polyhedral model can be built by StereoLithography without supports. If the answer is in the affirmative, then these algorithms also reports all such feasible build directions. There is a large body of experimental/heuristic work, of which we mention only a few for lack of space. The issue of optimizing supports (and surface finish) is raised by Frank and Fadel [10], and heuristics are given within the framework of an expert system. Bablani and Bagchi [4] quantify the notion of stair-stepping error and give an approach to compute it, assuming that the build direction is given. Allen and Dutta [1] give a heuristic for minimizing contact-area but do not analyze the running time or the quality of their approximation.

For lack of space, we will discuss only results (1)–(3) and parts of result (4) here, and will omit most details; a complete discussion can be found in [15].

2 Minimizing stair-stepping

Recall our problem: "*Given a polyhedron $\mathcal{P}$ with n vertices, find a direction $\mathbf{d}$ which minimizes the maximum error-triangle height*". (Figure 2.) Let L denote the slice thickness and let $h_f(\mathbf{d})$ denote the height of the error-triangle, $t_f(\mathbf{d})$, for facet f. Let $\theta'_f(\mathbf{d})$ be the angle between $\mathbf{d}$ and the normal, $\mathbf{n}_f$, to f, and let $\theta''_f(\mathbf{d})$ be the angle between $\mathbf{d}$ and $-\mathbf{n}_f$. Let $\theta_f(\mathbf{d}) = \min(\theta'_f(\mathbf{d}), \theta''_f(\mathbf{d}))$. It is easy to check that $h_f(\mathbf{d}) = L\cos\theta_f(\mathbf{d})$. We seek a direction $\mathbf{d}$ such that $\max_f h_f(\mathbf{d})$ is minimized, i.e., $\min_f \theta_f(\mathbf{d})$ is maximized.

Consider the set $S = \{\mathbf{n}_f \cap \mathbf{S}^2, -\mathbf{n}_f \cap \mathbf{S}^2 | f$ is a facet of $\mathcal{P}\}$. That is, S consists of the points where the facet normals and their negations intersect the unit-sphere $\mathbf{S}^2$. Note that S has $O(n)$ points or *sites*. We wish to find a direction $\mathbf{d}$, i.e., a point $\mathbf{d}$ on $\mathbf{S}^2$, such that the minimum angle between it and the sites is maximized. Define a *cap* on $\mathbf{S}^2$, with *pole* $\mathbf{d}$ and *radius* θ, as the set of all points on $\mathbf{S}^2$ that are at a distance of at most θ from $\mathbf{d}$, as measured along the surface of $\mathbf{S}^2$. Clearly, our problem is equivalent to finding the largest radius cap which does not contain any site in its interior, i.e., a largest empty cap; the pole of this cap is the desired optimal direction. The following properties of caps are easily shown:

1. Let c be the circle bounding a cap C and let $H(C)$ be the plane such that $c = H(C) \cap \mathbf{S}^2$. If C is empty, then all the sites in S lie on the same side of $H(C)$. Conversely, every such plane which intersects $\mathbf{S}^2$ corresponds to an empty cap.
2. The larger C is, the closer is $H(C)$ to the origin.
3. A largest empty cap must have at least three sites on its boundary.

By 1 and 2 above, we need to find a plane that is (a) closest to the origin and (b) has all the sites on one side of it. Let $CH(S)$ be the convex hull of S. By 3, it suffices to consider only the facets of $CH(S)$ when searching for a candidate plane. This follows because the plane containing any facet of $CH(S)$ must contain at least three sites and, moreover, all the sites of S lie on one side of this plane; on the other hand, the plane containing three or more co-planar sites that are not all on a facet of $CH(S)$ will have sites on both sides of it.

Therefore, our algorithm is as follows: We first compute the set S and then compute $CH(S)$. For each facet of $CH(S)$, we determine the plane containing the facet and find the one closest to the origin. We then compute the normal from the origin to this closest plane. The desired optimal direction $\mathbf{d}$ is the intersection of this normal and $\mathbf{S}^2$. The overall time is dominated by the $O(n \log n)$ time for the convex hull computation.

Theorem 1. *Let $\mathcal{P}$ be an n-vertex polyhedron. A direction $\mathbf{d}$ which minimizes the maximum error-triangle height can be found in time $O(n \log n)$.* $\square$

Remark 1 The algorithm makes no assumptions about $\mathcal{P}$ and hence works for any polyhedron. The algorithm does, however, assume that the slice thickness, L, is fixed. Consider building a rectilinear polyhedron P, i.e., one whose facets are all mutually orthogonal. If P is built in a direction $\mathbf{d}$ such that the facets are all either normal or parallel to $\mathbf{d}$, then there will be stair-stepping on those facets whose distance from the platform is not a multiple of L. (The height of the stair-step on these facets can be close to L.) Our algorithm would reduce this error by building P in a different, and, perhaps, less natural orientation. An interesting problem is to incorporate variable slice thicknesses in our algorithm.

3 Minimum-volume supports for a convex polyhedron

For a build direction $\mathbf{d}$, we call the facets of $\mathcal{P}$ whose outward normals make an angle greater than $90°$ with $\mathbf{d}$ the *back facets* of $\mathcal{P}$. Call the vertex v of $\mathcal{P}$ that is farthest away in direction $-\mathbf{d}$ the *extreme vertex* of $\mathcal{P}$. Thus, when $\mathcal{P}$ is built in direction $\mathbf{d}$, v rests on the platform and the facets requiring support are the back facets. The support volume is the volume of the polyhedron which is bounded below by the platform, above by the back facets, and on the sides by vertical facets that contain the edges on the boundary of the union of the back facets.

Our approach consists of partitioning 3-space into $O(n^2)$ regions, R, such that the back facets and the extreme vertex are the same for all directions $\mathbf{d} \in R$. This partition can be represented as an arrangement $\mathcal{A}$ on the unit-sphere $\mathbf{S}^2$. We generate a formula for the total support volume w.r.t. any $\mathbf{d} \in R$ and set up an optimization problem for finding the build direction which minimizes the total support volume within R. We solve this problem, in time $O(|R|)$, by using the Lagrangian Multipliers method [14] and exploiting the convexity of R. This implies that the total time for all regions is $O(n^2)$.

However, we must also set up the volume formula within each region R. Doing this in the straightforward way takes $O(n)$ time per region, hence $O(n^3)$ time overall. We circumvent this by visiting the regions of the arrangement in a certain order and updating the formula incrementally. For this, we rewrite the formula as two parts—one based mainly on the extreme vertex v for R and the other based on the back facets. We then show that after doing an $O(n^2)$-time precomputation, the formula in this new form can be updated incrementally in $O(n^2)$ total time as we visit the regions of $\mathcal{A}$.

Here is our approach in more detail: Using the method in [17], we partition 3-space into unbounded polyhedral regions called *cones*, each with its apex at the origin and such that for all directions within it the set of back facets is the same. Let $\mathcal{A}'$ be the arrangement obtained by intersecting these cones with $\mathbf{S}^2$. $\mathcal{A}'$ is essentially an arrangement of great circles on $\mathbf{S}^2$, and hence has size $O(n^2)$ and can be computed in time $O(n^2)$ using the algorithm given in [8].

Next using the algorithm of Bose *et al.* [6], we compute a second arrangement, $\mathcal{A}''$, which consists of portions of great circles (i.e., great arcs) on $\mathbf{S}^2$ such that all directions within a region of $\mathcal{A}''$ have the same extreme vertex. The desired arrangement, $\mathcal{A}$, is the intersection of $\mathcal{A}'$ and $\mathcal{A}''$. It can be computed in $O(n^2)$ time by extending the arcs of $\mathcal{A}''$ to great circles, intersecting the $O(n)$ great circles of $\mathcal{A}'$ and $\mathcal{A}''$, and then performing a depth-first search of the resulting arrangement to remove the edges that correspond to arc extensions.

We also need for each region $R \in \mathcal{A}$ the set of back facets and the extreme vertex. Rather than storing this information explicitly, which would take $O(n^3)$ total space, we compute it incrementally while updating the volume formula, as described in Section 3.4.

3.1 The volume formula

Given a region $R \in \mathcal{A}$, we determine a formula for the volume, $V(R)$, of the supports required by $\mathcal{P}$ for any direction $\mathbf{d} \in R$. Let $\mathbf{d} = x\mathbf{i} + y\mathbf{j} + z\mathbf{k}$ be any unit vector within R. Let $\mathbf{v}$ be the extreme vertex for R. Consider one of the back facets, say, f. Let the vertices of f be $P_0, P_1, \ldots, P_{m-1}$ in counterclockwise order looking from outside the polyhedron and let $\mathbf{n}_f$ be the unit-outer-normal to f. Let f project onto convex polygon $P_0', P_1', \ldots, P_{m-1}'$ on the plane passing through $\mathbf{v}$ and normal to $\mathbf{d}$. The volume of the support needed by f is then the volume of the polyhedron P_f shown in Figure 4.

Let the facets of P_f be $S_0, S_1, \ldots, S_{m+1}$, where S_0 is the top facet, S_1 is the bottom facet, and S_i, $2 \le i \le m+1$, is a side facet, i.e., $S_i = P_{i-2}P_{i-1}P_{i-1}'P_{i-2}'$. Let $\mathbf{Q}_j$ be any point on S_j, and let $\mathbf{N}_j$ be a unit outward normal vector to S_j. Then, using the formula in [12], the volume, $V_f(R)$, of P_f is given by $V_f(R) = \frac{1}{3} \sum_j (\mathbf{Q}_j \cdot \mathbf{N}_j) Area(S_j)$. In the full paper [15], we show:

$$V_f(R) = -\frac{1}{6}(\mathbf{P}_0 \cdot \mathbf{n}_f)(\mathbf{n}_f \cdot \boldsymbol{\Delta}_f) + \frac{1}{6}\sum_{i=1}^{m}(\mathbf{d} \cdot (\mathbf{P}_i \times \mathbf{P}_{i-1}))(\mathbf{d} \cdot (\mathbf{P}_i + \mathbf{P}_{i-1})) +$$

$$\frac{1}{2}(\mathbf{d} \cdot \boldsymbol{\Delta}_f)(\mathbf{d} \cdot \mathbf{v})$$

$$= V_f'(R) + \frac{1}{2}(\mathbf{d} \cdot \boldsymbol{\Delta}_f)(\mathbf{d} \cdot \mathbf{v}),$$

where $\boldsymbol{\Delta}_f = \sum_{i=0}^{m-1} \mathbf{P}_i \times \mathbf{P}_{i+1}$ is called the *area term* for f and $V_f'(R)$ denotes the part of $V_f(R)$ which is independent of $\mathbf{v}$. As we will see in Section 3.4, being able to decompose the formula in this way is crucial to the running time of the algorithm. Therefore, the total support volume, $V(R) = \sum_f V_f(R)$, associated with region $R \in \mathcal{A}$ is:

$$V(R) = \sum_f V_f'(R) + \frac{1}{2}(\mathbf{d} \cdot \boldsymbol{\Delta}(R))(\mathbf{d} \cdot \mathbf{v}),$$

where $\boldsymbol{\Delta}(R) = \sum_f \boldsymbol{\Delta}_f$ is the total area term for all the back facets associated with R.

3.2 The optimization problem

When we expand $V(R)$ we get an expression of the form $Ax^2 + By^2 + Cz^2 + Dxy + Eyz + Fxz + G$, where $A, B, C, \ldots, G$ are constants depending only on the P_i's of the different back facets of R and on the extreme vertex v of R.

Let $\mathbf{d}_R \in \mathbf{S}^2$ be a given point in R's interior. ($\mathbf{d}_R$ is computed when $\mathcal{A}$ is constructed.) For any great arc a bounding R, let $h(a)$ be the plane containing the corresponding great circle and let $\mathbf{n}_{h(a)}$ be a unit-normal for it. Note that $\mathbf{d}$ is in the interior of R if and only if $\mathbf{d} \cdot \mathbf{n}_{h(a)}$ and $\mathbf{d}_R \cdot \mathbf{n}_{h(a)}$ are both positive or both negative for each great arc a bounding R.

Thus our optimization problem within each region, R, is:

Minimize $\quad f(x, y, z) = Ax^2 + By^2 + Cz^2 + Dxy + Eyz + Fxz$

Subject to $\quad x^2 + y^2 + z^2 = 1 \quad$ (*Sphere Constraint.*)

$\qquad\qquad \mathbf{d} \cdot \mathbf{n}_{h(a)} \geq 0$ (resp. ≤ 0) if $\mathbf{d}_R \cdot \mathbf{n}_{h(a)} \geq 0$ (resp. ≤ 0) for each great arc a bounding R. (*Plane Constraints.*)

3.3 Solving the optimization problem

We use the method of Lagrangian Multipliers [14]. The optimization problem for a region R has $|R| + 1$ constraints, and, in general, could take $O(2^{|R|})$ time to solve. But in our case R is convex and this leads to an $O(|R|)$-time solution.

We proceed in three stages: (i) We first keep only the sphere constraint active and find the extreme points (i.e., minimum or maximum) over all of $\mathbf{S}^2$. (ii) Next, we take some arc a bounding R and make the corresponding plane constraint active as well. This gives extreme points lying on a's great circle. We repeat this for each great arc a bounding R. (iii) Finally, we consider arcs a and a' meeting at a vertex v of R and make the corresponding plane constraints active—thus making v an extreme point. Note that it is not necessary to make more than two plane constraints active since there is no point of R that is common to three great circles.

For lack of space, we discuss only stage (i) in some detail: The Lagrangian is $L(x, y, z, \lambda) = f(x, y, z) + \lambda(1 - x^2 - y^2 - z^2)$, for some parameter λ. The partial derivatives of L, w.r.t. each of x, y, and z, must be zero at an extreme point. This yields three linear equations in x, y, and z. The values of λ for which these three equations have non-trivial solutions can be found by solving a cubic equation in λ, given by:

$$\begin{vmatrix} 2A - 2\lambda & D & F \\ D & 2B - 2\lambda & E \\ F & E & 2C - 2\lambda \end{vmatrix} = 0.$$

For each such real-valued λ (there are at most three of them) we solve for x, y, and z, using any two of the three linear equations (the remaining one will depend on the two chosen) and the sphere constraint. This will yield, in general, two antipodal points on $\mathbf{S}^2$. If either of these lies in R, then we take this to be a candidate for the minimum value of $f(x, y, z)$. Clearly, all this can be done in $O(1)$ time. The remaining stages also take $O(1)$ time each, so that the optimization problem for R takes time $O(|R|)$. Summed over all regions R, this is $O(n^2)$.

3.4 Updating the volume formula incrementally

We precompute Δ_f for each facet $f \in \mathcal{P}$. We pick an initial region, $R_0 \in \mathcal{A}$, and compute its back facets, its extreme vertex and $V(R_0)$ and $\Delta(R_0)$, all in $O(n)$ time. Starting from R_0, we visit the regions of $\mathcal{A}$ in the order given by a depth-first search of the dual graph of $\mathcal{A}$. Suppose that the search visits region R' from region R. There are three cases:

1. A facet f which was a back facet for R ceases to be a back facet for R'.
2. A facet f which was not a back facet for R becomes a back facet for R'.
3. The extreme vertex v changes to v'.

In cases 1 and 2, the extreme vertex is unchanged while in case 3 the set of back facets is unchanged. In case 1, we obtain $V(R')$ from $V(R)$ by updating $V'_f(R)$, for the given f, set $\Delta(R') = \Delta(R) - \Delta_f$, and obtain the back facets of R' by removing f from R's set of back facets. All this takes $O(f)$ time. Case 2 is similar. In case 3, we update $V(R)$ to $V(R')$ by subtracting $\frac{1}{2}(\mathbf{d} \cdot \Delta(R))(\mathbf{d} \cdot \mathbf{v})$ and adding $\frac{1}{2}(\mathbf{d} \cdot \Delta(R'))(\mathbf{d} \cdot \mathbf{v}')$ (noting that $\Delta(R') = \Delta(R)$ here), and set v' to be the extreme vertex of R'.

It can be shown that f appears or disappears as a back facet only $O(n)$ times, so the total time for cases 1 and 2 is $O(\sum_{f \in \mathcal{P}} n|f|) = O(n^2)$. The total time for case 3 is clearly $O(n^2)$.

Theorem 2. *Let $\mathcal{P}$ be a convex polyhedron of n vertices. A build direction minimizing the total volume of supports needed by $\mathcal{P}$ can be computed in time $O(n^2)$.* $\square$

4 Minimum-contact-area supports for a convex polyhedron

Facet $f \in \mathcal{P}$ needs support w.r.t. $\mathbf{d}$ iff it is a back facet. Thus, the set of directions for which f needs support forms an open hemisphere, h_f, on $\mathbf{S}^2$, with pole $-\mathbf{n}_f$. If we associate with h_f a weight equal to the area of f, then our problem is to find a point on $\mathbf{S}^2$ covered by hemispheres, h_f, of minimum total weight. Let h_f^+ be the part of h_f above the equator of $\mathbf{S}^2$. Using *central projection* [18], we map h_f^+ to an open halfplane, ℓ_f^+, of the same weight, on the plane $z = 1$. We now need to find a face in the arrangement of the ℓ_f^+'s which is covered by halfplanes of minimum total weight. This can be done in $O(n^2)$ time and $O(n)$ space using topological sweep [9, pages 181–182]. Similarly for the parts of the h_f^+'s below the equator of $\mathbf{S}^2$.

Theorem 3. *Let $\mathcal{P}$ be an n-vertex convex polyhedron. A direction $\mathbf{d}$ which minimizes the total surface area of $\mathcal{P}$ in contact with supports can be found in time $O(n^2)$, using $O(n)$ space.* $\square$

4.1 A faster algorithm when building $\mathcal{P}$ on a facet

To build $\mathcal{P}$ on a facet f, we must choose $-\mathbf{n}_f$ as the build direction. Let h_f be the closed hemisphere on $\mathbf{S}^2$ with pole $-\mathbf{n}_f$. Then, a facet $f' \neq f$ will need support iff $\mathbf{n}_{f'}$ is not contained in h_f. Let $C = \{\mathbf{n}_f | f \text{ is a facet of } \mathcal{P}\}$ and let $C' = \{-\mathbf{n}_f | f \text{ is a facet of } \mathcal{P}\}$. Associate with each point $\mathbf{n}_f \in C$ a weight equal to f's area. Clearly, our problem is to find a point $-\mathbf{n}_f \in C'$ such that the total weight of the points $\mathbf{n}_{f'} \in C$ that are not in h_f is minimized, or, equivalently, the total weight of the points $\mathbf{n}_{f'} \in C$ that are in h_f is maximized.

We reformulate our problem as follows: We are given $r = O(n)$ weighted *blue points* and r *red points* on $\mathbf{S}^2$, corresponding to the points in C and C', respectively. Each red point is the pole of a closed *red hemisphere*. We wish to find the red hemisphere which contains blue points of maximum total weight.

Let $P = (p_1, p_2, p_3)$ be any blue point. Assume w.l.o.g. $p_2 \neq 0$ for any blue point. Let $H : z = ax + by$ be the bounding plane of a red hemisphere. We map H to the red point $H' = (a, b)$ and map P to the blue line $P' : y = (-p_1/p_2)x + (p_3/p_2)$, both in the plane. Next, using standard point-line duality [18], we map H' to a red line H'' and P' to a blue point P''.

Lemma 4. *(A): P is on or above H iff ($p_2 > 0$ and P'' is on or below H'') or ($p_2 < 0$ and P'' is on or above H'').*
(B): P is on or below H iff ($p_2 > 0$ and P'' is on or above H'') or ($p_2 < 0$ and P'' is on or below H''). $\square$

We refer to the red hemispheres that have their poles above the equator of $\mathbf{S}^2$ as *upper red hemispheres* and to those that have their poles below as *lower red hemispheres*. The blue points P that are contained in any upper red hemisphere all lie on or above the plane H for that hemisphere. Therefore, by Lemma 4(A), we must find a red line H'' such that the total weight of the blue points P'' with $p_2 > 0$ that are on or below H'' plus the total weight of the blue points P'' with $p_2 < 0$ that are on or above H'' is maximum. This is a halfplane range counting problem on weighted points in 2D.

We build two data structures D^+ and D^-: D^+ is the halfplane range counting structure in [16], built on the blue points P'' with $p_2 > 0$. D^- is a similar structure for the blue points P'' with $p_2 < 0$. For each upper red hemisphere with bounding plane H, we query D^+ with the lower halfplane of H'', query D^- with the upper halfplane of H'', and sum up the total weights returned. We handle the lower red hemispheres similarly, using Lemma 4(B).

Theorem 5. *Let $\mathcal{P}$ be an n-vertex convex polyhedron that is to be built on a facet. A build direction minimizing the total area of $\mathcal{P}$ that is in contact with supports can be found in time $O(n^{4/3}(\log n)^\gamma)$ time and space $O(n^{4/3}/(\log n)^{2\gamma})$, where $\gamma > 0$ is an arbitrarily small real number.*

Proof. Let m $(n \leq m \leq n^2)$ be a parameter. The total time taken is $O(m(\log n)^\delta)$, to build the data structure (where $\delta > 0$ is an arbitrarily small real [16]), plus $O(n/m^{1/2})$ for each of the $O(n)$ queries. The space is $O(m)$. (See [16].) We choose $m = n^{4/3}/(\log n)^{2\delta/3}$ and $\gamma = \delta/3$. $\square$

5 Length-optimal supports for a simple, non-convex polygon

Here we let $\mathcal{P}$ denote an n-vertex simple, non-convex polygon. We wish to find a build direction $\mathbf{d}$ which minimizes the total length of the boundary of $\mathcal{P}$ to which supports "stick". The main difficulty here is that some of the supports may actually be attached to other parts of $\mathcal{P}$, instead of being attached to the platform. (See Figure 3.) Our approach is similar to that used in [2] for a related problem.

An edge $e \in \mathcal{P}$ *needs support for direction* $\mathbf{d}$ iff the angle between $\mathbf{d}$ and e's outward normal, $\mathbf{n}_e$, is more than $90°$; in this case, the entire length of e needs support. However, even if the angle is at most $90°$, e may be *attached to supports for direction* $\mathbf{d}$ for some or all of its length. In Figure 5, edge f needs support, while edge $e = \overline{a_e b_e}$ is attached to supports for the indicated $\mathbf{d}$. Let $S_e(\mathbf{d})$ be the set of points on e that are attached to a support for direction $\mathbf{d}$.

Lemma 6. $S_e(\mathbf{d})$ *is (i) empty, (ii) consists of one segment on e, one of whose endpoints is an endpoint of e, (iii) consists of two segments on e, and each segment has an endpoint which is an endpoint of e, or (iv) is equal to the entire edge e.* $\square$

Let α_e be the angle between the positive x-axis and $\mathbf{n}_e$ and let $\phi_\mathbf{d}$ be the angle between the positive x-axis and $\mathbf{d}$. Let $\overline{a_e A_e(\mathbf{d})}$ and $\overline{b_e B_e(\mathbf{d})}$ be the segments alluded to in Lemma 6. If we shoot rays from $A_e(\mathbf{d})$ and $B_e(\mathbf{d})$, in direction $\mathbf{d}$, then these rays will (necessarily) hit vertices of $\mathcal{P}$; call these vertices $A'_e(\mathbf{d})$ and $B'_e(\mathbf{d})$, respectively. (See Figure 5.)

Denote by $l_e(\mathbf{d})$ the length of the supports associated with e: If e needs supports for direction $\mathbf{d}$, then $l_e(\mathbf{d})$ is the length of e; otherwise, if e is attached to supports for direction $\mathbf{d}$, then $l_e(\mathbf{d})$ is the total length of the segments in $S_e(\mathbf{d})$. We wish to find a $\mathbf{d}$ which minimizes $L_P(\mathbf{d}) = \sum_{e \in \mathcal{P}} l_e(\mathbf{d})$.

In the full paper [15], we show that $l_e(\mathbf{d}) = X_e + Y_e \tan(\alpha_e - \phi_\mathbf{d})$, where X_e and Y_e are (possibly negative) real numbers that only depend on e and the points $A'_e(\mathbf{d})$ and/or $B'_e(\mathbf{d})$.

If we change $\phi_\mathbf{d}$ slightly, then X_e and Y_e, hence the expression for $l_e(\mathbf{d})$, do not change; For some angles $\phi_\mathbf{d}$, however, X_e and Y_e will change. Call the corresponding directions *critical*. We wish to find these critical directions and thereby partition the interval $[0, 2\pi]$ of directions into subintervals, such that within each subinterval, I, the function $l_e(\mathbf{d})$ can be written as

$$l_e(\mathbf{d}) = X_e^I + Y_e^I \tan(\alpha_e - \phi_\mathbf{d}),$$

where X_e^I and Y_e^I are constant within I.

To find this partition, we follow [2] and define for each vertex v of $\mathcal{P}$, its *visibility cone*, $cone(v)$, as the cone with apex v and maximum angular range in which v can see to infinity. In [15], we show:

Lemma 7. *Let $\mathbf{D}_a$ be the set of directions determined by the bounding rays of the non-empty visibility cones. Let $\mathbf{D}_b$ contain for each edge e the two directions that are parallel to e. Then $\mathbf{D} = \mathbf{D_a} \cup \mathbf{D_b}$ is a set of $O(n)$ directions which includes all critical directions.* $\square$

Here is an outline of the algorithm (details can be found in [15]): First, we compute and sort $\mathbf{D}$ in $O(n \log n)$ time, using ray-shooting techniques to compute the visibility cones. Next, we pick the first subinterval I, compute $L_P(\mathbf{d}) = \sum_{e \in \mathcal{P}} (X_e^I + Y_e^I \tan(\alpha_e - \phi_\mathbf{d}))$, and minimize this function over I. Using standard techniques from calculus, this minimization involves computing the roots of a certain polynomial of degree $\Theta(n)$. Let $p(n)$ be the time to do this. Finally, we sweep over the $O(n)$ subintervals and, within each, we update the expression for $L_P(\mathbf{d})$ in $O(\log n)$ time, using ray-shooting techniques again, and perform a minimization. Thus the total time is $O(n \log n + np(n))$. If $\mathcal{P}$ is C-*oriented*, i.e., its sides have only a constant number, C, of different orientations, then the polynomial whose roots we wish to compute can be written so that it has degree $2C - 2 = \Theta(C)$. Under the reasonable assumption that the roots of this polynomial can be found in $O(1)$ time, the total time reduces to $O(n \log n)$.

Theorem 8. *For a simple, non-convex n-vertex polygon $\mathcal{P}$, a direction $\mathbf{d}$ which minimizes the total length of the supports that are in contact with $\mathcal{P}$ can be found in time $O(n \log n + np(n))$, where $p(n)$ is the time to compute the roots of a certain polynomial of degree $\Theta(n)$. If $\mathcal{P}$ is C-oriented, for some constant C, then the algorithm takes $O(n \log n)$ time.* $\square$

References

1. S. Allen and D. Dutta. Determination and evaluation of support structures in layered manufacturing. *Journal of Design and Manufacturing*, 5:153–162, 1995.
2. E.M. Arkin, Y.-J. Chiang, M. Held, J.S.B. Mitchell, V. Sacristan, S.S. Skiena, and T.-C. Yang. On minimum-area hulls. In *Proceedings of the 1996 European Symposium on Algorithms, LNCS 1136, Springer-Verlag, Berlin*, pages 334–348, 1996.
3. B. Asberg, G. Blanco, P. Bose, J. Garcia-Lopez, M. Overmars, G. Toussaint, G. Wilfong, and B. Zhu. Feasibility of design in stereolithography. In *Proceedings of the 13th FST&TCS Conference*, 1993.
4. M. Bablani and A. Bagchi. Quantification of errors in rapid prototyping processes and determination of preferred orientation of parts. In *Transactions of the 23rd North American Manufacturing Research Conference*, 1995.
5. P. Bose. *Geometric and computational aspects of manufacturing processes*. PhD thesis, School of Computer Science, McGill University, Montréal, Canada, 1995.
6. P. Bose, M. van Kreveld, and G. Toussaint. Filling polyhedral molds. In *Proc. 3rd Workshop Algorithms Data Structures*, volume 709 of *Lecture Notes in Computer Science*, pages 210–221. Springer-Verlag, 1993.

7. B. Chazelle et al. Application challenges to computational geometry. Technical Report 521, Princeton University, April 1996. (Also available at `http://www.cs.duke.edu/~jeffe/compgeom/taskforce.html`).

8. B. Chazelle, L.J. Guibas, and D.T. Lee. The power of geometric duality. *BIT*, 25:76–90, 1985.

9. H. Edelsbrunner and L. Guibas. Topologically sweeping an arrangement. *Journal of Computer and System Sciences*, 38:165–194, 1989.

10. D. Frank and G. Fadel. Preferred direction of build for rapid prototyping processes. In *Proceedings of the 5th International Conference on Rapid Prototyping*, pages 191–200, 1994.

11. A. Gajentaan and M. H. Overmars. On a class of $O(n^2)$ problems in computational geometry. *Computational Geometry: Theory & Applications*, 5:165–185, 1995.

12. R.N. Goldman. Area of planar polygons and volume of polyhedra. In J. Arvo, editor, *Graphics Gems II*, pages 170–171. Academic Press, 1991.

13. P. Jacobs. *Rapid Prototyping & Manufacturing: Fundamentals of StereoLithography*. McGraw-Hill, 1992.

14. D.G. Luenberger. *Introduction to Linear and Non-linear Programming*. Addison-Wesley, 1973.

15. J. Majhi, R. Janardan, M. Smid, and P. Gupta. On some geometric optimization problems in layered manufacturing. Technical Report TR–97–002, Dept. of Computer Science, University of Minnesota, 1997. Submitted.

16. J. Matoušek. Range searching with efficient hierarchical cuttings. *Discrete & Computational Geometry*, 10:157–182, 1992.

17. M. McKenna and R. Seidel. Finding the optimal shadows of a convex polytope. In *Proceedings of the 1st Annual ACM Symposium on Computational Geometry*, pages 24–28, 1985.

18. F.P. Preparata and M.I. Shamos. *Computational Geometry–An Introduction*. Springer–Verlag, 1988.

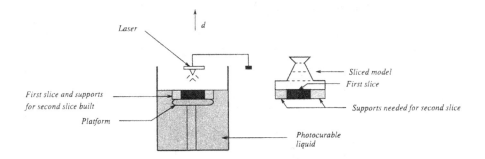

Fig. 1. The stereolithograpy apparatus.

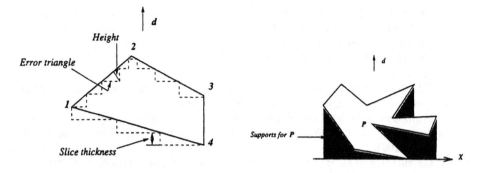

Fig. 2. Illustrating stair-stepping. **Fig. 3.** Supports for a non-convex polygon.

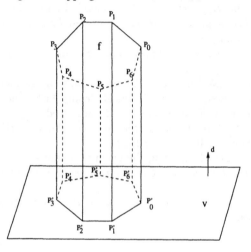

Fig. 4. Support polyhedron P_f for back facet f.

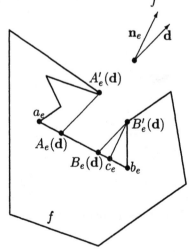

Fig. 5. Edge f (resp. $e = \overline{a_e b_e}$) needs (resp. is attached to) supports.

New TSP Construction Heuristics and Their Relationships to the 2-Opt

Hiroyuki Okano, Shinji Misono, and Kazuo Iwano

IBM Research, Tokyo Research Laboratory, 1623-14 Shimotsuruma, Yamato, Kanagawa-ken, Japan 242. E-mail: {okanoh,misono,iwano}@jp.ibm.com

Abstract. Construction heuristics for the traveling salesman problem (TSP), with the 2-Opt applied as a postprocess, are studied with respect to their tour lengths and computation times. In this study, the "2-Opt dependency," which indicates how the performance of the 2-Opt depends on the initial tours built by the construction heuristics, is analyzed. In accordance with the analysis, we devise a new construction heuristic, the *recursive selection with long edge preference* (RSL) method, which runs faster and produces shorter tours than the multiple-fragment method.

1 Introduction

The traveling salesman problem (TSP) is a representative NP-hard problem, and is known to have a wide range of practical applications [1]. Especially, for geometric TSP, in which distance is defined in L_2 or L_∞ metric, many efficient heuristics and their implementations have been discussed [2, 5]. Fast heuristics can be categorized into two classes according to their computation time: construction heuristics which build a tour from scratch, and local improvement heuristics which improve a given tour. The latter class is further divided into the 2-Opt (Fig. 1), 3-Opt, and Lin-Kernighan algorithms. We are interested in efficient heuristics for each class of time constraints. When we deal with on-the-fly route optimization in the drilling of printed-circuit boards (PCB), the available computation time and power are quite limited; thus, only a construction heuristic is used. For this class (the first class), we devised the *grand-tour* (GT) method [3], a variant of the random-addition (RA) method. The GT method generates a shorter tour and runs faster than the multiple-fragment (MF) method which Bentley regards as one of the best heuristics [2], and so it is satisfactory for the first class. When we have time to apply the 2-Opt to construction tours – we call this case the second class – however, we found that the final (2-Opt) tour of the GT method is longer than that given by the MF+2-Opt, which is regarded as the best pair of heuristics in the second class [2] (Fig. 2).

Figure 2 shows the results of various construction heuristics and their changes after application of the 2-Opt. Results before and after the 2-Opt for each heuristic are connected by lines. The vertical segments of the RA indicate the variances of its tour lengths. Tour lengths are shown as ratios to the results given by the Lin-Kernighan algorithm [8]. It is shown that the GT has the best performance

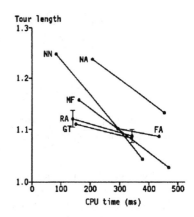

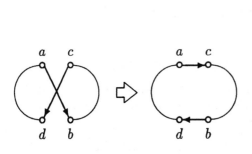

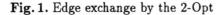

Fig. 1. Edge exchange by the 2-Opt

Fig. 2. Performances of various construction heuristics combined with the 2-Opt

in terms of tour length and CPU time among the *construction* heuristics examined here. When the GT is followed by the 2-Opt, however, the tours given by the GT+2-Opt become far longer than those given by the MF+2-Opt.

More generally, use of the 2-Opt improves the tour length less for the addition heuristics – the nearest-addition (NA) method, the farthest-addition (FA) method, the random-addition (RA) method, and the GT method – than for the nearest-neighbor type heuristics – the nearest-neighbor (NN) method and the MF method. Therefore, in this paper, we address the "2-Opt dependency," and devise a construction heuristic that outperforms the MF+2-Opt in terms of both computation time and tour length.

By the *2-Opt dependency* we mean the extent to which the *performance* of the 2-Opt depends on the initial tour created by the use of construction heuristics; here, the performance is measured in terms of execution time and a tour length. The reality of the 2-Opt dependency on construction heuristics was demonstrated by Bentley [2] and Reinelt [5], but no detailed analysis has yet been carried out. Perttunen [7] examined the 2-Opt dependency more closely, but focused only on the saving method [1]. In this paper, we conduct detailed analysis on the 2-Opt dependency on the construction heuristics by using our original measuring factors.

We use two factors: a *nearest-neighbor ratio* (an *nn ratio*), which measures how well a construction heuristic builds the local structure of a tour, and an *opt range*, which estimates the number of edges examined by each local search of the 2-Opt on the average. We show that the following assertions hold in most cases: (1) the higher a construction tour's nn ratio, the shorter the 2-Opt tour, and (2) the larger a construction tour's opt range, the greater the improvement given by the 2-Opt.

In accordance with this analysis, we propose a new construction heuristic, the *recursive selection with long edge preference* (RSL) method. In so doing, we modify the MF so that the nn ratio of the construction tour becomes higher, and at the same time the opt range becomes larger. We show through numerical experiments both for uniformly distributed instances and for TSPLIB instances that the RSL+2-Opt runs faster by about 20% than the MF+2-Opt, and finds a shorter tour than the MF+2-Opt. This is the first attempt to design construction heuristics with the aim of obtaining a shorter tour after combination with the 2-Opt, neglecting the length of construction tours.

This paper is organized as follows: in the next two subsections, the construction heuristics we use in this paper are described, and the 2-Opt dependency is defined. In Section 2, new measuring factors, the nn ratio and the opt range, are defined. In Section 3, three new construction heuristics we devised for the experiments are illustrated. One of these methods, the RSL method, is our main outcome of this study. In Section 4, we analyze the local structure in a tour by the use of the nn ratio. In Section 5, effects of long edges in a tour are analyzed by the use of the opt range, and the RSL method is shown to outperform the MF method. Finally, Section 6 summarizes the paper.

1.1 Experimental Setups

We discuss the 2-Opt dependency of some of existing construction heuristics whose efficient implementations for geometric TSP are described by Bentley [2]: three addition heuristics – the nearest-addition (NA) method, the farthest-addition (FA) method, and the random-addition (RA) method – and two nearest-neighbor type heuristics – the nearest-neighbor (NN) method and the multiple-fragment (MF) method. The NA, FA, and RA methods start with a subtour consisting of a single point, and insert the nearest (or furthest) points from a subtour (or randomly selected points) one by one into the place in the subtour that least increases the tour length. The addition heuristics use a parameter (called *AddRad* in [2]) in their procedures to determine the search radius of its neighborhood search, and it is set to 2.0 in our experiments. The NN method starts at an arbitrary point, visits the nearest unvisited points one by one, and constructs a tour. The MF method starts with each point as a fragment of a single point, and patches the closest pairs of fragments one by one without making points of degree three or small loops. For more detail descriptions of these heuristics, see [2, 3, 4]. For the GT method, see the description of the GN method in Section 3.

We follow Bentley [2] for the implementation details of the construction heuristics; we use k-d trees for near-neighbor searches and a priority queue for selecting the closest pairs of fragments. Tour lengths and CPU times are measured as the average over 100 instances of uniformly distributed points in a unit square. Tour lengths are normalized by the length of the Lin-Kernighan algorithm [8]. CPU times are measured on an IBM RS/6000 whose CPU is a PowerPC 604 112 MHz. The experiments shown below are for instances of size 1,000. We also experimented for instances of sizes up to 10,000, and observed the same results; thus, we omit to show them due to the space limitation.

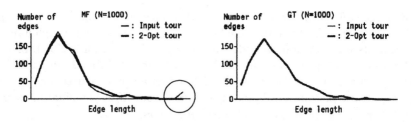

Fig. 3. Edge length distributions of the MF and the GT

1.2 The 2-Opt Dependency

As described before, the length of a 2-Opt tour depends strongly on which construction heuristic is used for building an initial tour. This is what we mean by the 2-Opt dependency, and it is expressed by slopes of lines in Fig. 2. For quantitative analysis of the 2-Opt dependency, we define the *slope* simply as:

$$\text{Slope} = \frac{\text{Improvement in tour length given by the 2-Opt}}{\text{CPU time of the 2-Opt}}.$$

In the following sections, tours before and after application of the 2-Opt will be referred to as construction tours and 2-Opt tours, respectively. Similarly, the names of factors may be used with the prefix construction or 2-Opt, which means that the factors are observed before or after the 2-Opt, respectively.

2 The Nearest-Neighbor Ratio and the Opt Range

2.1 The Nearest-Neighbor Ratio

In order to measure the extent to which the 2-Opt changes the structure of a tour, we first examine the edge length distributions. Figure 3 shows the edge length distributions of both the MF and the GT. The thin lines in Fig. 3 show the edge length distributions of construction tours, while the thick lines show those of 2-Opt tours. We cannot see shifts of the distributions leftward after the 2-Opt. Thus, in order to investigate shifts of short edges in detail and examine how the 2-Opt transforms the local structure of a tour, we introduce a new factor, the *nearest-neighbor (nn) ratio*.

The nn ratio is defined as the ratio of the number of *nn edges* to the total number of edges in a tour, where the nn edge is defined as an edge at least one of whose end points is the nearest neighbor of the other end point. For the same instance used in Fig. 3, the MF+2-Opt's nn ratio is 60%, and the GT+2-Opt's nn ratio is 54%. These values seem to reflect their 2-Opt tour lengths (Fig. 2). It is to be noted that the nn ratios of Lin-Kernighan tours are about 59% irrespective of which construction heuristic is used.

2.2 The Opt Range

When the 2-Opt exchanges a pair of edges, at least one of the resulting edges is shorter than either of the original pair of edges; thus, in choosing an edge (p_i, p_j) and another edge with which to exchange it, the 2-Opt examines only edges connected to each point in a circle centered at p_i with radius $d(p_i, p_j)$. Because of this characteristic of the 2-Opt, long edges in construction tours expand the size of the area in which the 2-Opt can affect the structure of tours.

Figure 3 shows that the construction tour of the MF contains larger number of long edges than the GT tour (note the circled part of Fig. 3), that means the 2-Opt can perform more efficiently for the MF tours than for the GT tours. In fact, the slope of the MF is steeper than that of the GT (Fig. 2).

The opt range of a tour is defined as the average value of $cover(i)$ over all points in the tour, where $cover(i)$ is defined as the number of points that lie in a circle centered at a point p_i with a radius given by the edge length $|e_i|$, excluding p_i and p_j where $e_i = (p_i, p_j)$. The average value is measured while a tour is scanned in one direction, where the tail of each edge e_i is always regarded as p_i. In the 2-Opt procedure, $cover(i)$ is the number of edges that are examined as candidates with which to exchange e_i. Thus, the opt range of a construction tour estimates the number of edges examined by each local search of the 2-Opt on the average, or the size of the area in which the 2-Opt can affect the structure of construction tours.

3 New Construction Heuristics

3.1 The Grand-Tour with NN Edge Preference (GN) Method [3]

The GN is a modification of the GT to increase the construction nn ratio. Its procedure is as follows:

1. Build a k-d tree for input points so that they are sorted in order of tree leaves, and make an *addition sequence* by selecting the sorted points in order of the van der Corput sequence [9] (a low discrepancy deterministic sequence).
2. Let the first point of the addition sequence be a subtour of a single point.
3. Select points p in the addition sequence to be inserted one by one.
4. Find all the points in a circle of radius $d(p, q) \times 2.0$ centered at p, where q is the nearest-neighbor point in the subtour from p, and choose one of the edges e connected to the points in the circle, with the following criteria:
   (a) The selection most greatly increases the number of the nn edges, or
   (b) The selection least increases the tour length.
5. Break the edge e, and insert p between the two end points of e.
6. Go to Step 3 until all the points in the addition sequence have been selected.

The only difference between the GT and the GN is criterion (a), which the GT does not have. Note that criterion (a) has priority over (b).

3.2 The Recursive Selection (RS) Method

The RS is a modification of the MF to increase the nn ratio. Its procedure is as follows:

1. Sort input points in increasing order of the distances between them and their nearest neighbors.
2. Select sorted point p one by one in the sorted order.
3. Connect p to its nearest neighbor, unless doing so creates points of degree three or small loops. In this case, simply cancel the selection.
4. Go to Step 2 until all the input points have been selected.
5. If there remain points of degree one or zero after all the input points have been selected, go to Step 1, assuming them to be the *input points.*

When we consider all the nn edges in a tour to obtain a graph, degrees of nodes in the graph turn out to be fairly small for uniformly distributed instances; the maximum degree is three and the average is 1.4 for the instance used in Fig. 3. Thus, the first execution of Step 2 effectively finds nearly maximum number of nn edges of an instance. The RS tour for the above instance includes 649 nn edges, which is less than the maximum only by two.

3.3 The RS with Long Edge Preference (RSL) Method

To increase the opt range, beginning from the fourth recursion of the RS procedure, change Step 2 as follows:

2. Select sorted point p one by one in the *reverse* order.

This modification of the RS can be implemented simply by scanning a sort table in reverse, after the fourth recursion. Note that, here, the first execution of the procedure is regarded as the first recursion. Note also that nn edges in an RS tour are all connected in the first recursion, and that subsequent recursions do not affect the resulting nn ratio.

4 Experiments on the NN Ratio

Figure 4 shows the tour lengths and the nn ratios of both construction and 2-Opt tours (the construction and 2-Opt nn ratios) for various construction heuristics with the dotted line (the *fitting line*) that best fits the data of 2-Opt tours. Results before and after application of the 2-Opt are connected by lines. The fitting line gives us one observation; the higher the 2-Opt nn ratio is, the shorter the 2-Opt tour is. We can also see from Fig. 4 that 2-Opt nn ratios have large variation (from 48% to 59%), depending strongly on the construction nn ratios. The nn ratio of the NA method, for example, gets increased from 42% to 48% by the 2-Opt, still its 2-Opt nn ratio is much smaller than that of the MF method. This is to be compared with the fact that the nn ratios of Lin-Kernighan tours concentrate around 59% and have small variation less than 3%. The 2-Opt, thus,

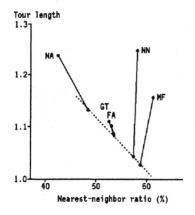
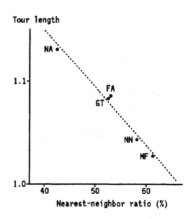
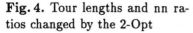

Fig. 4. Tour lengths and nn ratios changed by the 2-Opt

Fig. 5. 2-Opt tour lengths and
construction nn ratios

do not significantly change the nn ratio as the Lin-Kernighan algorithm does.
These observations suggest that there is a corelation between the construction
nn ratio and the 2-Opt tour length.

In Fig. 5, we plot the 2-Opt tour lengths and the construction nn ratios of
heuristics under consideration, and add a fitting line as before. This fitting line
clearly shows that the higher the *construction* nn ratio, the shorter the *2-Opt*
tour. In other words, one way to obtain a 2-Opt tour of shorter length is to realize
a construction tour with higher nn ratio. This observation motivates us to devise
two construction heuristics, the RS and GN methods, which are described in the
previous section.

We designed the RS (the GN resp.) by modifying the MF (the GT resp.) so
that the higher construction nn ratio is realized. We add the construction nn
ratios and the 2-Opt tour lengths of the RS and the GN to Fig. 4 and obtain
Fig. 7. It shows that the construction nn ratios of the RS and the GN are higher
than those of the MF and the GT by 6% and 16%, respectively.

None of the RS and the GN, however, improves the 2-Opt tour length significantly. The tour lengths of the RS and the GN are shorter than those of the
MF and the GT by 0.1% and 0.3%, respectively. Increasing the construction nn
ratio, on the other hand, has little influence on the slope (Fig. 6). Furthermore,
we can see a gap in 2-Opt tour length between two types of heuristics – nearest-
neighbor (RS, MF, and NN) and addition (GN, GT, NA, and FA) – irrespective
of the construction nn ratio (Fig. 7). The nn ratio, thus, is not a definite factor
to characterize the 2-Opt dependency; the nn ratio is a valid factor (for each
type of heuristic) to measure a tour's quality.

Although the construction nn ratio does not clarify the 2-Opt dependency, it
leads us to find heuristics of better performance. The RS method, when combined
with the 2-Opt, runs faster by 28.6% than the MF, and its 2-Opt tour length
is shorter by 0.1% than that of the MF (Fig. 6). This is because the number of
nearest-neighbor searches done by the RS is only 66.4% of the number done by

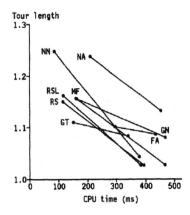
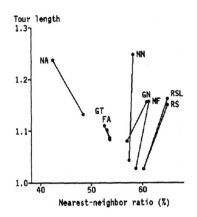

Fig. 6. Performances of the RS, GN, and RSL methods combined with the 2-Opt

Fig. 7. Tour lengths and nn ratios of the RS, GN, and RSL methods changed by the 2-Opt

the MF, and also because the RS does not require insert/delete operations to a heap as the MF does, and only scans the sorted list. The number of input points for each recursion of the RS, which is equal to the number of nearest-neighbor searches, for example, changes as follows: 1000, 681, 189, 52, 14, 4.

5 Experiments on the Opt Range

As described in Section 2, the opt range of a construction tour estimate the size of the area in which the 2-Opt can affect the structure of the construction tour. The opt range also measures the effect of long edges in a construction tour to the performance of the 2-Opt. In this section, we observe the opt ranges of various tours, and show that the opt range can measure the 2-Opt dependency.

Figure 9 shows the construction tours and the opt ranges of various construction heuristics with a fitting line. The smaller the opt ranges, the shorter the construction tours; this is easily understood, because the opt range reflects the average edge length. Figure 10 shows the slopes and the opt ranges of various construction heuristics with a fitting line. We can see almost the same trend as in Fig. 9, which means that the larger the opt ranges, the greater the slopes; that is, the greater the improvement that the 2-Opt gives.

Our next question is how to increase the slopes without affecting the lengths of the construction tours. Figures 9 and 10 show that the two objectives, decreasing the length of the construction tours and increasing the slopes, require a tradeoff. We must, therefore, seek a better tradeoff in order to improve the construction heuristics, instead of just changing the slopes. The opt range and the construction nn ratio, on the other hand, do not involve a tradeoff, because the opt range is mainly increased by long edges which are not the nn edges.

We modify the RS in such a way as to increase the opt range and keep the nn ratio high to devise the RSL method. Figure 6 shows that the slope of the

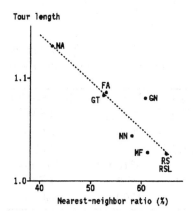

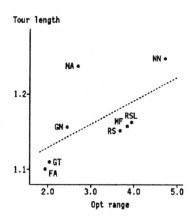

Fig. 8. 2-Opt tour lengths and construction nn ratios of the RS, GN, and RSL methods

Fig. 9. Construction tours and opt ranges of the RS, GN, and RSL methods

RSL+2-Opt is steeper than that of the MF+2-Opt. It also shows that the 2-Opt tour of the RSL is slightly shorter than that of the MF, while the construction tour of the RSL is longer than that of the MF. Note that the 2-Opt for the RSL can be computed faster than the 2-Opt for the MF or the RS, and that the RSL+2-Opt shows a consequent improvement over the MF+2-Opt of 18.0%, which is larger than that of the RS+2-Opt (15.6%). The RSL's improvement over the MF results from the differences in the numbers of nearest-neighbor searches they require and in their computation times for a quick-sort and a priority queue.

Figure 8 shows that the construction nn ratio of the RSL is exactly the same as that of the RS. Figures 9 and 10 show that the RSL succeeds in increasing the slope without greatly increasing the construction tour length. Table 1 shows that the RSL+2-Opt's advantage over the MF+2-Opt is the same for real instances from TSPLIB [6]. The RSL+2-Opt is faster than the MF+2-Opt by 21.6% on the average, and generates tours whose length are 4.5% in excess of the optimal values.

6 Conclusion

We have analyzed how and why the performance of the 2-Opt depends on the initial tour built by construction heuristics, and studied how to modify a construction heuristic to obtain a better performance when the 2-Opt is applied after it. For the analysis, we defined the slope as a quantitative evaluation of the 2-Opt dependency, and introduced two new measuring factors, the nn ratio and the opt range. We showed that the nn ratio measures a tour's quality, and that increase of the nn ratio of a construction tour improves the 2-Opt tour without affecting the slope greatly. We also showed that the opt range measures the 2-Opt dependency of a construction tour, and that increase of the opt range of a construction tour makes the slope steeper.

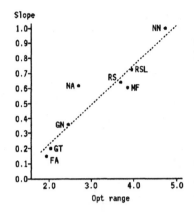

Fig. 10. Slopes and opt ranges of the RS, GN, and RSL methods

Table 1. Application to TSPLIB instances
(Percentages in excess of the optimal values)

Instance	MF+2-Opt	RSL+2-Opt
pcb3038	4.5%	4.2%
	2030 ms	1370 ms
pla7397	5.3%	4.5%
	6310 ms	5370 ms
pr2392	5.2%	4.7%
	1260 ms	1040 ms

Based on the analysis with these factors, we devised a new construction heuristic, the *recursive selection with long edge preference* (RSL) method, which is faster than the multiple-fragment (MF) method by about 20%, and finds a shorter tour than the MF method when the 2-Opt is used for postprocessing. This is one of the first attempts to analyze construction heuristics in conjunction with the improvement heuristic, and also the first to design construction heuristics so that they perform well when combined with the 2-Opt.

References

1. E. L. Lawler, J. K. Lenstra, A. H. G. Rinnooy Kan, and D. B. Shimoys, *The Traveling Salesman Problem*, John Wiley & Sons, New York, 1985.
2. J. L. Bentley, *Fast Algorithms for Geometric Traveling Salesman Problems*, ORSA Journal on Computing, Vol. 4, No. 4, pp. 387-411, 1992.
3. S. Misono and K. Iwano, *Experiments on TSP Real Instances*, IBM Research Report, RT0153, 1996.
4. H. Okano, S. Misono and K. Iwano, *New TSP Construction Heuristics and Their Relationships to the 2-Opt* IBM Research Report, RT0154, 1996.
5. G. Reinelt, *The Traveling Salesman, Computational Solutions for TSP Applications*, Springer-Verlag, Heidelberg, 1994.
6. G. Reinelt, *TSPLIB 1.1, Version of December 10, 1990*, Institut fuer Mathematik, Universitaet Augsburg, 1990.
7. J. Perttunen, *On the Significance of the Initial Solution in Traveling Salesman Heuristics*, Journal of the Operational Research Society, Vol. 45, No. 10, pp. 1131-1140, 1994.
8. S. Lin and B.W. Kernighan, *An Effective Heuristic Algorithm for the Traveling-Salesman Problem*, Operations Research, Vol. 21, pp. 498-516, 1973.
9. J.G. van der Corput, *Verteilungstunktionen I, II*, Nederal. Akad. Wetensh. Proc. Ser. B, 38, pp. 813-821, 1058-1066, 1935.

Pattern Matching in Hypertext

Amihood Amir* Moshe Lewenstein** Noa Lewenstein***
Georgia Tech   Bar-Ilan University   Bar-Ilan University
and
Bar-Ilan University

Abstract. A myriad of textual problems have been considered in the pattern matching field with many non-trivial results. Nevertheless, surprisingly little work has been done on the natural combination of pattern matching and hypertext. In contrast to regular text, hypertext has a non-linear structure and the techniques of pattern matching for text cannot be directly applied to hypertext.

Manber and Wu pioneered the study of pattern matching in hypertext and defined a hypertext model for pattern matching. Subsequent papers gave algorithms for pattern matching on hypertext with special structures – trees and DAGS.

In this paper we present a much simpler algorithm achieving the same complexity which runs on *any* hypertext graph. We then extend the problem to *approximate* pattern matching in hypertext, first considering hamming distance and then edit distance. We show that in contrast to regular text, it *does make a difference* whether the errors occur in the hypertext or the pattern. The approximate pattern matching problem in hypertext with errors in the hypertext turns out to be $\mathcal{NP}$-Complete and the approximate pattern matching problem in hypertext with errors in the pattern has a polynomial time solution.

1 Introduction

Most digital libraries today appear in *hypertext* form [15], with links between text and annotations, and, in multimedia libraries, between the text and pictures, video and voice. The World Wide Web is, perhaps, the most widely known hypertext system. In contrast, classical pattern matching (e.g. [6, 10]) has dealt primarily with unlinked textual files. Recently, there has been interest in non-standard matching, such as dictionary matching ([4, 3]) or dynamic indexing

---

* Department of Mathematics and Computer Science, Bar-Ilan University, 52900 Ramat-Gan, Israel, (972-3)531-8770; amir@cs.biu.ac.il; Partially supported by NSF grant CCR-95-31939, the Israel Ministry of Science and the Arts grants 6297 and 8560 and a 1997 Bar-Ilan University Internal Research Grant.

** moshe@cs.biu.ac.il.

*** noa@cs.biu.ac.il; Partially supported by the Israel Ministry of Science and the Arts grant 8560. This work is part of Noa Lewenstein's Ph.D. dissertation.

([8, 16]) but, surprisingly, there has been no concerted effort in analyzing string matching in hypertext.

In a pioneering paper by Manber and Wu [14] a first attempt is made to define pattern matching in hypertext. They suggest the concept of viewing a hypertext library as a general graph of unlinked files. For a formal definition, see section 2.

Akutsu [2], presented an algorithm that would be capable of pattern matching in a hypertext with an underlying tree structure. Kim and Park [11] also use the Manber and Wu model and present a general pattern matching algorithm. Their algorithm assumes that the hypertext links form an acyclic digraph.

We were motivated by a natural example of hypertext to seek a pattern matching algorithm in more general graphs. The application that motivated our research is the *Responsa Judaica library*. Hebrew Law has evolved over two thousand years and is recorded in the Bible, Mishna, Tosephta, Babylonian and Jerusalem Talmud, Maimonides, Shulchan Aruch, Tur, and thousands of published responsa. The links between the various works create a general hypertext digraph. In this context, one may want to search only a main text, but it is possible that a search is desired through the hyperlinks.

Some of the issues involved are discussed in a recent manuscript of Fraenkel and Klein [9]. An M.Sc. thesis in the Math and CS department at Bar-Ilan [5] deals with a hypertext system for the Babylonian Talmud. All existing pattern matching algorithms in the hypertext model ([14, 11] assume that the hypertext links form an acyclic digraph. In the Responsa environment, the text files present a general digraph.

In this paper we present the *first* algorithm for pattern matching in hypertext where the hypertext links form a general graph. The complexity of our algorithm is the same as that of the Kim and Park algorithm, $O(N + |E|m)$, where N is the overall size of the text, m is the pattern length and E is the edge set of the hypertext.

We extend this result to *approximate* matching in hypertext. We begin with the hamming distance as our metric. Some very surprising insights are achieved. In classical approximate pattern matching the error locations are *symmetric*. It does not matter if the errors occur in the text or in the pattern. In approximate matching in hypertext there are distinctly different cases.

This paper is the *first in the literature* to identify cases where the error location is not symmetric. We show that it is important to understand whether the mismatches occur in pattern or in the text.

We consider three flavors of the problem: with mismatches in the hypertext only, mismatches in the pattern only and mismatches in both. The first and third turn out to be $\mathcal{NP}$-Complete. For the second type, we present an algorithm that runs in $O(N\sqrt{m}+m\sqrt{N}+|E|m)$, where N is the overall size of the text, m the pattern length and E the edge set of the hypertext.

Finally, we consider edit distance. Once again, this comes in three variations and the variations allowing errors in the hypertext are $\mathcal{NP}$-Complete. If we allow errors in the pattern only, then it is polynomial. We present an algorithm of time complexity $O(Nm \log m + |E|m)$. This problem was already considered by Manber and Wu [14]. As previously mentioned, they assume that the underlying digraph is acyclic. Our algorithm works for an arbitrary graph.

2 Pattern Matching in Hypertext

Formally a *hypertext* above alphabet Σ is a triplet $H = (V, E, T)$ where (V, E) is a digraph and $T = \{T_v \in \Sigma^+ \mid v \in V\}$. If for every $v \in V$, $T_v \in \Sigma$ then we call the hypertext a *one-character hypertext*.

Let $v_1, ..., v_k$ be a path, possibly with loops, in (V, E) and let $W = T'_{v_1} T_{v_2} T_{v_3} ... T_{v_{k-2}} T_{v_{k-1}} T''_{v_k}$ be the concatenation of texts upon this path, where T''_{v_k} is a prefix of T_{v_k} and T'_{v_1} is a suffix of T_{v_1} beginning at location l of T_{v_1}. We say that W *matches* on the path $v_1, ..., v_k$ beginning at location l of v_1 or that $v_1 : l$ is a *W-match location*. In general we say that W *matches* in H. Note that for a one-character hypertext we may disregard the location since there is only one location per vertex.

The *Pattern Matching in Hypertext* problem is defined as follows:
INPUT: A pattern P and a hypertext H.
OUTPUT: All P-match locations.

Initially we solve the problem by transforming a given hypertext into a one-character hypertext. The transformation is done by taking every text T_v and splitting v into $|T_v|$ vertices (saving for each new vertex its origin). For the remainder of this section, we consider the hypertext $H = (V, E, T)$ to be a one-character hypertext. In the next section we discuss general hypertext graphs.

We would like to find all the P-match locations within the hypertext. Since our hypertext is a one-character hypertext, every path that $P = p_1 p_2 ... p_m$ matches upon is exactly m, the length of pattern P. To be precise, the path is of the form $v_1, v_2, v_3, ..., v_m$ such that $T_{v_1} = p_1$, $T_{v_2} = p_2$, ..., $T_{v_m} = p_m$. The idea is to create a digraph $G^{(P,H)} = (V^{(P,H)}, E^{(P,H)})$ that depends on the hypertext and the pattern, such that every path of length m in the digraph will represent a match in the hypertext. Corresponding to each vertex in the hypertext there will be m vertices in the digraph that will represent the m pattern locations. Similar to an edge in the hypertext, an edge in the digraph will represent two consecutive characters, but the edge will also represent their location in the pattern, i and $i + 1$. In addition, the edge represents a match between the hypertext character in the source of the edge and the ith pattern character.

Formally, we define the digraph in the following way:

$$V^{(P,H)} = \{v^i \mid v \in V, 1 \le i \le m\} \cup \{s, f\}$$
$$E^{(P,H)} = \{(s, v^1) \mid v \in V\} \cup \{(v^i, u^{i+1}) \mid (v, u) \in E, T_v = p_i\} \cup$$
$$\{(v^m, f) \mid T_v = p_m\}.$$

It is easy to see from the definition that $G^{(P,H)}$ is a DAG and the longest path in $G^{(P,H)}$ is of length $m + 1$ (the additional 1 is for the initial vertex s). For our algorithm we need the following lemma.

Lemma 1. *P matches on path $v_1, v_2, ..., v_m$ in H iff $s, v_1^1, v_2^2, ..., v_m^m, f$ is a path in $G^{(P,H)}$.*

Proof: Follows from the definition of $G^{(P,H)}$. $\square$

Algorithm for Pattern Matching in Hypertext

1. Construct a one-character hypertext $H' = (V', E', T')$ from the original hypertext $H = (V, E, T)$.
2. Construct $G^{(P,H')}$ from the pattern P and the hypertext H'.
3. Do a depth first search in $G^{(P,H')}$ starting from s. Denote by *true* every vertex v^i where there is a path from v^i to f.
4. For every vertex v in the hypertext, check v^1 in the digraph and if marked true announce the vertex and location in H corresponding to vertex v in H' as a P-match location.

Time: Building H' and $G^{(P,H')}$ takes time linear to the size of H', P and $G^{(P,H')}$. The size of $G^{(P,H')}$, where $H' = (V', E', T')$ is $O(m|E'|)$ and this is clearly the largest of the three. Step 3, can likewise be implemented in linear time. Step 4 takes $O(|V'|)$ time. The size of V' is at most $|V| + N$ and the size of E' is at most $|E| + N$ since for every new vertex in the one-character hypertext there is exactly one edge introduced. Therefore the overall time for the algorithm takes $O(m|E'|) = O(mN + m|E|)$. $\square$

Correctness: Follows from the construction of $G^{(P,H)}$ and Lemma 1. $\square$

3 Improved Algorithm for Pattern Matching in Hypertext

We now consider the case where a hypertext vertex may contain text of greater length than 1 (possibly even longer than the pattern). As in the previous section, we create a digraph to model the hypertext and the pattern, but in this case there will be two sets of m vertices for every vertex in the hypertext instead of one set. These two sets will model comparisons of subpatterns rather than comparisons of characters only. For a match of T_v from location i to location j in the pattern we set an edge from the ith vertex in the first set to jth vertex in the second set. (These edges are described in the first part of $E^{(P,H)}$ defined below.) We also need to consider a match beginning in the middle of the vertex's text or ending in the middle of the vertex's text. (These edges are described in the second and third part of $E^{(P,H)}$ defined below.)

We will use the following notation. Let x be a string and k an integer. Denote the k length suffix of x by $Suf(x,k)$ and the k length prefix of x by $Pref(x,k)$.

We now give a formal definition of the digraph. Let $H = (V, E, T)$ be a hypertext and $P = p_1 p_2 ... p_m$ be a pattern then $G^{(P,H)} = (V^{(P,H)}, E^{(P,H)})$ is defined in the following way.

$$V^{(P,H)} = \{\overline{v}^i \mid v \in V, 1 \le i \le m\} \cup \{\underline{v}^i \mid v \in V, 1 \le i \le m\} \cup \{s\}$$

$$E^{(P,H)} = \{(\overline{v}^i, \underline{v}^{i+|T_v|-1}) \mid |T_v| < m, 1 < i < m - |T_v| + 1, T_v = p_i...p_{i+|T_v|-1}\} \cup$$
$$\{(\overline{v}^1, \underline{v}^k) \mid Suf(T_v, k) = p_1...p_k, 1 \le k < \min\{m, |T_v|\}\} \cup$$
$$\{(\overline{v}^k, \underline{v}^m) \mid Pref(T_v, m - k + 1) = p_k...p_m,$$
$$\max\{1, m - |T_v| + 1\} < k \le m\} \cup$$
$$\{(\underline{v}^i, \overline{u}^{i+1}) \mid (v, u) \in E, 1 \le i < m\} \cup$$
$$\{(s, \overline{v}^1) \mid v \in V\}.$$

Lemma 2. *Let $v \in V$ and l be a location in T_v. If $|T_v| - l + 1 < m$ then there is a P-match location at $v : l$ in H iff there exists a path in $G^{(P,H)}$ beginning with $s, \overline{v}^1, \underline{v}^{|T_v|-l+1}$ and ending with $\underline{u}^m$ for some $u \in V$.*

Proof: Assume that there is a P-match location at $v : l$ in H with $|T_v| - l + 1 < m$. Let $v, v_2, ..., v_k$ be a path such that P matches upon it beginning at location l of v, i.e. $P = T'_v T_{v_2} T_{v_3} ... T_{v_{k-2}} T_{v_{k-1}} T''_{v_k}$. Since $|T'_v| = |T_v| - l + 1$, $(\overline{v}^1, \underline{v}^{|T_v|-l+1}) \in E^{(P,H)}$. It is also immediate that $(s, \overline{v}^1) \in E^{(P,H)}$. Moreover, since the prefix of length $|T''_{v_k}|$ of T_{v_k} matches at pattern location $m - |T''_{v_k}| + 1$, $(\overline{v}_k^{m-|T''_{v_k}|+1}, \underline{v}_k^m) \in E^{(P,H)}$. Also, for every $1 < i < k$, $(\overline{v}_i^{\Sigma_{j=1}^{i-1}|T_{v_j}|+1}, \underline{v}_i^{\Sigma_{j=1}^{i}|T_{v_j}|}) \in E^{(P,H)}$ because $T_{v_i} < m$ and T_{v_i} appears at pattern location $\Sigma_{j=1}^{i}|T_{v_j}| + 1$. Since $(v_i, v_{i+1}) \in E$, for $1 \le i < k$, it follows that $(\underline{v}_i^{|T_{v_i}|}, \overline{v}_{i+1}^{|T_{v_i}|+1}) \in E$. Therefore, we have a path that ends with v_k^m as required.

Conversely, a careful analysis of the construction of $G^{(P,H)}$ shows that any path from s to $\underline{u}^m$ for some $u \in V$ must be of the form $s, \overline{v}_1^1, \underline{v}_1^{i_1}, \overline{v}_2^{i_1+1}, \underline{v}_2^{i_2}, ..., \underline{v}_{k-1}^{i_{k-1}}, \overline{u}^{i_{k-1}+1}, \underline{u}^m$. Similar to the other direction if $i_1 = |T_v| - l + 1$ it is relatively straightforward to see that this represents a path beginning in v_1 upon which P matches at location l of v_1. $\square$

Algorithm for Pattern Matching in Hypertext

1. For every vertex v in H do standard pattern matching with pattern P and text T_v.

2. Based on the results of step 1, announce the internal matches as P-match locations.

3. Build $G^{(P,H)}$ from the pattern P and the hypertext H, using step 1 for the first three types of edges.

4. Do a depth first search in $G^{(P,H)}$ starting from vertex s saving at every vertex $w \in V^{(P,H)}$ true, if there is a path from w to $\underline{u}^m$ for some $u \in V$.

5. For every vertex v in the hypertext, check $\overline{v}^1$ in the digraph $G^{(P,H)}$ and if marked true do the following: For each $\underline{v}_j$ s.t. $(\overline{v}_1, \underline{v}_j) \in E^{(P,H)}$, if $\underline{v}_j$ is marked true, announce $v : |T_v| - j + 1$ a P-match location.

Correctness: Matches internal to a vertex will be detected in step 1 and announced in step 2. Now, let $v : l$ be a P-match location that crosses at least one hyperlink. Since it crosses a hyperlink it must be the case that $|T_v| - l + 1 < m$. The conditions of step 4 and step 5, where we set $j = |T_v| - l + 1$, together with Lemma 2 shows that we indeed find this P-match location and announce it in step 5.

Time: Step 1 takes $O(N + |V|m)$ time since pattern matching is linear for text and N is the overall text size. $O(|V|m)$ is charged for the vertices with text shorter than m. Step 2 is included in the complexity of the previous step. For step 3, note the size of $G^{(P,H)}$. There are $O(|V|m)$ vertices and there are $O(|E|m)$ edges. Constructing the edges is immediate for those with source s and those of the form $(\underline{v}^i, \overline{u}^{i+1})$. For the others we use the results of the pattern matching from step 1. A possible implementation can be done by slightly modifying the KMP algorithm. Therefore, the construction of $G^{(P,H)}$ is linear in its size. So, step 3 is $O(|E|m)$. Step 4 is, once again, linear in the size of $G^{(P,H)}$ and the time for step 5 is bounded by the size of $G^{(P,H)}$. Therefore, the algorithm runs in $O(N + |E|m)$ time.

4 Approximate Pattern Matching in Hypertext - Hamming Distance

Approximate pattern matching is one of the well-researched problems in pattern matching. Often text contains errors and searching for an exact match is not sufficient. In this case we may search for a closest solution or for all closest solutions or for solutions not exceeding k errors. The closest solution depends on what type of errors are considered. Mismatches is one of the most common errors and the number of mismatches between two equal length strings is called the *hamming distance. Approximate pattern matching with hamming distance* refers to the problem of finding the substring with minimum hamming distance from the pattern or finding all substrings with hamming distance less than a specified distance from the pattern.

Naturally, it would be interesting to investigate approximate pattern matching in hypertext using hamming distance as our metric. This needs some clarification. For approximate pattern matching in regular text the usual assumption is that mismatches occur in the text. Since the text and pattern both have linear structures it does not really matter whether the mismatches occur in the text or in the pattern. In hypertext this is not true. Consider a certain path in the hypertext passing through v, k times, where $k > 1$. If we change a character in that hypertext vertex we are changing all k instances on that path. On the other hand, if we are changing the characters in the pattern we can change each instance to a different character. From this reasoning it follows that the fewest number of changes in the pattern required so that the pattern matches may be different than the fewest number of changes required in the hypertext. Moreover, we can always change the pattern so that the pattern will match in the hypertext but it may be the case that we cannot change the hypertext so that the pattern will match.

In real life, applications vary and in some applications we naturally expect mismatches in the text, while in others we expect the mismatches in the pattern. Sometimes we may even expect errors to appear in both. Consider a hypertext similar to the internet in which information is often inserted quickly and erroneously. Here we would expect the mismatches to be in the text. On the other hand, a hypertext which is relatively constant and well checked such as the Responsa project which serves as a query system, is less prone to errors. Nevertheless, for large queries we may expect mismatches in the pattern. So that we actually have three versions for approximate pattern matching with hamming distance. In each of these versions the input is a pattern P and a hypertext $H = (V, E, T)$ and an error bound d.

> **Version 1:** Can we change at most d characters in the hypertext H so that P will match in H?
> **Version 2:** Can we change d characters in P so that P will match in H?
> **Version 3:** Can we change d characters in P and H so that P will match in H?

These questions are stated as decision questions but we may also ask, what is the fewest changes, min_d, necessary for P to match in H? Or, find all locations where P matches in the hypertext with min_d changes. Note that when the hypertext is a DAG all three versions boil down to the same version. In general for an arbitrary graph we shall see that versions 1 and 3 are $\mathcal{NP}$-Complete and version 2 is polynomial.

Proposition 1 *The Approximate hypertext-hamming distance problem - version 1 is $\mathcal{NP}$-Complete.*

Proof: Clearly the problem is in $\mathcal{NP}$. We will reduce from directed hamiltonian path.

Let $G = (V, E)$ be a directed graph in which we seek a hamiltonian path and let $n = |V|$. Construct a hypertext H setting $H = (V, E, T)$ where $T_v = \sigma_1$ for every $v \in V$. Take pattern $\sigma_1...\sigma_n$, where $\forall i \neq j$, $\sigma_i \neq \sigma_j$ and distance $d = n-1$. If we have a hamiltonian path then replacing the i-th vertex's text with σ_i will give us a match of P in the hypertext.

Conversely, if we can make text-changes in the hypertext (which, in our case, clearly, does not exceed n) so that P matches on some path then since $|T| = |V| = n$ and $\forall i \neq j$, $\sigma_i \neq \sigma_j$ it must be the case that the pattern matches on a hamiltonian path. $\qquad\qquad\qquad\qquad\qquad\qquad\qquad\qquad\qquad\qquad\qquad\qquad\qquad\square$

We can not use the same reduction for version 3, since any graph containing a cycle would have returned a match simply by changing all pattern characters to σ_1. However, we use a similar reduction.

Proposition 2 *The Approximate hypertext-hamming distance problem - version 3 is $\mathcal{NP}$-Complete.*

Proof: Clearly the problem is in $\mathcal{NP}$. We will reduce from directed hamiltonian cycle.

Let $G = (V, E)$ be a directed graph in which we seek a hamiltonian cycle and let $n = |V|$. Construct a hypertext H setting $H = (V, E, T)$ where $T_v = \sigma_1$ for every $v \in V$. Take pattern $\sigma_1...\sigma_n\sigma_1...\sigma_n$, where $\forall i \neq j$, $\sigma_i \neq \sigma_j$ and distance $d = n - 1$. If we have a hamiltonian cycle then replacing the i-th vertex's text with σ_i, for $i \geq 2$, making $n - 1$ changes altogether, will give us a match of P in the hypertext.

Conversely, assume we have a match of P in H with at most $n - 1$ changes on P and H together. Consider the path on which P matches in H after the changes were made, call it C, and let k be the number of different characters on C. Note, that it is always true that $k \leq n$ since there are only n characters in the whole hypertext. There are n different characters in the pattern. But since there are only k of these on C, at least $n - k$ of the pattern symbols must be changed. Since each symbol in the pattern appears twice we must make at least $2(n - k)$ changes in the pattern. Now, since we started with the same character in all vertices in the hypertext we must have made at least $k - 1$ changes in the text. So overall there are at least $2(n - k) + k - 1$ changes. Our distance is $d = n - 1$ and therefore it must be that $2(n-k)+k-1 \leq n-1$. This is equivalent to $n \leq k$ and since $n \geq k$ we have $k = n$. This means that after changes every vertex in the hypertext must contain a different character. This already accounts for the $n - 1$ allowed changes. So, the pattern must be in its original form.

C now serves for a hamiltonian cycle since we first visit the vertex labeled σ_1 and then the vertex labeled σ_2 and so on till we come back to the vertex labeled σ_1 (and then go for another round). $\qquad\qquad\qquad\qquad\qquad\qquad\qquad\square$

We now show a polynomial algorithm for version 2. The idea is similar to the algorithm of the previous section.

We build a digraph in a similar fashion to the digraph of section 3 but this time the digraph will be weighted. As in the previous construction, for each vertex in the hypertext we construct two sets of m vertices. All edges that were in the previous construction will also be in this digraph, all having weight zero. The weight zero expresses that there are no mismatches. We add edges that will capture the matches containing mismatches. The weight we assign to such an edge is the number of mismatches occuring between the corresponding subpattern and the vertex's text.

We denote the number of mismatches between two equal length strings x and y as $Ham(x,y)$.

Formally we define the directed weighted digraph $G^{(P,H)} = (V^{(P,H)}, E^{(P,H)})$, where H is a hypertext (V, E, T) and P is a pattern $p_1 p_2 ... p_m$, as follows:

$$V^{(P,H)} = \{\overline{v}^i \mid v \in V, 1 \le i \le m\} \cup \{\underline{v}^i \mid v \in V, 1 \le i \le m\} \cup \{s\}$$

$$
\begin{aligned}
E^{(P,H)} = &\{(\overline{v}^i, \underline{v}^{i+|T_v|-1}, w) \mid |T_v| < m, 1 < i \le m - |T_v| + 1, \\
&\qquad\qquad\qquad\qquad w = Ham(T_v, p_i...p_{i+|T_v|-1})\} \cup \\
&\{(\overline{v}^1, \underline{v}^k, w) \mid 1 \le k < m, w = Ham(Suf(T_v, k), p_1...p_k)\} \cup \\
&\{(\overline{v}^k, \underline{v}^m, w) \mid 1 < k \le m, w = Ham(Pref(T_v, m - k + 1), p_k...p_m)\} \cup \\
&\{(\underline{v}^i, \overline{u}^{i+1}, 0) \mid (v, u) \in E, 1 \le i \le m\} \cup \\
&\{s, \overline{v}^1, 0) \mid v \in V\}.
\end{aligned}
$$

Here we denoted edges as a triplet (u, v, w) where w is the weight of the edge from u to v. We will alternate between this notation and notation of an edge as (u, v) with a weight function over all edges $w(u, v)$.

Lemma 3. *Every path in $G^{(P,H)}$ from s to $\underline{v}^m$ for some $v \in V$ is of the form $s, \overline{v}_1^1, \underline{v}_1^{i_1}, \overline{v}_2^{i_1+1}, \underline{v}_2^{i_2}, ..., \underline{v}_{k-1}^{i_{k-1}}, \overline{v}^{i_{k-1}+1}, \underline{v}^m$, with $1 \le i_1 < i_2 < ... < i_{k-1} < m$.*

Proof: Let C be a path from s to some $\underline{v}^m$. The only edges leaving s enter vertices of the form $\overline{u}^1$ and all edges leaving vertices of the form $\underline{u}^i$ go to vertices of the form $\overline{w}^{i+1}$. All edges leaving vertices of the form $\overline{u}^i$ go to vertices of the form $\underline{u}^j$ for $j > i$. So the lemma follows. $\square$

Lemma 4. *Let P be a pattern of length m and $1 \le d \le m$. There exists a pattern P' of length m with $Ham(P, P') = d$ such that P' matches in H iff there exists a path in $G^{(P,H)}$ from s to v^m, for some vertex $v \in V$, with path weight d.*

Proof: Assume that there exists a P' for which the lemma's conditions holds. Let $v_1, ..., v_k$ be the path in H on which P' matches beginning from location l, i.e. $P' = T'_{v_1} T_{v_2} T_{v_3} ... T_{v_{k-2}} T_{v_{k-1}} T''_{v_k}$, where $|T'_{v_1}| = |T_{v_1}| - l + 1$. It is straightforward to check that the following path exists and has weight d:

$$s, \overline{v}_1^1, \underline{v}_1^{\frac{|T'_{v_1}|}{}}, \underline{v}_2^{\frac{|T'_{v_1}|+1}{}}, \underline{v}_2^{\frac{|T'_{v_1}|+|T_{v_2}|}{}}, ..., \overline{v}_i^{\frac{|T'_{v_1}|+\sum_{j=1}^{i-1}|T_{v_j}|+1}{}}, \underline{v}_i^{\frac{|T'_{v_1}|+\sum_{j=1}^{i}|T_{v_j}|}{}}, ...,$$
$$\overline{v}_k^{\frac{m-|T''_{v_k}|+1}{}}, \underline{v}_k^m.$$

Conversely, according to Lemma 3 every path from s to $\underline{v}^m$ is of the form $s, \overline{v}_1^1, \underline{v}_1^{i_1}, \overline{v}_2^{i_1+1}, \underline{v}_2^{i_2}, ..., \underline{v}_{k-1}^{i_{k-1}}, \overline{v}_k^{i_{k-1}+1}, \underline{v}_k^m$ such that $1 \leq i_1 < i_2 < ... < i_{k-1} < m$. An analysis of the construction of $G^{(P,H)}$ shows that $(s, \overline{v}_1^1)$ and $(\underline{v}_j^{i_j}, \overline{v}_{j+1}^{i_j+1})$ are edges with weight 0 and that $(\overline{v}_j^{i_{j-1}+1}, \underline{v}_j^{i_j})$ are edges with weight $Ham(T_{v_j}, p_{i_{j-1}+1}...p_{i_j})$ except for $(\overline{v}_1^1, \underline{v}_1^{i_1})$ and $(\overline{v}_k^{i_{k-1}+1}, \underline{v}_k^m)$ which have weight $Ham(Suf(T_{v_1}, i_1), p_1...p_{i_1})$ and $Ham(Pref(T_{v_k}, m-i_{k-1}), p_{i_{k-1}+1}...p_m)$ in correspondence.

Therefore, the overall weight $d = w(s, \overline{v}_1^1) + w(\overline{v}_1^1, \underline{v}_1^{i_1}) + w(\underline{v}_1^{i_1}, \overline{v}_2^{i_1+1}) + ... + w(\overline{v}_k^{i_{k-1}+1}, \underline{v}_k^m) = Ham(P, T'_{v_1}T_{v_2}T_{v_3}...T_{v_{k-2}}T_{v_{k-1}}T''_{v_k})$, where $T'_{v_1} = Suf(T_{v_1}, i_1)$ and $T''_{v_k} = Pref(T_{v_k}, m - i_{k-1})$. So, choosing $P' = T'_{v_1}T_{v_2}T_{v_3}...T_{v_{k-2}}T_{v_{k-1}}T''_{v_k}$ gives a pattern of length m that matches in H with $Ham(P, P') = d$. $\square$

The algorithm we will now display is assumed to output the minimum number of changes necessary to make P match. If desired, this can be adapted in a simple way to return P' that does match and is obtained by a minimum number of changes from P.

Algorithm for Approximate Pattern Matching with Hamming Distance in Hypertext

1. For each vertex v in H for which $|T_v| \geq m$, build an array of the hamming distance between pattern P and each substring of the text $\$^{m-1}T_v\$^{m-1}$, where $\$ \notin \Sigma$.
   For each vertex v in H for which $|T_v| \leq m$, build an array of the hamming distance between T_v, viewed as the pattern, and each substring of $\$^{|T_v|-1}P\$^{|T_v|-1}$, viewed as text, where $\$ \notin \Sigma$.

2. For each vertex with $|T_v| \geq m$ use the results of step 1 to find a substring of T_v of length m with least hamming distance from pattern P. Denote the distance $w(v)$.

3. Build $G^{(P,H)}$ from the pattern P and the hypertext H, using the results of step 1.

4. Do a depth first search in $G^{(P,H)}$ starting from vertex s finding the shortest path from s to $\underline{u}^m$ for some $u \in V$ and denote the length of the shortest (weighted) path w.

5. **Output:** $\min(\{w\} \cup \{w(v) \mid v \in V, |T_v| \geq m\})$.

Correctness: If the substring with shortest hamming distance from the pattern appears within a vertex then we will find it in step 2 and announce it in step 5. If the pattern crosses at least one hyperlink then by Lemma 4 there is a path of the form described in Lemma 4 with this weight. Since by Lemma 3 all paths from s to $\underline{v}^m$ for some $v \in V$ have this form which by Lemma 4 correspond to paths which model comparisons of P in the hypertext, it is sufficient to find the shortest path from s to some $\underline{v}^m$. This is exactly what we do in step 4. We implement it with a depth first search since $G^{(P,H)}$ is a DAG. In step 5 we announce the shortest.

Time: For each vertex, if $|T_v| > m$ then step 1 takes time $O(|T_v|\sqrt{m})$ [1, 12] and if $m > |T_v|$ then step 1 takes time $O(m\sqrt{|T_v|})$, for an overall $O(N\sqrt{m} + m\sqrt{N})$. Step 2 is linear in the size of the text within the vertices, i.e. $O(N)$. Step 3 is linear in the size of $G^{(P,H)}$, which has a vertex set of size $O(|V|m)$ and edge set of size $O(|E|m)$. Step 4 is linear in the size of $G^{(P,H)}$ and step 5 costs $O(|V|)$. So, we have an overall complexity of $O(N\sqrt{m} + m\sqrt{N} + |E|m)$.

5 Approximate Pattern Matching in Hypertext - Edit Distance

While it is true that mismatches are a common error in texts, often other errors occur such as accidently deleting a character or inserting a superfluous character. The minimal number of insertion, deletions and changes necessary to transfer one string into another is called the *edit distance* of these two strings. In this section we consider approximate pattern matching in hypertext with edit distance as our metric.

Similar to the previous section we have three versions to the problem and, not surprisingly similar results. We will first show that if the insertions, deletion and changes can be done only in the hypertext, then the problem is $\mathcal{NP}$-Complete. We then give an outline how to extend this result to the case when the errors occur both in the pattern and in the hypertext. Afterwards we present a polynomial algorithm for the case when errors occur in the pattern only. Note that the insertions or deletions are always on the characters whether in the pattern or in the hypertext, never on the structure of the hypertext. The underlying digraph always remains in its original form. Different from what we have seen up till now, it may be that after a deletion of a character of text in a vertex, the text is the empty word ϵ.

The *Approximate Hypertext Matching-Edit Distance in Hypertext* problem is defined as follows:
Input: A pattern P, a hypertext $H = (V, E, T)$ and a distance d.
Output: *True*, if with d error-corrections (insertions, deletions and changes) we can change the hypertext so that P matches in H. *False*, otherwise.

Proposition 3 *The Approximate Matching-Edit Distance in Hypertext problem is $\mathcal{NP}$-Complete.*

Proof: Clearly the problem is in $\mathcal{NP}$. We will reduce from directed hamiltonian path.

Let $G = (V, E)$ be a directed graph in which we seek a hamiltonian path and let $n = |V|$. Set $P = \$\sigma_1\#...\$\sigma_n\#$, where $\forall i \neq j$, $\sigma_i \neq \sigma_j$ and $\$, \# \notin \{\sigma_1, ..., \sigma_n\}$ and set the hypertext $H = (V, E, T)$ where $T_v = \$\#$ for every $v \in V$. Set the distance $d = n$. If we have a hamiltonian cycle then by inserting σ_i in between the $\$$ and the $\#$ of the i-th vertex will give us a match of P in the hypertext with n error-corrections.

Conversely, assume we have a match of P in H with at most n error-corrections on H. Since $\sigma_1, ..., \sigma_n$ do not appear in the hypertext before the error-corrections and must appear after the error-corrections and since $\forall i \neq j$, $\sigma_i \neq \sigma_j$ it must be the case that the n error-corrections are either changes or insertions intended for inserting $\sigma_1, ..., \sigma_n$ into the hypertext.

Consider the path C in the hypertext that P matches upon after the error-corrections. If this path is not hamiltonian then one of the following must be true: (a) there is a vertex that does not appear on C, (b) there is a vertex that appears twice, not including the first or last vertex on the path or (c) the first or last vertex appear a second time along the path.

If (case (a)) there is a vertex that does not appear on C then there must be a different vertex in which both σ_i and σ_{i+1} appear. But this other vertex must contain text $\sigma_i\#\$\sigma_{i+1}$. This cannot be, because at least $n-2$ error corrections are necessary for the other vertices and four corrections are necessary for the vertex containing $\sigma_i\#\$\sigma_{i+1}$.

If (case (b)) there is a vertex in C that appears twice, this vertex may not contain σ_j for any j. But then there must be a vertex with σ_i and σ_{i+1} which, as in case (a), is impossible. Case (c) is also impossible following similar reasoning. Therefore, C is a hamiltonian path. □

If we modify the definition of the problem to allow errors in the hypertext and the pattern then it is once again $\mathcal{NP}$-Complete. This can be proved reducing from the directed hamiltonian cycle problem with pattern $P = \$\sigma_1\#...\$\sigma_n\#\$\sigma_1\#... \$\sigma_n\#$, using a similar claim to Proposition 2 to show that there cannot be changes in the pattern and then the rest is similar to the proof of Proposition 3.

We now consider the version with errors in the pattern only. For this problem we will present a polynomial algorithm that, once again, extends on the previous ideas. First we will give a formal definition.

The *Approximate Hypertext Matching-Edit Distance in Pattern* problem is defined as follows:
Input: A pattern P and a hypertext $H = (V, E, T)$.
Output: The minimum d such that with d error-corrections (insertions, deletions and changes) in the pattern, P, it will match in H.

The best known algorithms [13] for approximate pattern matching with edit distance in regular text have complexity of $O(nm)$, where n is the text length and m is the pattern length. In a hypertext a vertex may contain $O(N)$, where

N is, as before, the overall size of the text in the vertices. So applying regular approximate pattern matching techniques would cost $O(Nm)$ for this vertex alone. Therefore, for simplicity, without sacrificing efficiency, we will turn the hypertext into a one-character hypertext.

As in section 2, for each vertex v in the hypertext the digraph contains the vertex set $\{v^i \mid 1 \leq i \leq m\}$, where v^i represents the comparison of T_v and p_i. The edges will be weighted in a fashion similar to the previous section. There will be edges with weight 0 to account for an exact match and edges with weight 1 to account for mismatch. We also add edges with weight 1 to account for insertion and deletion of a character.

We now describe the edges in the digraph. We have (a) edges describing an exact match of ith pattern character, (b) edges describing a mismatch at location i, (c) edges describing insertion of a character to the pattern before pattern location i, (d) edges describing deletion of the ith character of the pattern and (e) edges from the start vertex s to all the first location vertices.

Formally,

$$V^{(P,H)} = \{v^i \mid v \in V, 1 \leq i \leq m\} \cup \{s\} \cup \{f\}$$

$$
\begin{aligned}
E^{(P,H)} = (a)\ & \{(u^i, v^{i+1}, 0) \mid (u,v) \in E, T_u = p_i\}\ \cup \\
& \{(u^m, f, 0)\quad \mid T_u = p_m\}\ \cup \\
(b)\ & \{(u^i, v^{i+1}, 1) \mid (u,v) \in E, T_u \neq p_i\} \cup \\
(c)\ & \{(u^i, v^i, 1)\quad \mid (u,v) \in E,\ 2 \leq i \leq m\} \cup \\
(d)\ & \{(v^i, v^{i+1}, 1) \mid v \in V,\ 1 \leq i \leq m-1\} \cup \\
& \{(v^m, f, 1)\quad \mid T_v \neq p_m\} \cup \\
(e)\ & \{(s, v^1, 0)\quad \mid v \in V\}.
\end{aligned}
$$

Lemma 5. *Let P be a pattern of length m and $1 \leq d \leq m$. There exists a pattern P' formed by d error-corrections of P such that P' matches in H iff there exists a path in $G^{(P,H)}$ from s to f with path weight d.*

Proof: We leave the full proof for the journal version, but point out that the proof is similar to the proof of Lemma 4. $\square$

Algorithm for Approximate Hypertext Matching - Edit Distance in Pattern

1. Create a one-character hypertext $H' = (V', E', T')$ from H.
2. Build $G^{(P,H')}$ from the pattern P and the one-character hypertext H'.
3. Find the shortest path from s to f and announce its weighted length.

Correctness: Follows directly from Lemma 5.

Time: Step 1 takes time linear to H' and step 2 time linear to $G^{(P,H')}$. $|V'| = O(|V| + N) = O(N)$ and $|E'| = O(|E| + N)$. $G^{(P,H')}$ has vertex size $O(|V'|m) =$

$O(mN)$ and edge size $O(m|E'|) = O(mN + m|E|)$. With Dijkstra's algorithm implemented with Fibonacci heaps the third step can be done in $O(Nm \log m + |E|m)$ time [7].

References

1. K. Abrahamson. Generalized String Matching. *SIAM J. Computing* **16** (**6**), 1039-1051, 1987.
2. T. Akutsu. A Linear Time Pattern Matching Algorithm Between a String and a Tree. *Proceedings of the 4th Symposium on Combinatorial Pattern Matching*, 1-10, Padova, Italy, 1993.
3. A. Amir, M. Farach, R. Giancarlo, Z. Galil, and K. Park. Dynamic dictionary matching. *Journal of Computer and System Sciences*, 49(2):208–222, 1994.
4. A. Amir, M. Farach, R.M. Idury, J.A. La Poutré, and A.A Schäffer. Improved dynamic dictionary matching. *Information and Computation*, 119(2):258–282, 1995.
5. A. Aviad. HyperTalmud: a hypertext system for the Babylonian Talmud and its commentaries. Dept. of Math and Computer Science, Bar-Ilan University, 1993.
6. R.S. Boyer and J.S. Moore. A fast string searching algorithm. *Comm. ACM*, 20:762–772, 1977.
7. T.H. Cormen, C.E. Leiserson and R.L. Rivest. *Introduction to Algorithms* The MIT Press, Cambridge, Massachusetts, 1990.
8. P. Ferragina and R. Grossi. Optimal on-line search and sublinear time update in string matching. *Proc. 7th ACM-SIAM Symposium on Discrete Algorithms*, pages 531–540, 1995.
9. A. S. Fraenkel and S. T. Klein. Information Retrieval from Annotated Texts. TR-95-25, Dept. of Applied Math and Computer Science, The Weizmann Institute of Science, 1995.
10. D.E. Knuth, J.H. Morris, and V.R. Pratt. Fast pattern matching in strings. *SIAM J. Computing*, 6:323–350, 1977.
11. D.K. Kim and K. Park. String matching in hypertext. *6th Symposium on Combinatorial Pattern Matching*, Helsinki, Finland, 1995.
12. S. Rao Kosaraju. Efficient String Matching. Manuscript, 1987.
13. G. M. Landau and U. Vishkin. Fast parallel and serial approximate string matching. *Journal of Algorithms* **10** (**2**), 157-169, 1989.
14. U. Manber and S. Wu. Approximate string matching with arbitrary costs for text and hypertext. *IAPR Workshop on structural and syntactic pattern recognition*, Bern, Switzerland, 1992.
15. J. Nielsen. *Hypertext and Hypermedia*. Academic Press Professional, Boston, 1993.
16. S. C. Sahinalp and U. Vishkin. Efficient approximate and dynamic matching of patterns using a labeling paradigm. *Proc. of 37th Annual Symposium on Foundations of Computer Science*, 320-328, Burlington, Vermont, 1996.

Multiple Approximate String Matching *

Ricardo Baeza-Yates   Gonzalo Navarro

Department of Computer Science
University of Chile
Blanco Encalada 2120 - Santiago - Chile
{rbaeza,gnavarro}@dcc.uchile.cl

Abstract. We present two new algorithms for on-line multiple approximate string matching. These are extensions of previous algorithms that search for a single pattern. The single-pattern version of the first one is based on the simulation with bits of a non-deterministic finite automaton built from the pattern and using the text as input. To search for multiple patterns, we superimpose their automata, using the result as a filter. The second algorithm partitions the pattern in sub-patterns that are searched with no errors, with a fast exact multipattern search algorithm. To handle multiple patterns, we search the sub-patterns of all of them together. The average running time achieved is in both cases $O(n)$ for moderate error level, pattern length and number of patterns. They adapt (with higher costs) to the other cases. However, the algorithms differ in speed and thresholds of usefulness. We analyze theoretically when each algorithm should be used, and show experimentally that they are faster than previous solutions in a wide range of cases.

1   Introduction

Approximate string matching is one of the main problems in classical string algorithms, with applications to text searching, computational biology, pattern recognition, etc. Given a text of length n and a pattern of length m (both sequences over an alphabet Σ of size σ), and a maximal number of errors allowed, k, we want to find all text positions where the pattern matches the text up to k errors. Errors can be substituting, deleting or inserting a character. We use the term "error ratio" to refer to $\alpha = k/m$.

The solutions to this problem differ if the algorithm has to be on-line (i.e. the text is not known in advance) or off-line (the text can be preprocessed). In this paper we are interested in the first case, where the classical dynamic programming solution for a single pattern is $O(mn)$ running time [14].

In the last years several algorithms have improved the classical one. Some achieve $O(kn)$ cost by using the properties of the dynamic programming matrix [18, 8, 10, 19, 6]. Others filter the text to quickly eliminate uninteresting parts [17, 16, 7, 13, 5], some of them being sublinear on average for moderate α. Yet other approaches use bit-parallelism [2] to reduce the number of operations

---

* This work has been supported in part by FONDECYT grants 1950622 and 1960881.

[20, 22, 21, 4]. In [21] the search is modeled with a non-deterministic finite automaton, whose execution is simulated in parallel on machine words of w bits, achieving $O(kmn/w)$ time. In [4], we simulate the same automaton in a different way, achieving $O(n)$ time for small patterns. This algorithm is shown to be the fastest in that case (see also [3]).

The problem of approximately searching a set of patterns (i.e. the occurrences of anyone of them) has been considered only recently. A trivial solution is to do r searches, where r is the number of patterns. As far as we know, the only previous works on this problem are [11] and [13]. The first approach uses hashing to search many patterns with *one* error, being efficient even for one thousand patterns. The second one filters the text by counting matching positions, keeping many counters in a single computer word and updating them in a single operation.

In this work, we present two new algorithms that are extensions of previous ones to the case of multiple search. In Sections 2 and 3 we explain and extend [4]. In Section 4 we do the same for [5]. In Section 5 and 6 we analyze our algorithms and compare them against [11] and [13].

Although [11] allows to search for many patterns, it is limited to only one error. We allow any number of errors, and improve [11] when the number of patterns is not very large (say, less than 60). We improve [13] except for intermediate error ratios. We also improve the trivial algorithm (i.e. one separate search per pattern) when the error ratio is moderate. The extension of [5] is the fastest for small error ratios, while that of [4] adapts better to more errors.

2   Bit-Parallelism by Diagonals

In this section we review the main points of the algorithm [4]. We refer the reader to the original article for more details.

Consider the NFA for searching "this" with at most $k = 2$ errors shown in Figure 1 (for now disregard the "$(+ x)$"). Every row denotes the number of errors seen. The first one 0, the second one 1, and so on. Every column represents matching the pattern up to a given position. At each iteration, a new text character is considered and the automaton changes its states. Horizontal arrows represent matching a character (they can only be followed if the corresponding match occurs), vertical arrows represent inserting a character in the pattern, solid diagonal arrows represent replacing a character, and dashed diagonal arrows represent deleting a character of the pattern (they are empty transitions, since we delete the character from the pattern without advancing in the text). The loop at the initial state allows to consider any character as a potential starting point of a match. The automaton accepts a character (as the end of a match) whenever a rightmost state is active. Initially, the active states at row i ($i \in 0..k$) are those at the columns from 0 to i, to represent the deletion of the first i characters of the pattern, referred here as $pat[1..m]$.

Many algorithms for approximate string matching consist fundamentally in simulating this automaton by rows or columns. The dependencies introduced by the diagonal empty transitions prevent the parallel computation of the new

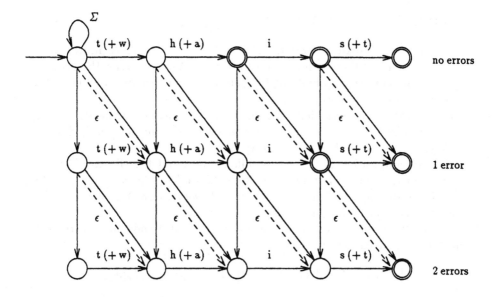

Fig. 1. An NFA for approximate string matching. Unlabeled transitions match any character.

values. In [4] we show that by simulating the automaton by diagonals, it is possible to compute all values in parallel.

Because of the empty transitions, once a state in a diagonal is active, all the subsequent states in that diagonal become active too, so we can define the minimum active row of each diagonal, D_i (diagonals are numbered by looking the column they start at). The new values for D_i ($i \in 1..m - k$) after we read a new text character c can be computed by

$$D_i' = \min(\ D_i + 1, \ D_{i+1} + 1, \ g(D_{i-1}, c)\ )$$

where $g(D_i, c) = \min(\ \{k + 1\} \cup \{\ j\ /\ j \geq D_i\ \wedge\ pat[i + j] == c\ \}\ )$

We use bit-parallelism to represent the D_i's in unary. With some pattern preprocessing, the parallel update is $O(1)$ cost and very fast in practice. A central part of this preprocessing is the definition of an m bits long mask $t[c]$, representing match or mismatch against the pattern, for each character c. The resulting algorithm is linear whenever the representation fits in a computer word (i.e. $(m - k)(k + 2) \leq w$, where w is the number of bits in a computer word).

Observe that the $t[]$ table mechanism allows more sophisticated searching: at each position of the pattern, we can allow not a single character, but a class of characters, at no additional search cost. It suffices to set $t[c]$ to "match" at position i for every $c \in pat[i]$. For example, we can search in case-insensitive by allowing each position to match the upper-case and lower-case versions of the letter. We use this property to allow multiple patterns.

3 Superimposed Automata

Suppose we have to search r patterns $P_1, ..., P_r$. We are interested in the occurrences of any one of them, with at most k errors. We can extend the previous bit-parallelism approach by building the automaton for each one, and then "superimpose" all the automata.

Assume that all patterns have the same length (otherwise, truncate them to the shortest one). Hence, all the automata have the same structure, differing only in the labels of the horizontal arrows.

The superimposition is defined as follows: we build the $t[]$ table for each pattern, and then take the bitwise-*or* of all the tables. The resulting $t[]$ table matches at position i with the i-th character of *any* pattern. We then build the automaton as before using this table.

The resulting automaton accepts a text position if it ends an occurrence of a much more relaxed pattern, namely

$$\{P_1[1], ..., P_r[1]\} \{P_1[2], ..., P_r[2]\} ... \{P_1[m], ..., P_r[m]\}$$

for example, if the search is for "this" and "wait", the string "whit" is accepted with *zero* errors. See Figure 1, this time paying attention to the two characters present at each horizontal transition.

For a moderate number of patterns, the filter is strict enough at the same cost of a single search. Each occurrence reported by the automaton has to be verified for all the involved patterns (we use [19] for this step).

If the number of patterns is so large that the filter does not work well, we partition the set of patterns into groups of r' patterns each, build the automaton of each group and perform $\lceil r/r' \rceil$ independent searches. The cost of this search is $O(r/r' \ n)$, where r' is small enough to make the number of verifications negligible. This r' always exists, since for $r' = 1$ we have a single pattern per automaton and no verification is needed.

If the length of the patterns does not allow to put their automata in single computer words (i.e. $(m - k)(k + 2) > w$), we partition the problem. We adapt the two partitioning techniques defined in [4].

3.1 Pattern Partitioning

The following lemma proved in [4] suggests a way to partition a large pattern.

Lemma: If $segm = Text[a..b]$ matches *pat* with k errors, and $pat = p_1...p_j$ (a concatenation of sub-patterns), then $segm$ includes a segment that matches at least one of the p_i's, with $\lfloor k/j \rfloor$ errors.

Thus, we can reduce the size of the problem if we divide the pattern in j parts, provided we search all the sub-patterns with $\lfloor k/j \rfloor$ errors. Each match of a sub-pattern must be verified to determine if it is in fact a complete match.

To perform the partition, we pick the smallest j such that the problem fits in a single computer word (i.e. $(\lceil m/j \rceil - \lfloor k/j \rfloor)(\lfloor k/j \rfloor + 2) \leq w$). We divide the pattern in j subpatterns as evenly as possible. The limit of this method is reached for $j = k + 1$, since in that case we search with zero errors. The resulting algorithm is qualitatively different and is described later.

Once we partition all the patterns, we are left with jr subpatterns to be searched with $\lfloor k/j \rfloor$ errors. We simply group them as if they were independent patterns to search with the general method. The only difference is that we have to verify the complete patterns when we find a possible occurrence. To avoid verifying all the patterns at each match, we try as much as possible to include subpatterns of the same patterns in the groups.

3.2 Automaton Partitioning

If the automaton does not fit in a single word, we can partition it using a number of machine words for the simulation.

The idea is as follows: once the (large) automata have been superimposed, we partition the automaton into a matrix of subautomata, each one fitting in a computer word. Those subautomata behave differently than the simple one, since they must propagate bits to their neighbors.

Once the automaton is partitioned, we run it over the text updating its subautomata. Observe, however, that it is not necessary to update all the subautomata. In the same spirit as [19], we work only on "active" diagonals.

The technique of grouping in case of a very relaxed filter is used here too. We use the heuristic of sorting the patterns and packing neighbors in the same group, trying to have the same first characters.

4 Exact Partitioning Extended to Multiple Patterns

We first briefly review the algorithm [5] (studied more in detail in [4, 3]). We refer the reader to the original articles for details.

A particular case of the lemma of Section 3.1 shows that if a pattern matches a text position with k errors, and we split the pattern in $k + 1$ pieces, then at least one of the pieces must be present with no errors in each occurrence (this is a folklore property which has been used several times [21, 12, 9]). Searching with zero errors leads to a completely different technique.

Since there are efficient algorithms to search for a set of patterns exactly, we partition the pattern in $k + 1$ pieces (of similar length), and apply a multipattern exact search for the pieces. Each occurrence of a piece is verified to check if it involves a complete match. If there are no too many verifications, this algorithm is extremely fast.

From the many algorithms for multipattern search, an extension of Boyer-Moore-Horspool-Sunday (BMHS) [15] gave the best results. We build a trie with the sub-patterns. At each text position we search the text that follows into the trie, until a leaf is found (match) or there is no path to follow (mismatch). The jump to the next text position is precomputed as the minimum of the jumps allowed in each sub-pattern by the BMHS algorithm.

Observe that we can easily add more patterns to this scheme. Suppose we have to search for r patterns $P_1, ..., P_r$. We cut each one into $k + 1$ pieces and search in parallel for *all* the $r(k+1)$ pieces. When a piece is found in the text, we use a classical algorithm to verify its pattern in the candidate area (this time we normally know which pattern to verify). As in the previous case, this constitutes a good filter if the number of patterns and errors is not too high. Unlike the previous case, grouping is of no use here, since there are no more matches in the union of patterns than the sum of the individual matches.

5 Analysis

We are interested in the restrictions that α and r must satisfy for each mechanism to be efficient in filtering most of the unrelevant part of the text. We are also interested in the complexity of the algorithms, especially in when they are linear on average and when they are *useful* (i.e. better than r sequential searches).

5.1 Superimposed Automata

Suppose that we search r patterns. As explained before, we can partition the set in groups of r' patterns each, and search each group separately (with its r' automata superimposed). The size of the groups should be as large as possible, but small enough for the verifications to be not significant. We analyze which is the optimal value for r' and which is the complexity of the search.

In [4] we prove that the probability of a given text position matching a random pattern with error ratio α is $O(a^m)$, where $a = 1/(\sigma^{1-\alpha}\alpha^{2\alpha}(1-\alpha)^{2(1-\alpha)})$. It is also proven that $a < 1$ whenever $\alpha < 1 - e/\sqrt{\sigma}$.

In this formula, $1/\sigma$ stands for the probability of a character crossing a horizontal edge of the automaton (i.e. the probability of two characters being equal). To extend this result, we note that we have r' characters on each edge now, so the above mentioned probability is $1 - (1 - 1/\sigma)^{r'}$, which is smaller than r'/σ. We use this upper bound as a pessimistic approximation (which stands for the case of all the r' characters being different, and is tight for $r' << \sigma$).

The algorithm is linear on average whenever the total cost of verifications is $O(1)$ per character. Since each verification costs $O(m^2)$ per pattern in the superimposed group, we pay $O(r'm^2)$ to verify the whole group. Thus, we want the probability of a verification to be $O(1/(r'm^2))$, which happens for $a < 1$.

To decide which is the optimal size of the groups, we state that the total cost of verifications must be at most one per character, since if it is larger we would prefer to make two separate searches with much less verifications. Hence, our r' satisfies $a^m = 1/(r'm^2)$, i.e.

$$r' = \left(\frac{\sigma^{m-k}\alpha^{2k}(1-\alpha)^{2(m-k)}}{m^2} \right)^{\frac{1}{m-k+1}}$$

which our algorithm uses to determine the correct grouping.

For m not too small, we can simplify and bound the above result to

$$r' = \sigma \alpha^{\frac{2\alpha}{1-\alpha}} (1-\alpha)^2 \left(1 + O\left(\frac{\log m}{m}\right)\right) \geq \frac{\sigma(1-\alpha)^2}{e^2} \left(1 + O\left(\frac{\log m}{m}\right)\right)$$

where the last step is valid because $1 \geq \alpha^{\frac{\alpha}{1-\alpha}} \geq e^{-1}$ for $0 \leq \alpha \leq 1$.

Since we partition in sets small enough to make the verifications not significant, the cost is simply $O(r/r' n)$, i.e. $e^2 rn/(\sigma(1-\alpha)^2) = O(rn/\sigma)$.

Observe that this means a linear algorithm for $r = O(\sigma)$ (taking the error ratio as a constant), and that for $\alpha > 1 - e/\sqrt{\sigma}$, the cost is $O(rn)$, not better than the trivial solution (i.e. $r' = 1$ and hence no superimposition occurs).

Pattern Partitioning We have now jr patterns to search with $\lfloor k/j \rfloor$ errors. The error level is the same for subproblems (recall that the subpatterns are of length m/j). Since we group as many subpatterns of the same pattern as possible, we have that if our sets have size r', they have sub-patterns of at most $1 + \lceil r'/j \rceil$ distinct patterns. Therefore, the cost of a verification is $O(m^2 r'/j)$ and our equation to define r' is $a^{m/j} = j/(r'm^2)$, i.e.

$$r' = \left((j/m^2)^j \sigma^{m-k} \alpha^{2k} (1-\alpha)^{2(m-k)} \right)^{\frac{1}{m-k+j}}$$

To approximate this value we expand j, whose formula is obtained in [4]. We rephrase it as $j = (m-k)d$, with $d = d(\alpha, w) = \left(1 + \sqrt{1 + \frac{w\alpha}{1-\alpha}}\right)/w = O(1/\sqrt{w})$. The order is valid because $1 - \alpha \geq e/\sqrt{\sigma}$ (past this point the verifications are too many even with no superimposition). We then have $r' = (d/m)^{\frac{d}{d+1}} \sigma^{\frac{1}{d+1}} \alpha^{\frac{2\alpha}{(1-\alpha)(d+1)}} (1-\alpha)^{\frac{d+2}{d+1}}$, and the total cost is $O(jr/r' n)$, i.e.

$$\frac{d^{\frac{1}{d+1}} m^{\frac{2d+1}{d+1}} rn}{\sigma^{\frac{1}{d+1}} \alpha^{\frac{2\alpha}{(1-\alpha)(d+1)}} (1-\alpha)^{\frac{1}{d+1}}} = \frac{dm\,rn}{\sigma \alpha^{\frac{2\alpha}{1-\alpha}}(1-\alpha)} \left(1 + O\left(\frac{1}{\sqrt{w}}\right)\right) = O\left(\frac{m}{\sigma\sqrt{w}} rn\right)$$

which is linear for $r = O(\sigma\sqrt{w}/m)$ (the order is valid for constant α), and is *useful* when the total cost is less than rn, i.e. $dm < \sigma\alpha^{\frac{2\alpha}{1-\alpha}}(1-\alpha)$. This condition that can be pessimistically bounded by $\alpha \leq 1 - e^2 m/(\sigma\sqrt{w})$.

Automaton Partitioning The analysis for this case is similar to the simple one, except because each step of the large automaton takes time proportional to the total number of subautomata. In [4] we show that this number is $O(k(m-k)/w)$. Therefore, the cost formula is

$$\frac{e^2}{(1-\alpha)^2\sigma} \frac{k(m-k)}{w} rn = \frac{e^2 m^2 \alpha}{\sigma w(1-\alpha)} rn = O\left(\frac{m^2}{\sigma w} rn\right)$$

which is linear in n for $r = O(\sigma w/m^2)$, and *useful* for $\alpha \leq \sigma w/((em)^2 + \sigma w)$.

5.2 Exact Partitioning

In [4] we analyze this algorithm as follows. Except for verifications, the search time is linear (e.g. by using an Aho-Corasick machine [1] although, as mentioned before, we use an algorithm which is faster in practice). Hence, we are interested in the cases where there are few verifications.

Since we cut the pattern in $k + 1$ pieces, they are of length $\lfloor m/(k + 1) \rfloor$ and $\lceil m/(k + 1) \rceil$. On average, half of the subpatterns have each length. The probability of each piece matching is $1/\sigma^{\lfloor m/(k+1) \rfloor}$ (the case $1/\sigma^{\lceil m/(k+1) \rceil}$ is exponentially smaller than this one, so we disregard it). Ignoring equal pieces, the probability of any piece matching is $(k + 1)/(2\sigma^{\lfloor m/(k+1) \rfloor})$.

We can easily extend that analysis to the case of multiple search, since we have now $r(k+1)$ pieces of the same length. Hence, the probability of verifying is $r(k+1)/(2\sigma^{\lfloor \frac{m}{k+1} \rfloor})$. This must be $O(1/m^2)$ (i.e. the cost of one verification, since in this algorithm we know which pattern to verify) to ensure that verifications do not affect on average the linearity of the search. This happens pessimistically for $\alpha \leq 1/(3 \log_\sigma m + \log_\sigma r)$ (although roundoffs must be observed in practice). The method is better than r sequential searches for $\alpha \leq 1/3 \log_\sigma m$.

6 Experimental Results[2]

We experimentally study our algorithms and compare them against previous work. We tested with 1 Mb of random text ($\sigma = 30$) and lower-case English text. The patterns were randomly generated in the first case and randomly selected from the the same text (at the start of non-stopwords) in the second case. We use a Sun SparcStation 4 running Solaris 2.4, with 32 Mb of RAM, and $w = 32$. Each data point was obtained by averaging the Unix's user time over 10 trials.

We first show the degree of parallelism achieved by our different algorithms, in terms of the ratio between the parallel version and r applications of the same single-pattern algorithm. Figure 2 shows the behavior in terms of r, and Figure 3 in terms of k. We observe that for low error ratio the parallel versions take 0.3 to 0.6 of their sequential versions, and that exact partitioning works better for more patterns, while the others quickly stabilize. On the other hand, only automaton partitioning works well for a moderate number of errors, while the other two quickly become worse than their single-pattern counterparts. We tried other m and r values with very similar results (omitted here for lack of space).

We compare now our algorithms against Muth-Manber [11] and Navarro [13]. Figure 4 shows a comparison for $k = 1$ and varying r. Exact partitioning is the best for a moderate number of patterns (i.e. near 60 patterns, except for the bad case of short English patterns). For more patterns, Muth-Manber is better. Pattern and automaton partitioning are better than Muth-Manber for a small number of patterns (10-15) and better than Navarro on random text.

Figure 5 shows a comparison (excluding Muth-Manber because $k > 1$) for fixed r and varying k. Exact partitioning is the fastest algorithm for low error

---

[2] We thank Robert Muth and Udi Manber for their implementation of [11].

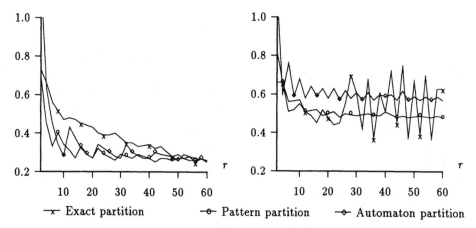

Fig. 2. Parallelism ratio for $m = 10$ and $k = 1$. The left plot is for random text ($\sigma = 30$), the right plot for English text.

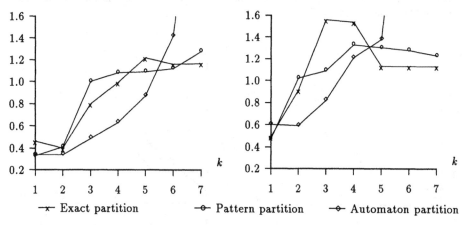

Fig. 3. Parallelism ratio for $m = 10$ and $r = 15$. The left plot is for random text ($\sigma = 30$), the right plot for English text.

ratios. For moderate error ratios the verifications make it useless and Navarro becomes the best algorithm. When the error ratio increases further, automaton partitioning becomes the best choice. The usefulness of this last algorithm ends for a large number of errors, when it becomes similar to r sequential searches.

7 Concluding Remarks

We are working on a number of heuristic optimizations to our algorithms, for instance

- If the patterns have different length, we truncate them to the shortest one when superimposing automata. We can select cleverly the substrings to use,

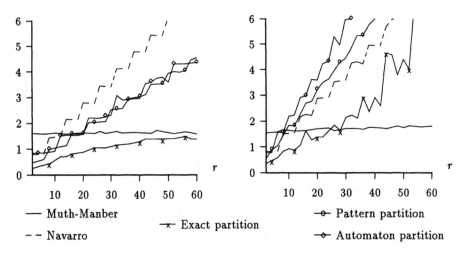

Fig. 4. Times in seconds for $m = 10$ and $k = 1$. The left plot is for random text ($\sigma = 30$), the right plot for English text.

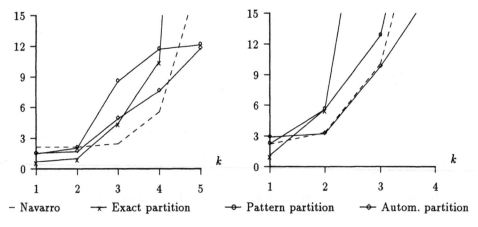

Fig. 5. Times in seconds for $m = 10$ and $r = 15$. The left plot is for random text ($\sigma = 30$), the right plot for English text.

since having the same character at the same position in two patterns improves the filtering mechanism.

- We used simple heuristics to group subpatterns in superimposed automata. These can be improved to maximize common letters too.

- We are free to partition each pattern in $k+1$ pieces as we like in exact partitioning. This is used to minimize the expected number of verifications when the letters of the alphabet do not have the same probability of occurrence (e.g. in English text). We have an $O(m^3)$ dynamic programming algorithm to select the best partition. The same can be done in superimposed automata.

References

1. A. Aho and M. Corasick. Efficient string matching: an aid to bibliographic search. *CACM*, 18(6):333–340, June 1975.
2. R. Baeza-Yates. Text retrieval: Theory and practice. In *12th IFIP World Computer Congress*, volume I, pages 465–476. Elsevier Science, Sept. 1992.
3. R. Baeza-Yates and G. Navarro. A fast heuristic for approximate string matching. In N. Ziviani, R. Baeza-Yates, and K. Guimarães, editors, *Proc. of WSP'96*, pages 47–63, 1996. ftp://ftp.dcc.uchile.cl/pub/users/gnavarro/wsp96.2.ps.gz.
4. R. Baeza-Yates and G. Navarro. A faster algorithm for approximate string matching. In *Proc. of CPM'96*, pages 1–23, 1996. ftp://ftp.dcc.uchile.cl/pub/users/gnavarro/cpm96.ps.gz.
5. R. Baeza-Yates and C. Perleberg. Fast and practical approximate pattern matching. In *Proc. CPM'92*, pages 185–192. Springer-Verlag, 1992. LNCS 644.
6. W. Chang and J. Lampe. Theoretical and empirical comparisons of approximate string matching algorithms. In *Proc. of CPM'92*, pages 172–181, 1992. LNCS 644.
7. W. Chang and E. Lawler. Sublinear approximate string matching and biological applications. *Algorithmica*, 12(4/5):327–344, Oct/Nov 1994.
8. Z. Galil and K. Park. An improved algorithm for approximate string matching. *SIAM J. of Computing*, 19(6):989–999, 1990.
9. D. Greene, M. Parnas, and F. Yao. Multi-index hashing for information retrieval. In *Proc. FOCS'94*, pages 722–731, 1994.
10. G. Landau and U. Vishkin. Fast parallel and serial approximate string matching. *J. of Algorithms*, 10:157–169, 1989.
11. R. Muth and U. Manber. Approximate multiple string search. In *Proc. of CPM'96*, pages 75–86, 1996.
12. E. Myers. A sublinear algorithm for approximate keyword searching. *Algorithmica*, 12(4/5):345–374, Oct/Nov 1994.
13. G. Navarro. Approximate string matching by counting. Submitted, ftp://ftp.dcc.uchile.cl/pub/users/gnavarro/count.ps.gz, 1997.
14. P. Sellers. The theory and computation of evolutionary distances: pattern recognition. *J. of Algorithms*, 1:359–373, 1980.
15. D. Sunday. A very fast substring search algorithm. *CACM*, 33(8):132–142, Aug. 1990.
16. E. Sutinen and J. Tarhio. On using q-gram locations in approximate string matching. In *Proc. of ESA'95*. Springer-Verlag, 1995. LNCS 979.
17. T. Takaoka. Approximate pattern matching with samples. In *Proc. of ISAAC'94*, pages 234–242. Springer-Verlag, 1994. LNCS 834.
18. E. Ukkonen. Algorithms for approximate string matching. *Information and Control*, 64:100–118, 1985.
19. E. Ukkonen. Finding approximate patterns in strings. *J. of Algorithms*, 6:132–137, 1985.
20. A. Wright. Approximate string matching using within-word parallelism. *Software Practice and Experience*, 24(4):337–362, Apr. 1994.
21. S. Wu and U. Manber. Fast text searching allowing errors. *CACM*, 35(10):83–91, Oct. 1992.
22. S. Wu, U. Manber, and E. Myers. A sub-quadratic algorithm for approximate limited expression matching. *Algorithmica*, 15(1):50–67, 1996.

Applied Computational Geometry – Abstract

David P. Dobkin *

Department of Computer Science, Princeton University, Princeton NJ 08 544, USA,
dpd@cs.princeton.edu

Abstract. Computational Geometry has been a thriving research area for the past 20 years. During that time, the field has grown from a handful of researchers working on a small set of problems to a full-blown research area with multiple conferences and journals and many hundreds of researchers. The initial motivation for the field was to develop algorithms that would find application in practice in other fields.

In this talk, I will trace some of the original history of the field to see where it came from. Next, I demonstrate some application areas where computational geometry has been applied. These applications are largely drawn from computer graphics and visualization. Techniques that apply to sampling problems as well as progressive refinements of meshes will be shown. In addition, algorithms for graph layout will be considered.

These applications will be presented as case studies. Some begin with a problem for which solution techniques need to be developed. Others begin with a technique and problems are found to which the technique applies. In each case, there is an underlying implementation that justifies the work. Difficulties of the implementation will also be discussed.

---

* This work has been supported in part by NSF Grant CCR-9643913 and by the US Army Research Office under Grant DAAH04-96-1-0181

Checking the Convexity of Polytopes and the Planarity of Subdivisions* (Extended Abstract)

O. Devillers[1]   G. Liotta[2]   F. P. Preparata[3]   R. Tamassia[3]

[1] INRIA, BP 93, 06902 Sophia Antipolis Cedex, France
odevil@sophia.inria.fr
[2] Dipartimento di Informatica e Sistemistica, Università di Roma "La Sapienza"
via Salaria 113, 00198 Roma, Italy
liotta@dis.uniroma1.it
[3] Center for Geometric Computing, Department of Computer Science
Brown University, Providence, RI 02912–1910, USA
franco@cs.brown.edu   rt@cs.brown.edu

Abstract. This paper studies the problem of verifying the correctness of geometric structures. We design optimal checkers for convex polytopes in two and higher dimensions, and for various types of planar subdivisions, such as triangulations, Delaunay triangulations, and convex subdivisions. Our checkers are simpler and more general than the ones previously described in the literature. Their performance is studied also in terms of the degree, which characterizes the arithmetic precision required.

1   Introduction

The development of checkers for geometric structures is justified by the expectation that it is easier to evaluate the quality of the output than the correctness of the algorithm producing it, and is further motivated by the increasing availability of geometric software on the Internet (see, e.g., [1]), and by the emerging client-server distributed models of geometric computing over the Web (see, e.g., [2]). Mehlhorn *et al.* [19] identify three fundamental features of a good checker: *correctness*, *simplicity*, and *efficiency*.

In this paper, we consider checkers for subdivisions in two and higher dimensions. In particular, we consider two-dimensional planar subdivisions and convex polytopes in a fixed dimension d. The subdivision to be checked can be either a primitive structure, or a derived structure computed from a primitive one (e.g., the Voronoi diagram of a set of sites in the plane). More formally, we consider checkers whose input consists of:

- a geometric graph Γ (i.e., a graph with coordinates assigned to its vertices), which is claimed to induce a subdivision S;
- an optional primitive structure P, from which S is claimed to be derived;

---

* Research supported in part by the U.S. Army Research Office under grant DAAH04-96-1-0013, by the National Science Foundation under grant CCR-9423847, and by the EC ESPRIT Long Term Research Project ALCOM-IT under contract 20244.

- an optional *certificate* C, provided to facilitate the task of the checker; and
- a predicate $\mathcal{P}$ stating a property of S.

The task of the checker is to verify that a subdivision S induced by geometric graph Γ and satisfying predicate $\mathcal{P}$ does indeed exists. The checker either *accepts* Γ, or *rejects* Γ producing evidence that no such subdivision S exists.

Consider the following two examples:

- Γ is a graph with (x, y) coordinates assigned to the vertices; P is a set of points in the plane; C is a circular ordering of the edges incident on each vertex of Γ; and predicate $\mathcal{P}$ states that S is the Voronoi diagram of P.
- Γ is a graph with (x, y, z) coordinates assigned to the vertices; P and C are not defined; and predicate $\mathcal{P}$ states that S is a convex 3D polytope.

Clearly, the availability of a suitable certificate may be crucial to the efficiency of the checker. We consider three scenarios for the type of checker available:

Arbitrary-Certificate Scenario: This is the most favorable scenario for the checker. Namely, the checker can specify an arbitrary certificate C to be provided as additional input. The size $|C|$ of the certificate should be $O(|S|)$ if S is itself primitive, and $O(|S|+|P|)$ if S is derived from a primitive structure P. Also, if S is derived from P, the certificate should be computable as a byproduct of an optimal algorithm that constructs S from P without using additional asymptotic time or space. A checker that operates within the Arbitrary-Certificate Scenario is called an *A-checker*.

Consider, for example, the problem of verifying that a polygon Γ is the convex hull S of a set P of primitive points in the plane. A useful certificate C would consist, for each point p of P, either of a vertex of Γ coincident with p, or of a triplet a, b, c of vertices of Γ such that p is contained in the triangle $\triangle(a, b, c)$. It has been shown that there exists an A-checker using certificate C for verifying that a polygon Γ is the convex hull S of the points of a set P in linear time $O(|G|+|P|)$ (see, e.g., [21]).

Topology-Certificate Scenario: In this intermediate scenario, the checker has available a certificate C that describes the (claimed) topology of the subdivision S. A checker that operates within the Topology-Certificate Scenario is called a *T-checker*.

In the previous example of the planar convex-hull verification, a certificate C for the Topology-Certificate Scenario would consist simply of the circular sequence of the vertices of polygon Γ. With this certificate, it is possible to perform in linear time only a partial verification that polygon Γ is the convex hull S of the points of P. Namely, one can verify in time $O(|G|)$ that Γ is a convex polygon (see Section 3). However, verifying that the points of P that are not vertices of Γ are interior to Γ requires time $\Omega(\log |G|)$ per point.

For the more complex problem of verifying that a 3D geometric graph Γ realizes a convex polytope S, a certificate C for the Topology-Certificate Scenario would be a data structure that describes a planar embedding of Γ, e.g., circularly-sorted adjacency lists.

No-Certificate Scenario: This is the least favorable scenario for the checker. Namely, no certificate C is available to the checker, which must perform the verification using only geometric graph Γ. A checker that operates within the No-Certificate Scenario is called an *N-checker.*

The Arbitrary-Certificate Scenario follows the program checking paradigm pioneered by Blum and Kannan [4]. On the negative side, it requires that the algorithms constructing the subdivision be modified to produce the specified certificate. On the positive side, A-checkers are often faster and simpler to implement than other types of checkers, and their correctness is usually easily established. Sullivan *et al.* [21] show A-checkers for planar convex hull, sorting, and shortest path algorithms. A-checkers for d-dimensional convex hulls are also discussed in [19].

The Topology-Certificate Scenario requires a natural type of certificate, which many algorithms for constructing subdivisions are likely to produce by default. Mehlhorn *et al.* [19] present a T-checker for certifying the convexity of a d-dimensional polytope and mention T-checkers for several other geometric structures, including Delaunay triangulations and Voronoi diagrams.

The No-Certificate Scenario is likely to occur in various application contexts, such as CAD models, where 3D subdivisions are represented as "polygon soups", i.e., collections polygons with no topological information [11]). Also, the availability of N-checkers is important when one tries to incorporate in a program modules developed by others, whose source code may be difficult to understand or modify.

This paper contains a systematic study of checkers for various types of planar subdivisions and for convex polytopes in two and higher dimensions. We advocate a new requirement that good geometric checkers should satisfy, and present simple and efficient N-checkers for several structures for which only A-checkers and T-checkers have been so far designed. Specifically, our contributions can be summarized as follows.

1. As an additional measure of effectiveness for a checker, we adopt the notion of *degree* [5, 16, 17], which takes into account the number of bits required by the checker to carry out error-free computations. A good checker should have degree no higher than the problem at hand allows. We give lower bounds on the degree of checkers for planar subdivisions and convex polytopes, and present optimal-degree checkers.

2. We present a new T-checker for convex polytopes that is simpler than the one given in [19]. Our T-checker works in any dimension and recursively reduces the verification of a d-dimensional polytope to the verification of an associated $(d-1)$-dimensional polytope. It is optimal with respect to both the time complexity and the degree. The design of our T-checker for convex polytopes reveals new combinatorial and geometric properties that may be of independent interest.

3. We present linear-time optimal degree T-checkers for triangulations, convex

subdivisions, and general planar subdivisions. Such checkers use as subroutines elementary graph algorithms and do not require to test the planarity of the underlying input graph.

4. Extending the above results on T-checkers, we present linear-time optimal degree N-checkers for triangulations and convex subdivisions. This solves significant special cases of an open problem mentioned by Kirkpatrick [14] on the existence of an $o(n \log n)$ algorithm to verify the planarity of a geometric graph.

5. As a further application, we give linear-time optimal-degree N-checkers for Delaunay triangulations, locally minimum-weight triangulations, and Delaunay diagrams. Finally, we give a linear-time optimal-degree N-checker for three-dimensional convex polytopes.

Near the completion of our investigations, we became aware of two ongoing projects on the design of T-checkers for planar subdivisions, including triangulations and convex subdivisions. A manual [18] describing the functionality of C++ functions that implement T-checkers for Delaunay triangulations, Voronoi diagrams and convex planar subdivisions is available from Mehlhorn's Web page. A manuscript in progress [20] contains characterizations of triangulations and convex planar subdivisions similar to those of the present paper. There is therefore an undeniable shared objective between our research and that of Mehlhorn *et al.* Any minor overlaps of independently obtained results are largely offset by the fact that our proof techniques are different from theirs, and our checkers work also within the No-Certificate Scenario, while maintaining the same efficiency as the T-checkers of [18, 20].

The rest of this paper is organized as follows. Preliminaries and lower bounds on the degree of checkers are in Section 2. Our T-checker for convex polytopes is presented in Section 3. Section 4 is devoted to T-checkers for triangulations, convex subdivisions, and general planar subdivisions. N-checkers are studied in Section 5. Finally, open problems are given in Section 6. Due to space limitations, detailed proofs have been omitted in this extended abstract.

2 Preliminaries

We start with definitions for geometric graphs, ordered graphs, and planarity. We then recall the notion of degree of geometric algorithms, introduced in [16, 17] (a related concept is defined in [5]). Finally, we present lower bounds on the degree of checkers for convex polytopes and planar subdivisions.

A d-dimensional *geometric graph* is a graph drawn with straight-line edges in d-dimensional space, i.e., a graph whose vertices have d-dimensional coordinates. In the following, we often denote with Γ a geometric graph, and with G its underlying combinatorial structure. To simplify the notation and when the context is not misleading, we may denote with G both the geometric graph and its underlying combinatorial structure.

A two-dimensional geometric graph Γ is *planar* if it has no crossing edges, i.e., any two edges of Γ intersect only at a common vertex. For every planar graph G, there exists a planar geometric graph Γ with underlying graph G, i.e., every planar graph admits a planar straight-line drawing (see, e.g., [10]). However, a geometric graph with an underlying planar graph is not necessarily planar.

A planar geometric graph Γ determines a *planar subdivision* S, i.e., a partition of the plane into regions called faces. Planar subdivision S is said to be *induced* by Γ. A planar subdivision is *convex* if the boundary of each face is a convex polygon. A planar subdivision S induced by geometric graph Γ is *maximal* if there is no other planar geometric graph Γ' such that Γ is a subgraph of Γ'. In a maximal planar subdivision, the boundary of each internal face is a triangle, and the boundary of the external face is a convex polygon. A maximal planar subdivision is also called a *triangulation*.

A graph G is *ordered* (or G has an *ordering*) if for each vertex v of G, a circular ordering of the edges incident on v is given. The ordering of a graph is usually denoted with Ψ. A two-dimensional geometric graph Γ has a natural ordering associated with it, called *the ordering of Γ*, given by the clockwise circular sequence of the edges incident on each vertex. An ordering Ψ of a graph induces a set of directed *circuits*, where the edge (v, w) following (u, v) in a circuit is the successor of (u, v) in the circular ordering of the edges incident on v. Every edge of the graph is traversed by exactly two circuits, once in each direction.

For a geometric graph Γ and its ordering Ψ, we call *outer circuit* the circuit induced by Ψ that contains the first edge following clockwise a horizontal rightward ray emanating from the topmost vertex of Γ, and *internal circuits* the remaining circuits induced by Ψ. If Γ is planar, then the circuits induced by Ψ are the boundaries of the faces of the subdivision S induced by Γ. In particular, the outer circuit is the boundary of the external face traversed clockwise, and the internal circuits are the boundaries of the internal faces traversed counterclockwise. The ordering Ψ of an ordered graph G is *planar* if there exists a planar geometric graph Γ whose underlying graph is G and whose ordering is Ψ. A planar ordering of G is associated with a planar (topological) embedding of G. A graph G admits a planar ordering only if it is planar. Also, the number of distinct planar orderings (topological embeddings) of a planar graph can be super-exponential.

An algorithm has *degree d* if its test computations involve the evaluation of multivariate polynomials of arithmetic degree at most d [16, 17]. We state two lower bounds on the degree of the checkers that will be studied in the rest of the paper. The proof is omitted in this extended abstract.

Theorem 1. *A checker for d-dimensional convex polytopes has degree at least d.*

Theorem 2. *A checker for planar subdivisions has degree at least 2.*

3 T-checkers for Convex Polytopes

I this section, we describe the design of a T-checker for convex polytopes in a fixed dimension d. The input to the checker is a geometric graph Γ in E^d, and a certificate describing the topology of the polytope supposedly induced by Γ.

Because of its appeal to intuition, we consider the question for $d \leq 3$ (specifically for $d = 3$) and indicate later how to extend the result to higher dimensions. A 3-dimensional polytope has the topology of a planar graph and, without loss of generality, we may assume that all faces are triangles (simplicial polytope). In addition, for each edge a condition of convexity (local convexity) must be satisfied. Specifically for edge $p_1 p_2$ ($p_i = (x_i, y_i, z_i)$), with incident facets $p_1 p_2 p_3$ and $p_1 p_2 p_4$, the simplex $(p_1 p_2 p_3 p_4)$ (see Figure 1(a)) lies in the interior half-space of each of the two hyperplanes supporting the facets. This is equivalent to specifying that the sign of the determinant Δ_{1234} shown in Figure 1(b) is positive.

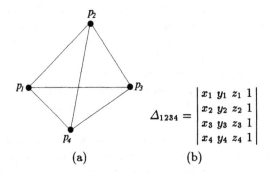

$$\Delta_{1234} = \begin{vmatrix} x_1 & y_1 & z_1 & 1 \\ x_2 & y_2 & z_2 & 1 \\ x_3 & y_3 & z_3 & 1 \\ x_4 & y_4 & z_4 & 1 \end{vmatrix}$$

(a)   (b)

Fig. 1. (a) The simplex $(p_1 p_2 p_3 p_4)$. (b) Determinant Δ_{1234}.

It has been shown by Mehlhorn *et al* [19] that a locally convex plane-bounded surface (without boundary) is the surface of a convex polytope if and only if any ray from a point q in E^d lying on the negative side of all facets intersects the interior of exactly one facet. Therefore if a ray intersects the interiors of more than one facet, then the surface is not a convex polytope.

We now discuss an equivalent but simpler criterion. We will prove the validity of our criterion by showing its equivalence to the criterion of Mehlhorn *et al.* We say that an edge e of a 3-dimensional polytope Γ is a *seam edge* if a vertical (i.e., parallel to the z-axis) plane by e leaves both facets incident on e in the same closed half space and the interior of the upper facet in the open half space.

Lemma 3. *If polytope Γ is locally convex at each edge, then the subgraph Γ' of Γ induced by the seam edges of Γ is a collection of cycles.*

Assume now that there is a point q, on the interior side of each facet, such that a ray r from q intersects the interior of more that one facet, say $f_1, f_2, \ldots, f_s$. Without loss of generality we discuss $s = 2$. Consider the vertical half-plane α containing r and the intersection of α with the surface defined by Γ, and let p_i be the intersection of r with f_i. Points p_1 and p_2 belong each to a closed curve in α (since Γ is without boundary). Starting from p_i and proceeding on this curve by increasing z we shall traverse a polygonal curve. Since this curve is bounded, it will have a point of maximum z in the half-plane α, denoted u_i. Analogously, if we proceed from p_i by decreasing z we will reach a point of minimum z in α, denoted b_i. If u_i does not belong to the vertical line by q, then it admits a horizontal supporting in α; otherwise the angle formed by the ascending polygonal line at u_i with the vertical by q is ≤ 0. Analogously, b_i either admits a horizontal supporting plane or the angle of the polygonal line is ≥ 0. We conclude that traversing the subchain $u_i b_i$ downwards we reach a point s_i which is the first to admit a vertical supporting plane, i.e., s_i belongs to the seam.

If we now project the seam and point q on the horizontal plane, point q projects to q', half-plane α projects to a ray by q', and each s_i projects to a point on this ray. Since intersection is preserved by projection, we conclude that any ray from q' intersects the projected seam σ' as many times as a ray from q intersects the surface of Γ.

We now show that for a locally convex polytope, the criterion of [19] translates into an analogous criterion applied to the projected seams, i.e., the projected seam edges must form a single cycle. The following lemma enables us to reduce by one the dimensionality of the criterion.

Lemma 4. *Let Γ be a locally convex polytope and let σ' be the collection of polygons in the xy-plane obtained by projecting the seam edges of Γ along the z-axis. Each cycle of σ' is locally convex.*

Sketch of Proof. Let $|w_1, w_2, w_3.w_4|$ denote the determinant whose i-th row is $[x(w_i), y(w_i), z(w_i), 1]$; $|w_1, w_2, w_3|$ is analogously defined. Let (u_1, v) and (v, u_s) be two seam edges incident on v and obviously consecutive in their cycle. If these two edges are also consecutive around v, then facet (u_1, v, u_2) trivially projects to a (locally convex) triangle. So for $s \geq 3$ let $(u_2, v), \ldots, (u_{s-1}, v)$ be the intermediate edges clockwise around v (clockwise for an external observer). For $s=3$, by the local convexity test applied to edge (u_1, u_3) we conclude that segment (u_1, u_3) is internal to the polytope. For $s > 3$ the local convexity at (u_3, v) yields $|u_2, v, u_3, u_1| \geq 0$ and the local convexity at (u_4, v) yields $|u_3, v, u_4, u_2| \geq 0$. The latter is equivalent to $|u_2, v, u_3, u_4| \geq 0$. Now, for any real $0 \leq \alpha \leq 1$, we have:

$$\alpha|u_2, v, u_3, u_1| + (1 - \alpha)|u_2, v, u_3, u_4| \geq 0$$

i.e., $|u_2, v, u_3, \alpha u_1 + (1 - \alpha)u_4| \geq 0$, which shows that segment (u_1, u_4) is internal to the polytope (since any of its points is a convex combination of u_1 and u_4). Iterating the argument on $u_5, \ldots, u_s$, we conclude that segment (u_1, u_s) is internal to the polytope, i.e., $|u_1, v, u_s, u_j| \geq 0$ for any $2, \ldots, s - 1$. We observe that $|u_1, v, u_s, u_j|/|u_1, v, u_s|$ is the signed distance of point u_j from plane (u_1, v, u_s). Since u_j is above this plane, $|u_1, v, u_s, u_j| \geq 0$ implies $|u_1, v, u_s| > 0$ and establishes the lemma.

Therefore the previous process can be applied to σ' in the plane. Now the seam (referred to as 2-seam, for 2-dimensional seam) is formed by the vertices of σ' having a supporting line parallel to the y-axis. Specifically, $v \in \sigma'$ belongs to the 2-seam if the mentioned supporting line leaves both edges incident on v in the same closed half-plane, and the interior of the upper edge in the open half-plane. We project all seam vertices to the x-axis and conclude that Γ is a convex polytope if and only if the latter projection consists of the boundary of a one-dimensional convex set, that is, it consists of exactly *two* points.

We observe that the test for membership of an edge in the seam is a subcomputation of its test for local convexity. Namely, let $\Delta_{1234} > 0$ be the condition for local convexity on edge (p_1p_2) (see Figure 2).

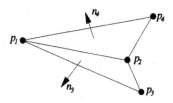

Fig. 2. The outward normals n_4 and n_3.

The outward normal n_4 to $p_1p_2p_4$ has z-component

$$\begin{vmatrix} x_2 - x_1 & y_2 - y_1 \\ x_4 - x_1 & y_4 - y_1 \end{vmatrix} = \begin{vmatrix} x_1 & y_1 & 1 \\ x_2 & y_2 & 1 \\ x_4 & y_4 & 1 \end{vmatrix} = \Delta_{124}$$

and n_3 has z-component

$$-\begin{vmatrix} x_1 & y_1 & 1 \\ x_2 & y_2 & 1 \\ x_3 & y_3 & 1 \end{vmatrix} = -\Delta_{123}$$

Therefore, if we evaluate Δ_{1234} by expansion according to the z-column, we obtain (for free) the minors required to test vertical support ($\text{sign}(\Delta_{124}) = \text{sign}(\Delta_{123})$).

Our T-checker for 3-dimensional convex polytopes is shown in Fig. 3. The certificate describing the topology is used in the local convexity test. We maintain two counters L and R, and, for every vertex v of Γ, a pointer $seam(v)$ to another vertex of Γ. Function $seamvertex(u, v, w)$ tests for seam membership in two dimensions and, depending upon whether v is not a seam vertex, is a left seam vertex, or is a right seam vertex, it returns 0, 1, or 2, respectively. Specifically, v is a right seam vertex either if both u and w are to its left, or if it lies above the vertex with the same abscissa while the other vertex lies to its left. A left seam vertex is analogously characterized.

The above approach can be generalized to higher dimensions. Let Γ be a simplicial polytope in d dimensions. A j-facet of Γ is a simplex defined by j linearly independent points; a conventional facet is a d-facet. Given points $p_1, \ldots, p_j$ we

```
L := 0; R := 0;
foreach vertex v do seam(v) := nil
foreach edge e = pq do
        if e is not locally convex
            then "reject"
            else if e is a seam edge
                then if seam(p) = nil
                        then seam(p) := q
                        else b := seamvertex(q, p, seam(p))
                             if b = 1 then L := L + 1
                     if seam(q) = nil
                        then seam(q) := p
                        else b := seamvertex(p, q, seam(q))
                             if b = 2 then R := R + 1

if L = 1 and R = 1
    then "accept"
    else "reject"
```

Fig. 3. T-checker for 3-dimensional convex polytopes.

let $[p_1 \ldots p_j]$ denote the $j \times (d+1)$ matrix whose i-th row is $(x_1^{(i)}, \ldots, x_d^{(i)}, 1)$; $\|p_1 \ldots p_j\|$ is the determinant of the leftmost j columns of $[p_1 \ldots p_j]$.

The initial local convexity test applies to a $(d-1)$-facet. Specifically, let $(d-1)$-facet F be defined by $p_1, \ldots, p_{d-1}$, indexed so that $\|p_1 \ldots p_{d-1}\| > 0$. F is shared by two d-facets respectively defined by $p_1, \ldots, p_{d-1}, p_d$ and $p_1, \ldots, p_{d-1}, p_{d+1}$, for two additionally independent points p_d and p_{d+1}. We say that F is *locally convex* if $\|p_1 \ldots p_{d-1} p_d\| > 0$ and $\|p_1 \ldots p_{d-1} p_d p_{d+1}\| > 0$. We also say that f belongs to the d-seam if $\|p_1 \ldots p_{d-1} p_d\| > 0$ and $\|p_1 \ldots p_{d-1} p_{d+1}\| \leq 0$. A $(d-1)$-facet in the d-seam is easily shown to belong to a hyperplane parallel to the x_d-axis. If we project all d-seam $(d-1)$-facets along the x_d-axis we obtain a $(d-1)$-dimensional polytope Γ', which can be shown to inherit the property of local convexity. The seam test for Γ' now involves membership of $(d-2)$-facets in a $(d-1)$-seam. Each $(d-1)$-facet F in the d-seam shares a $(d-2)$-facet with $(d-1)$ adjacent $(d-1)$-facets, all belonging to the d-seam. Let F' be one such $(d-2)$-facet (for $d = 3$ F' is a point), which is visited twice as all d-seam facets are visited; through the obvious use of a pointer the two sharing $(d-1)$-facets are linked to each other, and F' can be tested for membership in the $(d-1)$-seam. The process proceeds recursively until the 1-seam is reached, at which level a counter is available for incrementing when appropriate.

The results presented in this section are summarized in the following theorem.

Theorem 5. *There exists an optimal T-checker for convex polytopes in any fixed dimension d that runs in linear time and has degree d.*

4 T-checkers for Planar Subdivisions

In this section we consider T-checkers for planar subdivisions that take as input a two-dimensional geometric graph Γ and the ordering Ψ of Γ, and verify whether Γ induces a planar subdivision satisfying a certain predicate (e.g., convex faces).

A building block of such a T-checker is an algorithm that tests whether an ordering Ψ of a graph G is planar. A linear-time algorithm for answering the question was given by Kirkpatrick in [14]. His algorithm considers the circuits induced by Ψ and adds to G a new vertex v_c for each induced circuit c and a new edge (v_c, w) for each vertex w of c. The resulting augmented graph G^* is planar if and only if the ordering Ψ is planar. Thus, the planarity of Ψ can checked by running a planarity-testing algorithm (e.g., [13]) on G^*. Besides its theoretical interest, however, this algorithm may not be the most suited for practical applications, since the implementation of a linear-time planarity testing algorithm is complex and requires sophisticated data structures.

We show that there exists a much simpler solution to the problem of testing in linear time whether an ordering Ψ of a connected graph G is planar. Our algorithm follows from basic results in planarity theory [12]. Namely, we determine the circuits induced by Ψ and check whether their number C is equal to $E - V + 2$, where V and E are the number of vertices and edges of G, respectively. Clearly, this takes linear time.

Lemma 6. *Let Γ be a two-dimensional connected geometric graph with a planar ordering Ψ. If Γ is not planar (i.e., it has crossing edges), then at least one circuit of Γ induced by Ψ is not planar (i.e., it is a self-intersecting polygon).*

By Lemma 6, verifying whether a connected geometric graph Γ with a planar ordering is a planar can be reduced to testing separately whether the circuits induced by Ψ are simple (i.e., not self-intersecting) polygons. Lemma 6 also provides an alternative and shorter proof of a result of [9] (Lemma 4.5), as the following corollary shows.

Corollary 7. *Let Γ be a connected two-dimensional geometric graph. If the ordering Ψ of Γ is planar, all the internal circuits induced by Ψ are triangles, and the outer circuit induced by Ψ is a convex polygon, then Γ is planar and induces a triangulation.*

Observe, however, that Corollary 7 cannot be easily extended to other types of planar subdivisions. For example, Figure 4 shows a triconnected geometric graph (obtained by removing two edges from a triangulation) that has a planar ordering but is itself not planar.

Lemma 8. *A connected two-dimensional geometric graph Γ with ordering Ψ is planar and induces a triangulation if and only if: (i) ordering Ψ is planar; (ii) all the internal circuits induced by Ψ are triangles; and (iii) the outer circuit induced by Ψ is a convex polygon.*

The input to our T-checker for triangulations is a two-dimensional geometric graph Γ with n vertices and the ordering Ψ of Γ. The T-checker constructs the circuits induced by Ψ and then performs the following sequence of 6 simple

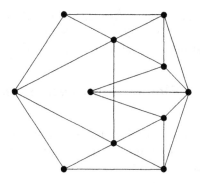

Fig. 4. A triconnected geometric graph whose ordering is planar, but which is itself not planar.

checks. It fails as soon as one of the checks fails, and succeeds if all the checks succeed. Checks 1 and 6 have degree 2. The other checks have degree 1.

1. Check that Ψ is the ordering of Γ.
2. Check that ordering Ψ is planar.
3. Check that Γ is connected.
4. Check that Γ has no more than $3n - 6$ vertices.
5. Check that all internal circuits induced by Ψ are triangles.
6. Check that the outer circuit induced by Ψ is a convex polygon.

Theorem 9. *There exists an optimal T-checker for planar triangulations that runs in linear time and has degree 2.*

Lemma 8 can be generalized to convex planar subdivisions.

Lemma 10. *A two-dimensional connected geometric graph Γ with ordering Ψ is planar and induces a convex planar subdivision if and only if: (i) ordering Ψ is planar; and (ii) all the circuits induced by Ψ are convex polygons.*

Theorem 11. *There exists an optimal T-checker for convex planar subdivisions that runs in linear time and has degree 2.*

As shown in Fig. 4, there exist nonplanar geometric graphs with a planar ordering and an underlying triconnected graph. In order to verify whether such geometric graphs are planar, the strategy suggested by Lemma 6 is to check the simplicity of the circuits induced by the ordering are simple polygons. Testing the simplicity of a polygon with k vertices can be done in $O(k \log k)$ by a simple optimal-degree sweep-line algorithm [3] or in $O(k)$-time as an application of the more elaborate (and suboptimal in terms of degree) triangulation algorithm by Chazelle [6]. This discussion can be summarized as follows.

Lemma 12. *A connected geometric graph Γ with ordering Ψ is planar (and thus induces a planar subdivision) if and only if: (i) ordering Ψ is planar; and (ii) all the circuits induced by Ψ are simple polygons.*

Theorem 13. *There exists a T-checker for planar subdivisions that runs in linear time and has degree 3. Alternatively, there exists a T-checker that runs in $O(n \log n)$ time and has degree 2, where n is the number of vertices.*

5 N-checkers for Planar Subdivisions and Convex Polytopes

In this section, we present N-checkers for planar subdivisions. Our N-checkers compute the ordering of the input geometric graph and then use one of the T-checkers presented in Section 4.

Theorem 14. [14] *Let Γ be a geometric graph with n vertices such that its underlying graph G is planar and has $\lambda(G)$ distinct planar orderings. There exists an algorithm that either computes the ordering of Γ in $O(n + \log \lambda(G))$ time or fails. If it fails, then the ordering of Γ is not planar.*

Lemma 15. *The algorithm of Theorem 14 has degree 2.*

Our N-checker for planar triangulations exploits Theorem 14, Lemma 15, and the fact that the underlying graph G of a triangulation has $\lambda(G) = 2$. Let Γ be a geometric graph with n vertices. By Theorem 14, we can compute the ordering of Γ in $O(n)$ time or else we can conclude that Γ is not a triangulation (we reach this conclusion either when the running time of Kirkpatrick's algorithm exceeds $O(n)$ or when Kirkpatrick's algorithm fails in $O(n)$-time). Once the ordering of Γ is computed, we apply the T-checker of Theorem 9.

Theorem 16. *There exists an optimal N-checker for planar triangulations that runs in linear time and has degree 2.*

Theorem 17. *There exists an N-checker for Delaunay triangulations that runs in linear time and has degree 4.*

Sketch of Proof. First, we verify that Γ is planar and induces a triangulation S using the T-checker of Theorem 16. To verify that S is a Delaunay triangulation, it suffices to check whether for every triangle $T = \Delta(a, b, c)$ of S, the disk through a, b, c contains any of the opposite vertices in the triangles sharing one edge with T. Clearly, this can be done in linear time. Also, the above in-circle test can be executed with a degree 4 algorithm (see, e.g. [16, 17]).

A *locally minimum-weight triangulation* is a triangulation such that for every edge shared by two triangles $\Delta(a, b, c)$ and $\Delta(a, b, d)$, edge bd is the shortest diagonal of the quadrilateral with vertices a, b, c, d. Locally minimum-weight triangulations have been extensively studied for their relationship to minimum-weight triangulations (see, e.g., [15]).

Theorem 18. *There exists an optimal N-checker for locally minimum-weight triangulations that runs in linear time and has degree 2.*

The optimal time complexity of the N-checkers of Theorems 16, 17, and 18 relies on the fact that $\lambda(G) = 2$ for the underlying graph of a triangulation.

The next lemma, shows that for convex planar subdivisions, $\lambda(G)$ is simply exponential.

Lemma 19. *Let G be the underlying graph of a convex planar subdivision. Then the number $\lambda(G)$ of topologically distinct planar orderings of G is $O(2^n)$, where n is the number of vertices of G.*

Sketch of Proof. The proof is based on the results of [7, 8], where planar graphs that admit a planar straight-line drawing with convex faces are characterized.

By combining Theorem 14, Lemma 15, Lemma 19, and Theorem 11 we obtain the following.

Theorem 20. *There exists an optimal N-checker for convex planar subdivisions that runs in linear time and has degree 2.*

Theorem 21. *There exists an N-checker for Delaunay diagrams that runs in linear time and has degree 4.*

Theorem 22. *There exists an optimal N-checker for three-dimensional convex polytopes that runs in linear time and has degree 3.*

6 Open Questions

Several questions remain open. Among them, we consider especially relevant the following.

1. Design effective T-checkers and N-checkers for other types of geometric graphs, such as Gabriel graphs, relative neighborhood graphs, and β-skeletons graphs.

2. Design a simple T-checker for planar subdivisions that runs in linear time and has degree 2.

3. Extend Theorem 22 on optimal N-checkers for 3-dimensional convex polytopes to any fixed dimension d.

Acknowledgments

We would like to thank Giuseppe Di Battista, David Kirkpatrick, Michael Goodrich, and Luca Vismara for useful discussions.

References

1. N. Amenta. Directory of computational geometry software.
   http://www.geom.umn.edu/software/cglist/.

2. G. Barequet, S. S. Bridgeman, C. A. Duncan, M. T. Goodrich, and R. Tamassia. Classic computational geometry with GeomNet. In *Proc. ACM Symp. Computational Geometry*, 1997.

3. J. L. Bentley and T. A. Ottmann. Algorithms for reporting and counting geometric intersections. *IEEE Trans. Comput.*, C-28:643–647, 1979.

4. M. Blum and S. Kannan. Designing programs that check their work. *J. ACM*, 42(1):269–291, Jan. 1995.

5. C. Burnikel. *Exact Computation of Voronoi Diagrams and Line Segment Intersections*. Ph.D thesis, Universität des Saarlandes, Mar. 1996.

6. B. Chazelle. Triangulating a simple polygon in linear time. *Discrete Comput. Geom.*, 6:485–524, 1991.

7. G. Di Battista, R. Tamassia, and L. Vismara. On-line convex planarity testing. In *Graph-Theoretic Concepts in Computer Science (Proc. WG 94)*, Lecture Notes in Computer Science. Springer-Verlag, 1994.

8. G. Di Battista, R. Tamassia, and L. Vismara. On-line convex planarity testing. Technical Report CS-95-26, Dept. Computer Science, Brown Univ., 1995. `ftp://ftp.cs.brown.edu/pub/techreports/95/cs95-26.ps.Z`.

9. G. Di Battista and L. Vismara. Angles of planar triangular graphs. *SIAM J. Discrete Math.*, 9(3):349–359, 1996.

10. I. Fary. On straight lines representation of planar graphs. *Acta Sci. Math. Szeged.*, 11:229–233, 1948.

11. S. Gottschalk, M. C. Lin, and D. Manocha. OBB-tree: A hierarchical structure for rapid interference detection. *Comput. Graph.*, 1996. Proc. SIGGRAPH '96.

12. F. Harary. *Graph Theory*. Addison-Wesley, Reading, MA, 1972.

13. J. Hopcroft and R. E. Tarjan. Efficient planarity testing. *J. ACM*, 21(4):549–568, 1974.

14. D. G. Kirkpatrick. Establishing order in planar subdivisions. *Discrete Comput. Geom.*, 3:267–280, 1988.

15. C. Levcopoulos and A. Lingas. On approximation behavior of the greedy triangulation for convex polygons. *Algorithmica*, 2:175–193, 1987.

16. G. Liotta, F. P. Preparata, and R. Tamassia. Robust proximity queries: an illustration of degree-driven algorithm design. *SIAM J. Computing*. to appear.

17. G. Liotta, F. P. Preparata, and R. Tamassia. Robust proximity queries: an illustration of degree-driven algorithm design. In *Proc. ACM Symp. Computational Geometry*, 1997.

18. K. Mehlhorn and S. Näher. *Checking Geometric Structures*, December 1996. Manual.

19. K. Mehlhorn, S. Näher, T. Schilz, S. Schirra, M. Seel, R. Seidel, and C. Uhrig. Checking geometric programs or verification of geometric structures. In *Proc. 12th Annu. ACM Sympos. Comput. Geom.*, pages 159–165, 1996.

20. K. Mehlhorn, S. Näher, T. Schilz, S. Schirra, M. Seel, R. Seidel, and C. Uhrig. Checking geometric programs or verification of geometric structures. Manuscript, 1997.

21. G. F. Sullivan, D. S. Wilson, and G. M. Masson. Certification of computational results. *IEEE Transactions on Computers*, 44(7):833–847, 1995.

Voronoi Diagrams for Polygon-Offset Distance Functions*

Gill Barequet[1], Matthew T. Dickerson[2], and Michael T. Goodrich[1]

[1] Center for Geometric Computing, Dept. of Computer Science, Johns Hopkins
University, Baltimore, MD 21218. E-mail: [barequet|goodrich]@cs.jhu.edu
[2] Department of Mathematics and Computer Science, Middlebury College,
Middlebury, VT 05753. E-mail: dickerso@middlebury.edu

Abstract. In this paper we develop the concept of a *polygon-offset* distance function and show how to compute the respective nearest- and
furthest-site Voronoi diagrams of point sites in the plane. We provide
optimal deterministic $O(n(\log n + \log m) + m)$-time algorithms, where
n is the number of points and m is the complexity of the underlying
polygon, for computing compact representations of both diagrams.

Keywords: Voronoi diagrams, medial axis, distance function, offset,
convexity, geometric tolerancing.

1 Introduction

The Voronoi diagram of a set $S \subset \mathbb{R}^2$ is a powerful tool for handling many geometric problems dealing with distance relationships. Voronoi diagrams have been
used extensively, for example, for solving nearest-neighbor, furthest-neighbor,
and matching problems in many contexts. The underlying distance function is
typically either the usual Euclidean metric, or more generally a distance function
based upon one of the L_p metrics. There has also been some interesting work
done using convex distance functions, which are extensions of the scaling notion
for circles to convex polygons. Nevertheless, we feel that defining distance in
terms of an offset from a polygon is more natural than scaling in many applications, including those dealing with manufacturing processes. This is because the
relative error of the production tool is independent of the location of the produced feature relative to some artificial reference point (the "origin"). Therefore
it is more likely to allow (and expect) local errors bounded by some tolerance,
rather than errors scaled around some (arbitrary) center.

In this paper we investigate distance functions based on *offsetting* convex
polygons, where the distance is measured along the infinitely extended medial-
axis of such a polygon. While the scaling operation shifts each edge of the polygon *proportionally* to its distance from the origin, the offset operation shifts all
the edges by the *same amount*. Offset polygons are therefore not homothetic
copies of the original polygon (unless the original polygon is regular). We are
interested in the investigation of basic properties of polygon-offset distance functions, with particular attention paid to how they may be used in the definition
and computation of Voronoi diagrams.

* Work on this paper by the first and the third authors has been supported in part by
the U.S. ARO under Grant DAAH04-96-1-0013. Work by the second author has been
supported in part by the National Science Foundation under Grant CCR-93-1714.
Work by the third author has been supported also by NSF grant CCR-96-25289.

1.1 Related Previous Work

We are not familiar with any prior work on defining distance in terms of offset polygons, nor in methods for defining Voronoi diagrams in terms of such functions. Minkowski was the first to study the related notion of *convex distance functions*. He showed, for example, that distance can be defined in terms of a scaling of a convex polygon, and that while such functions do not in general define metrics, they exactly characterize the "distance" functions satisfying the triangle inequality. Chew and Drysdale [CD] show that one can define nearest-neighbor Voronoi diagrams using convex distance functions. They give an $O(nm \log n)$-time method for constructing such diagrams for a set S of n points in the plane, with distance defined by a scaling of an m-edge convex polygon. This is actually quite close to optimal, as they show the Voronoi diagram can be of size $\Theta(nm)$. This work was generalized even further [Kl, KMM, KW, MMO, MMR], showing how to define Voronoi diagrams in a very abstract setting. They also give randomized incremental constructions for this abstract setting that can be applied to nearest- and furthest-neighbor Voronoi diagrams for convex distance functions. The running times of these constructions are expected to be $O(nm \log n)$, but it is possible to use their approaches to construct "compact" Voronoi diagram representations in $O(n \log n \log m + m)$ expected time. For the case of nearest-neighbor diagrams this was improved by McAllister, Kirkpatrick, and Snoeyink [MKS], who give a deterministic $O(n(\log n + \log m) + m)$-time method for constructing a compact Voronoi diagram complete enough for answering nearest-neighbor queries for a given convex distance function in $O(\log n + \log m)$ time. They do not address furthest-neighbor diagrams, however, nor do they address the case when distance is defined by offsets of a convex polygon.

Aichholzer and Aurenhammer present in [AA] an algorithm for constructing the so-called *straight skeleton* of a collection of polygonal chains, which identifies with the medial-axis of a convex polygon. They mention the use of the skeleton as a distance function and give a sketch [ibid., Figure 4] of a Voronoi diagram of two straight segments. They take the set of polygonal chains which define the distance function to also be the set of sites for which the Voronoi diagram is sought. We, however, make the distinction between the convex polygon which defines the distance function and the set of points, sites in the diagram. Aichholzer et al. [AAAG] further study the properties and applications of the straight skeleton of a simple polygon. They remark that the skeleton of a polygon "is no Voronoi-diagram-like structure." Unlike in the cited work the constructions presented in this paper are Voronoi diagrams of *point sets* and not of polygons; the polygon only defines the underlying distance function. We are therefore concerned with a very different set of problems and constructions.

1.2 Our Results

In this paper we formally develop the concept of a convex polygon-offset distance function and explore its properties. One such interesting property is the fact that polygon-offset distance functions do *not* in general satisfy the triangle inequality. This, of course, follows from the contra-positive of Minkowski's characterization theorem, but we provide a simple constructive proof. Nevertheless, we show that convex polygon-offset distance functions satisfy all the topological properties for abstract Voronoi diagrams [Kl, KMM, KW, MMO, MMR]. Finally, given a set S

of n points in the plane, we show how to deterministicly construct compact representations of nearest- and furthest-site Voronoi diagrams for S with respect to an offset distance defined by an m-edge convex polygon in $O(n(\log n + \log m) + m)$ time. We apply the tentative prune-and-search paradigm of Kirkpatrick and Snoeyink [KS], as utilized by McAlister et al. [MKS], to the polygon-offset distance function. Like the method of McAlister et al., our nearest-neighbor diagram is based upon using this approach to design a non-trivial generalization of Fortune's plane-sweep algorithm [Fo], whereas our furthest-neighbor diagram algorithm is based upon applying tentative prune-and-search to a generalization of Rappaport's space-sweeping algorithm [Ra].

2 Preliminaries

2.1 Convex Polygon-Offset Distance Functions

We first define the *offset* of a convex polygon. For simplicity of expression in this extended abstract we define the offset so as to be piecewise-linear, and we show in the full version that this can be extended to the usual definition of an offset (when, for example, an outer offset is made up of alternating line segments and circular arcs).

The *outer* ε-offset of P is obtained by translating each edge $e \in P$ by ε in a direction orthogonal to e and by extending it so as to meet the translations of the edges neighboring to e. The edge e' is trimmed by the lines parallel and at distance ε (outside of P) of the neighboring edges of e. The *inner* offset is defined in a similar way: For each edge $e \in P$ we construct a line parallel to it and at distance ε on the inside of P. If we increase ε continuously we observe that the edge e'' (the inner offset of e) gets smaller continuously and may even "disappear". This happens when the neighboring offset lines meet "before" they intersect with the line that corresponds to e''.

Figure 1 shows inner and outer offsets of a convex polygon for some value of ε. We adopt the notation of [BBDG] and denote the inner and outer ε-offset polygons of P by $I_{P,\varepsilon}$ and $O_{P,\varepsilon}$, respectively. In addition, the notation $O_{P,-\varepsilon}$ is used as a synonym for $I_{P,\varepsilon}$, in particular when the sign of ε is unknown.

As long as we increase ε (for inner offsets) more and more edges of $I_{P,\varepsilon}$ "disappear" in a well-defined order. By increasing the value of ε we actually move all the vertices of P along (and inward) the medial-axis of P. When two neighboring vertices of P meet at a junction of the medial axis the corresponding edge disappears and the two vertices are merged into one vertex. This process continues and at some point we are left with a triangle which in turn eventually collapses into a point for some value of ε. We denote this point as the *center* of P.

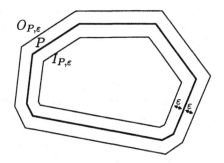

Fig. 1. Inner and outer offsets

Figure 2 illustrates the medial-axis-based offset operation. In case P contains two parallel edges which also define the maximum width of P, $I_{P,\varepsilon}$ becomes (for some value of ε) a segment that can no more be offset inwards. Let ε_0 be the value of ε for which I_{P,ε_0} degenerates into a point or a segment. ε_0 is also the Euclidean distance from the center to the nearest edge of P. The outer offset behaves more "normally": no edges appear or disappear.

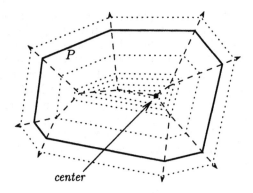

Fig. 2. Offset and the medial-axis

We now define the convex polygon-offset distance function $\mathcal{D}_P$ between two points p and q. Let d be the Euclidean distance from the center to the closest edge in the offset of P centered at p which first touches q. Then $\mathcal{D}_P(p,q) = d/\varepsilon_0$. This gives us a similar normalization where $\mathcal{D}_P(p,p) = 0$, and $\mathcal{D}_P(p,q) = 1$ if q is on the unit polygon.

The polygon-offset distance function has several interesting properties. In general, if the polygon P is not regular then the polygon-offset distance function $\mathcal{D}_P$ is not symmetric. Thus it is not a metric. Moreover, it does not even obey the triangle inequality, as we show below. Nevertheless, we show that the polygon-offset distance function $\mathcal{D}_P$ *can* be used as a basic distance function in a well-defined Voronoi diagram. We will denote the nearest- and furthest-site Voronoi diagrams with respect to $\mathcal{D}_P$ by $\mathcal{V}_P^n$ and $\mathcal{V}_P^f$, respectively.

2.2 Basic Properties

For measuring the distance from some point p to another point q we center P at p and offset it until it hits q. We say that the edge of P that hits q *dominates* it. We show later in this section how the dominating edge may change along some direction. Let us consider what happens as we move points farther from p, the center of P, in some direction defined by a ray $\vec{pq}$. When $\vec{pq}$ crosses an edge of the medial-axis, this corresponds to switching from the dominance of one edge to the dominance of another edge, as each region of the plane subdivision induced by the (infinite-extension of the) medial axis corresponds to an edge. Figure 3 illustrates this situation: The

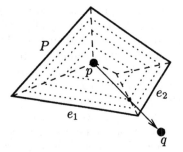

Fig. 3. Dominating edges

direction $\vec{pq}$ (p is positioned at the center of P) is first dominated by $e_1 \in P$ (in the inner offsets of P the edge e_2 does not exist yet), and then it crosses an edge of the medial-axis and becomes dominated by e_2.

We now give some more specific relationships between $\mathcal{D}_P$ and the Euclidean distance E. Refer to Figure 4. Assume that the dominating edge e is moving in a direction ℓ_e from the center point p (so that ℓ_e is perpendicular to e). In moving from p to q along this "wave" e, denote the angle between ℓ_e and $\overrightarrow{pq}$ by θ. A unit move of e in the direction of $\overrightarrow{pq}$ corresponds to an offset of $\cos(\theta)$ along ℓ_e. Intuitively, if the offset from P is growing along a moving "wave" e, the point of contact is like a "surfer" moving on this wave

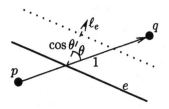

Fig. 4. $\mathcal{D}_P$ and the Euclidean distance

in the direction ℓ_e. Increasing the offset by 1 increases the Euclidean distance in the direction of $\overrightarrow{pq}$ by $1/\cos\theta$. When $\theta = 0$, ℓ_e and $\overrightarrow{pq}$ have the same "speed". As the absolute value of θ increases, a unit offset of an edge outward results in an increasing Euclidean distance. Conversely, as θ increases, a unit Euclidean distance corresponds to a decreasing distance with respect to $\mathcal{D}_P$ along ℓ_e, until it vanishes at $\theta = \pi/2$. Intuitively, the "speed" along $\overrightarrow{pq}$ is inversely proportional to the charged $\mathcal{D}_P$ distance (the offset). At $\theta = \pi/2$ the charge (offset) is 0 and hence the "speed" is infinite.

This explains why the "speed" along any direction v can only decrease. Consider the situation where v crosses an edge of the medial-axis. That is, it moves from a region that corresponds to some edge $e_1 \in P$ to the region that corresponds to another edge e_2. The edge of the medial-axis is the bisector of e_1 and e_2. Elementary geometry shows that v forms a larger angle with the normal to e_1 than that with the normal to e_2 and therefore its "speed" decreases, that is, the offset charge is increased. Hence we have:

Theorem 1. *If p, s, q are colinear points in this order, then $\mathcal{D}_P(p, s) + \mathcal{D}_P(s, q) \leq \mathcal{D}_P(p, q)$ and equality holds if and only if every point on the edge pq is dominated by the same edge of P when P is centered at p.*

Corollary 2. *A polygon-offset distance function does **not** fulfill the triangle inequality.*

Minkowski [KN, p. 15, Theorem 2.3] proved that given a scaled distance function based on a shape S, the triangle inequality holds if and only if S is convex.

3 Nearest-Site Voronoi Diagram

Let us now address the construction of a nearest-site Voronoi diagram. We begin by showing that this notion of distance fits the unifying approach of Klein [KW, Kl]. Klein proposes to replace the *distance* notion by *bisecting curves*. Each pair of sites are separated by curves which divide the plane into two portions each corresponding to one site. The Voronoi region of some site s is the intersection of all the s-portions defined by bisecting curves between s and all the other sites.

3.1 Properties of $\mathcal{D}_P$

In order to show that the polygon-offset distance function is valid for defining Voronoi diagrams, we need to precisely define the nearest-site Voronoi diagram in this context: The Voronoi cell that corresponds to a point p consists of all the points x in the plane for which $\mathcal{D}_P(x,p) \leq \mathcal{D}_P(x,q)$ for all $q \neq p$. Note that this definition measures the distance from the points x and not from the sites! If we reverse the direction of measuring the distance, then we obtain a different Voronoi diagram: that obtained with the former definition using a reflected copy of P.

Klein makes use of the following three properties of *metrics*:
1. The topology induced is the same as that induced by the Euclidean metric;
2. The distance between every pair of points is invariant under translations;
3. Distances are additive along every straight line.
In fact, the third property can be replaced by a weaker property:

3'. Distances are monotonically increasing along every straight line.

Theorem 3. $\mathcal{D}_P$ *has the properties 1, 2, and 3' above. Moreover,*

1. *The bisector of every pair of points consists of disjoint simple curves.*
2. *Curves portions of different bisectors intersect a finite number of times.*

We next follow the approach of [KW] and show additional properties of a polygon-offset distance function and the respective nearest-site Voronoi diagram.

Theorem 4. *Every cell of $\mathcal{V}_P^n$ is connected and $|\mathcal{V}_P^n| = O(nm)$.*

A distance function is called *complete* if it is uniquely defined for every ordered pair of points. The next theorem establishes an important intermediate-position property for the convex polygon-offset distance function.

Theorem 5. $\mathcal{D}_P$ *is complete and for every pair of points p,q in the plane there exists a point $r \notin \overline{pq}$ such that $\mathcal{D}_P(p,r) + \mathcal{D}_P(r,q) = \mathcal{D}_P(p,q)$.*

Proof. Center the polygon P at p. Sweeping the whole plane with $O_{P,\varepsilon}$ by ranging ε from $-\varepsilon_0$ to $+\infty$ we can find the unique ε_1 for which $q \in \partial O_{P,\varepsilon_1}$. Hence $\mathcal{D}_P$ is complete. For the second property we observe that if the whole segment $\overline{pq}$ is dominated by some edge $e_i \in P$ (where the normal to e_i and $\vec{pq}$ form the angle θ_i), then all the points $r \in \overline{pq}$ fulfill the claimed equality (since then $\mathcal{D}_P$ is merely the Euclidean distance multiplied by a constant—$\cos(\theta_i)$). This is not the case when $\overline{pq}$ is dominated by more than one edge of P. Assume that $\overline{pq}$ is first (at the vicinity of p) dominated by the edge e_i and finally (at the vicinity of q) dominated by e_j. We have already shown in Section 2.2 that $\cos(\theta_i) < \cos(\theta_j)$ and that there exists a point $s \in \overline{pq}$ near q such that $\mathcal{D}_P(p,s) + \mathcal{D}_P(s,q) < \mathcal{D}_P(p,q)$. It is fairly easy to find another point s' in the plane for which $\mathcal{D}_P(p,s') + \mathcal{D}_P(s',q) > \mathcal{D}_P(p,q)$: select, for example, a point s' whose distance from $\overline{pq}$ is twice the diameter of P. Now define a function $f(x) = \mathcal{D}_P(p,x) + \mathcal{D}_P(x,q)$. Since $\mathcal{D}_P$ is a continuous distance function, the function $f(x)$ is continuous too. According to the mean-value theorem there exists a point r in the plane for which $f(r) = \mathcal{D}_P(p,q)$ and the claim follows. $\qquad\square$

Note that the theorem does not require r to belong to $\overline{pq}$. The next theorem establishes a crucial extension property. (Its proof is similar to that of Theorem 5 and omitted in this extended abstract.)

Theorem 6. *For every pair of points p, q in the plane there exists a point $r \neq q$ such that $\mathcal{D}_P(p, q) + \mathcal{D}_P(q, r) = \mathcal{D}_P(p, r)$.*

Theorem 7. *Every cell of $\mathcal{V}_P^n$ is simply-connected.*

Proof. The topology induced by $\mathcal{D}_P$ is the same as that of the Euclidean metric (by Theorem 3). $\mathcal{D}_P$ is complete, for every pair of points p, q in the plane there exists a point $r_1 \notin \overline{pq}$ such that $\mathcal{D}_P(p, r_1) + \mathcal{D}_P(r_1, q) = \mathcal{D}_P(p, q)$ (by Theorem 5) and a point $r_2 \neq q$ such that $\mathcal{D}_P(p, q) + \mathcal{D}_P(q, r_2) = \mathcal{D}_P(p, r_2)$ (by Theorem 6). This is sufficient for applying Theorem 4.1 of [KW] and obtaining the claim. □

3.2 Computing the Diagram

We define (following [KW]) the $(\alpha, \overline{\alpha})$-support as follows. A distance function D has an $(\alpha, \overline{\alpha})$-support (for $\alpha \neq \overline{\alpha} \in [0, \pi]$) if it fulfills the following condition: Let p, q be a pair of points in the plane, where $\overrightarrow{pq}$ is of slope α, and let ℓ be a line of slope $\overline{\alpha}$ that passes through q. Then all points r that satisfy $D(p, r) \leq D(p, q)$ lie on the same side of ℓ as p (see Figure 5).

Lemma 8. *There exist angles $\alpha \neq \beta$ s.t. $\mathcal{D}_P$ has both $(\alpha, \overline{\alpha})$- and $(\beta, \overline{\beta})$-support.*

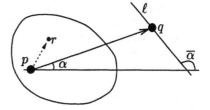

Fig. 5. $(\alpha, \overline{\alpha})$-support

Proof. There exist three edges $e_i, e_j, e_k \in P$ that never disappear in inner offsets of P (in a degenerate case mentioned earlier there are only two such parallel edges). Without loss of generality assume that the normal vectors to e_i and e_j have slopes in $[0, \pi]$. Consider the edge e_i and the normal to it ℓ_i. Set α to the slope of ℓ_i and $\overline{\alpha}$ to the slope of e_i. It is trivial to verify that $\mathcal{D}_P$ has an $(\alpha, \overline{\alpha})$-support. Similarly set β to the slope of ℓ_j and $\overline{\beta}$ to the slope of e_j. Obviously $\alpha \neq \beta$ and the claim follows. □

Now we apply the algorithmic framework of [KW] to establish the following:

Theorem 9. *Let S be a set of n points in the plane. A compact representation of $\mathcal{V}_P^n(S)$ can be computed in $O(n \log n \log m + m)$ expected time.*

Proof. The needed topological properties of $\mathcal{D}_P$ have already been established. This is sufficient for applying Theorem 5.1 of [KW] and obtaining the claim for a constant m. The extra $\log m$ factor comes from the cost of the primitive function (computing a Voronoi vertex), which is called (due to the randomized divide-and-conquer paradigm) $O(n \log n)$ times. Bisecting curves are then represented implicitly by the relative positions of the sites that define them. □

4 Compact Nearest-Site Polygon-Offset Vor. Diagram

We now extend the work of McAllister, Kirkpatrick, and Snoeyink [MKS]. They describe in detail an algorithm for constructing a *compact* nearest-site Voronoi diagram where the underlying scaled distance function is defined by a convex polygon. The compact diagram simplifies the full Voronoi diagram by maintaining a coarse "dual" of it: For each vertex of the full diagram, the compact diagram maintains a set of *spokes* (minimum-length segments from the vertex to the sites), and each polygonal site is replaced by the convex hull of vertices of the site in which spokes occur. (See [MKS, p. 81, Fig. 4] for an illustration.) This allows the complexity of the compact diagram to be $O(n)$ instead of $O(nm)$, where m is the complexity of P.

To save space in this extended abstract, we assume the familiarity of the reader with [MKS] and sketch how to generalize the compact Voronoi diagram to a convex polygon-offset distance function. In order to do that we have to: (1) Make sure that the geometric properties of the compact Voronoi diagram are preserved when we change the distance function to convex polygon-offset; and (2) Verify that we can compute the same information by using the same amount of time when we make this change.

For the first goal we need only note that for every $\varepsilon_1 > \varepsilon_2$ the convex polygon O_{P,ε_1} fully contains the convex polygon O_{P,ε_2} and that two offsets of the same polygon intersect at most twice [BBDG, Theorem 1] (the second property is crucial for applying the *tentative prune-and-search* technique of [KS]). This suffices to prove the spoke properties (Lemmas 2.2–2.6 of [MKS]) and the correctness of the plane-sweep algorithm (Lemmas 3.1–3.4 and Theorems 3.5, ibid.) We omit here the details for lack of space. For the second goal we prove the following theorem:

Theorem 10. *Allowing $O(m)$-time preprocessing, $\mathcal{D}_P(p,q)$ can be computed in $O(\log m)$ time for every pair of points p, q in the plane.*

Proof. We observe first how P is used for measuring regular convex-distance between points in the plane. A *binary search* is performed (in $O(\log m)$ time) in the (cyclic) ordered list of edge-slopes of P for finding the edge e that hits $\overrightarrow{pq}$. Then the distance is easily determined. For a polygon-offset distance function we preprocess P in $O(m)$ time and build an appropriate (tree-like) data-structure that represents the medial-axis of P. Each segment of the data-structure represents a vertex (or several consecutive vertices) of P and each region is attributed by the dominating edge. The data-structure should answer, given a vector $\overrightarrow{pq}$ positioned such that p coincides with the center of P, in which region of the medial-axis q falls. This region (which corresponds to a pair of vertices of P) defines the edge of P that dominates q. The elegance of this approach is in that "disappearing" edges are reflected by regions that correspond to pairs of non-consecutive vertices along P. The query thus requires $O(\log m)$ time. □

Thus we are able to replicate the claims of [MKS] regarding the running time of the primitive operations (Lemmas 3.13–3.15). In particular, the function **vertex**(ABC) (given three sites A, B, and C, it returns the Voronoi vertex around which they appear counter-clockwise) can still be implemented so that it runs in $O(\log n + \log m)$ time. This function is called $O(n)$ times (due to the sweep paradigm), resulting in an $O(n(\log n + \log m) + m)$-time method.

5 Furthest-Site Voronoi Diagram

In this section we show how to construct the furthest-site Voronoi diagram with respect to $\mathcal{D}_P$. For this purpose we show that $\mathcal{D}_P$ fits the framework of Mehlhorn, Meiser, and Rasch [MMR] which follows Klein's unifying approach [KW, Kl]. Mehlhorn et al. show that under some conditions the complexity of the furthest-site Voronoi diagram is $O(n)$, where n is the number of sites.

First we adopt some terminology of [MMR]. Let p, q be a pair of points in the plane. The *dominant* set $M_P(p, q)$ contains all the points that are closer to p than to q with respect to $\mathcal{D}_P$.

Let $S = \{p_i | 1 \le i \le n\} \in \mathbb{R}^2$ be a set of n points. The family $\mathcal{M} = \{M(p_i, p_j) | 1 \le i \ne j \le n\}$ is called a *dominant* system if for all $p_i, p_j \in S$:
1. $M(p_i, p_j)$ is open and nonempty.
2. $M(p_i, p_j) \cap M(p_j, p_i) = \emptyset$ and $\partial M(p_i, p_j) = \partial M(p_j, p_i)$.
3. $\partial M(p_i, p_j)$ is homeomorphic to the open interval $(0, 1)$.

Theorem 11. $\mathcal{M}_P$ *(the family $\mathcal{M}$ with respect to $\mathcal{D}_P$) is a dominant system.*

Assume that for the nearest-site Voronoi diagram every portion of a bisector $\partial M(p_i, p_j)$ is put in the cell of $\min(i, j)$.

A dominance system is called *admissible* if it satisfies in addition:
4. Bisectors intersect finitely-many time.
5. For all $S' \subset S, S' \ne \emptyset$ and for all reordering of indices of points in S:
   (a) Every Voronoi cell is connected and has a nonempty interior.
   (b) The union of all the Voronoi cells is the entire plane.
A dominant system which fulfills only properties 4 and 5b is called *semi-admissible*.

Theorem 12. $\mathcal{M}_P$ *is admissible.*

The proof of Theorem 12 is straightforward (part of the theorem was proven in a previous section) and is omitted in this version of the paper.

We now consider $\mathcal{M}_P^*$, the "dual" of $\mathcal{M}_P$, in which the dominance relation as well as the ordering of the points are reversed.

Theorem 13. *[MMR, Lemma 1] If $\mathcal{M}_P$ is semi-admissible then so is $\mathcal{M}_P^*$. Moreover, the (so-called furthest-site) Voronoi diagram that corresponds to $\mathcal{M}_P^*$ is identical to the (nearest-site) Voronoi diagram that corresponds to $\mathcal{M}_P$.*

Note that admissibility is not preserved when moving to the dual of the dominance system. This corresponds to the fact that cells in the furthest-site Voronoi diagram may well be disconnected.

We have thus shown that the furthest-site Voronoi diagram (with respect to $\mathcal{D}_P$) can be defined in terms of a dominance system (by moving to its dual). Therefore the results of [Kl, KW, KMM] apply to both nearest- and furthest-site Voronoi diagrams. Furthermore, we can benefit from all the results of [MMR] on this diagram, namely, that it is a tree, and that we can compute it by a randomized algorithm in $O(n \log nK)$ time, where K is the time needed by a primitive function which considers 5 sites. Since computing the intersection between two bisectors requires $O(\log m)$ time, this primitive requires also this amount of time in our setting. Hence the overall expected time required for computing the furthest-site Voronoi diagram (with respect to $\mathcal{D}_P$) is $O(n \log n \log m + m)$.

We have followed the approach of [KMM] only for proving the properties of the Furthest-site Voronoi diagram. Our real goal is to obtain an optimal deterministic algorithm for computing the diagram. For this purpose we adopt the plane-sweep (in 3-space) approach of Rappaport [Ra] which follows Fortune's algorithm [Fo], in which axis-parallel cones are emanating from the sites and the plane-sweep detects their intersections and produces the corresponding Voronoi vertices. The major detail that we need to note is that in our setting we have "polyhedral" cones whose intersection can be computed in $O(\log m)$ time. Thus we spend $O(n \log n + m)$ time in preprocessing and $O(n(\log n + \log m))$ time in the sweep (n events, for each we spend $O(\log m)$ time for computing and $O(\log n)$ time for queue operations), resulting in an optimal $\Theta(n(\log n + \log m) + m)$-time algorithm. We again store the diagram compactly, representing bisectors implicitly via the relative positions of the two sites that define them.

6 Conclusion

We develop in this paper the notion of a polygon-offset distance function. This is an extremely important notion for tolerancing in manufacturing. We note not only that this is not a metric, it does not even fulfill the triangle inequality. We describe in detail how to compute the nearest and furthest-site Voronoi diagrams with respect to this distance function. In a companion paper [BBDG] we use these diagrams for solving a tolerancing problem: given a set S of points in the plane and a convex polygon P, find the minimum ε and the respective translation τ for which the ε-*offset annulus* of $\tau(P)$ covers S.

Acknowledgements. The authors wish to thank J. Snoeyink and J. Bose for helpful discussions on Voronoi diagrams. We also thank V. Mirelli for introducing us the concept of polygon offsets and their applications.

References

[AA]   O. AICHHOLZER AND F. AURENHAMMER, Straight skeletons for general polygonal figures in the plane, *Proc. 2nd COCOON*, 1996, 117-126, *LNCS 1090*, Springer Verlag.

[AAAG]   O. AICHHOLZER, D. ALBERTS, F. AURENHAMMER, AND B. GÄRTNER, A novel type of skeleton for polygons, *J. of Universal Computer Science* (an electronic journal), 1 (1995), 752–761

[BBDG]   G. BAREQUET, A. BRIGGS, M. DICKERSON, AND M.T. GOODRICH, Offset-polygon annulus placement problems, these proceedings, 1997.

[CD]   L.P. CHEW AND R.L. DRYSDALE, Voronoi diagrams based on convex distance functions, Technical Report PCS-TR86-132, Dept. of Computer Science, Dartmouth College, Hanover, NH 03755, 1986; Short version: *Proc. 1st Symp. on Comp. Geom.*, 1985, 324–244.

[Fo]   S. FORTUNE, A sweepline algorithm for Voronoi diagrams, *Algorithmica*, 2 (1987), 153–174.

[KN]   J.L. KELLEY AND I. NAMIOKA, Linear Topological Spaces, Springer Verlag, 1976.

[KS]   D. KIRKPATRICK AND J. SNOEYINK, Tentative prune-and-search for computing fixed-points with applications to geometric computation, *Fund. Informaticæ*, 22 (1995), 353–370.

[Kl]   R. KLEIN, Concrete and abstract Voronoi diagrams, *LNCS 400*, Springer Verlag, 1989.

[KMM]   R. KLEIN, K. MEHLHORN, AND S. MEISER, Randomized incremental construction of abstract Voronoi diagrams, *Comp. Geometry: Theory and Applications*, 3 (1993), 157–184.

[KW]   R. KLEIN AND D. WOOD, Voronoi diagrams based on general metrics in the plane, *Proc. 5th Symp. on Theoret. Aspects in Comp. Sci.*, 1988, 281–291, *LNCS 294*, Springer Verlag.

[MKS]   M. MCALLISTER, D. KIRKPATRICK, AND J. SNOEYINK, A compact piecewise-linear Voronoi diagram for convex sites in the plane, *D&CG*, 15 (1996), 73–105.

[MMO]   K. MEHLHORN, S. MEISER, AND Ó'DÚNLAING, On the construction of abstract Voronoi diagrams, *Discrete & Computational Geometry*, 6 (1991), 211-224.

[MMR]   K. MEHLHORN, S. MEISER, AND R. RASCH, Furthest site abstract Voronoi diagrams, Technical Report MPI-I-92-135, Max-Planck-Institut für Informatik, Saarbrücken, Germany.

[Ra]   D. RAPPAPORT, Computing the furthest site Voronoi diagram for a set of disks, *Proc. 1st Workshop on Algorithms and Data Structures, LNCS*, 382, Springer Verlag, 1989, 57-66.

Randomized Algorithms
for that
Ancient Scheduling Problem

Steve Seiden*

Department of Information and Computer Science
University of California
Irvine, CA 92697-3425

Abstract. The problem of scheduling independent jobs on m parallel machines in an online fashion was introduced by Graham in 1966. While the deterministic case of this problem has been studied extensively, little work has been done on the randomized case. For $m = 2$ an algorithm achieving a competitive ratio of $\frac{4}{3}$ was found by Bartal, Fiat, Karloff and Vohra. These same authors show a matching lower bound. Chen, van Vliet and Woeginger, and independently Sgall, have shown a lower bound which converges to $\frac{e}{e-1}$ as m goes to infinity. Prior to this work, no randomized algorithm for $m > 2$ was known. A randomized algorithm for $m \geq 3$ is presented. It achieves competitive ratios of 1.55665, 1.65888, 1.73376, 1.78295 and 1.81681, for $m = 3, \ldots, 7$ respectively. These competitive ratios are less than the best deterministic lower bound for $m = 3, 4, 5$ and less than the competitive ratio of the best deterministic algorithm for $m \leq 7$.

1 Introduction

Consider the following problem: we have m machines. Jobs arrive periodically and are scheduled on a machine. The duration of each job is known when it arrives, and each job is performed equally well on any machine. There are no precedence constraints or release times. Scheduling occurs online; each job is irrevocably assigned to a machine before the next job arrives. By the *height* of a machine, we mean the sum of the sizes of the jobs assigned to that machine. The cost or *makespan* of a schedule is the maximum height of any machine. For a given job sequence σ let $\text{cost}_A(\sigma)$ be the cost incurred by an algorithm A on σ. Let $\text{cost}(\sigma)$ be the cost of the optimal (offline) schedule for σ. A scheduling algorithm A is *c-competitive* if

$$\text{cost}_A(\sigma) \leq c \cdot \text{cost}(\sigma),$$

for all job sequences σ. If the algorithm is randomized, then we use $\text{E}[\text{cost}_A(\sigma)]$ instead of $\text{cost}_A(\sigma)$ in the above definition. The constant c is called the *competitive ratio* of A. The goal is to minimize the competitive ratio.

---

* Email: sseiden@acm.org. Author's address after August 1997: Institut für Mathematik, TU Graz, Steyrergasse 30, A-8010 Graz, Austria

This problem originates with Graham [9], who shows that the deterministic greedy algorithm, which he calls LIST, is

$$2 - \frac{1}{m}$$

competitive. Faigle, Kern and Turàn show that LIST is optimal for $m \in \{2, 3\}$ and show a lower bound of

$$1 + \frac{1}{\sqrt{2}} \approx 1.7071,$$

for $m \geq 4$. Galambos and Woeginger [8] present an algorithm which improves the upper bound to

$$2 - \frac{1}{m} - \epsilon_m.$$

However, this bound converges to 2 as m grows. Bartal, Fiat, Karloff and Vohra [3] show an algorithm which is

$$2 - \frac{1}{70} \approx 1.9857$$

competitive for all m. This was considered to be a major breakthrough, which revitalized interest in the problem. Karger, Phillips and Torng [11] improved this bound by giving an algorithm which achieves a competitive ratio which is less than $2 - 1/m$ for $m \geq 6$ and which is at most 1.945 for all m. Chen, van Vliet and Woeginger [6] show deterministic lower bounds of 1.7310 for $m = 4$ and 1.8319 for sufficiently large m. They also give an algorithm which achieves a competitive ratio of

$$\max \left\{ \frac{4m^2 - 3m}{2m^2 - 2}, \frac{2(m-1)^2 + \sqrt{1 + 2m(m-1)} - 1}{(m-1)^2 + \sqrt{1 + 2m(m-1)} - 1} \right\}.$$

This algorithm gives the best known competitive ratio for small m. Bartal, Karloff, and Rabani [4] show a lower bound of 1.837 for $m \geq 3454$. Recently, Albers [1] has discovered an algorithm which is 1.923-competitive for all $m \geq 2$. She also shows that no algorithm is better than 1.852-competitive for $m \geq 80$.

Little is known about the randomized version of this problem. For $m = 2$ an algorithm achieving a competitive ratio of $\frac{4}{3}$ is presented by Bartal, Fiat, Karloff and Vohra [3]. These same authors show a matching lower bound for $m = 2$ and a lower bound of 1.4 for $m = 3$. Chen, van Vliet and Woeginger [5], and independently Sgall [12], show a general lower bound of

$$1 + \frac{1}{\left(\frac{m}{m-1}\right)^m - 1},$$

which converges to $\frac{e}{e-1} \approx 1.5819$ as m goes to infinity. Note that this bound matches the bound of Bartal et $al.$ for $m = 2$.

Prior to this work, no randomized algorithm for $m > 2$ has been shown to perform better than the best deterministic algorithm. We present an algorithm which we call LINEAR INVARIANT. LINEAR INVARIANT performs better than the best deterministic algorithm for $m \leq 7$. In addition, it performs better than the best deterministic lower bound for $m = 3, 4, 5$. Our results are summarized in Table 1. In the table we also present bounds on the performance of an algorithm which we call CONSTANT INVARIANT. CONSTANT INVARIANT was the predecessor of LINEAR INVARIANT. Both algorithms are generalizations of the two machine algorithm of Bartal *et al.* [3]. Since the competitive ratio upper bound we are able to show for CONSTANT INVARIANT is higher than that of LINEAR INVARIANT, we describe CONSTANT INVARIANT here, but omit our analysis. The

	Randomized	CONSTANT INVARIANT		LINEAR INVARIANT		Deterministic	
m	L.B.	L.B.	U.B.	L.B.	U.B.	L.B.	U.B.
2	1.33333	1.33333	1.33334	1.33333	1.33334	1.50000	1.50000
3	1.42105	1.52576	1.55871	1.51383	1.55665	1.66666	1.66667
4	1.46285	1.62962	1.67951	1.60731	1.65888	1.73101	1.73334
5	1.48738	1.68750	1.75134	1.64859	1.73376	1.74625	1.77084
6	1.50352	1.72800	1.79799	1.65750	1.78295	1.77301	1.80000
7	1.51496	1.75925	1.83040	1.65178	1.81681	1.79103	1.82292

Table 1. Performance of CONSTANT INVARIANT and LINEAR INVARIANT.

deterministic upper and lower bounds for small m are from [6]. The randomized lower bound for $m = 2$ is from [3], while the randomized lower bound for $m \geq 3$ is from [5, 12].

2 The LINEAR INVARIANT Algorithm

We begin by defining notation and describing how LINEAR INVARIANT schedules each job.

Given a schedule, we call the machine in this schedule with greatest height (breaking ties arbitrarily) the *tall* machine. We define similarly the *second-shortest* and *short* machines. The algorithm either places J on the short machine or on the second-shortest machine.

After n jobs have been scheduled, there are $N = 2^n$ schedules $\pi_1, \ldots, \pi_N$ that the algorithm might have chosen. The algorithm keeps track of each of these schedules, and of the probability with which it has chosen each schedule. (The number of schedules which are remembered can be reduced to n, using the technique described in [3].) In schedule π_i, we let x_i, y_i and z_i be the heights of the tall, second-shortest and short machines, respectively. We let p_i be the probability of π_i (we describe how the p_i's are chosen later). We define

$$x = \sum_i p_i \cdot x_i, \qquad y = \sum_i p_i \cdot y_i, \qquad z = \sum_i p_i \cdot z_i.$$

The expected makespan is x. We may assume, without loss of generality, the sum of the sizes of all jobs is 1. We refer to this as the *normalizing assumption*.

LINEAR INVARIANT schedules each job on the two shortest machines, maintaining the invariant

$$x \geq \alpha z.$$

where α is a real constant greater than 1. The intuition behind this invariant is that the algorithm should keep the load on the short machine low in order to 'be prepared' for the possibility of a large job arriving.

This is accomplished as follows: When a new job J arrives, let z_{short} and x_{short} be the new values of z and x, respectively, if J is scheduled on the short machine in every schedule. Let z_{2nd} and x_{2nd} be the new values of z and x if J is scheduled on the second-shortest machine in every schedule. If $x_{short} \geq \alpha z_{short}$ then J is scheduled on the short machine in every schedule. Otherwise, note that $z_{2nd} = z$ and that $x_{2nd} \geq x$. Therefore $x_{2nd} \geq \alpha z_{2nd}$. Let

$$p = \frac{x_{2nd} - \alpha z_{2nd}}{x_{2nd} - \alpha z_{2nd} - (x_{short} - \alpha z_{short})}.$$

It is not hard to see that $p \in [0, 1]$. We schedule J on the short machine with probability p, and on the second-shortest machine with probability $1 - p$. Thus the invariant is maintained at each step.

Note that LINEAR INVARIANT is a generalization of the 2-machine algorithm of Bartal *et al.* [3], since for $m = 2$ we maintain $x = 1 - z \geq \alpha z$ which implies $z \leq 1/(\alpha + 1)$. Their algorithm maintains $z \leq 1/3$.

3 The CONSTANT INVARIANT Algorithm

The predecessor of LINEAR INVARIANT was an algorithm which we call CONSTANT INVARIANT. CONSTANT INVARIANT schedules each job on the two shortest machines, maintaining the invariant

$$z \leq \gamma,$$

where γ is a positive real constant less than $\frac{1}{m}$. CONSTANT INVARIANT is a generalization of the $m = 2$ algorithm of Bartal, Fiat, Karloff and Vohra [3].

Since LINEAR INVARIANT outperforms CONSTANT INVARIANT for $m \geq 3$, we do not present an analysis of CONSTANT INVARIANT here. However, we note that the analysis is very similar to that of LINEAR INVARIANT.

4 An Upper bound for LINEAR INVARIANT

In this section, we prove an upper bound on the competitive ratio of LINEAR INVARIANT. This is accomplished by showing that, when each job is scheduled, a second invariant involving the optimal offline cost is maintained.

For $m = 3, \ldots, 7$ let α and c be as defined by Table 2. Let

m	c	α
3	1.55665	1.806865
4	1.65888	2.040258
5	1.73376	2.123240
6	1.78295	2.113960
7	1.81681	2.103110

Table 2. Values of c and α.

$$G(a) = \max \left\{ \frac{1}{m}, \frac{(m-1)a+1-c}{m-c} \right\}.$$

Note that G is monotone non-decreasing and that $G(a) = 1/m$ for $a \le c/m$ and $G(1) = 1$. We show that the algorithm maintains the invariant that the optimal offline cost is at least $G(x)$. This implies that the competitive ratio is at most c since $x/c \le G(x)$ for $0 \le x \le 1$. This is illustrated for the case $m = 3$ in Figure 1. It is sufficient to show that the optimal offline cost is at least $g(x)$

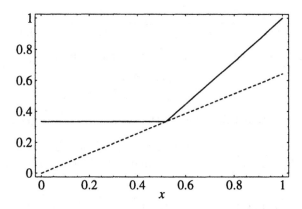

Fig. 1. The functions $G(x)$ (solid) and x/c (dashed) for $m = 3$.

where

$$g(a) = \frac{(m-1)a+1-c}{m-c}.$$

since the optimal offline cost is always at least $\frac{1}{m}$.

Let x and x' be LINEAR INVARIANT's expected makespan before and after J is scheduled, respectively. Let s be the size of J. We prove that the optimal offline cost is at least $G(x')$ by induction. If J is the first job, under the normalizing assumption $x' = 1$ and the optimal offline cost is 1. Therefore, the basis holds since $G(1) = 1$.

Otherwise job J is either scheduled deterministically on the short machine, or is scheduled probabilistically. In the later case we have $x' = \alpha z'$, where z' is the expected height of the short machine after the job is scheduled. The following lemma handles this case.

Lemma 1. *When $x' = \alpha z'$ the optimal offline cost is at least $G(x')$.*

Proof. Under the normalizing assumption we have $z' \leq (1 - x')/(m - 1) = (1 - \alpha z')/(m-1)$ and so $z' \leq 1/(m-1+\alpha)$. Therefore we have $x' \leq \alpha/(m-1+\alpha)$. One can verify that $\alpha/(m - 1 + \alpha) \leq c/m$ for the values in Table 2. As noted above, $G(x') = 1/m$ for $x' \leq c/m$ and the optimal offline cost is always at least $1/m$. □

Lemma 2. *When $s \geq 1$ the optimal offline cost is at least $G(x')$.*

Proof. Since $s \geq x_i - z_i$ for all i we have $x' = z + s$. The optimal offline cost is s. Under the normalizing assumption, we need to show $g(x'/(1 + s)) \leq s/(1 + s)$. First we note that $x \geq \alpha z$ implies $z \leq 1/(\alpha + m - 1)$. We have

$$g\left(\frac{x'}{1+s}\right) = g\left(\frac{z+s}{1+s}\right) \leq g\left(\frac{1/(\alpha+m-1)+s}{1+s}\right).$$

We show that

$$\frac{s}{s+1} - g\left(\frac{1/(\alpha+m-1)+s}{1+s}\right) = \frac{\alpha c + mc - c - \alpha - 2m + 2}{(m-c)(\alpha+m-1)(1+s)} \geq 0.$$

The denominator is positive, so this is true if and only if

$$c \geq \frac{2m - 2 + \alpha}{\alpha + m - 1}.$$

It is easily verified that this holds for the values of c and α given. □

So for the remainder of the analysis we consider the case where J is scheduled deterministically on the short machine and $s < 1$. Under these assumptions we have

$$x' = \sum_i p_i \max\{x_i, z_i + s\} = \sum_{x_i - z_i > s} p_i \cdot x_i + \sum_{x_i - z_i \leq s} p_i(z_i + s).$$

Define

$$A = \{i \mid x_i - z_i > s\}, \qquad B = \{i \mid x_i - z_i \leq s\},$$

$$P = \sum_{i \in A} p_i,$$

$$x_A = \frac{1}{P}\sum_{i \in A} p_i \cdot x_i = E[x_i \mid i \in A],$$

$$z_B = \frac{1}{1-P}\sum_{i \in B} p_i \cdot z_i = E[z_i \mid i \in B].$$

Then we have

$$x' = Px_A + (1 - P)(z_B + s).$$

We define x_B, z_A and y_A analogously. We define x_{opt} to be the optimal offline makespan. Our approach is to formulate a non-linear program in terms of these variables, which shows the inductive hypothesis to be true. To accomplish this, we need to show the variables obey various constraints.

By definition of z_A, z_B and P we have $z = Pz_A + (1 - P)z_B$. We define

$$x_{min} = Px_A + (1 - P)\frac{1 - z_B}{m - 1},$$
$$x_{max} = Px_A + (1 - P)(1 - (m - 1)z_B),$$

and note that $x_{max} \geq x \geq x_{min}$. By definition of the algorithm we have

$$x_{max} \geq x \geq \alpha z.$$

It is easily seen that $x_{opt} \geq \frac{1}{m}(1 + s)$ and that $x_{opt} \geq s$. In order to complete the inductive step, we derive four additional lower bounds on x_{opt}. The first two are given in the following two lemmas.

Lemma 3.

$$x_{opt} \geq x_A - \frac{1 - x_A - z_A}{m - 2}.$$

Proof. Note that $y_A \leq (1 - x_A - z_A)/(m - 2)$. There exists a schedule i such that $x_i - y_i \geq x_A - y_A$. Let S be the size of the largest job ever to arrive. Note that S is a lower bound on the optimal offline makespan. Since every job is scheduled on the short or second shortest machine, we have

$$x_i - y_i \leq S,$$

for all i. □

Lemma 4.

$$x_{opt} \geq g(x_{min}).$$

Proof. By the inductive hypothesis, the optimal makespan is at least $g(x)$. Note that $x \geq x_{min}$ and that g is monotone increasing, and so $g(x) \geq g(x_{min})$. □

For a given schedule, call the most recent job on a machine the *top job* for that machine.

There exists a schedule k such that $x_k \geq x_A$. If several such schedules exist, we pick k to be the least indexed one. Let J^* be the top job on the tall machine of schedule k. Let u be the size of J^*. Our third lower bound is $x_{opt} \geq u$. We now show various constraints on u.

Note that

$$u \geq x_k - y_k \geq x_k - \frac{1 - x_k}{m - 2} \geq x_A - \frac{1 - x_A}{m - 2}.$$

Further note that $u \le x_B \le 1 - (m-1)z_B$, since J^* must appear in every schedule. Let z_{old} be the value of z before J^* was scheduled. We have

$$z_{\text{old}} \le Pz_A + (1-P)(x_B - u)$$
$$\le Pz_A + (1-P)(1 - (m-1)z_B - u).$$

Define $z_{\text{old-max}} = Pz_A + (1-P)(1 - (m-1)z_B - u)$. Let v be the total size of all jobs which arrived after J^*. In schedule k any job scheduled after J^* is not scheduled on the tall machine since J^* is the top job on this machine. At the time J^* was scheduled the value of y_k was at least $x_k - u$. We therefore have that

$$v \le 1 - x_k - z_k - (m-2)(x_k - u) + z_k$$
$$= 1 - (m-1)x_k + (m-2)u$$
$$\le 1 - (m-1)x_A + (m-2)u.$$

Define $v_{\text{max}} = 1 - (m-1)x_A + (m-2)u$.

Lemma 5. *Either $x_{\text{min}} - v_{\text{max}} \le \alpha z$ or*

$$u \ge x_A - \frac{1 - x_A}{m-1}.$$

Proof. Suppose J^* was scheduled probabilistically. Let x^* and z^* be the values of x and z immediately after J^* was scheduled. Then $x^* = \alpha z^*$. We have

$$x_{\text{min}} - v_{\text{max}} \le x - v \le x^* = \alpha z^* \le \alpha z.$$

Otherwise, J^* was scheduled on the short machine of every schedule. We have

$$u \ge x_k - z_k \ge x_k - \frac{1 - x_k}{m-1} \ge x_A - \frac{1 - x_A}{m-1}.$$

$\square$

The next two lemmas provide our fourth lower bound.

Lemma 6. *If $z < (m-1)/(\alpha m(m-1) + m)$ under the normalizing assumption, and there is a schedule i such that $y_i - z_i \ge \Delta$ then there are at least $m-1$ jobs of size at least Δ.*

Proof. Let $\bar{J}$ be the most recently scheduled job which is not on the short machine of schedule i. Let $a = |\bar{J}|$. Note that whatever machine $\bar{J}$ is on, it is the top job on that machine. Further note that the height of the machine on which $\bar{J}$ is scheduled is at least y_i.

Suppose first that $\bar{J}$ was scheduled deterministically. Then it was placed on the short machine in every schedule. Therefore $a \ge y_i - z_i \ge \Delta$. Before $\bar{J}$ was scheduled, the two shortest machines had height at most z_i. Each of the other $m-2$ machines in schedule i has some top job. Each of these jobs was scheduled before $\bar{J}$ and was scheduled on either the shortest or second shortest machine. Each of these $m - 2$ top jobs therefore has size at least $y_i - z_i \geq \Delta$. The total is $m - 1$ jobs of size at least Δ.

Suppose now that $\bar{J}$ was scheduled probabilistically. Let $\bar{x}$ and $\bar{z}$ be the values of x and z immediately after $\bar{J}$ was scheduled. Then $\bar{x} = \alpha\bar{z}$. Since all jobs which came after $\bar{J}$ are on the short machine of schedule i, under the normalizing assumption their total size is at most $1/m$. We have

$$\frac{1 - 1/m - \bar{z}}{m - 1} \leq \bar{x} = \alpha\bar{z},$$

which implies

$$\frac{m - 1}{\alpha m(m - 1) + m} \leq \bar{z} \leq z.$$

$\square$

Lemma 7. *If* $z_{\text{old-max}} \leq (1 - u - v_{\text{max}})(m - 1)/(\alpha m(m - 1) + m)$ *then*

$$x_{\text{opt}} \geq 2\min\{s, u, \Delta\},$$

where $\Delta = mz_B - v_{\text{max}} - 1 + u$.

Proof. There exists a schedule j such that $z_j \geq z_B$. Before J^* was scheduled, the short machine of this schedule had height at most $x_j - u \leq 1 - (m - 1)z_j \leq 1 - (m - 1)z_B$. The height of the second shortest machine was at least $z_j - v \geq z_B - v_{\text{max}}$. Therefore, before J^* was scheduled, schedule j had $y_j - z_j \geq \Delta$. Since $z_{\text{old-max}} \leq (1 - u - v_{\text{max}})(m - 1)/(\alpha m(m - 1) + m)$, by Lemma 6 there were $m - 1$ jobs of size at least Δ. Consider these $m - 1$ jobs, job J^* and the current job. We have a total of $m + 1$ jobs of size at least $\min\{s, u, \Delta\}$. Two of these jobs must appear on the same machine in any schedule. $\square$

To complete the induction, we show that the following non-linear program has solution at most 0. Maximize

$$g\left(\frac{x'}{1 + s}\right)(1 + s) - x_{\text{opt}}, \tag{1}$$

subject to

$$x_{\text{opt}} \geq 2\min\{s, u, \Delta\}$$

$$\text{if } z_{\text{old-max}} \leq \frac{(m-1)(1-u-v_{\max})}{(\alpha m(m-1)+m)}, \tag{2}$$

$$u \geq x_A - \frac{1-x_A}{m-1} \qquad \text{if } x_{\min} - v_{\max} > \alpha z, \tag{3}$$

$$x_{\text{opt}} \geq \frac{1}{m}(1+s), \qquad x_{\text{opt}} \geq s, \tag{4,5}$$

$$x_{\text{opt}} \geq g(x_{\min}), \qquad x_{\text{opt}} \geq x_A - \frac{1-x_A-z_A}{m-2}, \tag{6,7}$$

$$x_{\text{opt}} \geq u, \qquad x_{\max} \geq \alpha z, \tag{8,9}$$

$$u \geq x_A - \frac{1-x_A}{m-2}, \qquad u \leq 1-(m-1)z_B, \tag{10,11}$$

$$0 \leq s, \qquad s \leq 1, \tag{12,13}$$

$$0 \leq P, \qquad P \leq 1, \tag{14,15}$$

$$\frac{1}{m} \leq x_A, \qquad x_A \leq 1, \tag{16,17}$$

$$0 \leq z_A, \qquad z_A \leq \frac{1}{m}, \tag{18,19}$$

$$0 \leq z_B, \qquad z_B \leq \frac{1}{m}, \tag{20,21}$$

over variables x_A, z_A, z_B, P, s, u and x_{opt}.

We break this non-linear program into 5 cases which have no conditional constraints, namely:

1. $z_{\text{old-max}} > (m-1)(1-u-v_{\max})/(\alpha m(m-1)+m)$.
2. $x_{\min} - v_{\max} > \alpha z$.
3. $\min\{s, u, \Delta\} = s$ and $x_{\min} - v_{\max} \leq \alpha z$.
4. $\min\{s, u, \Delta\} = u$ and $x_{\min} - v_{\max} \leq \alpha z$.
5. $\min\{s, u, \Delta\} = \Delta$ and $x_{\min} - v_{\max} \leq \alpha z$.

For each value of $m \in \{3, 4, 5, 6, 7\}$ and the values of c and α given in Table 2, we have shown that each of the 5 non-linear programs has solution at most 0. This was accomplished using a computer program written using `Mathematica`. Details appear in Section 6.

5 Lower Bounds for LINEAR INVARIANT

We now show several lower bounds on the competitive ratio of LINEAR INVARIANT. The analysis differs for $m = 3$ and $m \geq 4$. For $m \leq 5$, we are able to show a lower bound over all choices of α. For $m \geq 6$ we are able to show partial results. We present here the bounds, without proof. Proofs shall appear in the full version.

The first bound we show applies to all $\alpha > 1$, but is relatively weak:

Lemma 8. *For $m \geq 3$, if* LINEAR INVARIANT *is c-competitive then*

$$c \geq L_1(\alpha) = \begin{cases} \dfrac{m+(m-1)\alpha}{1+(m-1)\alpha} & \text{for } 1 \leq \alpha < A, \\[2ex] \dfrac{2(m-1)\alpha^2 + m\alpha}{(1+\alpha)(1+\alpha m - \alpha)} & \text{for } A \leq \alpha, \end{cases}$$

where

$$A = \frac{m - 1 + \sqrt{(m-1)(5m-1)}}{2(m-1)}.$$

Lemma 9. *For $m = 3$, if* LINEAR INVARIANT *is c-competitive then*

$$c \geq L_2(\alpha) = \frac{7 + 13\alpha + 8\alpha^2 + 8\alpha^3 - 2B}{(\alpha - 1)^2(1 + \alpha)},$$

$$B = \sqrt{(1 + \alpha)(1 + 2\alpha)(2 + \alpha^2)(3 + 10\alpha + 5\alpha^2)},$$

for $1.288 \leq \alpha \leq 2.953$.

The following theorem combines the bounds of the previous two lemmas to get a lower bound for $m = 3$.

Theorem 10. *For $m = 3$, if* LINEAR INVARIANT *is c-competitive then $c \geq$* 1.51383.

The next lemma provides a fairly strong bound for general m, but only holds for α in a limited range.

Lemma 11. *For $m \geq 4$, if* LINEAR INVARIANT *is c-competitive then*

$$c \geq L_3(\alpha) = \begin{cases} 1 + \dfrac{m-2}{m-1} & \textit{for} \quad 1 \ \leq \alpha < \ \frac{m-1}{m-2}, \\[2mm] \dfrac{\alpha(m^2 - 3) + 3(m-1)(m-2)}{(m-1)^2(1+\alpha)} & \textit{for} \ \frac{m-1}{m-2} \leq \alpha < \ C, \\[2mm] \dfrac{\alpha(2m-3)(m\alpha - \alpha + 2m - 4)}{(m-1)(1+\alpha)(m\alpha - \alpha + m - 3)} & \textit{for} \ \ C \ \leq \alpha < \ \frac{2(m-2)}{m-3}, \end{cases}$$

where

$$C = \frac{3m - 5 + \sqrt{12m^4 - 72m^3 + 165m^2 - 174m + 73}}{2(m-1)(m-2)}.$$

By combining Lemmas 8 and 11 get the following theorem.

Theorem 12. *For $m \geq 4$, if* LINEAR INVARIANT *is c-competitive then*

$$c \geq \min\{L_3(C), L_1(2(m-2)/(m-3))\}.$$

The actual values obtained from this theorem are displayed in Table 1.

6 Solving the Non-linear Programs

We explain how to solve the non-linear programs of Section 4. The key is that each of the non-linear programs to be solved has the following properties:

- The constraints and objective function of each of these programs are linear for a fixed P.
- The objective function of each is bounded above: $g(x'/(1+s))(1+s) - x_{opt}$ is a rational function with one singularity at $s = -1$. We know that $s > 0$ and that each of the other variables is bounded.
- When P is considered a constant, each variable in the resulting linear program has a coefficient which is a polynomial function (possibly constant) of P.

We describe an algorithm for solving such non-linear programs.

Let $\mathcal{N}$ be a nonlinear program with a variable P such that $\mathcal{N}$ is linear for fixed P. Call the value of this non-linear program val($\mathcal{N}$). Call the linear program which results when P is fixed $\mathcal{N}(P)$. Let V be the set of variables of $\mathcal{N}(P)$. Let $d = |V|$ and n be the dimension and number of constraints of $\mathcal{N}(P)$, respectively. From the theory of linear programming, we know that the maximum of $\mathcal{N}(P)$ is achieved at some vertex of the feasible region of $\mathcal{N}(P)$. At a vertex of $\mathcal{N}(P)$, at least d constraints are satisfied with equality. Let S be a subset of the constraints of $\mathcal{N}(P)$ of size d. There are at most $\binom{n}{d}$ such subsets. Let $E(S)$ be the set S with each constraint changed to an equality. Let $\mathcal{N}(S, P)$ be the linear program obtained from $\mathcal{N}(P)$ by changing each constraint in S to be an equality. When P is considered to be a constant, $E(S)$ is a set of d simultaneous linear equations in d variables. We say that $E(S)$ has a unique solution if we can solve $E(S)$ symbolically (using for instance Gaussian elimination) and get a value for each $v \in V$ which is a function only of P. For each vertex of $\mathcal{N}(P)$ there exists some S such that the set of equations $E(S)$ has a unique solution, i.e., at least d constraints are satisfied with equality at a vertex. Let $\mathcal{N}(S)$ be the non-linear program obtained from $\mathcal{N}(S, P)$ by treating P as a variable again. Let val($\mathcal{N}(S)$) be the value of this non-linear program. It is easily seen that

$$\text{val}(\mathcal{N}) = \max_S \text{val}(\mathcal{N}(S)).$$

We therefore use the following algorithm to find val($\mathcal{N}$).

1. For each subset of constraints S of size d:
   (a) Apply symbolic Gaussian elimination to $E(S)$ to find a value for each $v \in V$ as a function of P. If Gaussian elimination fails to assign a value to each $v \in V$, go to the next subset S.
   (b) Assign each variable its value in the above solution.
   (c) Solve the resulting one dimensional non-linear program.
2. Return the maximum over all one dimensional non-linear programs.

We now describe how each one dimensional non-linear program is solved. Let $f(S, P)$ be the objective function of $\mathcal{N}(S, P)$. Note that each coefficient of a variable in $\mathcal{N}(S, P)$ is a polynomial function of P. Gaussian elimination uses only addition, subtraction, multiplication and division. Therefore $f(S, P)$ is a rational function of P. Hence $f(S, P)$ is continuous except at some set $\{P_0, \ldots, P_i\}$ of singularities. If $f(S, P)$ approaches infinity at some singularity P_j, then P_j cannot be in the feasible region of the program, because we know that the objective function of our original non-linear program is bounded above. Therefore, the maximum of $f(S, P)$ will occur at a point where

$$\frac{df(S, P)}{dP} = 0,$$

or at a point where some constraint is satisfied with equality. We enumerate such points, evaluate the objective function at each of them where all constraints are satisfied, and return the maximum.

We implemented the above algorithm using **Mathematica**, and used it to verify that the non-linear programs of Section 4 have solution at most 0. Our implementation solves several thousand one dimensional non-linear programs for each case. In order to ensure the correctness of our results, all intermediate values are stored symbolically. (Floating point approximations are not used in intermediate calculations.) The only difficulty is in comparing a given algebraic number a (a potential maxima) to 0. In order to show that a is negative we show that $a < -\epsilon$ for some small positive real number ϵ. Potential maxima which are not thus shown to be negative are stored, and the set of all such potential maxima is output at the end of computation. This output is hand checked. To speed computation, rational values of c and α were used. These values are not the 'best possible' values, but are all within 10^{-6} of the best possible.

7 Conclusions

LINEAR INVARIANT is a special case of an algorithm which we conjecture to be $(2 - \epsilon)$-competitive for all m. To describe this algorithm, we need a more general notation. Let h_i be the expected height of the ith tallest machine. In terms of our previous notation $h_1 = x$ and $h_m = z$. Let

$$H_k = \sum_{j=m-i+1}^{m} h_i.$$

For some $k = k(m)$, the algorithm schedules each job on the $k + 1$ shortest machine or the shortest, maintaining the invariant

$$h_1 \geq \alpha \frac{H_k}{k}.$$

Note that with $k(m) = 1$ we have LINEAR INVARIANT. Also note the similarity with the deterministic m-machine algorithm of Bartal et al. [3]. (They set $k(m) =$

$0.445m$.) It is our conjecture that with $k(m) = am$ for some $a \in (0, 1)$, this algorithm is $(2 - \epsilon)$-competitive for some positive ϵ.

It would also be interesting to investigate the algorithm which maintains

$$\frac{H_m - H_k}{m - k} \geq \alpha \frac{H_k}{k}.$$

This is a randomized version of Albers' algorithm [1].

8 Acknowledgment

Thanks to Sandy Irani for many helpful discussions. Thanks to the creators of **Mathematica**. Thanks to Gerhard Woeginger, Howard Karloff and Jiří Sgall for their correspondence.

References

1. ALBERS, S. Better bounds for online scheduling. In *Proc. 29th ACM Symposium on Theory of Computing (STOC)* (1997), pp. 130–139.
2. BARTAL, Y., FIAT, A., KARLOFF, H., AND VOHRA, R. New algorithms for an ancient scheduling problem. In *Proceedings of the Twenty-Fourth Annual ACM Symposium on the Theory of Computing* (May 1992), pp. 51–58.
3. BARTAL, Y., FIAT, A., KARLOFF, H., AND VOHRA, R. New algorithms for an ancient scheduling problem. *Journal of Computer and System Sciences 51*, 3 (Dec 1995), 359–366.
4. BARTAL, Y., KARLOFF, H., AND RABANI, Y. A better lower bound for on-line scheduling. *Information Processing Letters 50*, 3 (May 1994), 113–116.
5. CHEN, B., VAN VLIET, A., AND WOEGINGER, G. A lower bound for randomized on-line scheduling algorithms. *Information Processing Letters 51*, 5 (September 1994), 219–22.
6. CHEN, B., VAN VLIET, A., AND WOEGINGER, G. New lower and upper bounds for on-line scheduling. *Operations Research Letters 16*, 4 (November 1994), 221–230.
7. FAIGLE, U., KERN, W., AND TURÀN, G. On the performance of on-line algorithms for partition problems. *Acta Cybernetica 9*, 2 (1989), 107–19.
8. GALAMBOS, G., AND WOEGINGER, G. An online scheduling heuristic with better worst case ratio than Graham's list scheduling. *SIAM Journal on Computing 22*, 2 (April 1993), 349–355.
9. GRAHAM, R. Bounds for certain multiprocessing anomalies. *Bell Systems Technical Journal 45* (1966), 1563–1581.
10. KARGER, D., PHILLIPS, S., AND TORNG, E. A better algorithm for an ancient scheduling problem. In *Proc. 5th Annual ACM-SIAM Symposium on Discrete Algorithms (SODA)* (1994), pp. 132–140.
11. KARGER, D., PHILLIPS, S., AND TORNG, E. A better algorithm for an ancient scheduling problem. *Journal of Algorithms 20*, 2 (March 1996), 400–430.
12. SGALL, J. *On-Line Scheduling on Parallel Machines*. PhD thesis, Carnegie-Mellon University, 1994.

Optimal Parallel Algorithms for Proximate Points,* with Applications
(Extended Abstract)

Tatsuya Hayashi[1], Koji Nakano[1], and Stephan Olariu[2]

[1] Department of Electrical and Computer Engineering, Nagoya Institute of Technology, Showa-ku, Nagoya 466, Japan

[2] Department of Computer Science, Old Dominion University, Norfolk, VA 23529, USA

Abstract. Consider a set P of points in the plane sorted by x-coordinate. A point p in P is said to be a *proximate point* if there exists a point q on the x-axis such that p is the closest point to q over all points in P. The *proximate points problem* is to determine all proximate points in P. We propose optimal sequential and parallel algorithms for the proximate points problem. Our sequential algorithm runs in $O(n)$ time. Our parallel algorithms run in $O(\log \log n)$ time using $\frac{n}{\log \log n}$ Common-CRCW processors, and in $O(\log n)$ time using $\frac{n}{\log n}$ EREW processors. We show that both parallel algorithms are work-time optimal; the EREW algorithm is also time-optimal. As it turns out, the proximate points problem finds interesting and highly nontrivial applications to pattern analysis, digital geometry, and image processing.

1   Introduction

Let P be a set of points in the plane sorted by x-coordinate. A point p in P is termed a *proximate point* if there exists a point q on the x-axis such that p is the closest point to q over all the points in P. The *proximate points problem* is to determine all proximate points in P. We propose optimal sequential and parallel algorithms for the proximate points problem. The sequential algorithm runs in $O(n)$ time, being therefore optimal. The parallel algorithms run in $O(\log \log n)$ time using $\frac{n}{\log \log n}$ Common-CRCW processors and in $O(\log n)$ time using $\frac{n}{\log n}$ EREW processors. Both these algorithms are work-time optimal; in fact, the EREW algorithm turns out to also be time-optimal. The proximate points problem has interesting and quite unexpected applications to problems in pattern recognition, digital geometry, shape analysis, compression, decomposition, and image reconstruction. We now summarize the main applications of our algorithm for the proximate points problem.

- A very simple algorithm for the convex hull of a set of n planar points sorted by x-coordinate running in $O(\log n)$ time using $\frac{n}{\log n}$ EREW processors and in $O(\log \log n)$ time using $\frac{n}{\log \log n}$ Common-CRCW processors. Our algorithm has the same performance as those of [1, 2, 3], being much simpler and more intuitive.
- Algorithms for the Voronoi map, the distance map, the maximal empty figure and the largest empty figure of a binary image of size $n \times n$. The *Voronoi map* assigns each pixel the position of the nearest black pixel. The *distance map* assigns each pixel the distance to the nearest black pixel. An *empty circle* of the image is a circle filled with white pixels. The *maximal empty circle* is an empty circle included in no other circle. The *largest empty circle* is an empty circle with the largest radius.

* Work supported in part by NSF grant CCR-9522093, by ONR grant N00014-97-1-0526, and by a grant from the Casio Science Promotion Foundation.

- A work optimal algorithm for the Euclidean distance map of a binary image of size $n \times n$ running in $O(\log \log n)$ time using $\frac{n^2}{\log \log n}$ Common-CRCW processors or in $O(\log n)$ time using $\frac{n^2}{\log n}$ EREW processors. We also show that the distance map of various metrics, including the well-known L_k metric ($k \geq 1$), can be computed in the same manner. Fujiwara *et al.* [5] presented a work-optimal algorithm running in $O(\log n)$ time and using $\frac{n^2}{\log n}$ EREW processors and in $O(\frac{\log n}{\log \log n})$ time using $\frac{n^2 \log \log n}{\log n}$ Common-CRCW processors. As we see it, our algorithm has three major advantages over Fujiwara's algorithm. First, the performance of our algorithm for the CRCW is superior; second, our algorithm applies to a large array of distance metrics; finally, our algorithm is much simpler and more intuitive.

- An algorithm for the maximal empty circle and for the largest empty circle of an $n \times n$ binary image running in $O(\log n)$ time using $\frac{n^2}{\log n}$ EREW processors and in $O(\log \log n)$ time using $\frac{n^2}{\log \log n}$ Common-CRCW processors. This algorithm is applicable to various other figures including circles, squares, diamonds, n-gons.

Due to page stringent limitations, the applications to pattern analysis and image processing will be discussed in the journal version of this work. The reader can refer to the companion technical report [6] available from http://cs.odu.edu/~olariu.

2 The proximate points problem: a first look

For a point p in the plane we write $p = (x(p), y(p))$ in the obvious way. We let $d(p, q)$ denote the Euclidean distance between the points p and q.

Throughout this section we consider a set $P = \{p_1, p_2, \ldots, p_n\}$ of n points above the x-axis sorted by increasing x-coordinate. Let I_i, ($1 \leq i \leq n$), be the locus of all the points q' on the x-axis for which $d(q', p_i) \leq d(q', p_j)$, i.e., $q' \in I_i$ if and only if p_i is the closest point to q' over all points in P. Elementary geometry confirms that every set I_i is an interval. Accordingly, $I_1, I_2, \ldots, I_n$ are the *proximate intervals* of P. Notice that some of these intervals may be empty. In case the interval I_i is non-empty, we say that p_i is a *proximate point* of P. A point q on the x-axis is the *boundary* between p_i and p_j if $d(p_i, q) = d(p_j, q)$. Figure 1 illustrates an example of proximate intervals and the Voronoi diagram of P. Clearly, the Voronoi diagram partitions the x-axis into proximate intervals. In Figure 1, p_1, p_2, p_4, p_6, and p_7 are proximate points; the others are not. Observe that the leftmost and rightmost points of P are always proximate points.

For three points p_i, p_j, p_k with $x(p_i) < x(p_j) < x(p_k)$, we say that p_j is *dominated* by p_i and p_k if p_j is not a proximate point of $\{p_i, p_j, p_k\}$. Note that one processor can determine in $O(1)$ time whether p_j is *dominated* by p_i and p_k. A point p_i of P is a proximate point if and only if no pair of points in P dominates p_i.

Let P be as above and let p be a point to the right of P. We wish to compute the proximate intervals of $P \cup \{p\}$. Assume, wlog, that all points in P are proximate points, and let $I_1, I_2, \ldots, I_n$ be the proximate intervals of P. Further, let $I'_1, I'_2, \ldots, I'_n, I'_p$ be the proximate intervals of $P \cup \{p\}$ and refer to Figure 2. There exists a *unique* point p_i, called the *contact point* between P and p, such that

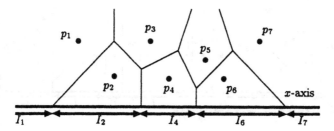

Fig. 1. *Voronoi diagram and proximate intervals*

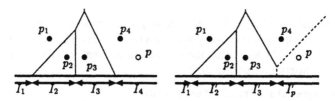

Fig. 2. *Addition of p to* $P = \{p_1, p_2, p_3, p_4\}$.

1. for every j, $(1 < j < i)$, p_j is not dominated by p_{j-1} and p. Moreover, $I'_j = I_j$ and p_j is a proximate point of $P \cup \{p\}$;
2. p_i is not dominated by p_{i-1} and p, and the boundary between p_i and p is in I_i; the left part of I_i, separated by the boundary, is I'_i. The right part of the x-axis is I'_p;
3. for every j, $(i < j \leq n)$, p_j is dominated by p_{j-1} and p. Moreover, I'_j is empty and p_j is not a proximate point of $P \cup \{p\}$.

Next, suppose that P is partitioned into subsets $P_L = \{p_1, p_2, \ldots, p_n\}$ and $P_R = \{p_{n+1}, p_{n+2}, \ldots, p_{2n}\}$. We are interested in updating the proximate intervals in the process or merging P_L and P_R. Let $I_1, I_2, \ldots, I_n$ and $I_{n+1}, I_{n+2}, \ldots, I_{2n}$ be the proximate intervals of P_L and P_R, respectively. We assume, wlog, that all these proximate intervals are nonempty. Let $I'_1, I'_2, \ldots, I'_{2n}$ be the proximate intervals of $P = P_L \cup P_R$, and refer to Figure 3. There exist *unique* proximate points $p_i \in P_L$ and $p_j \in P_R$, called the *contact points* between P_L and P_R, such that

1. for every k, $(1 < k < i)$, p_k is not dominated by p_{k-1} and p_j. Moreover, $I'_k = I_k$ and p_k is a proximate point of P;
2. p_i is not dominated by p_{i-1} and p_j, and the boundary between p_i and p_j is in I_i. The left part of I_i separated by the boundary is I'_i;
3. for every k, $(i < k \leq n)$, p_k is dominated by p_{k-1} and p_j. Moreover, the interval I'_k is empty and p_k is not a proximate point of P;
4. for every k, $(n < k \leq j)$, p_k is dominated by p_i and p_{k+1}. Moreover, the interval I'_k is empty and p_k is not a proximate point of P;
5. p_j is not dominated by p_i and p_{j+1}, and the boundary between p_i and p_j is in I_j. The right part of I_j separated by the boundary is I'_j;
6. for every k, $(j < k < 2n)$, p_k is not dominated by p_i and p_{k+1}. Moreover, $I'_k = I_k$ and p_k is a proximate point of P.

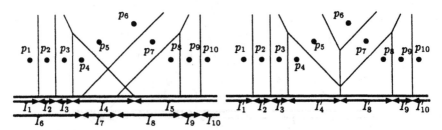

Fig. 3. *Illustrating the merging of two sets of proximate intervals.*

Next, using the two observations above, we propose a simple $O(n)$-time sequential algorithm for finding the proximate points of a set $P = \{p_1, p_2, \ldots, p_n\}$ with $x(p_1) < x(p_2) < \cdots < x(p_n)$. The algorithm uses a stack that, when the algorithm terminates, contains all the proximate points in P.

Algorithm Sequential-Proximate-Points;
1 **for** $i := 1$ **to** n **do**
2   **while** the stack has two or more points **do**
3   **begin**
4       $p' :=$ the top of the stack;
5       $p'' :=$ the next to the top of the stack;
6   **if** p' is dominated by p'' and p_i **then** pop p' from the stack
7   **else** exit while loop
8   **end**;
9   push p_i
10 **end**;

An easy inductive argument shows that at the end of the i-th iteration of the **for** loop, the stack contains all proximate points in $\{p_1, p_2, \ldots, p_i\}$. Once a point is removed from the stack it will never be considered again. Thus, we have

Lemma 1. *The task of finding the proximate points of a set of n points sorted by x-coordinate can be performed in $O(n)$ sequential time.*

3 Parallel algorithms for the proximate points problem

Consider a set $P = \{p_1, p_2, \ldots, p_n\}$ of points sorted by x-coordinate. For every point p_i we use three indices c_i, l_i, and r_i of p defined as:

1. $c_i = \max\{j \mid j \leq i$ and p_j is an proximate point$\}$;
2. $l_i = \max\{j \mid j < c_i$ and p_j is an proximate point$\}$;
3. $r_i = \min\{j \mid j > c_i$ and p_j is an proximate point$\}$.

Note that $l_i < c_i \leq i < r_i$ holds and there is no proximate point p_j such that $l_i < j < c_i$ or $c_i < j < r_i$. If $c_i = i$ then p_i is a proximate point, as shown in Figure 4.

Next, we are interested in finding the contact point between the set P and a point p to the right of P. We assume that for every i, $(1 \leq i \leq n)$, the indices c_i, l_i, and r_i are given and that m, $(m \leq n)$, processors are available.

non-proximate points

proximate points

p_i

p_{l_i} p_{c_i} p_{r_i}

Fig. 4. *Illustrating l_i, c_i, and r_i for p_i*

Algorithm Find-Contact-Point

Step 1 Extract a sample $S(P)$ of size m consisting of the points $p_{c_1}, p_{c_{\frac{n}{m}+1}}, p_{c_{2\frac{n}{m}+1}}, \cdots$ in P. For every k, $(k \geq 0)$, check whether the point $p_{c_{k\frac{n}{m}+1}}$ is dominated by $p_{l_{k\frac{n}{m}+1}}$ and p, and whether $p_{r_{k\frac{n}{m}+1}}$ is dominated by $p_{c_{k\frac{n}{m}+1}}$ and p. If $p_{c_{k\frac{n}{m}+1}}$ is not dominated but $p_{r_{k\frac{n}{m}+1}}$ is dominated, then $p_{c_{k\frac{n}{m}+1}}$ is the desired contact point.

Step 2 Find k such that the point $p_{r_{k\frac{n}{m}+1}}$ is not dominated by $p_{c_{k\frac{n}{m}+1}}$ and p, and $p_{c_{(k+1)\frac{n}{m}+1}}$ is dominated by $p_{l_{(k+1)\frac{n}{m}+1}}$ and p.

Step 3 Execute recursively this algorithm for the set of points $P' = \{p_{r_{k\frac{n}{m}+1}}, p_{r_{k\frac{n}{m}+1}+1},$ $p_{r_{k\frac{n}{m}+1}+2}, \ldots, p_{l_{(k+1)\frac{n}{m}+1}}\}$ to find the contact point.

Since the set P' contains at most $l_{(k+1)\frac{n}{m}+1} - r_{k\frac{n}{m}+1} + 1 \leq \frac{n}{m} - 1$ points, the depth of the recursion is $O(\frac{\log n}{\log m})$. Thus, we have

Lemma 2. *The task of finding the contact point between a set P of n points in the plane sorted by x-coordinate and a point p to the right of P can be performed in $O(\frac{\log n}{\log m})$ time using m CREW processors.*

Next, consider two sets $P_L = \{p_1, p_2, \ldots, p_n\}$ and $P_R = \{p_{n+1}, p_{n+2}, \ldots, p_{2n}\}$ of points in the plane such that $x(p_1) < x(p_2) < \cdots < x(p_{2n})$. Assume that for every i the indices c_i, l_i and r_i are given and that m processors are available. The following algorithm finds the contact points of P_L and P_R.

Algorithm Find-Contact-Points-Between-Sets

Step 1 Extract $\sqrt{m}$ sample points $S(P_L) = \{p_{c_1}, p_{c_{2\frac{n}{\sqrt{m}}+1}}, p_{c_{3\frac{n}{\sqrt{m}}+1}}, \ldots\}$ from P_L. Using Find-Contact-Point and $\sqrt{m}$ of the processors available, determine for each sample point $p_{c_{k\frac{n}{\sqrt{m}}+1}}$, the corresponding contact point $q_{c_{k\frac{n}{\sqrt{m}}+1}}$ in P_R.

Step 2 For each k, $(0 \leq k \leq \sqrt{m} - 1)$, check whether the point $p_{c_{k\frac{n}{\sqrt{m}}+1}}$ is dominated by $p_{l_{k\frac{n}{\sqrt{m}}+1}}$ and $q_{c_{k\frac{n}{\sqrt{m}}+1}}$, and whether the point $p_{r_{k\frac{n}{\sqrt{m}}+1}}$ is dominated by $p_{c_{k\frac{n}{\sqrt{m}}+1}}$ and $q_{c_{k\frac{n}{\sqrt{m}}+1}}$. If $p_{c_{k\frac{n}{\sqrt{m}}+1}}$ is not dominated, yet $p_{r_{k\frac{n}{\sqrt{m}}+1}}$ is, output $p_{c_{k\frac{n}{\sqrt{m}}+1}}$ and $q_{c_{k\frac{n}{\sqrt{m}}+1}}$ as the desired contact points.

Step 3 Find k such that the point $p_{r_{k\frac{n}{\sqrt{m}}+1}}$ is not dominated by $p_{c_{k\frac{n}{\sqrt{m}}+1}}$ and $q_{c_{k\frac{n}{\sqrt{m}}+1}}$, yet $p_{c_{(k+1)\frac{n}{\sqrt{m}}+1}}$ is dominated by $p_{l_{(k+1)\frac{n}{\sqrt{m}}+1}}$ and $q_{c_{(k+1)\frac{n}{\sqrt{m}}+1}}$.

Step 4 Execute recursively this algorithm for $P'_L = \{p_{r_{k\frac{n}{\sqrt{m}}+1}}, p_{r_{k\frac{n}{\sqrt{m}}+1}+1}, p_{r_{k\frac{n}{\sqrt{m}}+1}+2}, \ldots,$ $p_{l_{(k+1)\frac{n}{\sqrt{m}}+1}}\}$ and P_R to find the contact points.

By Lemma 2, Step 1 can be takes $O(\frac{\log n}{\log m})$ time on the CREW model. Steps 2 and 3 run, clearly, in $O(1)$ time. Since P'_L contains at most $\frac{n}{\sqrt{m}} - 1$ points, the depth of recursion is $O(\frac{\log n}{\log m})$. Thus, we have

Lemma 3. *The task of finding the contact points between the sets $P_L = \{p_1, p_2, \ldots, p_n\}$ and $P_R = \{p_{n+1}, p_{n+2}, \ldots, p_{2n}\}$ of points in the plane such that $x(p_1) < x(p_2) < \cdots < x(p_{2n})$ can be performed in $O(\frac{\log^2 n}{\log^2 m})$ time using m CREW processors.*

We now show how to compute the proximate points of a set P of n points in the plane sorted by x-coordinate in $O(\log \log n)$ time on the Common-CRCW. Assume that n processors are available. The idea is simple: first, we determine for every i, the indices c_i, l_i, and r_i. Next, we retain the points p_i for which $c_i = i$.

Algorithm Find-Proximate-Points
Step 1 Partition the set P into $n^{1/3}$ subsets $P_0, P_1, \ldots, P_{n^{1/3}-1}$ such that for every k, $(0 \le k \le n^{1/3} - 1)$, $P_k = \{p_{kn^{2/3}+1}, p_{kn^{2/3}+2}, \ldots, p_{(k+1)n^{2/3}}\}$. For every point p_i in P_k, $(0 \le k \le n^{1/3} - 1)$, determine the indices c_i, l_i, and r_i local to P_k.
Step 2 Compute the contact points of each pair of sets P_i and P_j, $(0 \le i < j \le n^{1/3} - 1)$, using $n^{1/3}$ of the processors available. Let $q_{i,j} \in P_i$ denote the contact point between P_i and P_j.
Step 3 For every P_i, find the rightmost contact point p_{rc_i} among all the points $q_{i,j}$ with $j < i$ and find the leftmost contact point p_{lc_i} over all points $q_{i,j}$ with $j > i$. Clearly, $x(p_{rc_i}) = \max\{x(q_{i,j}) \mid j < i\}$ and $x(p_{lc_i}) = \min\{x(q_{i,j}) \mid j > i\}$.
Step 4 For each set P_i, the proximate points lying between rc_i and lc_i (inclusive) are proximate points of P. Update each c_i, l_i, and r_i.

Clearly, Step 2 runs in $O(\frac{\log^2 n^{2/3}}{\log^2 n^{1/3}}) = O(1)$ time. Step 3 runs in $O(1)$ time as well. The updating of the indices c_i, l_i, and r_i in Step 4 can be performed in $O(1)$ time. We only discuss c_i. In each P_i, the value of c_j, $(rc_i < j \le lc_i)$, is not changed. For all the points p_j with $lc_i < j$, the value of c_j must be changed to lc_i. For all points p_j with $j < rc_i$, the value of c_j is changed to lc_{i-1}, if P_{i-1} has an proximate point. However, if P_{i-1} does not contain a proximate points, we have to find the nearest subset that does. For this, first check whether each P_i has a proximate point. Next, we determine in $O(1)$ time P_k such that $k = \max\{j \mid j < i \text{ and } P_j \text{ contains a proximate point}\}$. Thus, Step 4 can be done in $O(1)$ time. Note that the depth of the recursion is $O(\log \log n)$. Thus, we have

Lemma 4. *An instance of size n of the proximate points problem can be solved in $O(\log \log n)$ time using n Common-CRCW processors.*

Next, we show that the number of processors can be reduced by a factor of $\log \log n$ without increasing the running time. The idea is as follows: begin by partitioning the set P into $\frac{n}{\log \log n}$ subsets $P_1, P_2, \ldots, P_{\frac{n}{\log \log n}}$ each of size $\log \log n$. Next, using algorithm **Sequential-Proximate-Points** find the proximate points within each subset in $O(\log \log n)$ sequential time and, in the process, remove from P all the points that are not proximate points. For every i, $(1 \le i \le \frac{n}{\log \log n})$, let $\{p_{i,1}, p_{i,2}, \ldots\}$ be proximate points in the set P_i.

At this moment, run **Find-Proximate-Point** on $P_1 \cup P_2 \cup \cdots \cup P_{\frac{n}{\log \log n}}$. Since n processors are needed to update the indices c_i, l_i, and r_i in $O(1)$, we will proceed slightly differently. The idea is the following: while executing the algorithm, some of the proximate points will cease to be proximate points. To maintain this information efficiently, we use *ranges* $[L_1, R_1], [L_2, R_2], \ldots, [L_{\frac{n}{\log \log n}}, R_{\frac{n}{\log \log n}}]$ such that for each P_i, $\{p_{i,L_i}, p_{i,L_i+1}, \ldots, p_{i,R_i}\}$ are the current proximate points. While executing the algorithm, P_i may contain no proximate points. To find the neighboring proximate points, we use the pointers $L'_1, L'_2, \ldots, L'_{\frac{n}{\log\log n}}$ and $R'_1, R'_2, \ldots, R'_{\frac{n}{\log\log n}}$ such that

- $L'_i = \max\{j \,|\, j < i$ and the set P_j contains a proximate point$\}$,
- $R'_i = \min\{j \,|\, j > i$ and the set P_j contains a proximate point$\}$.

By using this strategy, we can find the contact point between a point and P in $O(\frac{\log n}{\log m})$ time using m processors as discussed in Lemma 2. Thus, the contact points between two subsets can be found in the same manner as in Lemma 3. Finally, the task of updating L_i, R_i, L'_i, and R'_i in Step 4 can be done in $O(1)$ time by using $\frac{n}{\log\log n}$ processors. To summarize, we have the following result.

Theorem 5. *An instance of size n of the proximate points problem can be solved in $O(\log\log n)$ time using $\frac{n}{\log\log n}$ Common-CRCW processors.*

Finally, we show that algorithm Find-Proximate-Points can be implemented efficiently on the EREW-PRAM. For this recall that one step of an m-processor CRCW can be simulated by an m-processor EREW in $O(\log m)$ time [7]. Consequently, Steps 2, 3, and 4 can be performed in $O(\log n)$ time using n EREW processors, as the CRCW performs these steps in $O(1)$ time using n processors. Let $T_{EREW}(n)$ be the worst-case running time on the EREW. Then, the recurrence describing the EREW time complexity becomes $T_{EREW}(n) = T_{EREW}(n^{\frac{2}{3}}) + O(\log n)$, confirming that $T(n) \in O(\log n)$. Consequently, we have:

Lemma 6. *An instance of size n of the proximate points problem can be solved in $O(\log n)$ time using n EREW processors.*

Using, essentially, the same idea as for the Common-CRCW, we can reduce the number of processors by a factor of $\log n$ without increasing the computing time. Thus, we have

Theorem 7. *An instance of size n of the proximate points problem can be solved in $O(\log n)$ time using $\frac{n}{\log n}$ EREW processors.*

Further, it is not hard to see that Theorems 5 and 7 hold for the case of any L_k metric with $k \geq 1$.

4 Lower Bounds

The main goal of this section is to show that the running time of the Common CRCW algorithm for the proximate points problem cannot be improved while retaining work-optimality. This, in effect, will prove that our Common-CRCW algorithm is work-time optimal. We then show that our EREW algorithm is time-optimal.

The work-optimality of both algorithms is obvious; every point must be accessed to solve the proximate points problem, thus, $\Omega(n)$ work is required of any algorithm solving the problem. Our arguments rely, in part, on the following well known result [7, 8].

Lemma 8. *The task of finding the minimum (maximum) of n real numbers requires $\Omega(\log\log n)$ time on the CRCW provided that $n\log^{O(1)} n$ processors are available.*

Obviously, even if all the input numbers are non-negative, the task still requires $\Omega(\log\log n)$ time. Further, we rely on the following classic result of Cook et al. [4].

Lemma 9. *The task of finding the minimum (maximum) of n real numbers requires $\Omega(\log n)$ time on the CREW (therefore, also on the EREW) even if infinitely many processors are available.*

We shall reduce the task of finding the minimum of a collection A of n non-negative $a_1, a_2, \ldots, a_n$ to the proximate points problem. In other words, we will show that the minimum finding problem can be converted to the proximate points problem in $O(1)$ time.

For this purpose, let $a_1, a_2, \ldots, a_n$ be an arbitrary input to the minimum problem and refer to Figure 5. We construct a set $P = \{p_1, p_2, \ldots, p_{2n}\}$ of points in the plane by setting for every i, $(1 \le i \le n)$, $p_i = (i, \sqrt{a_i + 4n^2 - i^2})$ and $p_{i+n} = (i + n, \sqrt{a_i + 4n^2 - (i+n)^2})$. Notice that this construction guarantees that the points in P are sorted by x-coordinate and that for every i, $(1 \le i \le n)$, the distance between the point p_i and the origin is exactly $\sqrt{a_i + 4n^2}$. The set P that we just constructed has the following property.

Lemma 10. *a_j is the minimum of A if and only if both points p_j and p_{j+n} are proximate points of P.*

Proof. Assume that a_j is the minimum of A. Consider the circle C of radius is $\sqrt{a_j + 4n^2}$ centered at the origin. Clearly, the points a_j and a_{j+n} are on C, while all the other points are outside C. Therefore, there exists a small real number $\epsilon > 0$ such that a_j is the closest points of $(-\epsilon, 0)$ over all points in P, and a_{j+n} is the closest points of $(\epsilon, 0)$. Thus, both p_j and p_{j+n} are proximate points of P. Further, the proximate intervals for p_j and p_{j+n} are adjacent. Thus, no point p_i with $k < i < k+n$ can be a proximate point.

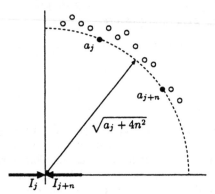

Fig. 5. *Illustrating the construction of P*

Lemma 10 guarantees that we can determine the minimum of A once the proximate points of P are known. Now the conclusion follows immediately from Lemma 8. Thus, we have the following important result.

232

Theorem 11. *Any algorithm that solves an instance of size n of the proximate points problem on the CRCW must take $\Omega(\log\log n)$ time provided that $n\log^{O(1)} n$ processors are available.*

Using exactly the same construction we obtain the following lower bound for the CREW.

Theorem 12. *Any algorithm that solves an instance of size n of the proximate points problem on the CREW (also on the EREW) must take $\Omega(\log n)$ time even if an infinite number of processors are available.*

Obviously, the EREW algorithm presented in Section 3 solving the proximate points problem in $O(\log n)$ time and optimal work runs, within the same resource bounds, on the CREW-PRAM. By Theorem 11 the corresponding CREW algorithm is also time-optimal.

Further, in the case of the L_k metric, for every i, $(1 \leq i \leq n)$, the points $p_i = (i, \sqrt[k]{a_i + (2n)^k - i^k})$ and $p_{i+n} = (i + n, \sqrt[k]{a_i + (2n)^k - (i+n)^k})$ allow us to find the minimum of A. Thus, the above results hold for the L_k metric.

5 Application to the convex hull problem

Let $P = \{p_1, p_2, \ldots, p_n\}$ be a planar set of n points sorted by x-coordinate. We assume, with no loss of generality that for every i, $x(p_i)^2 \leq y(p_i)$. The line segment $p_1 p_n$ partitions the convex hull of P into the *lower hull*, lying below the segment, and the *upper hull*, lying above it. In this section we focus on the computation of the lower hull only, the computation of the upper hull being similar.

Referring to Figure 6 let $Q = \{q_1, q_2, \ldots, q_n\}$ be the set of n points obtained from P by setting for every i, $q_i = (x(p_i), \sqrt{y(p_i) - x(p_i)^2})$. The following result captures the relationship between P and Q.

Lemma 13. *For every j, $(1 \leq j \leq n)$, the point p_j is an extreme point of the lower hull of P if and only if q_j is a proximate point of Q.*

Proof. If $j = 1$ or $j = n$, then p_j is an extreme point and so q_j is a proximate point. Thus, the lemma is correct for $j = 1$ and $j = n$. For a fixed j, $2 \leq j \leq n-1$, let i and k be arbitrary indices such that $1 \leq i < j < k \leq n$. Let b_i and b_k be the boundaries between q_i and q_j and between q_j and q_k, respectively. Notice that $d(q_i, b_i) = d(q_j, b_i)$ implies that $(x(p_i) - x(b_i))^2 + y(p_i) - x(p_i)^2 = (x(p_j) - x(b_i))^2 + y(p_j) - x(p_j)^2$. Thus, we have $2x(b_i) = \frac{y(p_i)-y(p_j)}{x(p_i)-x(p_j)}$. Similarly, we have $2x(b_k) = \frac{y(p_k)-y(p_j)}{x(p_k)-x(p_j)}$.

Notice that the slopes of the segments $p_i p_j$ and $p_k p_j$ equal $2x(b_i)$ and $2x(b_k)$, respectively. It follows that the point p_j lies below the segment $p_i p_k$ if and only if q_j is not dominated by q_i and q_k. In other words, the point p_j is an extreme point of the lower hull if and only if q_j is a proximate point of P.

Lemma 13 suggests the following algorithm for determining the extreme points of the lower hull of $P = \{p_1, p_2, \ldots, p_n\}$.

Algorithm Find-Lower-Hull

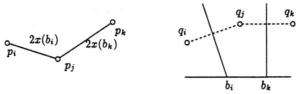

Fig. 6. *Illustrating the transformation of P into Q*

Step 1 Compute the minimum $Y = \min\{y(p_i) - x(p_i)^2 \mid 1 \le i \le n\}$ and construct the set $Q = \{q_1, q_2, \ldots, q_n\}$ such that for every i, $(1 \le i \le n)$, $q_i = (x_i, \sqrt{y(p_i) - x(p_i)^2 - Y})$.

Step 2 Determine the proximate points of Q. Having done that, select p_i as an extreme point in the lower hull whenever q_i is a proximate point.

Clearly, the minimum Y is used to avoid that the argument of the square root be negative. The minimum finding can be done in $O(\log n)$ time using $\frac{n}{\log n}$ EREW processors and in $O(\log \log n)$ time using $\frac{n}{\log \log n}$ CRCW processors [7]. Step 2 can be completed by using the algorithms of Theorems 5 and 7. Thus, we have

Theorem 14. *The task of determining the convex hull of n points sorted by x-coordinate can be performed in $O(\log \log n)$ time using $\frac{n}{\log \log n}$ Common-CRCW processors and in $O(\log n)$ time using $\frac{n}{\log n}$ EREW processors.*

We now state, without proof, two important results implying the optimality of our convex hull algorithm. The proofs can be found in [6].

Theorem 15. *The task of finding the convex hull of n sorted points requires $\Omega(\log \log n)$ time on the CRCW-PRAM provided that $n \log^{O(1)} n$ processors are available.*

Theorem 16. *The task of finding the convex hull of n sorted points requires $\Omega(\log n)$ time on the CREW even if infinitely many processors are available.*

References

1. O. Berkman, B. Schieber, and U. Vishkin. A fast parallel algorithm for finding the convex hull of a sorted point set. *Int. J. Comput. Geom. Appl.*, 6(2):231–241, June 1996.
2. D. Z. Chen. Efficient geometric algorithms on the EREW PRAM. *IEEE Trans. Parallel Distrib. Syst.*, 6(1):41–47, January 1995.
3. W. Chen, K. Nakano, T. Masuzawa, and N. Tokura. Optimal parallel algorithms for computing convex hulls. *IEICE Transactions*, J74-D-I(6):809–820, September 1992.
4. S. A. Cook, C. Dwork, and R. Reischuk. Upper and lower time bounds for parallel random access machines without simultaneous writes. *SIAM J. Comput.*, 15:87–97, 1986.
5. A. Fujiwara, T. Masuzawa, and H. Fujiwara. An optimal parallel algorithm for the Euclidean distance maps. *Information Processing Letters*, 54:295–300, 1995.
6. T. Hayashi, K. Nakano, and S. Olariu. Optimal parallel algorithms for finding proximate points, with applications. Technical Report TR97-01, Hayashi Laboratory, Department of Electrical and Computer Engineering, Nagoya Institute of Technology, February 1997.
7. J. JáJá. *An Introduction to Parallel Algorithms*. Addison-Wesley, 1992.
8. L. G. Valiant. Parallelism in comparison problem. *SIAM J. Comput.*, 4(3):348–355, 1975.

An Efficient Algorithm for Shortest Paths in Vertical and Horizontal Segments

David Eppstein *   David W. Hart *

Abstract. Suppose one has a line segment arrangement consisting entirely of vertical and horizontal segments, and one wants to find the shortest path from one point to another along these segments. Using known algorithms one can solve this in $O(n^2)$ time and in $O(n^2)$ space. We show that it is possible to find a shortest path in time $O(n^{1.5} \log n)$ and space $O(n^{1.5})$. Furthermore, if only one path endpoint is known in advance, it is possible to preprocess the arrangement in the same time and space and then find shortest paths for query points in time $O(\log n)$.

1 Introduction

Consider the problem of finding the shortest path from one location to another in a large city. Many cities have two sets of parallel streets, where the streets of one set are perpendicular to the streets of the other. These streets may have sections closed or there may be points where the streets do not go through.

Geometrically, we can represent the city streets as line segments, with endpoints where the streets do not go through. What we have is an arrangement of horizontal and vertical line segments. We would like to find a shortest path from one point on some segment to another point on another segment.

There are many other routing problems that fit this model, such as routing pipes and wiring in a building, routing utility lines in a city, and routing connections on a circuit board.

One field in particular where this problem has application is VLSI design. In VLSI design, one has a set of cells, which represent logic elements in the circuit and which have fixed locations. In between cells, one has vertical and horizontal channels where one can route connections. The cost of the connection is the path length, so one wants to minimize path lengths in this arrangement of vertical and horizontal segments.

De Rezende et al. [8] study a related problem of finding a rectilinear shortest path from one point to another in a region where there are obstacles consisting of disjoint, axis parallel rectangles. The constraint is that the path must not pass through the rectangles. For this problem, the authors give a solution that takes $O(n \log n)$ preprocessing time, given the path starting point, and $O(\log n)$ time per query for each path ending point.

Their problem is less general, requiring that the obstacles be disjoint rectangles. For the problem presented here of an arrangement of segments, we make

---

* Department of Information and Computer Science, University of California, Irvine, CA 92697-3425. Email to eppstein@ics.uci.edu and dhart@ics.uci.edu.

no assumptions about what obstacles or other constraints end each segment; we allow any set of vertical and horizontal segments.

The problem discussed here is also related to other problems in computational geometry. The first is, given an arrangement consisting of line segments with $O(1)$ different orientations, find a shortest path between two points. We have not found an easy way to extend the methods of this paper to solve this problem. The second, more general problem is, given any arrangement of lines, find a shortest path between two points in less than $O(n^2)$ time and space. Both of these problems remain open. Bose et al. [2] describe a method for approximating the shortest path in a line arrangement.

Returning to the problem discussed in this paper: one can solve this using known algorithms as follows. Given n line segments, we can compute the arrangement for these segments in time $O(n^2)$. The arrangement has $O(n^2)$ vertices. An algorithm by Klein, et al. [5] computes shortest paths in a planar graph in linear time. Using this gives a time of $O(n^2)$ to find a shortest path.

We would like to improve this time, but of perhaps greater concern is that this approach also takes $O(n^2)$ space. We show here that one can compute shortest paths faster and in less space. We describe an algorithm that computes a path in time $O(n^{1.5} \log n)$ and space $O(n^{1.5})$.

We also look at the case where only one path endpoint is known in advance. We show that one can preprocess the segments in the same time and space bounds, and one can then find a shortest path for query points in time $O(\log n)$.

2 The Basic Concept

The intuition for what our algorithm does is easy to describe. Consider first where the undesirable $O(n^2)$ bound arises. We have only n lines, but these produce as many as $n^2/4$ vertices at intersection points. For the number of intersection points to be large, each of the n lines must intersect most of the other n lines. If this were not the case, then there would be few intersection points, and we could solve the problem quickly.

In an arrangement of vertical and horizontal segments, if one has most vertical lines intersecting most horizontal lines, the arrangement contains many regions like the one shown in Figure 1, where all vertical lines intersect all horizontal lines. Suppose there is a path starting point at one side of this region and a path destination point on another side. The algorithms described above look at paths going through all n^2 vertices and edges. But clearly there are very many different paths in this grid that are all equally good. An algorithm that checks all these paths is doing a tremendous amount of unnecessary work.

2.1 Algorithm Overview

It is possible to avoid looking at many of the vertices and edges in the arrangement. Consider a grid of lines and the rectangular shape consisting of the four outermost segments. If the path starts at one side of this rectangle and leaves

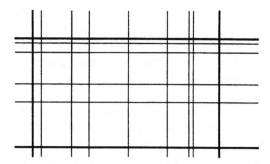

Fig. 1. A region where all vertical segments intersect all horizontal segments, defined here to be a *complete grid*. The bold lines are *boundary segments* and all other lines are *interior segments*.

from a side with different orientation, then a path through the interior of the rectangle has corresponding to it an equally short path traveling only on the outermost lines. If a path starts at one side of this rectangle and leaves from the opposite side, it may be necessary to travel through the interior of the rectangle, but all of this travel can be on a single segment.

Thus, the algorithm described here constructs a graph that eliminates most of the intersection points and the corresponding, numerous edges, but the graph produced has a shortest path equal to the shortest path in the original arrangement.

The algorithm finds regions where all vertical lines intersect all horizontal lines. It adds the outermost segments—the boundary of a rectangle—to the graph, including vertices at all intersections of these lines with other lines. It adds lines crossing the rectangle but omits vertices at the intersections of these lines with each other; this eliminates many of the intersection points found in the arrangement.

2.2 Definitions

Suppose one can identify a rectangular region where every vertical line intersects every horizontal line. This region is what we term a *complete grid* (see Figure 1). A paper by Chan and Chin [3] uses a similar concept of complete grids. When our algorithm finds a complete grid, it adds all the line segments to the graph but leaves out every intersection point in the interior of the rectangle.

More precisely, the algorithm determines which four line segments are outermost. These are what we term *boundary lines*. Every intersection of a line with a boundary line is included as a vertex of the graph. Every intersection point cuts line segments into pieces; these are edges of the graph.

Lines in the complete grid other than boundary lines are what we term *interior lines*. The algorithm does not include, as vertices of the graph, the intersection points of interior lines with other interior lines. Each interior line becomes three edges: the two fragments outside the rectangle and the one subsegment that crosses the interior of the rectangle (see Figure 1).

A special case occurs if a path endpoint is located on a line segment of the complete grid. In this case, the algorithm treats this segment just like a boundary segment, including as vertices in the graph all intersection points of this segment with any other segments.

One of the difficulties making this approach work is finding, in a complex arrangement of segments, regions that fit the description of a complete grid. The solution we describe uses a technique similar to that used in a paper by Bern, Dobkin, and Eppstein [1]. The algorithm finds and enlarges complete grids by looking in vertical strips of the arrangement that are $\sqrt{n}$ vertical lines wide.

It makes use of the fact that there are few $(2n)$ segment endpoints in the arrangement. The only interruption of a complete grid that can occur is the start or end of a line segment. The algorithm enlarges grids until an endpoint interrupts its progress. The algorithm then adds the complete grid to the graph, and continues by finding the next complete grid.

This produces a graph representing the arrangement, but unlike the naively constructed graph, this is not a planar graph. The previously mentioned algorithm by Klein et al. [5] works only for planar graphs. Thus, we apply Dijkstra's shortest path algorithm to our graph.

2.3 Partitioning the Rectangle

Before discussing the algorithm in detail, we discuss an abstraction that will aid precise mathematical analysis. Consider a rectangular region R of the plane defined by the extreme x and y values of all segment endpoints (and thus enclosing all segments). Imagine that all vertical line segments extend the full length of this rectangle. Choose vertical lines at intervals of $\sqrt{n}$, including the left and right boundaries of the rectangle. These lines partition the rectangle into vertical strips. The vertical line segments on the boundary of each strip could be considered to belong to the strips on either side; we assign them to the strip on their left.

Now imagine that, for each vertical strip, we sweep a horizontal line down the strip. Every time the sweep line intersects the endpoint of a vertical line, we add to the partition of R a horizontal segment across the strip. If the sweep line crosses a horizontal line extending all the way across the strip, we do nothing. If the line crosses a horizontal line segment with an endpoint in the strip, we add another partitioning horizontal segment.

Doing this for each strip of the rectangle gives a partition of R into subrectangles, where the interiors of the subrectangles are disjoint. The subrectangles will correspond to complete grids. The rectangles include and intersect at their boundary lines. Taking the intersection of each subrectangle with the set of line segments gives a set of subsegments. Since the union of the subrectangles equals rectangle R, the union of all subsegments in the subrectangles equals the set of line segments.

3 Algorithm

The first step in the algorithm is to create an events list. We combine the lists of horizontal and vertical lines and sort them by y coordinate, where both endpoints of the vertical line segments count as events.

We choose vertical line segments that are separated by $\sqrt{n}$ vertical line segments. The vertical lines corresponding to these segments partition the rectangle into vertical strips. We will start with the leftmost vertical strip. We include any vertical line segment that is on a strip boundary as part of the vertical strip for which it is the right hand boundary. The algorithm processes all vertical strips going from left to right.

The computations we do for a vertical strip are as follows. We have a horizontal sweep line travel down the vertical strip, stopping for each event on the event list. We will build up a set of line segments corresponding to a subrectangle in our previously described partition of rectangle R, and the line segments will satisfy the definition given earlier of a complete grid. The set of segments is initially empty at the top of each vertical strip.

One event type is a horizontal line. If the horizontal line does not intersect the vertical strip, the algorithm does nothing. If the horizontal line segment extends the full width of the vertical strip, we add the line segment to a list of horizontal line segments in the current complete grid. If the horizontal line segment has one or both endpoints in the interior of the vertical strip, we define this segment to be a *free segment*, which is not part of the complete grid. This ends the complete grid, and we call the graph construction routine, providing the set of lines in the complete grid and the free segment as the parameters.

The event encountered by the sweep line may be an endpoint of a vertical line segment. If so, we call the graph construction routine with the set of lines in the complete grid as the parameter. In this case, there is no free segment.

After adding a complete grid, we reset the current list of lines. Doing this means setting the list of horizontal lines to null, and for the vertical lines, finding all line segments in the vertical strip that both intersect the horizontal sweep line and extend below it. This becomes the current set of lines for the next complete grid.

We continue moving the sweep line until the next event is reached or we reach the end of the vertical strip. At the end of the strip, we call our routine for adding the current complete grid to the graph. We then start the sweep down the next vertical strip.

3.1 Adding a Complete Grid to the Graph

The part of the algorithm described above finds sets of lines that fit our definition of a complete grid. In this section we describe how the algorithm constructs a part of the graph containing all line segments in the complete grid.

We use an adjacency list representation for the graph, and for each vertex we record its Cartesian coordinates. To add a complete grid, the algorithm constructs the graph by adding subsegments from the grid going from left to right and top to bottom.

At any time in the construction of the graph, line segments may have subsegments that are represented in the graph and other subsegments that are to be added to the graph. Adding a complete grid to the graph may add subsegments from a segment already represented in the graph, or it may add entirely new subsegments. Here we describe low level operations that are used in constructing the graph.

The low level operations depend on what is already in the graph. Vertices in the graph may be one of two types. The first type is a vertex at the intersection of segments. These vertices, once placed in the graph, are never changed except to add edges. The second type of vertex is a subsegment endpoint; it is not at an intersection. If a colinear subsegment is to be added to the graph (that is, both subsegments are part of the same segment), the algorithm always moves the existing endpoint by changing the coordinates associated with it. The vertex becomes the endpoint of the new, longer subsegment.

It may happen that a new vertex is at an intersection of two segments, and both segments have subsegments already in the graph. If both subsegments in the graph end in vertices that are not at intersections, one has a choice of which existing vertex to move. The algorithm picks one vertex. Since the other subsegment should be lengthened to end at this intersection, the vertex that is its endpoint is deleted and replaced by the newly moved vertex.

Still considering a new vertex at an intersection, if exactly one of the subsegments ends at an intersection point, the other vertex is moved. If both subsegments end at intersection points, then a new vertex is added at the new intersection point.

In summary, the low level operations for placing a vertex in the graph are to add, move, or move and delete vertices.

4 Algorithm Running Time

Here we analyze the running time of the algorithm. This has three components: the time for finding complete grids, the time for constructing a graph from the complete grids, and the time for applying Dijkstra's shortest paths algorithm to the graph. To determine how long Dijkstra's algorithm takes, we must bound the number of vertices and edges in the constructed graph.

4.1 Time to Find Complete Grids

The time for finding complete grids includes the time for sorting the events list and the time for passing a sweep line down $\sqrt{n}$ vertical strips processing events. Each event is simply to check whether endpoints are in the vertical strip, so each takes $O(1)$ time. There are $O(n)$ events processed in each vertical strip, giving $O(n^{1.5})$ time spent looking at events.

For certain events, the algorithm adds or removes lines from the two linked lists of lines representing the complete grid. Horizontal lines are encountered in sorted order, so adding them to a sorted linked list takes constant time. This happens at most n times in a vertical strip, giving a bound of $O(n^{1.5})$ total time for these operations.

Encountering vertical segment endpoints requires adding or removing vertical lines from the sorted linked list of vertical segments. The list may be as long as $\sqrt{n}$, and these events happen at most $2n$ times total in the algorithm. The total time for these operations is thus $O(n^{1.5})$.

We conclude that the time for finding complete grids is $O(n^{1.5})$.

4.2 Graph Construction Time

Here we look at the time it takes to construct the graph. We have the following lemma.

Lemma 1. *The time for constructing the graph is $O(n^{1.5} + |V|)$.*

Proof sketch. In every step the algorithm adds subsegments to the graph, and to do this it adds new vertices, moves existing vertices, and deletes existing vertices. The addition, deletion, or move each take constant time once one has accessed the nearest vertices. The algorithm accesses vertices in the graph in order from left to right and top to bottom, so the algorithm takes constant time to access vertices.

We now bound the total time of constructing the graph by charging the time of the low level operations (except moves) to vertices of the graph.

If we add a vertex, we charge this constant time to the vertex. For deletions: any time we delete a vertex, it is because we have placed a vertex at the intersection of two subsegments. This vertex is never itself deleted or moved. We thus charge the cost of deleting a vertex to the vertex that stays in the graph, and we also transfer the cost accrued to the deleted vertex (accrued when it was added to the graph) to the vertex that stays in the graph.

We next look at moving a vertex. Instead of charging this cost to vertices, we count the total number of moves possible. One can show that the total number of move operations on all vertices is $O(n^{1.5})$.

We have shown that the time for finding the complete grids is $O(n^{1.5})$, the time for moving vertices as we construct the graph is $O(n^{1.5})$, and the time for adding and deleting vertices is $O(|V|)$, proving the lemma. □

4.3 A Bound on the Number of Vertices and Edges

We bound the total number of vertices and edges in the graph by determining how many intersection points the algorithm adds for all complete grids and free lines. The only other vertices in the graph are $O(n)$ segment endpoints.

In a vertical strip, a complete grid (and possibly a free line) is added to the graph whenever the algorithm encounters an endpoint of a line. Complete grids are also added when the algorithm reaches the bottom of a vertical strip. There are only $2n$ line endpoints and $\sqrt{n}$ vertical strips, so the algorithm adds complete grids at most $2n + \sqrt{n}$ times.

A complete grid has (at most) 4 boundary lines. The horizontal boundary lines intersect no more than $\sqrt{n}$ vertical lines, so the complete grid horizontal boundary lines add at most this many intersection points. The vertical boundary lines can intersect up to n horizontal lines, but in fact the sum of intersections for all vertical boundary lines for all complete grids in the vertical strip is at most $2n$. This is because at any point in the strip there are only two vertical boundary lines; except for the left/right pair, vertical boundary lines in the same strip do not overlap.

Since there are 2 path endpoints, they add at most 2 segments for which all intersection points are added to the graph. Free lines (always horizontal) can intersect up to $\sqrt{n}$ vertical lines each.

All intersection points of segments with other segments are accounted for in these computations since there are no interior to interior line intersection points.

We add at most one free line each time we add a complete grid. The number of vertices added for horizontal segments for each complete grid is at most $3\sqrt{n}$.

Let n_i be the number of complete grids in vertical strip i. Then the total number of intersection points in strip i is bounded by $3\sqrt{n}\, n_i + 2n$. The total number of intersection points in all $\sqrt{n}$ strips (ignoring those added by segments containing path endpoints) is bounded by

$$\sum_{1 \le i \le \sqrt{n}} (3\sqrt{n}\, n_i + 2n)$$

We stated above that $\sum_i n_i \le 2n + \sqrt{n}$. This gives us a bound on intersection points of

$$3\sqrt{n}(2n + \sqrt{n}) + 2n\sqrt{n}$$

Adding the intersections produced by boundary lines containing the path endpoints, we get

$$|V| \le 3\sqrt{n}(2n + \sqrt{n}) + 2n\sqrt{n} + 2n$$
$$= O(n^{1.5})$$

This lets us finish the work of bounding the time for constructing the graph with the following theorem.

Theorem 2. *This algorithm takes $O(n^{1.5})$ time to construct a graph.*

The number of edges in the constructed graph is bounded by 4 times the number of intersection points plus (in case there are no intersection points) the number of line segments; this gives $O(n^{1.5})$ edges. Applying Dijkstra's single source shortest paths algorithm to this graph, we compute a shortest path in time

$$O(|V|\log|V| + |E|)$$
$$= O(n^{1.5}\log(n^{1.5}) + n^{1.5})$$
$$= O(n^{1.5}\log n)$$

We state this as a theorem.

Theorem 3. *The constructed graph requires $O(n^{1.5})$ space, and a shortest path in it can be computed in time $O(n^{1.5}\log n)$.*

5 Paths in the Constructed Graph

Here we prove that the shortest path in the original arrangement G has an equally short path in the constructed graph G'.

Lemma 4. *The constructed graph G' contains edges representing all line segments in the the arrangement G and vertices for all intersection points except interior to interior line intersections.*

This follows from the manner of constructing G', and the proof is omitted.

Theorem 5. *Suppose one is given an arrangement of vertical and horizontal segments G and points s and t in it. If G' is the graph constructed by the algorithm described above, then G' contains a path P' between s and t with weight equal to the shortest path P in G.*

Proof. We first need to explain some terminology. We say that a path *crosses through* a vertex v if it goes from one line segment to a different line segment through intersection point v. If the path goes through a vertex but continues on what, in the arrangement, is the same line segment, then we say that the path *does not cross through* the vertex.

Let P be a shortest path in the arrangement G. By our lemma, G' contains all segments found in G. If P crosses through only vertices found in G', then the same path exists in G', so we have nothing to prove.

Suppose instead that P crosses through some vertex (or vertices) not found in G'. Then, since G' does not contain a vertex at this intersection point, a path cannot go from one line segment to another, so the same path cannot exist in G'. Let S be the set of vertices that P crosses through that are not in G'. Choose one vertex v from S.

By the lemma above, G' contains every vertex contained in G except vertices at the intersection of interior lines with interior lines. Since the vertex v is not in G', it is at an intersection of interior lines. Note also that the omitted vertices are always at intersections of interior line segments from the same complete grid. Thus we can say that v belongs to a complete grid.

In the complete grid, the outermost line segments are designated as boundary segments. The four (at most) boundary segments define a rectangle; we denote this rectangle, including its interior, by r. (If there are fewer than four boundary segments, there is no enclosed region and so no interior.)

In the remainder of the proof we consider three cases, depending upon whether the path endpoints s and t are inside the rectangle r bounded by boundary segments. These are (1) both path endpoints s and t are exterior to or on the boundary of r, (2) one of s or t is in r, and (3) both path endpoints are in r.

Case 1 Suppose that both path endpoints s and t are exterior to or on the boundary of rectangle r. Consider vertex v from set S: by the definitions of boundary lines and interior lines, v is inside r.

Since neither s nor t is inside rectangle r, there must be some first point p_1 (starting from endpoint s) where P intersects the boundary of r and some last point p_2 where P ends travel on the boundary of r, never again intersecting r. (These points may be s or t.)

The first possibility is that one of these crossing points (say WLOG p_1) is on a horizontal boundary line h_1 and the other point p_2 is on a vertical boundary line v_2. Then there is no shorter path from p_1 to p_2 than $\overline{p_1 h_1 v_1 p_2}$. (We describe paths by listing both points and line segments on the path.) Thus we can replace the part of P from p_1 to p_2 with this subpath and we have at least as good a path as P.

This new path does not travel through the interior of r, and in particular it does not cross through vertex v. The number of interior to interior intersection points crossed through (the vertices of S) has decreased by at least one.

The second possibility is that both of the intersection points are on vertical boundary lines v_1 and v_2. The intersection at vertex v involves some horizontal line h_1. No path from p_1 to p_2 that travels through any point of h_1 is shorter than $\overline{p_1 v_1 h_1 v_2 p_2}$.

Thus we can replace the subpath of P from p_1 to p_2 with this path and have a new path that is at least as good. The new path reduces the number of cross-through vertices by at least one.

Case 2 Suppose that exactly one path endpoint, say s, is in the interior of rectangle r. Since t is exterior to r, starting at s the path must cross the rectangle boundary at least once. Let p_1 be the last point (this may be t) where P intersects the boundary of rectangle r before reaching t.

Suppose WLOG that p_1 is on a vertical boundary line v_1. If path endpoint s is on a horizontal segment h_1, then no path from p_1 to s is shorter than $\overline{p_1 v_1 h_1 s}$. Replacing the subpath of P going from p_1 to s with this subpath gives a new path that is at least as good as P. This new path reduces the size of the set of cross-through vertices S by at least one.

Still supposing that the boundary line containing p_1 is vertical line v_1, now suppose that path endpoint s is on a vertical segment v_2. Since the path endpoint is on this segment, the algorithm we described treats this segment as if it were a boundary segment; that is, all intersection vertices of segment v_2 with other lines are contained in the constructed graph G'.

The intersection point v is at an intersection involving some horizontal line h_1. There is no shorter path from p_1 to v than $\overline{p_1 v_1 h_1 v}$ and no shorter path from v to s than $\overline{v h_1 v_2 s}$. Replacing the subpaths $\overline{p_1 v}$ and $\overline{v s}$ in P with these new subpaths gives a path that is at least as good as P and does not cross through vertex v. Thus, the number of vertices in the set S of cross through vertices is decreased by at least one.

Case 3 Both path endpoints are in rectangle r. Remembering that any segment holding an endpoint has all intersection points with other segments included in G', this case is easy and is omitted.

All cases For any path in G, we use the above arguments to find, for any path that crosses through an interior to interior intersection vertex v, an equally good path that eliminates this cross-through vertex using a subpath contained entirely in G'. We can do this iteratively on P, removing cross through vertices, until we have a path that is at least as good and does not cross through interior to interior vertices. This path can be represented in G', completing the proof. □

This gives us our main result:

Theorem 6. *The algorithm described here computes a shortest path in an arrangement of horizontal and vertical segments in time $O(n^{1.5} \log n)$ and space $O(n^{1.5})$.*

6 Query points

Suppose one has a shortest path problem as described above, but one only knows the location of one path endpoint in advance. We show here that it is possible to preprocess the arrangement in time $O(n^{1.5} \log n)$ and space $O(n^{1.5})$ and then compute shortest paths for query points in time $O(\log n)$.

The algorithm begins by constructing a graph as described above, where we have only one known path endpoint. It then computes the shortest paths from the known endpoint to all vertices in the graph using Dijkstra's single source shortest paths algorithm.

To find the shortest path for a query point t, the algorithm must find vertices in the constructed graph that are close to the query point. In the worst case, the algorithm must check paths from 6 selected vertices in the graph.

Suppose the query point is at the intersection $\overline{p_1 t}$ of two segments. If the intersection point is not the intersection of two interior segments, then the intersection point is a vertex of the constructed graph, and the already computed shortest path to the vertex is correct (as discussed below). If the intersection point is the intersection of two interior lines, then the algorithm finds the four vertices where these two interior lines intersect the boundary lines. It computes the path length to t from each of these four points and chooses the best of these. This is a shortest path in the arrangement.

Suppose the query point is on a single segment and so is not at an intersection. If this point is not in the interior of the grid boundary rectangle, one finds the two vertices that are nearest to the point on the segment and chooses the best path from these vertices.

The most complex case occurs when the point is interior to a boundary rectangle and not at an intersection. First one needs to look at the two vertices where the segment containing t intersects the boundary. Next, one needs to find two segments perpendicular to this segment that are closest to t on each side of t. These two segments intersect the boundary at four points. This gives six vertices total. One finds the best paths from these six vertices to t (in $O(1)$ time each) and chooses the best of these. This gives a shortest path in the arrangement.

If t is located in the interior of the boundaries of a complete grid and s is also interior to this rectangle, then one treats it as a special case; one can compute the shortest path in $O(1)$ time once one has found "close" segments.

We must describe, then, how one quickly finds the "close" vertices described above. The algorithm builds a lookup table for vertical lines and lookup tables for each complete grid.

For the set of vertical lines we have the following. We have an array of vertical lines sorted by x coordinate. For each vertical line, we store a (variable sized) array containing a list, sorted by y coordinate, of all complete grids that intersect that line (even though the segment, as opposed to the line, may not pass through a complete grid). Then, given the coordinates of the query point, we can find the vertical line (or nearest vertical line) in time $O(\log n)$ and find the complete grid on that line in time $O(\log n)$.

In a vertical strip there are $\sqrt{n}$ vertical lines, and we define n_i to be the number of complete grids in the strip. As stated previously, $\sum_i n_i \leq 2n + \sqrt{n}$. Then the space for this data structure is

$$\sum_i \sqrt{n}\, n_i$$
$$= O(n^{1.5})$$

For each complete grid, we maintain two sorted arrays: one for horizontal boundaries and one for vertical boundaries. These each contain coordinates paired with pointers to vertices in the graph that are located at that x or y coordinate. Thus, once one finds the complete grid containing the query point, one can find the "closest" vertices—the ones needed by the computations described above—in $O(\log n)$ time. The number of pointers maintained is less than the number of vertices in the graph, so the space is $O(n^{1.5})$.

To see that the computations described here actually find a shortest path, we need to prove the following.

Lemma 7. *The shortest paths computed by Dijkstra's algorithm for all vertices in the constructed graph are shortest paths to these vertices in the arrangement.*

This does require proof, since the previous proofs all assume that t is known in advance, and the segments containing t are treated specially. The proof is easy and is omitted.

Theorem 8. *There is a shortest path from s to t that passes through one of the (at most) 6 vertices described above.*

Proof sketch. If t is not in the interior of the boundaries of a complete grid, then it is either at an intersection point, for which the shortest path has already been computed, or it is between two vertices, and the shortest path passes through one of these vertices.

Suppose that t is in the interior of the region bounded by grid boundary segments. Assume also that s is not in the interior. Then a shortest path from s to t must pass through the boundary. Let p_1 be the first point where the path from s intersects the boundary. Then traveling from p_1 to one of (whichever gives the best path) the six vertices described above (all on the boundary) and taking the obvious shortest path from that vertex to t gives a path that is as good as any other path traveling through p_1 to t.

If s is in the interior of the rectangle, one treats this as a special case and computes the shortest path in $O(1)$ time once t has been located. □

7   Concluding remarks

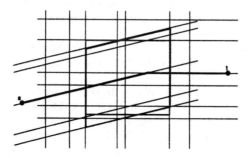

Fig. 2. An example with 3 line orientations where a shortest path (bold path in figure) does not have an equally short path in the constructed graph of overlapping, different-orientation grids.

It would be nice if the algorithm described here could be generalized to find shortest paths in an arrangement containing segments with $O(1)$ different orientations. However, a direct extension of the methods here to more line orientations does not appear to work.

The most direct way to extend the ideas of this paper (to 3 orientations for example) is as follows. Choose one line orientation and choose partitioning lines at $\sqrt{n}$ intervals. Find complete grids with respect to lines at a second orientation and build a graph from these. Then find complete grids with line segments of the third orientation and add these to the graph.

One must decide for this approach whether interior segments of grids from one orientation should have vertices at intersections with interior segments from different orientation, overlapping grids. The answer has to be that these intersections are not added as vertices; otherwise, the graph will have too many vertices. But as shown in Figure 2, there exist shortest paths in the arrangement for which no equally short path can be found in a graph constructed this way.

The factor $O(n^{1.5})$ in our results may be induced by the solution method rather than the problem. It is an open question whether one can find shortest paths in vertical and horizontal segments more quickly and in less space.

References

1. M. Bern, D. Dobkin, and D. Eppstein. Triangulating polygons without large angles. *International Journal of Computational Geometry & Applications* 5 (1995) 171-192.
2. P. Bose, W. Evans, D. Kirpatrick, M. McAllister, and J. Snoeyink. Approximating shortest paths in arrangements of lines. *Proceedings of the 8th Canadian Conference on Computational Geometry* (1996) 143-148.
3. W.-T. Chan and F. Y. L. Chin. Efficient algorithms for finding disjoint paths in grids. *Proceedings of the 8th ACM-SIAM Symposium on Discrete Algorithms* (1996) 454-463.
4. J. Hershberger and S. Suri. Efficient computation of Euclidian shortest paths in the plane. *Proceedings of the 24th Annual Symposium on Foundations of Computer Science* (1993) 508-517.
5. P. Klein, S. Rao, M. Rauch, and S. Subramanian. Faster shortest-path algorithms for planar graphs. *Proceedings of the 26th Annual ACM Symposium on the Theory of Computing* (1994) 27-37.
6. D.T. Lee, C.D. Yang, and C.K. Wong. Rectilinear paths among rectilinear obstacles. *Discrete Applied Mathematics* 70 (1996) 185-215.
7. F. Preparata and M. Shamos. *Computational Geometry.* Springer-Verlag, 1985.
8. P.J. de Rezende, D.T. Lee, and Y.F. Wu. Rectilinear shortest paths in the presence of rectangular barriers. *Discrete & Computational Geometry* 4 (1989) 41-53.

On Geometric Path Query Problems*

Danny Z. Chen, Ovidiu Daescu, and Kevin S. Klenk**

Department of Computer Science and Engineering, University of Notre Dame, Notre Dame, IN 46556, USA, E-mail: {chen,odaescu,kklenk}@cse.nd.edu

Abstract. In this paper, we study several geometric path query problems. Given a scene of disjoint polygonal obstacles with totally n vertices in the plane, we construct efficient data structures that enable fast reporting of an "optimal" obstacle-avoiding path (or its length, cost, directions, etc.) between two arbitrary query points s and t that are given in an on-line fashion. We consider geometric paths under several optimality criteria: L_p length, number of edges (called *links*), monotonicity with respect to a certain direction, and some combinations of length and links. Our methods are centered around the notion of *gateways*, a small number of easily identified points in the plane that control the paths we seek. We present solutions for the general cases based upon the computation of the *minimum size visibility polygon* for query points. We also give better solutions for several special cases based upon new geometric observations.

1   Introduction

Geometric path planning problems can be stated as follows: Given a d-dimensional space scattered with obstacles, find a path P connecting two points s and t such that P satisfies certain constraints (e.g., obstacle-avoiding, fixed orientations, bounded curvature) and that P optimizes certain combinations of criteria (e.g., L_p length, number of links, turning angle). Geometric path problems play an important role in many application areas, such as geographical information systems, operations research, robotics, transportation, and VLSI design. Also, they have significant connections with other fundamental topics in computational geometry, such as convexity, visibility, Voronoi diagrams, and geometric optimization. A great deal of research has been done on solving various geometric path planning problems, especially for the planar and 3-space settings (see [9, 11] for a survey).

In this paper, we study *geometric path query problems*: Given a scene of disjoint polygonal obstacles with totally n vertices in the plane, construct an efficient data structure that enables a fast reporting of an "optimal" obstacle-avoiding path (or its length, cost, directions, etc.) between two arbitrary query points s and t that are given in an on-line fashion. Geometric path query problems often arise in situations where good routes between many pairs of locations are desired.

The optimality of geometric paths is criterion-dependent. We consider geometric paths under several criteria: L_p length, number of edges (called *links*), monotonicity with respect to a certain direction, and some combinations of length and links. The corresponding optimal paths are called, respectively, shortest paths, minimum-link paths, monotone paths, and optimal bicriteria paths.

---

* This research was supported in part by the National Science Foundation under Grant CCR-9623585.

** This research was supported in part by the William D. Mensch, Jr. Summer Fellowship.

Interestingly, much work on geometric path planning has focused on the *single-source* versions of the problems (i.e., one of the two endpoints of a desired path is fixed while the other is arbitrary). In particular, a breakthrough has recently taken place on computing the planar single-source Euclidean shortest paths, for which Hershberger and Suri [7, 8] gave an optimal $O(n \log n)$ time algorithm. However, only a few results are known for planar geometric path *query* problems (with the endpoints of a desired path being both arbitrary), especially for the general case in which the plane is scattered with *multiple* obstacles (see Mitchell [11] for a survey).

Ideally, it would be highly desirable to come up with data structures for various geometric path queries that have a reasonable time and space bounds for their construction and a polylogarithmic time for queries on the lengths/costs of the sought paths. However, it appears that straightforward solutions for such data structures might require high time and space complexities for their construction and storage, and thus would be impractical. Therefore, we pursue in this paper a less ambitious but quite general approach, called the *visibility-sensitive approach*: For two arbitrary query points s and t, let Q_s (resp., Q_t) denote the (possibly unbounded) polygonal region of the plane that is visible from s (resp., t), called the *visibility polygon* of s (resp., t). Design a geometric path data structure that supports an $O(\min\{|Q_s|, |Q_t|\} \times \log n)$ query time on path lengths/costs. Clearly, $O(1) \leq \min\{|Q_s|, |Q_t|\} \leq O(n)$.

This approach enables us to build data structures for several geometric path query problems in the following construction bounds:

- Euclidean shortest paths among polygonal obstacles: $O(n^2 \log n)$ time and $O(n^2)$ space.
- Approximate minimum-link paths among polygonal obstacles: The problem is to compute an obstacle-avoiding approximate minimum-link path between two query points. The path we obtain has at most two more links than the optimal path. $O(n((|E| + ln)^{2/3} n^{2/3} l^{1/3} \log^{3.11} n + |E| \log^3 n))$ time and $O(n^2)$ space, where E is the visibility graph of the obstacle scene and l is the longest link length among all the minimum-link paths between any two obstacle vertices.
- Monotone paths among convex polygonal obstacles: The problem is to find a direction d such that there is an obstacle-avoiding path between two query points that is monotone to d (it was shown in [2] that for convex polygonal obstacles, such a direction always exists). $O(n^2 \log n)$ time and $O(n^2)$ space.

Although the visibility-sensitive approach does not achieve a polylogarithmic query time, it has several significant advantages: (i) The approach is quite general and is applicable to a variety of geometric path query problems. (ii) The querying procedures used by this approach can be much more efficient (even in the worst case) and are much simpler than the corresponding best known algorithms for computing an optimal path between two points. Note that $|Q_s|$ and $|Q_t|$ can differ from each other dramatically and can be much smaller than $O(n)$. Hence, using $\min\{|Q_s|, |Q_t|\}$ in practice can save considerable time. In comparison, an Euclidean shortest path (resp., minimum-link path, monotone path among convex polygonal obstacles) is computed in $O(n \log n)$ (resp., $O(|E|\alpha^2(n) \log n)$, $O(|E| + n \log n)$) time and $O(n \log n)$ (resp., $O(|E|)$, $O(|E|)$) space [1,2,7,8,12], and the algorithm to perform the computation is complicated to implement. (iii) The data structures need to be built only once, in an off-line fashion, and are more expensive than the corresponding single-source solutions only by a factor of n or less. Therefore, this approach could be useful in various situations where on-line geometric path queries need to be processed frequently (e.g., database systems).

We also present efficient techniques for solving several interesting special cases of the above geometric path query problems. By exploiting the geometric structures of the special cases, we construct much better data structures for these cases (e.g., with a polylogarithmic query time) than the ones for the corresponding general cases:

- Optimal rectilinear bicriteria paths among rectilinear polygonal obstacles: Three kinds of rectilinear bicriteria paths are considered: (1) paths with the minimum number of links among all shortest paths between two points, (2) shortest paths among all minimum-link paths between two points, and (3) minimum-cost paths between two points where the cost is a nondecreasing function of the number of links and the length of a path. For single-source versions of these problems, we build data structures in $O(n \log^{3/2} n)$ time and $O(n \log n)$ space, with an $O(\log n)$ query time on path lengths/costs. For arbitrary query versions, we build data structures in $O(n^2 \log^2 n)$ time and space, with an $O(\log^2 n)$ query time on path lengths/costs.
- Shortest A-distance paths among polygonal obstacles: A-distance paths are paths whose edges can be in arbitrary directions but whose edge lengths are measured based on $|A|$ given orientations [17]. We build a data structure for shortest A-distance path queries between arbitrary points in $O(|A|^2 n^2 \log^2 n)$ time and space, with an $O(|A|^2 \log^2 n)$ length query time.
- Monotone paths among convex polygonal obstacles: The version of the problem here is to decide whether there is an obstacle-avoiding path between two query points that is monotone to a *specified* direction d. We build data structures for two cases of the problem: (1) the direction d is fixed ($O(n \log n)$ time, $O(n)$ space, $O(\log n)$ query time), and (2) d is given as part of the query input ($O(n^2 \log n)$ time, $O(n^2)$ space, $O(\log n)$ query time).

Note: In all our query results above, an actual path (in addition to its length/cost) can be output in an additional $O(L)$ time, where L is the number of edges of the reported path.

2 Overview of general paradigm

The main idea that we use for solving geometric path query problems is: For a query point s, identify a set W_s of special points in the plane that "control" the desired paths to another query point t. We call these special points the *gateways* of s. Given W_s, the desired paths between s and t that go through a point in W_s can be obtained from the single-source data structures (with each point in W_s being the source of such a data structure). For example, for Euclidean shortest paths, minimum-link paths, and monotone paths, it is well-known that a desired path between s and t passes through a vertex of the visibility polygon Q_s that is also an obstacle vertex, unless s is visible from t [2, 7, 8, 12]. However, for this idea to work well, several difficulties must be overcome. (*i*) W_s must have a reasonably small size, because this will lead to an (at least) $O(|W_s| \times f(n))$ query time, where $f(n)$ is the time for a single-source query. (*ii*) It must be possible to quickly identify the gateway points of W_s for s. (*iii*) It must be known how to build single-source query data structures. (*iv*) The set W of points that include *all* gateway points for *any* arbitrary query points cannot be too big, since for each point of W, it needs to build a single-source query data structure.

As shown in the subsequent sections, several different forms of the above general paradigm emerge in our solutions to the studied geometric path query problems, due to the specific geometric structures of each of these problems. For example, the visibility-sensitive approach can use many gateways, the optimal bicriteria rectilinear path query technique uses $O(\log n)$ gateways, and the monotone path query methods use $O(1)$ gateways.

3 The visibility-sensitive approach

In this section, we present a general technique, called the *visibility-sensitive approach*, for various geometric path queries. This approach is based on the computation of visibility polygons of query points. The visibility polygon Q_s of a point s is represented as a counterclockwise sequence of vertices of the obstacle scene that are visible from s when scanning a ray around s that originates at s. Using this technique, we achieve a time of $O(\min\{|Q_s|, |Q_t|\} \times \log n)$ for queries on path lengths/costs, where s and t are two arbitrary query points. We will show how this approach can be applied to solving three path query problems among polygonal obstacles: (1) Euclidean shortest paths, (2) approximate minimum-link paths, and (3) monotone paths (among convex polygonal obstacles).

We use the *visibility complex* of Pocchiola and Vegter [14] to compute the visibility polygons of the query points s and t. The visibility complex of an obstacle scene is a two-dimensional cell complex whose faces consist of collections of rays with constant forward and backward views. The combinatorial complexity of the visibility complex is proportional to the number k of obstacle bitangents lying in free space, and the visibility complex can be constructed in $O(n \log n + k)$ time and $O(k)$ space, where $k = O(n^2)$ (see [14] for more details).

Based on the results in [14], the visibility polygon Q_s of s can be constructed in $O(|Q_s| \log n)$ time using the method of crossing faces. The computation of Q_s is based on the fact that the image of the curve in the visibility complex describing the movement of a ray emanating from s in the obstacle scene crosses the left (right) boundary of a face of the visibility complex in at most one point. Using the method of crossing faces, we spend $O(\log n)$ time to traverse a face by doing a binary search on the sequence of edges on the boundary of that face.

We compute the visibility complex of the scene and augment the underlying space of the visibility complex so that a point location can be done in $O(\log n)$ time and a ray shooting in an arbitrary direction to locate the initial face of the visibility complex to which the point belongs can also be done in $O(\log n)$ time [13].

To compute $\min\{|Q_s|, |Q_t|\}$, we cannot afford to obtain both Q_s and Q_t for query points s and t by simply using [14], since Q_s and Q_t can have much different sizes. Based on the approach in [14], we find the visibility polygon with the smaller size among Q_s and Q_t as follows: (1) Locate the faces of the visibility complex to which s or t belongs. (2) Traverse the face for each query point, and simultaneously compute the vertices of Q_s and the vertices of Q_t, in an incremental and alternate fashion: Compute a vertex of Q_s, then compute a vertex of Q_t, then the next vertex of Q_s, then the next vertex of Q_t, and so on. (3) Stop Step 2 when one of Q_s and Q_t is obtained completely.

Without loss of generality (WLOG), we assume $|Q_s| = \min\{|Q_s|, |Q_t|\}$. Using this incremental and alternate approach, we obtain the visibility polygon Q_s in $O(|Q_s| \log n)$ time. Our visibility-sensitive approach then uses the vertices of Q_s that also are obstacle vertices as the gateways in processing geometric path queries on any two points s and t, as to be shown next.

3.1 Euclidean shortest paths

Our method for processing Euclidean shortest path queries among polygonal obstacles is based upon [8], which computes a single-source *shortest path map* encoding shortest path information from a fixed source point to all other points in the plane. The map is built in $O(n \log n)$ time and space, and requires $O(n)$ space to store. Once the map is available, single-source shortest path length queries can be answered in $O(\log n)$ time each.

3.2 Approximate minimum-link paths and monotone paths

For approximate minimum-link path queries among polygonal obstacles, we use the single-source minimum-link path maps of Mitchell, Rote, and Woeginger [12]. For monotone path queries among convex polygonal obstacles, we make use of the single-source *monotone path map* of Arkin, Connelly, and Mitchell [2]. Our data structures and query procedures for these path queries have similar components and steps as those for Euclidean shortest path queries. Hence, we can process a query on path links/direction between any two query points s and t in $O(\min\{|Q_s|, |Q_t|\} \times \log n)$ time.

4 Optimal rectilinear paths

In this section, we present efficient data structures and algorithms for processing various optimal bicriteria rectilinear path queries among rectilinear obstacles in the plane. In particular, we focus on three types of path queries between query points s and t: (1) a path with the minimum number of links among all the shortest paths from s to t, (2) the shortest path among all the minimum-link paths from s to t, and (3) the minimum-cost path from s to t where the cost is a nondecreasing function of the number of links and the length of a path. We will refer to the target optimal bicriteria paths for any of these problems simply as *optimal bicriteria* paths or *optimal* paths for short.

Our main results are as follows: For single-source queries, we construct data structures in $O(n \log^{3/2} n)$ time and $O(n \log n)$ space, with an $O(\log n)$ query time on path costs. For arbitrary path queries, we construct data structures in $O(n^2 \log^2 n)$ time and space, with an $O(\log^2 n)$ query time on path costs.

4.1 Reduced visibility graph

A key structure we need is the reduced visibility graph. This graph was first introduced by Clarkson, Kapoor, and Vaidya [6], and has since been used by quite a few efficient algorithms for L_1 path problems [4–6, 10, 18]. We briefly review this graph below (see [4, 6, 10] for more details). Given a set of rectilinear polygonal obstacles, a reduced visibility graph $G = (V(G), E(G))$ includes all obstacle vertices as well as a set of *Steiner points* as its vertices. The Steiner points are introduced recursively, as follows: Use a vertical (resp., horizontal) cut line to partition the obstacle vertices into two halves, and project the visible obstacle vertices onto the cut line, resulting in Steiner points on the cut line; the problem then is recursively solved on each of the two halves of obstacle vertices.

G includes edges between an obstacle vertex and its $O(\log n)$ projected points on the cut lines, and between consecutive Steiner points on each cut line. As a result, G has $O(n \log n)$ vertices and edges. It is well-known that G so obtained contains rectilinear shortest paths between obstacle vertices.

Yang, Lee, and Wong [18] recently showed that the graph $G = (V(G), E(G))$ also contains paths that are *homotopic* to optimal paths between points in the plane that are represented as vertices in $V(G)$. Two paths are homotopic to each other if one can be continuously "dragged" (i.e., deformed) to become the other without crossing any points in $V(G)$. This knowledge of the existence of a homotopic path in G allows a search to be performed for a preliminary path that can be converted to an optimal path by some segment dragging operations [18].

The method in [18] for solving the three optimal path problems involves running Dijkstra's algorithm on G with some modifications to the advancing process. As an advance is made in the graph search, a vertical or horizontal segment is appended to the end of the computed path. This last segment of a computed path is subject to a dragging operation that tries to reduce the number of links on the path. When the destination is reached in this manner, the result is an optimal path. Note that [18] computes optimal paths only between points in the plane that are vertices in $V(G)$, and hence it does not solve our significantly more difficult single-source and arbitrary query problems.

4.2 Segment dragging to arbitrary query points

We would like to extend the segment dragging approach in [18] to handle segment dragging operations not only from vertices to vertices in $V(G)$ but also from vertices in $V(G)$ to any query points. Further, to achieve the claimed $O(\log n)$ single-source query time, we need to be able to: (i) reduce the number of vertices in $V(G)$ (the gateways) from which a segment needs to be dragged to a query point to $O(\log n)$, and (ii) perform each segment dragging from a gateway to a query point in $O(1)$ time.

We use the techniques introduced in [4] to determine the gateways for each query point. It was shown in [4] that the set of gateways $V_g(q)$ for each point q in the plane has size $|V_g(q)| = O(\log n)$ and can be computed in $O(\log n)$ time. While these gateway techniques allow a quick report of the gateways, [4] does not provide methods for segment dragging from the gateways to a query point.

A key problem thus is how to check in constant time whether a segment w (the segment of an optimal paths that enter a gateway of $V_g(q)$) can be dragged to its query point q. To solve this problem, we make use of the geometric structure of the gateway region. The *gateway region* $\mathcal{R}(q)$ of a point q is the area enclosed by an ordered interconnection of the gateways in $V_g(q)$. We have the following lemma on $\mathcal{R}(q)$:

Lemma 1. *The gateway region $\mathcal{R}(q)$ of every point q is rectilinearly convex and contains no obstacle points in its interior. Further, for any point $z \in \mathcal{R}(q)$, both points (z_x, q_y) and (q_x, z_y) are in $\mathcal{R}(q)$.*

Since the interior of each gateway region contains no obstacle points and hence no vertices in $V(G)$, it allows us to extend the segment dragging methods of [18] to handling segment dragging from a gateway to an arbitrary query point for that gateway.

We now discuss several cases of our segment dragging operations to illustrate the idea. Consider the situation where a gateway g of $V_g(q)$ is in the first quadrant of a point q (the other cases are similar). We need to do a segment dragging to q from each of the set of $O(1)$ line segments $W(g)$ entering g from either the horizontal or vertical direction. Note that we will only be discussing the cases when a segment $w \in W(g)$ enters g from *outside* of the gateway region $\mathcal{R}(q)$ since the other cases can be handled in a straightforward manner.

If the segment $w \in W(g)$ is fixed (i.e., it cannot be dragged further due to its proximity to other obstacles and has hence been removed from further dragging consideration), then the path extending from g to q will consist of a segment e_1 collinear to w followed by a segment e_2 perpendicular to w, thereby connecting g and q with a path $P(g,q)$ of length $length(P(g,q))$ and with one extra link. If w is floating, we must attempt to drag it into position thereby possibly eliminating this extra link.

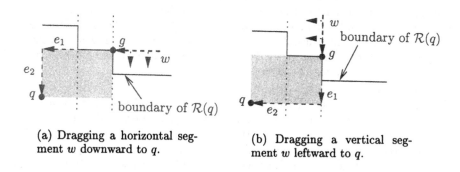

(a) Dragging a horizontal segment w downward to q.

(b) Dragging a vertical segment w leftward to q.

Fig. 1. Dragging a segment w to q. The shaded region is within $\mathcal{R}(q)$ and hence, obstacle-free.

Assume that the segment w enters g horizontally (the case shown in Figure 1(a)). We need to perform a segment dragging to determine whether w can be dragged downward to q_y thereby eliminating e_2 and connecting g and q with a single segment e_1, as follows. First check for the first vertex $v \in V(G)$ (referred to as the *hit-vertex*) that would be hit first if w were dragged downward. This checking can be done efficiently since w is collinear with an edge in $E(G)$ and we can use a data structure similar to one that is used in [18] to store the necessary information for the checking. If the hit-vertex of w is below q_y or such a hit-vertex does not exist, then we can drag w downward to q_y because we are sure that there are no obstacle points inside the gateway region (i.e., e_1 can be dragged downward as well). Another case in which the segment w enters g vertically (see Figure 1(b)) can be handled in a similar way (with leftward segment dragging).

If in both the two cases above, w cannot be dragged to eliminate a link, then w should be dragged as far as possible and then be fixed. In this situation, both segments e_1 and e_2 are used to connect g to q, adding one more link to the resulting path.

5 Monotone path query problems

In this section, we study monotone path queries between two arbitrary query points among disjoint convex polygonal obstacles in the plane. Monotonicity is a useful geometric property and monotone paths find applications in separability and part assembling/disassembling problems [2, 16]. Arkin, Connelly, and Mitchell [2] gave interesting algorithms to several monotone path problems (some are on arbitrary polygonal obstacles). Our work is based on some of their ideas and techniques.

We have considered in Section 3 the more general problem of finding a direction d to which there is a monotone path between two query points. Here we focus on the version of the problem in which d is given: Determining whether there is a path between two query points that is monotone to a *specified* direction d. We consider two cases of this problem. Case 1 is to process monotone path queries among convex polygonal obstacles with respect to a *fixed* direction d. Case 2 is on monotone path queries among convex polygonal obstacles with d being specified as part of the query input. Our data structure for Case 1 takes $O(n \log n)$ time and $O(n)$ space to build, and supports an $O(\log n)$ time query. Our data structure for Case 2 is built in $O(n^2 \log n)$ time and $O(n^2)$ space, and supports an $O(\log n)$ time query. We also give solutions to some other cases of the monotone path query problem.

5.1 Case 1

Case 1 can be stated as: "Given two query points s and t, is there a monotone path with respect to a fixed direction d?" WLOG, assume that d is the horizontal direction. We first construct the vertical trapezoidal decomposition VD of the obstacle-free region of the plane. We then obtain a directed graph G_{VD} from VD in the following manner: Each vertex u of G_{VD} corresponds to exactly one trapezoid T_u of VD, and two vertices u and v of G_{VD} are connected by a directed edge from u to v if and only if their corresponding trapezoids T_u and T_v share a common vertical boundary segment in VD and T_u is to the right of T_v. The following is a key observation whose proof is omitted here:

Lemma 2. *Suppose the graph G_{VD} is obtained as described above from a set of polygonal obstacles with n vertices, with each obstacle being monotone to the horizontal direction. Then G_{VD} is an embedded st-graph with $O(n)$ vertices.*

We construct a structure for processing transitive closure queries on the st-graph G_{VD} (for properties and transitive closure algorithms for embedded st-graphs, see [15]). We also create a planar point location structure for the trapezoidal decomposition VD. Our data structure as described above can be easily built in $O(n \log n)$ time and $O(n)$ space.

5.2 Case 2

Case 2 can be stated as: "Given two query points s and t and a direction d, does there exist a d-monotone path between s and t?" We make use of the single-source monotone path map in [2] and let each obstacle vertex be the source of such a map. After an $O(|E| + n \log n)$ time and $O(|E|)$ space preprocessing, the map for each obstacle vertex takes $O(n \log n)$ time and $O(n)$ space to build, leading to an overall $O(n^2 \log n)$ time and $O(n^2)$ space construction for our data structure.

Arkin, Connelly, and Mitchell [2] introduced the monotone path map $MPM(s)$, with a source point s, for a scene consisting of convex polygonal obstacles.

$MPM(s)$ is a subdivision of the plane such that for any point t in a face of the subdivision, the interval of directions (called the *cone*), with respect to each of which there is a monotone path from s to t, has a common structure. In fact, it was proved in [2] that among convex polygonal obstacles, the directions for which a monotone path exists from any s to any t always form a single non-empty cone. Although this subdivision may have a quadratic size, it can be represented implicitly as two linear-size subdivisions; let these be $S(s)$ and $S'(s)$ (we mainly focus on $S(s)$, since $S'(s)$ is similar). It was shown that the set of directions which may influence the structure of $S(s)$, called *critical directions*, correspond to the directions of common tangents between pairs of obstacles [2].

To construct $S(s)$ (resp., $S'(s)$), we start with a critical direction d and obtain a path $P(d^{\perp})$ (resp., $P(-d^{\perp})$) by shooting a ray leftward (resp., rightward) from the point s in direction $d^{\perp}$ (resp., $-d^{\perp}$) that is orthogonal to d. If this ray goes to infinity without hitting an obstacle, then that is the end of the path $P(d^{\perp})$. If it hits an obstacle W, then we proceed by letting $P(d^{\perp})$ follow a d-monotone path along the boundary of W, until we reach a vertex of W from which we can shoot a ray again in the $d^{\perp}$ (resp., $-d^{\perp}$) direction. Note that all the segments of $P(d^{\perp})$ that are not completely on the boundary of an obstacle are in direction $d^{\perp}$. In [2], it was shown that, after an $O(|E| + n\log n)$ time preprocessing, $S(s)$ can be constructed in $O(n\log n)$ time by sweeping the path $P(d^{\perp})$ through the plane (i.e., by continuously changing the direction d). In this sweeping process, the topology of the path may change at a critical direction and thus $S(s)$ actually consists of a collection of paths.

The following observations are important to our data structure.

Lemma 3. *Let $V(S(s))$ be the set of $O(n)$ obstacle vertices on the paths of $S(s)$ such that for each vertex $v \in V(S(s))$, a ray shooting occurs from v (in a certain direction) to construct a path of $S(s)$. Then the vertices in $V(S(s))$ form a tree $T(S(s))$ based on their relation along the paths of $S(s)$.*

We call a vertex $v \in V(S(s))$ a *special vertex* if its ray shooting goes to infinity. These special vertices play an important role in the query process.

Lemma 4. *Each special vertex v of the tree $T(S(s))$ is associated with a cone $cone(v)$ of directions that is bounded by two semi-infinite segments of the paths of $S(s)$, such that the following hold: (1) For two different special vertices u and v of $T(S(s))$, $cone(u) \cap cone(v) = \emptyset$, (2) the union of $cone(v)$ for all special vertices v of $T(S(s))$ is 2π, and (3) the left-to-right order of the special vertices of $T(S(s))$ corresponds to the order in which their cones are encountered by the direction of the sweeping algorithm for constructing $S(s)$.*

Let $d^{\perp}$ be the leftward direction orthogonal to a direction d. Based on Lemmas 3 and 4, for any direction d, we can obtain the path $P(d^{\perp})$ starting at s, as follows: (1) Find the special vertex u of the tree $T(S(s))$ such that $d^{\perp} \in cone(u)$ (this can be done by a binary search of $d^{\perp}$ in the cones of the special vertices of $T(S(s))$), and (2) trace the path from u to s along the tree $T(S(s))$, and include appropriate boundary portions of the obstacles visited in the tracing.

Actually, tracing the whole path $P(d^{\perp})$ can be too expensive for our query procedure and is not necessary. What we really need is the ability to perform binary search on the tree path of $T(S(s))$ from u to s. This will be possible if we preprocess $T(S(s))$ so that the so called *i-th level ancestor queries* [3] on any node of $T(S(s))$ can be handled in $O(1)$ time each. This preprocessing takes $O(n)$ time and space by Berkman and Vishkin's algorithm [3].

From the above discussion, we construct our data structure by doing the following. For each obstacle vertex v, compute the two subdivisions $S(v)$ and $S'(v)$ for the monotone path map $MPM(v)$, obtain the trees $T(S(v))$ and $T(S'(v))$, and preprocess $T(S(v))$ and $T(S'(v))$ for the i-th level ancestor queries.

References

1. N. M. Amato, M. T. Goodrich, and E. A. Ramos. Computing faces in segment and simplex arrangments. In *Proceedings of the 27th Annual Symposium on the Theory of Computing*. ACM, 29 May–1 June 1995.
2. E. M. Arkin, R. Connelly, and J. S. B. Mitchell. On monotone paths among obstacles, with applications to planning assemblies. In *Proceedings of the 5th Annual Symposium on Computational Geometry*, pages 334–343. ACM, June 1989.
3. O. Berkman and U. Vishkin. Finding level-ancestors in trees. Technical Report UMIACS-TR-91-9, University of Maryland, 1991.
4. D. Z. Chen, K. S. Klenk, and H.-Y. T. Tu. Shortest path queries among weighted obstacles in the rectilinear plane. In *Proceedings of the Eleventh Annual Symposium on Computational Geometry*, pages 370–379. ACM, 5–7 June 1995.
5. K. L. Clarkson, S. Kapoor, and P. M. Vaidya. Rectilinear shortest paths through polygonal obstacles in $O(n \log^{3/2} n)$ time. Unpublished manuscript.
6. K. L. Clarkson, S. Kapoor, and P. M. Vaidya. Rectilinear shortest paths through polygonal obstacles in $O(n(\log n)^2)$ time. In *Proceedings of the 3rd Annual Symposium on Computational Geometry*, pages 251–257. ACM, June 1987.
7. J. Hershberger and S. Suri. Efficient computation of Euclidean shortest paths in the plane. In *Proceedings of the 34th Annual Symposium on Foundations of Computer Science*, pages 508–517. IEEE, 3–5 Nov. 1993.
8. J. Hershberger and S. Suri. An optimal algorithm for Euclidean shortest paths in the plane. Manuscript, Feb. 1996.
9. J. Hershberger and S. Suri. Shortest path problems. Manuscript, 1996.
10. D. T. Lee, C. D. Yang, and T. H. Chen. Shortest rectilinear paths among weighted obstacles. *International Journal of Computational Geometry & Applications*, 1(2):109–124, 1991.
11. J. S. B. Mitchell. Shortest paths and networks. To appear in the *CRC Handbook on Computational Geometry*.
12. J. S. B. Mitchell, G. Rote, and G. Woeginger. Minimum-link paths among obstacles in the plane. *Algorithmica*, 8:431–459, 1992.
13. M. Pocchiola and G. Vegter. Pseudotriangulations: Theory and applications. In *Proceedings of the Twelfth Annual Symposium on Computational Geometry*, pages 291–300. ACM, 24–26 May 1996.
14. M. Pocchiola and G. Vegter. The visibility complex. *Internatational Journal of Computational Geometry & Applications*, 6(3):279–308, 1996.
15. S. Sairam, R. F. Cohen, R. Tamassia, and J. S. Vitter. Fully dynamic techniques for reachability in planar sT-graphs. Technical Report CS-91-02, Department of Computer Science, Brown University, Providence, Rhode Island, 02912-1910, Dec. 1990.
16. G. T. Toussaint. Movable separability of sets. In G. T. Toussaint, editor, *Computational Geometry*, volume 2 of *Machine Intelligence and Pattern Recognition*, pages 335–375. Elsevier, Amsterdam, 1985.
17. P. Widmayer, Y. F. Wu, and C. K. Wong. On some distance problems in fixed orientations. *SIAM J. Comput.*, 16(4):728–746, Aug. 1987.
18. C.-D. Yang, D. T. Lee, and C. K. Wong. Rectilinear path problems among rectilinear obstacles revisited. *SIAM J. Comput.*, 24(3):457–472, June 1995.

On-line Scheduling with Hard Deadlines (Extended Abstract)

Sally A. Goldman* and Jyoti Parwatikar** and Subhash Suri***

Dept. of CS, Washington University, St. Louis MO 63130, USA
{sg, jp, suri}@cs.wustl.edu

Abstract. We study non-preemptive, online admission control in the hard deadline model: each job must be either serviced prior to its deadline, or be rejected. Our setting consists of a single resource that services an online sequence of jobs; each job has a *length* indicating the length of time for which it needs the resource, and a *delay* indicating the maximum time it can wait for the service to be started. The goal is to maximize total resource utilization. We obtain a series of results, under varying assumptions of job lengths and delays.

1 Introduction

Online scheduling is an important problem area, with diverse applications. In this paper, we consider a scheduling framework where *jobs* have a *hard deadline*, meaning any job not serviced by its deadline is lost. Jobs arrive online, and must be scheduled non-preemptively. While much of the classical work on scheduling has been done in an offline setting, online scheduling is becoming especially important in high-speed network applications. The scheduling with hard deadlines models some interesting problems in providing Quality-of-Service (QoS) in shared packet-switched networks, as well as high bandwidth multimedia applications. For instance, in a packet-switched network such as the Internet, multiple traffic streams pass through a network of switching nodes (routers). Each node implements a *service discipline* (scheduling algorithm) for forwarding the incoming packets to outbound links. Due to the highly variable traffic rates (from few bits to several megabits per second) and packet sizes (from few bytes to several kilobytes), simple schedulers such as FIFO and round robin can fail miserably and fail to provide the bandwidth fairness and worst-case latency bounds. Packets in a latency-sensitive traffic stream such as video must be delivered within a small window, or else be dropped—the latency window corresponds to the maximum permissible delay at a router. (In these applications, delivering packets late has a worse effect than simply dropping the packets.) Similarly, in such emerging applications as video-on-demand, a useful model for requests is where users specify a window of time during which the delivery is acceptable.

* Supported in part by NSF Grant CCR-9110108 and NSF National Young Investigator grant CCR-9357707 with matching funds provided by Xerox PARC and WUTA.
** Research supported in part by Defense Advanced Research Projects Agency under contract DABT-63-95-C-0083.
*** Research supported in part by National Science Foundation Grant CCR-9501494

Job Lengths	Delay Type	Competitive Ratio
1	Arbitrary	2
1	Min. Delay 1	1.5
$\{1, k\}$	Arbitrary	4
$\{1, k\}$	Uniform	$\left(1 + \frac{\lceil k \rceil}{k}\right)$
$\{1, 2, 2^2, \ldots, 2^c\}$	Arbitrary	$3(c+1)$
$\{1, 2, 2^2, \ldots, 2^c\}$	Uniform	$2.5(c+1)$
$[1, 2^c]$	Arbitrary	$6(c+1)$
$[1, 2^c]$	"Uniform"	$5(c+1)$

Table 1. Summary of Results. We require that k is a real number greater than 1 and that c is a known integer.

Online scheduling in these applications can be used for admission control, accepting some requests and rejecting others with the goal of maximizing the total resource utilization. The requests or packets must be serviced *online* in these applications since the future arrival sequence is generally not known. We use competitive analysis[4] to measure the quality of our algorithms [12, 10, 6]; we consider both deterministic as well as randomized algorithms.

1.1 Our Results

In earlier work, admission control algorithms have allowed preemption [8, 9], however, newer technologies seem to favor non-preemption. In particular, high speed networks based on packet switching technology (e.g. Asynchronous Transfer Mode) are *connection oriented*, meaning that resources are reserved during a call set up, and the overhead of this setup makes preemption highly undesirable. ATM networks are well-suited for latency-sensitive real-time traffic, such as video, voice, and multimedia. Consequently, there has been considerable recent work on non-preemptive admission control [1, 11, 3, 2, 4, 7]. Our work is most closely related to, and a generalization of, Lipton and Tomkins [11]. Our main results are summarized in Table 1. (By appropriate scaling, we can assume that the shortest job has length one. Uniform delay means that jobs of the same length have the same delay. The competitive ratios in all cases except for the first two are *expected*—the algorithms for the unit length jobs are deterministic, while all others are randomized.)

The unit length jobs may seem artificial but, in the networking context, they correspond to packet lengths, and therefore are well-suited to ATM where all packets have the same fixed size (53 bytes). Thus the special case of all jobs having the same length under the arbitrary delay model is of great interest. Also, it seems quite reasonable in this setting to require that there is a minimum delay. For other networking protocols and in the example of a video server, the jobs will have different lengths. As the table shows, for these settings we are able

---

[4] We consider the *oblivious* competitive ratio in which the input sequence is selected independently of the random choices of the algorithm [5].

to get stronger results by enforcing uniform delays; this seems like a reasonable way to treat equal-length jobs.

When all jobs have unit length, we prove that the (deterministic) strategy of choosing the *available* job with the earliest deadline first is strongly 2-competitive. (Note that this does not mean that jobs are scheduled in the earliest deadline order—a future job may have deadline sooner than an already scheduled job.) Strong-competitiveness means that no deterministic algorithm can do better in a worst-case. If *randomized* algorithms are permitted, then we can show a lower bound of 4/3 for the *expected* competitive ratio. When all the jobs have equal length and the *minimum* delay is at least the job length, then we prove that our greedy algorithm is 3/2-competitive.

When jobs have one of the two *known* lengths (1 and $k > 1$), we give a randomized algorithm that is 4-competitive. Additionally, if the delays are *uniform*, meaning that equal-length jobs have the same maximum delay, we give a randomized algorithm that is $\left(1 + \frac{\lceil k \rceil}{k}\right)$-competitive. For k an integer, this algorithm is strongly 2-competitive. This generalizes the result of Lipton and Tomkins [11], who consider jobs with *no delays*, and settles an open question of their paper. There are instances (a large number of jobs arriving at roughly the same time, but each with a large allowable delay) where the *optimal solution with delays* is significantly better than that with no delays. While Feldmann et. al [7] show that there are worst-case inputs where delay does not help, we believe that in practice delay will help increase the resource utilization significantly. Finally, we extend our results to jobs of multiple lengths.

1.2 Previous Work

Our work is most closely related to the work of Lipton and Tomkins [11] and Awerbuch, Bartal, Fiat, and Rósen [3]. The paper of Lipton and Tomkins [11] considers scheduling *without* delays, and we extend their work by considering several models of delays. If all delays are set to zero, our results achieve the same performance as [11]. Finally, our results can be combined with the methods of Awerbuch et. al [3] to handle routing on a tree network or the situation in which the bandwidth of the requests can vary.

The paper by Awerbuch, Bartal, Fiat, and Rósen [3] considers admission control for tree networks. Again, no delays are allowed, meaning a request must be either scheduled immediately or rejected. They present a general technique called "classify-and-randomly-select" that randomly selects a bandwidth b, length ℓ, and benefit f, all of which are powers of 2. The algorithm then rejects all requests that do *not* have bandwidth between b and $2b$, length between ℓ and 2ℓ, and benefit between f and $2f$. They first give an $O(\log n)$-competitive algorithm for the case of a tree network of n nodes where all calls use the maximum bandwidth, have infinite length, and equal benefits. Then by using the classify-and-randomly-select paradigm they allow any of the other parameters to vary with a multiplicative increase of $\log \Delta$, where Δ is the ratio of the largest to smallest value for the parameter being varied. While the classify-and-randomly-select algorithm is simple and has provably good worst-case performance, in practice it seems unlikely that one would want an admission control algorithm that rejects all calls whose length is less than ℓ or greater than 2ℓ. In real-life, a rejected user is quitelikely to immediately re-issue another request, forcing poor behavior. Allowing users to specify a maximum delay would eliminate such behavior. Our algorithm is also more natural for real-life use since it does not pre-select the lengths of the requests to accept.

Finally, Feldmann et. al [7] consider scheduling jobs *with delays* for a network that is a linear array of n nodes. They show that in some cases fixed length delays do not help. When delay is at most a constant multiple of the job length, they give request sequences where the competitive ratio between the amount of time the resource is used by an on-line algorithm with delays to the amount of time the resource is used by an off-line algorithm with no delays is $\Omega(\lg n)$. They also consider infinite delays, and try to optimize either the total completion time for all jobs, or the maximum delay between a job's arrival and its start of service. They present a $O(\log n)$-competitive greedy strategy for these measures on a n-node tree for requests of arbitrary bandwidth.

2 Preliminaries

Scheduling Model. A job (or call) J consists of a triple of positive real numbers $\langle a_J, |J|, w_J \rangle$, where a_J is the arrival time, $|J|$ is the length, and w_J is the maximum wait time for job J. That is, job J must be started during the interval $[a_J, a_J + w_J)$. A problem instance is a finite set $\mathcal{S}$ of jobs to be scheduled. We say that a schedule $\sigma \subseteq \mathcal{S}$ is *feasible* if no two jobs are running at the same time. We define the *gain* of a schedule σ to be $\sum_{J \in \sigma} |J|$. That is, it is the amount of time in which the resource is scheduled. (If clients are charged a rate per minute of usage, then the gain accurately reflects the profit of the schedule.) In the *uniform delay model*, the delay of each job is a function of the job's length. The only property of the uniform delay model used in our proofs is that jobs of the same length have the same delay. We also consider the *arbitrary delay model* in which each job can specify its own delay without any restrictions.

Method of Analysis. Our scheduling algorithms use randomization and thus we study their expected performance. We use $G_A(\mathcal{S})$ to denote the expected gain of algorithm A on a problem instance $\mathcal{S}$. The gain of algorithm A on a particular job $J \in \mathcal{S}$ is defined as $G_A(J) = \sum_{\sigma} \Pr[J \in \sigma] \cdot |J|$, where the probability that σ occurs is with respect to the algorithm A. We let $G_A(\mathcal{S}) = \sum_{J \in \mathcal{S}} G_A(J)$. When the algorithm being studied is clear, we just use $G(J)$ and $G(\mathcal{S})$. We use standard competitive analysis [12, 10, 6] to evaluate our algorithms. Namely, let σ_* be an optimal solution. We say that algorithm A is *c-competitive* if $\forall \mathcal{S}$: $c \cdot G_A(\mathcal{S}) \geq |\sigma_*|$. An algorithm A is *strongly c-competitive* if it is c-competitive and there exists no c'-competitive algorithm for $c' < c$.

Additional Definitions. Each job in $\mathcal{S}$ is either (1) scheduled, (2) virtually scheduled, or (3) rejected. The job only runs if it is scheduled. A virtually scheduled job J does not run itself, but prevents any job of length $\leq |J|$ from running during a period of length $|J|$. Thus, virtually scheduling a job J holds the resource for a longer job (length $\geq 2|J|$) with a short wait time that may arrive during the interval when J is virtually scheduled.

Fig. 1. In the left drawing J_j^σ blocks $J_i^{\sigma*}$, and in the right one J_j^σ covers $J_i^{\sigma*}$. If $J_j^\sigma = J_i^{\sigma*}$ then by our definition of blocking, J_j blocks J_i.

A job can run at different times in σ and σ_*, so we use J_i^σ to denote J_i as run or virtually run in σ, and likewise use $J_i^{\sigma*}$ to denote J_i as run in σ_*. Abusing the notation slightly, we also use J_i^σ and $J_i^{\sigma*}$ as times at which J_i started in σ and σ_*, respectively. We say that a scheduled or virtually scheduled job J_j^σ *blocks* job $J_i^{\sigma*}$ if $J_j^\sigma \le J_i^{\sigma*} < J_j^\sigma + |J_j|$. Job J_j^σ *covers* $J_i^{\sigma*}$ if $J_j^\sigma < J_i^{\sigma*}$ and $J_j^\sigma + |J_j| > J_i^{\sigma*} + |J_i|$. (See Figure 1.)

Basic Proof Structure. Given a schedule σ produced by algorithm A, we associate jobs in σ with jobs in the optimal schedule σ_*. For jobs $I \in \sigma$ and $J \in \sigma_*$, we let $A_\sigma(I, J)$ denote the portion of job I assigned to job J. The total gain associated with J for the schedule σ is defined as $gain(\sigma, J) = \sum_{I \in \sigma} A_\sigma(I, J)$. For $\mathcal{C}$ a set of possible schedules, we define $E_{\mathcal{C},A}[gain(J)] = \sum_{\sigma \in \mathcal{C}} \Pr[\sigma \text{ occurs}] \cdot gain(\sigma, J)$. When the algorithm A is clear from the context, we simply use $E_{\mathcal{C}}[gain(J)]$. Finally, when $\mathcal{C}$ is the set of all legal schedules for the jobs in $\mathcal{S}$, we use $E[gain(J)]$.

Our assignments for schedule σ will have the following two properties: (1) $\sum_{J \in \sigma_*} A_\sigma(I, J) \le |I|$, and (2) for each $J \in \sigma_*$, $E[gain(J)] \ge |J|/r$ where $r > 0$. Property (1) is easily enforced. In order to achieve Property (2), we show that any schedule produced by our algorithm belongs to one of k cases. We prove that for each job $J \in \sigma_*$ and each case $\mathcal{C}_i$, $E_{\mathcal{C}_i}[gain(J)] \ge |J|/r$. It follows that $G_A(\mathcal{S}) \ge \sum_{J \in \sigma_*} \frac{|J|}{r} = \frac{G_{opt}(\mathcal{S})}{r}$.

Properties (1) and (2) imply that our algorithm is r-competitive. Although a similar proof structure was used by Lipton and Tomkins, the assignment from jobs in $\mathcal{S}$ to the jobs in σ_* are more complicated here because we allow delays. In particular, a job can be run at different times (potentially overlapping) in different legal schedules (including σ and σ_*), which complicates our proofs.

3 Unit Length Jobs

When no delays are allowed and all jobs have the same length, the greedy strategy of scheduling the jobs in the order of arrival is easily shown to be optimal. The problem becomes more complicated if arbitrary delays are allowed *even if all jobs have the same length*. The following theorem establishes lower bounds on the competitive ratio achievable by any online algorithm for this case. Without loss of generality, assume that all jobs have unit length.

Theorem 1. *Consider the task of scheduling jobs of unit length where each job can specify an arbitrary delay. No deterministic algorithm can be c-competitive for $c < 2$, and no randomized algorithm can be c-competitive for $c < 4/3$.*

Proof. Let scenario S_1 consists of two jobs: J_1 arrives at time 0 and has a wait time of 1.25, and J_2 arrives at time .25 and has zero wait time. Scenario S_2 also has two jobs: J_1 is same as before, but J_2 arrives at time 1 and has zero wait time. Suppose a scheduling algorithm A schedules J_1 at time $t = 0$ with probability p. (If A is deterministic, then $p \in \{0, 1\}$.) Then $G_A(S_1) = p \cdot 1 + (1 - p) \cdot 2 = 2 - p$, and $G_A(S_2) = p \cdot 2 + (1 - p) \cdot 1 = 1 + p$. The optimal schedule in each case has gain of 2. Thus, the competitive ratio for A is at least

$$\max \left(\frac{G_{opt}(S_1)}{G_A(S_1)}, \frac{G_{opt}(S_2)}{G_A(S_2)} \right) = \max \left(\frac{2}{2 - p}, \frac{2}{1 + p} \right) = \frac{2}{\min(2 - p, \ 1 + p)}.$$

If A is deterministic, we get $\min\{2 - p, 1 + p\} = 1$, and if A is randomized, we get $\min\{2 - p, 1 + p\} \geq 3/2$. □

Let Greedy denote the following algorithm. At any time t, let $Q(t)$ be the set of all available jobs, meaning jobs J such that $a_J \leq t < a_J + w_J$. If the resource is free at time t and $Q(t) \neq \emptyset$, we schedule the job in $Q(t)$ that minimizes $(a_J + w_J) - t$; that is, the job with the smallest wait time remaining. Greedy is strongly 2-competitive among all deterministic algorithms (proof omitted).

Theorem 2. *When jobs have unit lengths and arbitrary delays, Greedy is a strongly 2-competitive deterministic algorithm.*

We now consider when the minimum wait time is at least 1, the length of the jobs. This proof uses a different technique than that used in our other proofs.

Theorem 3. *When jobs have unit lengths and wait times of at least one, Greedy is a $\frac{3}{2}$-competitive deterministic algorithm.*

Proof Sketch. Consider the schedule σ produced by Greedy, and call the periods during which the resource is continuously in use the *busy periods* of σ. Label these periods as $\pi_1, \pi_2, \ldots, \pi_m$, and let b_i and e_i, respectively, denote the times at which π_i begins and ends. Partition the job sequence S into classes $S_1, S_2, \ldots, S_m$, where S_i consists of exactly those jobs that *arrived* during the period $[b_i, e_i)$. Observe that Greedy schedules only jobs from S_i during π_i, and the job queue is empty at $t = e_i$.

We consider an arbitrary busy period π_i, and partition the jobs in S_i scheduled by σ_* into three classes: (1) $L_i \subseteq S_i$, the jobs scheduled in σ_* *after* the time e_i; (2) $R_i \subseteq S_i$, jobs in σ_* that are *not* in σ; (3) jobs common to σ and σ_* during π_i. Use the mnemonic "lazy" for L and "rush" for R: jobs in L_i have their deadlines past e_i, while all the jobs in R_i have their deadlines before e_i—otherwise the job queue of σ wouldn't be empty at e_i. Let l_i and r_i denote the cardinality of L_i and R_i, and let $k_i = |\pi_i|$ be the length of π_i in units of jobs. Then, it is clear that the total length of the jobs in S_i scheduled by σ_* cannot exceed either $k_i + l_i$ or $k_i + r_i$. Therefore, the competitive ratio of σ is $\min \left(\frac{k_i + l_i}{k_i}, \frac{k_i + r_i}{k_i} \right)$.

If $l_i \leq k_i/2$, then the competitive ratio during π_i is 3/2, and the proof is complete. Otherwise, we show that $r_i \leq k_i/2$ must hold. Consider a rush job J_r in σ_*. We claim that J_r cannot overlap *two* lazy jobs in σ. Suppose it did, and let

L and L' be the two jobs overlapping J_r. Since π_i is a busy period, jobs L and L' run consecutively, and let $t, t+1, t+2$ denote the time at the start of L, the end of L, and the end of L'. Since the job J_r has a *minimum delay* of 1, it must be eligible to run either at t or at $t+1$ (if it arrives during $(t, t+1)$, then it can run at $t+1$, and if arrives before t, then it can run at t). Since J_r is a rush job, its deadline expires before e_i, while both L and L' have their deadlines after e_i, we get a contradiction that *Greedy* schedules jobs in the order of earliest deadline first. Thus, we have the inequality $r_i \leq k_i - l_i$. Since $l_i > k_i/2$, we get $r_i < k_i/2$, which gives the desired upper bound of $3/2$ on the competitive ratio. Since the ratio holds for all busy periods of σ, it holds for the entire schedule. □

4 Jobs of Two Lengths

We present a 4-competitive algorithm when each job can have one of two lengths, 1 and k (for any real $k > 1$). If the delays are uniform, rather than arbitrary, then we show that our algorithm is $\left(1 + \frac{\lceil k \rceil}{k}\right)$-competitive. We refer to the length-1 jobs as *short jobs*, and length-k jobs as *long jobs*. Let the short jobs be $S_1, \ldots, S_m$ in their order of arrival; similarly, let the long jobs be labeled $L_1, \ldots, L_n$, where L_i arrives after L_{i-1}. We use J_i to denote a job that could be of either length. Note that in this result, the competitive ratio does *not* depend on the ratio between the longest and shortest job.

Our scheduling algorithm, Schedule-Two-Lengths uses a queue Q_1 (respectively, Q_k) for short jobs (respectively, long jobs) that have arrived and are still waiting to be scheduled. Within each queue, the jobs are given priorities in decreasing order of the last time at which the job can be started. Whenever the resource is not in use (i.e. either unscheduled or virtually scheduled) and Q_k is not empty, we schedule the highest priority job from Q_k. If Q_k is empty and Q_1 is not empty then with probability $1/2$ the short job of the highest priority from Q_1 is scheduled and otherwise it is virtually scheduled. To simplify our proofs, if a short job is virtually scheduled, we remove it from Q_1 even if its wait time has not expired. Clearly, in practice one would leave it on Q_1 allowing it to still be scheduled. It is easily seen that this optimization can only improve the resource utilization. Also we assume that, if possible, two jobs in σ_* are scheduled so that the one with the earlier deadline is first.

We now prove that Schedule-TwoLengths is 4-competitive. Our proof is presented in a more general form than necessary to ease the transition to the case of arbitrary length jobs (which is given in the next section). For the schedule σ, we define the following graphs. For jobs of length ℓ (which is either 1 or k) we construct a graph $G_\ell = (V_\ell, E_\ell)$ as follows[5]. For each job of length ℓ there is a vertex in V_ℓ. If J_i^σ blocks $J_j^{\sigma*}$ for $|J_i| = |J_j| = \ell$, then the directed edge (J_i, J_j) is placed in E_ℓ. (Note that a virtually scheduled short job can block another short job.) Because each job can block only one other, the graph G_ℓ consists of a set of *chains* where a chain is a path $\langle (v_1, v_2), (v_2, v_3), \ldots, (v_{n-1}, v_n) \rangle$. We refer to

---

[5] The graph G_ℓ is defined with respect to a schedule σ. For ease of exposition we use the notation G_ℓ versus $G_\ell(\sigma)$

v_1 as the *head* of the chain, and v_n as the *tail*. A chain is a *singleton chain* if it contains a single vertex. The assignments $A_\sigma(\cdot, \cdot)$ are defined as follows:

Assignment 1: For $(S_i, S_j) \in E_1$ where S_i is scheduled in σ, let $A_\sigma(S_i, S_j) = A_\sigma(S_i, S_i) = 1/2$.

Assignment 2: Let L_i be a long job. For each short job $S_j \in \sigma_*$ covered or blocked by L_i^g, let $A_\sigma(L_i, S_j) = 1/4$.

Assignment 3: For $(L_i, L_j) \in E_k$, let $A_\sigma(L_i, L_j) = k/2$.

Assignment 4: The remaining portion of L_i is assigned to itself.

Since a long job can cover at most k short jobs, it is easily seen that for each job $J \in \sigma$, $\sum_{J_j} A_\sigma(J, J_j) \leq |J|$. To prove that Schedule-Two-Lengths is 4-competitive we show that for all jobs J in σ_* the expected value of the assignment to J is at least $|J|/4$.

Theorem 4. Schedule-Two-Lengths *is 4-competitive under arbitrary delays.*

Proof Sketch. We show that $E[gain(J)] \geq |J|/4$, for each $J \in \sigma_*$.

Case 1: J is a short job $S_j \in \sigma_*$. Let t be the time when S_j starts in σ_*. We partition the set of feasible schedules based on the following conditions:

C_{1a}: A short job S_i is considered (removed from Q_1 and a coin flipped) during the interval $(t-1, t]$. With probability $1/2$, S_i is scheduled thus blocking S_j. Since $A_\sigma(S_i, S_j) = 1/2$, $E_{C_{1a}}[gain(S_j)] \geq (1/2)(1/2) = 1/4$.

C_{1b}: A long job L_i is considered during the interval $(t - k, t]$. Since L_i runs with probability 1 and $A_\sigma(L_i, S_j) = 1/4$, $E_{C_{1b}}[gain(S_j)] \geq 1/4$.

C_{1c}: If neither of the above occur, then both queues must be empty at time t. However, S_j runs in σ_* at time t and thus it must have been removed from Q_1 because it was considered earlier. Thus with probability $1/2$, S_j runs in which case $A_\sigma(S_j, S_j) = 1/2$. So $E_{C_{1c}}[gain(S_j)] \geq (1/2)(1/2) = 1/4$.

Case 2: J is a long job $L_j \in \sigma_*$. Let t be the time when L_j starts in σ_*. We partition the set of schedules based on the following conditions:

C_{2a}: A short job S is considered during the interval $(t - 1, t]$. Q_k must be empty when S was considered, yet by time t, job L_j arrives. Let L be the first long job to arrive after S was considered. (Job L could be L_j or a different long job that arrives before L_j.) With probability $1/2$, job S is virtually scheduled and thus L runs in σ when it arrives. Since $A_\sigma(L, L_j) = k/2$, $E_{C_{2a}}[gain(S_j)] \geq (1/2)(k/2) = k/4$.

C_{2b}: A long job L_i is considered during the interval $(t - k, t]$. Since L_i runs with probability 1 and $A_\sigma(L_i, L_j) = k/2 \geq k/4$, $E_{C_{2b}}[gain(L_j)] \geq \frac{k}{4}$.

C_{2c}: If neither of the above occur, then both queues must be empty at time t. However, L_j runs in σ_* at time t and thus must have been removed from Q_k when considered earlier. The maximum amount of L_j is assigned to other jobs when L_j covers k short jobs ($k/4$ assigned) and blocks a long job ($k/2$ assigned). Thus $A_\sigma(L_j, L_j) \geq k - (k/2 + k/4) = k/4$. $\qquad\square$

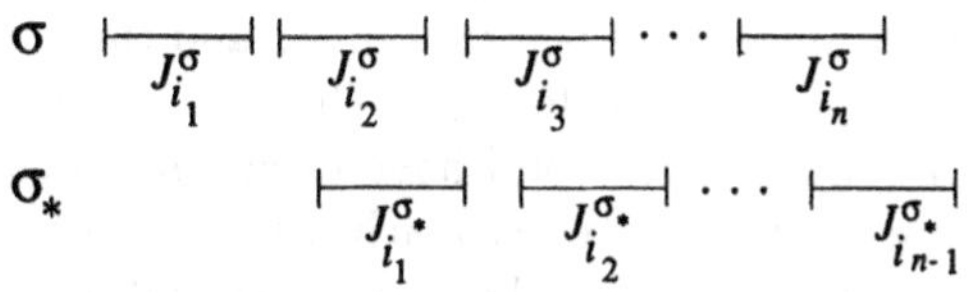

Fig. 2. A backward chain. Note that jobs of other lengths can be interleaved within the chain. Also the jobs in σ could be scheduled or virtually scheduled.

We now consider the model of uniform delays: all jobs in the same class (short or long) have the same delay. In this case, we can assume that σ_* schedules jobs of the same class in the order of their arrival: if J and J' are two same length jobs in σ_*, then equal delay implies that they can be ordered by their arrival time. Observe that σ also schedules the jobs in this order. We say a chain is a *backward chain* if each job blocks one that arrives before it. Thus in a backward chain jobs appear in the reverse order of arrival. (See Figure 2.)

Theorem 5. Schedule-Two-Lengths *is* $\left(1 + \frac{[k]}{k}\right)$*-competitive in the uniform delay model.*

Proof Sketch. Let $\alpha = \left(1 + \frac{[k]}{k}\right)$. The assignments $A_\sigma(\cdot, \cdot)$ are defined as:

Assignment 1: For $(S_i, S_j) \in E_1$ where S_i is scheduled in σ, $A_\sigma(S_i, S_j) = 1$.

Assignment 2: For each $S_j \in \sigma_*$ covered or blocked by the long job $L_i \in \sigma$, $A_\sigma(L_i, S_j) = 1/\alpha$.

Assignment 3: For $(L_i, L_j) \in E_k$, $A_\sigma(L_i, L_j) = k/\alpha$.

Assignment 4: For each job J, all unassigned portions of J are assigned to job J', the first job from σ_* in the chain of G_1 or G_k that includes J.

First, notice that $\sum_{J_j} A_\sigma(J, J_j) \leq |J|$, for each job $J \in \sigma$. Short jobs clearly satisfy the constraint. A long job can cover at most k short jobs and block one long job, and hence the total assigned value from Assignments 2 and 3 is $2k/\alpha$, which is at most k since $\alpha \geq 2$. We now prove that $E[gain(S_j)] \geq |S_j|/\alpha$ for each $S_j \in \sigma_*$.

Case 1: We consider when $S_j \in \sigma_*$ is a short job. As in the last proof, let t be the time $S_j^{\sigma_*}$.

$\mathcal{C}_{1a}$: A short job S_i is considered during the interval $(t-1, t]$. Here $E_{\mathcal{C}_{1a}}[gain(S_j)] = 1/2 \geq 1/\alpha$.

$\mathcal{C}_{1b}$: A long job L_i is considered during the interval $(t-k, t]$. Since $A_\sigma(L_i, S_j) = 1/\alpha$, $E_{\mathcal{C}_{1b}}[gain(S_j)] \geq 1/\alpha$.

$\mathcal{C}_{1c}$: Both queues are empty at time t and thus S_j was considered earlier. We now argue that S_j is the head of a backwards chain in G_1. If S_j does not block a short job in σ_* then S_j is a singleton chain. Suppose that S_j

blocks $S_{j'}$. Then since $S_{j'}$ is before S_j in σ_* and there are uniform delays, we know that $S_{j'}$ arrived before S_j. Thus $S_{j'}$ must be considered in σ before S_j (or otherwise it would have been considered instead of S_j at time S_j^σ). So we must eventually reach a short job S_i that is considered in σ yet blocks no short job. Thus S_T is the tail of the chain with head S_j. Since only Assignment 4 applies for S_T, $A_\sigma(S_T, S_j) = 1$. S_T runs with probability $1/2$, and thus $E_{\mathcal{C}_{1c}}[gain(S_j)] \geq 1/2 \geq 1/\alpha$.

Case 2: We consider when $L_j \in \sigma_*$ is a long job. Let t be time $L_j^{\sigma_*}$.

$\mathcal{C}_{2a}$: A short job S is considered during the interval $(t-1, t]$, and thus Q_k was empty when S was considered. Let L be the first long job to arrive after S was considered. With probability $1/2$, job S is virtually scheduled and thus L runs in σ when it arrives. L cannot block or cover any short jobs since there is less than one unit between its arrival time and t. Since we have uniform delays, L's deadline must have passed by time t or otherwise σ_* would have scheduled it instead of L_j. Thus $L \notin \sigma_*$, and hence there is a chain in G_k with head L followed by L_j. Since $L \notin \sigma_*$, all of the unassigned portions of L are assigned to L_j. Thus $A_\sigma(L, L_j) = k$ and hence $E_{\mathcal{C}_{2a}}[gain(S_j)] \geq (1/2)k = k/2 \geq k/\alpha$.

$\mathcal{C}_{2b}$: A long job L_i is considered during $(t - k, t]$. $E_{\mathcal{C}_{2b}}[gain(L_j)] \geq k/\alpha$.

$\mathcal{C}_{2c}$: Both queues are empty at time t and thus L_j was considered earlier. Using an argument as in Case 1c, it follows that L_j is the head of a backwards chain in G_k. Let L_T be the tail of the chain. The only portions of L_T not assigned to L_j are the $1/\alpha$ units assigned to each of the at most $\lceil k \rceil$ short jobs covered or blocked by L_T. Thus $E_{\mathcal{C}_{2c}}[gain(S_j)] \geq$

$$k - \frac{\lceil k \rceil}{\alpha} = k - \lceil k \rceil \left(\frac{k}{k + \lceil k \rceil} \right) = k \left(\frac{k}{k + \lceil k \rceil} \right) = \frac{k}{\alpha}. \qquad \square$$

The lower bound of 2 given by Lipton and Tomkins [11] holds here (since all delays could be zero), thus we immediately obtain the following corollary.

Corollary 6. Schedule-Two-Lengths *is strongly 2-competitive in the uniform delay model when the ratio, k, between the length of a long job and the length of a short job is an integer.*

5 Arbitrary Length Jobs

In this section we consider jobs that can have any length, but the maximum length of a job is known to the algorithm; as usual, we assume that the shortest job has length 1. We first consider the case when job lengths are powers of 2, namely, $1, 2, 4, \ldots, 2^c$, for some known constant c. While our algorithm is similar to the Marriage Algorithm of Lipton and Tomkins [11], our analysis is completely different. Note that the classify-and-randomly-select paradigm [3] in this case would just pick one randomly selected length to schedule. Our approach, in contrast, seems much more reasonable in practice: though there is a bias in favor of longer jobs, shorter length jobs also have a chance of being scheduled. The following lower bound was shown by Lipton and Tomkins [11].

Schedule-With-General-Delays
    Initialize $Q_0, \ldots, Q_c$ to be the empty queues
    When job J arrives
        Let 2^ℓ be the length of J (for $0 \le \ell \le c$)
        If the resource is not sched, or if J' is virtually sched for $|J'| \le 2^{\ell-1}$
            With prob $1/(c+1-\ell)$ schedule J, otherwise virtually schedule J
        Else Place J in Q_ℓ
    When a scheduled or virtually scheduled job finishes
        Let Q_ℓ be the non empty queue for the largest ℓ possible
        Let J be the highest priority job in Q_ℓ
        Remove J from Q_ℓ
        With prob $1/(c+1-\ell)$ schedule J, otherwise virtually schedule J

Fig. 3. Our admission control algorithm for when the jobs have lengths of $1, 2, 4, \ldots, 2^{c-1}$, or 2^c for a known constant c.

Theorem 7. *[11] Let $\Delta = 2^c$ be the ratio between the longest and shortest lengths. There is a lower bound of $\Omega(\log \Delta) = \Omega(c)$ for the randomized competitive ratio in the model with no delays.*

Following Schedule-Two-Lengths we maintain a queue Q_ℓ for jobs of length 2^ℓ. We favor the longer jobs in that (1) the probability of scheduling a job of length 2^ℓ is $1/(c+1-\ell)$, and (2) whenever the resource is available we schedule or virtually schedule the longest job available. During the period a job J is virtually scheduled, it prevents jobs of the same length or shorter jobs from running, but not longer ones. Our algorithm is shown in Figure 3.[6]

We now prove that this algorithm is $3(c+1)$-competitive. As in Schedule-Two-Lengths we use a graph for jobs of each length. Instead of using $G_{2^\ell} = (V_{2^\ell}, E_{2^\ell})$ for the graph corresponding to jobs of length 2^ℓ we use $G_\ell = (V_\ell, E_\ell)$ for $0 \le \ell \le c$. For each job of length 2^ℓ there is a vertex in V_ℓ. If a scheduled or virtually scheduled job J_i^σ blocks $J_j^{\sigma\bullet}$ where $|J_i| = |J_j| = 2^\ell$ then the edge (J_i, J_j) is placed in E_ℓ. For a given schedule σ, we make the following assignments based on G_ℓ (for $0 \le \ell \le c$). Unless otherwise given, the length of job J_i is 2^{ℓ_i}.

Assignment 1: If job J_i covers J_j (so $\ell_i > \ell_j$), then $A_\sigma(J_i, J_j) = 2^{\ell_j}/3$.
Assignment 2: If job J_i blocks J_j for $\ell_i \ge \ell_j$, then $A_\sigma(J_i, J_j) = 2^{\ell_i}/3$.
Assignment 3: The remaining portion of J_i is assigned to itself.

Since the sum of the lengths of the jobs covered by J_i is at most 2^{ℓ_i}, it follows that $\sum_{J_j \in \sigma_\bullet} A_\sigma(J_i, J_j) \le |J_i|$, for each $J_i \in \sigma$.

---

[6] For ease of exposition, we assume that jobs are removed from the queue when their wait time expires without explicitly including such checks in our pseudo-code. If a job is virtually scheduled, we remove it from the queue even if its wait time has not expired. Again, in practice one would leave such a job on the queue.

Theorem 8. Schedule-General-Delays *is* $3(c+1)$*-competitive if job lengths are in the set* $\{1, 2, 4, \ldots, 2^c\}$, *for some known constant* c.

Proof Sketch. We prove that $E[gain(J)] \geq \frac{|J|}{3(c+1)}$ for each $J \in \sigma_*$. Let J_i be the last job considered in σ before time $t = J_j^{\sigma *}$. We consider the following :

$\mathcal{C}_1$: $\ell_i < \ell_j$ **and** J_i **is considered during the interval** $(t - 2^{\ell_i}, t]$. Let J be the first job of length ℓ_j to arrive after J_i is considered. In the worst case, our algorithm might consider a job of length $\ell_i, \ell_i + 1, \ldots, \ell_j - 1$ before considering J. For J to run, the other jobs considered (at most one per length) must be virtually scheduled and then J must be scheduled. Thus:
$$\text{Prob } J \text{ scheduled} \geq \left(1 - \frac{1}{c+1-\ell_i}\right) \cdots \left(1 - \frac{1}{c+1-(\ell_j-1)}\right)\left(\frac{1}{c+1-\ell_j}\right)$$
$$= \frac{1}{c+1-\ell_i} \geq \frac{1}{c+1}$$
The above argument can be applied to show that any job considered would be scheduled with probability at least $1/(c+1)$. Since $A_\sigma(J, J_j) = 2^{\ell_j}/3$,
$E_{\mathcal{C}_1}[gain(J_j)] \geq \frac{2^{\ell_j}}{3(c+1)}$.

$\mathcal{C}_2$: $\ell_i \geq \ell_j$ **and** J_i **is considered during the interval** $(t - 2^{\ell_i}, t]$. Since the probability that J_i runs is at least $1/(c+1)$, and $A_\sigma(J_i, J_j) = 2^{\ell_i}/3$,
$E_{\mathcal{C}_2}[gain(J_j)] \geq \frac{2^{\ell_i}}{3(c+1)} \geq \frac{2^{\ell_j}}{3(c+1)}$.

$\mathcal{C}_3$: **All queues are empty at time** t. This case occurs if neither of the above do. Since Q_j is empty, J_j was considered earlier and ran with probability $\geq 1/(c+1)$. At most $2^{\ell_j}/3$ of J_j is assigned by Assignment 1, and $2^{\ell_j}/3$ of J_j is assigned to the job it blocks (if any) by Assignment 2. Thus $A_\sigma(J_j, J_j) \geq 2^{\ell_j} - (2^{\ell_j}/3 + 2^{\ell_j}/3) = 2^{\ell_j}/3$. Thus $E_{\mathcal{C}_3}[gain(J_j)] \geq \frac{2^{\ell_j}}{3(c+1)}$. $\square$

Corollary 9. Schedule-General-Delays *is* $6(c+1)$*-competitive if jobs have lengths between 1 and* 2^c, *for some known constant* c.

Proof Sketch. The only modification needed in **Schedule-General-Delays** is to treat job J like a job of length $2^{\lceil \lg |J| \rceil}$. Two key observations used in the proof are that (1) each job $J \in \sigma$ has true length that is at least $1/2$ of $2^{\lceil \lg |J| \rceil}$, and (2) if job $J \in \sigma_*$ is shorter than $2^{\lceil \lg |J| \rceil}$ then we can apply the excess that has been assigned to J to any other jobs (or portions thereof) that run in σ_* between $J^{\sigma *}$ and $J^{\sigma *} + 2^{\lceil \lg |J| \rceil}$. $\square$

We now consider the special case when there are uniform delays.

Theorem 10. Schedule-General-Delays *is* $2.5(c+1)$*-competitive in the uniform delay model when jobs have lengths in the set* $\{1, 2, 4, \ldots, 2^c\}$, *for some known constant* c.

Proof Sketch. We use the following assignments:

Assignment 1: If job J_i covers J_j (so $\ell_i > \ell_j$), then $A_\sigma(J_i, J_j) = 2^{\ell_j}/2.5$.

Assignment 2: If job J_i blocks J_j for $\ell_i \geq \ell_j$, then $A_\sigma(J_i, J_j) = 2^{\ell_i}/2.5$.

Assignment 3: For each job J_i, all unassigned portions of J_i are assigned to job J', where J' is the first job from σ_* in the chain G_{ℓ_i}.

Let J_i be the last job considered in σ before time $t = J_j^{\sigma_*}$. We consider the following cases.

$\mathcal{C}_1$: $\ell_i < \ell_j$ **and** J_i **is considered during the interval** $(t - 2^{\ell_i}, t]$. Let J be the first job of length ℓ_j to arrive after J_i is considered. As in the proof of Theorem 8, the probability that J runs is $\geq 1/(c+1)$. Since there are uniform delays J's deadline must have expired by time t (or σ_* would have scheduled it). Thus $J \notin \sigma_*$, and hence J_j is in a chain with head J followed by J_j. So, in Assignment 3, all excess from J is assigned to J_j. Finally, since J can cover jobs of length at most $2^{\ell_i} \leq 2^{\ell_j - 1}$ and blocks J_j, it follows that $A_\sigma(J, J_j) \geq 2^{\ell_j} - \frac{2^{\ell_j - 1}}{2.5} = 2^{\ell_j}/1.25$. Thus $E_{\mathcal{C}_1}[gain(J_j)] \geq \frac{2^{\ell_j}}{1.25(c+1)} \geq \frac{2^{\ell_j}}{2.5(c+1)}$.

$\mathcal{C}_2$: $\ell_i \geq \ell_j$ **and** J_i **is considered during the interval** $(t - 2^{\ell_i}, t]$. Since $A_\sigma(J_i, J_j) = 2^{\ell_j}/2.5$, $E_{\mathcal{C}_2}[gain(J_j)] \geq \frac{2^{\ell_i}}{2.5(c+1)} \geq \frac{2^{\ell_j}}{2.5(c+1)}$.

$\mathcal{C}_3$: **All queues are empty at time** t. Thus J_j must be considered earlier. As in the two job case, J_j is the head of a singleton chain with tail J_T. The maximum amount from J_T assigned to other jobs occurs when it covers jobs of lengths 2^{ℓ_j} and blocks a job of length $2^{\ell_j - 1}$. Thus $A_\sigma(J_T, J_j) \geq 2^{\ell_j} - \left(\frac{2^{\ell_j} - 2^{\ell_j - 1}}{2.5}\right) = \frac{2 \cdot 2^{\ell_j}}{5}$, and hence $E_{\mathcal{C}_3}[gain(J_j)] \geq \frac{2 \cdot 2^{\ell_j}}{5(c+1)} = \frac{2^{\ell_j}}{2.5(c+1)}$. $\quad\square$

Corollary 11. Schedule-General-Delays *is* $5(c+1)$*-competitive when scheduling jobs between lengths 1 and* 2^c *for some known constant* c *where the delay for each job* $J \in S$ *is a function of* $\lceil \lg|J| \rceil$.

6 Concluding Remarks

We have presented upper and lower bounds on the competitive ratio for non-preemptive, online admission control in the hard deadline model.

There are many interesting open questions raised by our work. In the model of unit length jobs with arbitrary delays, we have a lower bound (for randomized algorithms) of 4/3 on the competitive ration. Yet the best algorithm we've given is Greedy, which is a deterministic algorithm that is 2-competitive. Can randomization be used to obtain a better result? Can the lower bound be improved? Another interesting open question is to study the off-line problem of scheduling unit length jobs with arbitrary delays in the hard deadline model: Is there a polynomial time algorithm to find an optimal solution, or is this a NP-hard problem? While there is significant work on off-line scheduling, none addresses the hard deadline model.

While we have lower bounds to demonstrate that the competitive ratios of Schedule-TwoLengths and Schedule-With-General-Delays are asymptotically tight (in both the uniform and arbitrary delay models), the constants may not be tight. Thus another open problem is to try to develop matching upper and lower bounds on the competitive ratio for the various settings we considered.

Finally, another direction of study is to associate an additional *payoff* parameter with each job, which is the amount that a job is willing to pay for being scheduled. (In our current work, the implicit payoff was always equal to the length of the job.) Under this model a job that has a very short delay could provide a high payment to increase the chance that it is scheduled.

Acknowledgments. We thank Ben Gum who participated in the early discussions of this work. We also thank Michael Goldwasser and Eric Torng for many helpful discussions. Finally, we thank David Mathias for his comments on an earlier draft of this paper.

References

1. B. Awerbuch, Y. Azar, and S. Plotkin. Throughput-competitive on-line routing. In *Proc. of the 34th Ann. Symp. on Foundations of Comp. Sci.*, pp. 32–41. IEEE, Computer Society Press, Los Alamitos, CA, 1993.

2. B. Awerbuch, Y. Azar, S. Plotkin, and O. Waarts. Competitive routing of virtual circuits with unknown durations. In *Proc. of the 5th Ann. ACM-SIAM Symp. on Discrete Algs.*, pp. 321–327. SIAM, San Fransico, CA, 1994.

3. B. Awerbuch, Y. Bartal, A. Fiat, and A. Rósen. Competitive non-preemptive call control. In *Proc. of the 5th Ann. ACM-SIAM Symp. on Discrete Algs.*, pp. 312–320. SIAM, San Fransico, CA, 1994.

4. B. Awerbuch, R. Gawlick, T. Leighton, and Y. Rabani. On-line admission control and circuit routing for high performance computing and communication. In *Proc. of the 35th Ann. Symp. on Foundations of Comp. Sci.*, pp. 412–423. IEEE, Computer Society Press, Los Alamitos, CA, 1994.

5. S. Ben-David, A. Borodin, R. Karp, G. Tardos, and A. Wigderson. On the power of randomization in on-line algorithms. *Algorithmica*, 11(1):2–14, 1994.

6. A. Borodin, N. Linial, and M. Saks. An optimal online algorithm for metrical task systems. *Communications of the ACM*, 39(4):745–763, 1992.

7. A. Feldmann, B. Maggs, J. Sgall, D. Sleator, and A. Tomkins. Competitive analysis of call admission algorithms that allow delay. Technical Report CMU-CS-95-102, School of Computer Science, CMU, Pittsburgh, PA 15213, 1995.

8. J. Garay and I. Gopal. Call preemption in communication networks. In *Proc. of INFOCOM '92*, pp. 1043–1050. IEEE, Computer Society Press, Los Alamitos, CA, 1992.

9. J. Garay, I. Gopal, S. Kutten, Y. Mansour, and M. Yung. Efficient on-line call control algorithms. In *Proc. of the 2nd Israel Symp. on Theory and Computing Systems*, pp. 285–293. IEEE, Computer Society Press, Los Alamitos, CA, 1993.

10. A. Karlin, M. Manasse, L. Rudolpoh, and D. Sleator. Competitive snoopy caching. *Algorithmica*, 3(1):79–119, 1988.

11. R. Lipton and A. Tomkins. Online interval scheduling. In *Proc. of the 5th Ann. ACM-SIAM Symp. on Discrete Algs.*, pp. 302–311. SIAM, San Fransico, CA, 1994.

12. D. Sleator and R. Tarjan. Amortized efficiency of list update and paging rules. *Communications of the ACM*, 28(2):202–208, 1985.

Load Balanced Mapping of Data Structures in Parallel Memory Modules for Fast and Conflict-Free Templates Access [*]

Sajal K. Das[1] and M. Cristina Pinotti[2]

[1] Dept of Computer Sciences, Univ of North Texas, Denton, TX 76203, U.S.A.
[2] IEI, Consiglio Nazionale delle Ricerche, Via S. Maria, 46, 56126 Pisa, ITALY

Abstract. Conflict-free memory access is one of the important factors for the overall performance of a multiprocessor system in which the available memory is partitioned into several modules. Even if there is no contention in the processor-memory interconnection path, conflicts may still occur when two or more processors attempt to gain access to a single memory module or a memory location within a module. With a goal to achieve higher memory bandwidth, in this paper we resolve access conflicts at the level of memory modules. In particular, we deal with the problem of evenly mapping a data structure, called *host*, into as few distinct memory modules as possible to guarantee that subsets of distinct host nodes, called *templates*, can be accessed simultaneously in a conflict-free manner.

Since trees are among the most frequently used data structures in numerous applications, we propose a simple algebraic function based on the node indices for assigning the nodes of a k-ary tree to the memory modules in such a way that each subtree of a given height and arity can be accessed without conflicts. The assignment is direct, load balanced and also optimal in terms of the number of modules required. We also investigate conflict-free access to d-dimensional subcubes (Q_d) of n-dimensional hypercubes (Q_n), where Q_n represents a set of items indexed with n-digit addresses and accesses will be made to subsets of items differing in any arbitrary d-digit positions. With the help of the coding theory, we propose a novel approach to solve the subcube access problem. Codes with minimum distance $d \geq 2$ play a crucial role in our applications. We prove that any occurrence of a subcube $Q_s \subset Q_n$, for $0 \leq s \leq d-1$, can be accessed without conflicts using $\lceil \frac{2^n}{M} \rceil$ memory modules, by associating Q_n with a linear code C of length n, size M and minimum distance d. Associating the hypercube nodes with perfect or maximum distance separable (MDS) codes, our problem is solved optimally both in terms of the number of memory modules required and load balancing per module. These codes can be easily modified (without node relocation) according to the change in the size of the host or the number of available memory modules.

[*] This work is partially supported by Texas Advanced Technology Program grant TATP-003594031, and by Progetto 40% "Efficienza di Algoritmi e Progetto di Strutture Informative".

1   Introduction

We study the problem of mapping a subset of nodes in a data structure onto distinct memory modules so that they can be accessed in parallel. A data structure can be represented as a *host* graph $G = < V, E >$, where the node-set V represent the data to be mapped onto the memory modules and the edge-set represents logical adjacencies. A *template* $T \subseteq G$ is defined as a subgraph which is frequently accessed at run time. (Note here that a template is defined irrespective of its occurrence or *instance* in G.) Thus, the *conflict-free (memory) access problem* is formulated as the assignment of the host nodes to as *few memory modules* as possible such that the nodes of any instance of a specified template T in the host G are assigned to distinct memory modules. In this way, T can be accessed in a conflict-free manner. A trivial lower bound on the number of memory modules is the number of nodes in the template, which however is not tight even for very simple data structures due to the overlapping of the template instances in the host structure. Another requirement of the assignment is *balanced load* among the modules, that is each module should store almost the same number of host nodes.

For arbitrary host graphs and templates, the conflict-free access problem is computationally hard since a special case of it can be reduced to graph coloring. Therefore, a reasonable way to tackle this problem is by restricting the hosts to special structures. So far, significant attention has been paid in accessing arrays containing such templates as rows, columns, diagonals, submatrices, distributed blocks, and so on. For recent references, see [2, 9, 11]. Most of these solutions using an optimal number of memory modules are based on combinatorial objects like magic squares, Latin squares or circulant matrices.

Compared with arrays, mapping trees has received less attention, although various kinds of trees are among the most frequently used data structures in numerous applications. The reported literature on subtree access includes [7, 12]. These solutions either use a non-optimal number of memory modules or only allow the access to restricted templates. The approach due to Das and Sarkar [2] for k-ary subtree access in a k-ary tree uses an optimal number of memory modules (equal to the template size). However, it is recursive in nature and suffers from the drawback that there is no direct mapping by which one can derive the memory module assigned to a given node.

To understand how a solution to the conflict-free template access problem can be useful, refer to Das, Pinotti and Sarkar [4] who recently proposed elegant solutions to the conflict-free access to any *path* template in a heap data structure. This approach led to the development of the first load balanced and optimal mapping of priority queues on the hypercube-based, distributed-memory architecture [3]. Another application is range query in a B-tree modeling an external file system. Here, a template instance is a combination of paths and subtrees – the nodes along the subpath originating from the root to the first key in the given range and the nodes of a subtree rooted at such a key. Thus, paths and subtrees are fundamental templates in k-ary trees. More recently, Auletta et al. [1] have generalized our approach [4], so as to allow multiple template (both subtrees and path) access in a tree.

In this paper, for the k-ary trees as host, we design schemes for accessing various kinds of templates including t-ary subtrees where $2 \leq t \leq k$, levels or sublevels, and any combination of these basic templates. We also define a concept called *oriented* templates, which require exactly the same number of memory modules as the template size, to guarantee conflict-freeness. Also, all the assignments are scalable with respect to the host size, and the templates in k-ary trees can support addition of new levels at the bottom of the host without node relocation in memory modules. (For solution to the subtree access in binomial trees, refer to [5].)

We also study the subcube access problem in hypercubes. To the best of our knowledge, there exist no known results in the literature for subcube access. These templates have important applications in practice. For example, consider a data base of compressed files, each stored with a binary signature (i.e., a word either in the Boolean space $\{0,1\}^n$) into the disks of a parallel I/O subsystem. A typical query may require to retrieve all the files whose signatures match in at least $n - d$ digit positions with a given template signature. For faster query processing, the number of I/O transfers between the secondary and the main memories needs to be minimized. This problem can be elegantly solved (in a single I/O transfer) by formulating it as a conflict-free subcube access problem. Here each signature is interpreted as the address of a node in an n-dimensional binary hypercube, Q_n, and the signatures sought by a query are spread over some d-dimensional subcube, Q_d.

For the hypercube host, we apply various codes as an appropriate hashing function to decluster the host nodes. A crucial role is played by the codes with the larger *minimum distance*. Associating a binary hypercube Q_n with a linear code C of length n, size M, and minimum distance d, we show that any occurrence of an s-dimensional subcube $Q_s \subset Q_n$, for $0 \leq s \leq d - 1$, can be accessed in a conflict-free manner using $\lceil \frac{2^n}{M} \rceil$ memory modules.

2 t-ary Subtree Templates in k-ary Trees

Let the complete k-ary tree of height w, denoted by H_w^k, be the *host* and let $T_v^t \subseteq H_w^k$ be a *template*, where $2 \leq t \leq k$ and $0 \leq v \leq w$. Let the jth node at level i of the host H_w^k be denoted by (i, j), where the levels are counted from the root (at level 0) to the leaves and the nodes at each level from left to right, with the leftmost node counted as 0. Observed that two nodes (i, j) and (i', j') of H_w^k can belong to the same template T_v^t, for $2 \leq t \leq k$, if their lowest common ancestor lies at level l such that $max\{i, i'\} - l \leq v$, it follows:

Lemma 1. *The conflict-free access to any occurrence of $T_v^t \subseteq H_w^k$, with $2 \leq t \leq k$ and $0 \leq v \leq w$, requires $\mathcal{M} \geq \dfrac{k^{v+1} - 1}{k - 1}$ memory modules.*

In the following, a procedure for an optimal assignment of the host nodes is given. It is based on an easy mapping which leads to a well balanced memory load. Figure 1 shows the memory assignment for conflict-free access to any subtree $T_2^t \subset H_3^3$, where $2 \leq t \leq 3$.

Procedure Subtree-Coloring /* Conflict-free access to the template $T_v^t \subseteq H_w^k$ */

$M[(0,0)] = 0;$

$$M[(i,j)] = \left(\sum_{r=0}^{i-1} (j_r + 1) k^{i-1-r} \right) \bmod \frac{k^{v+1} - 1}{k - 1}, \text{ for } (i,j) \in H_w^k, \quad j = \sum_{r=0}^{i-1} j_r k^r, \text{ and } i \geq 1.$$

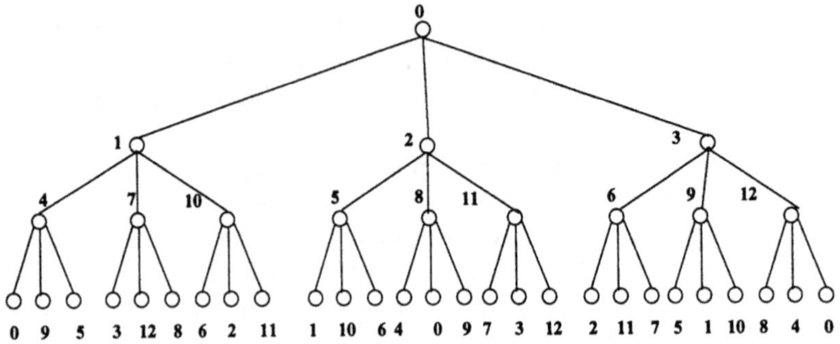

Fig. 1. Mapping of H_3^3 to access conflict-free any subtree T_2^t of height 2, for $t \leq 3$.

The proof of correctness of this assignment is stated below, and its proof is omitted here due to space limitations.

Theorem 2. *Any collection of $\mathcal{M}$ arbitrary nodes $(i,j) \in H_w^k$ whose addresses have v consecutive digits $(j_s j_{s+1} \cdots j_{s+v-1})$ spanning all the configurations in $\{0, 1, \ldots, k-1\}^v$ can be accessed in a conflict-free manner.*

The fact that any set of $\mathcal{M}$ nodes of the host at the same level $i \geq v + 1$ which are at mutual distance k^r, for $0 \leq r \leq i - v$, satisfies Theorem 2, therefore sublevels, subtrees as well as their combinations can be accessed with a minimum number of conflicts.

Note that the arity, t, of the template plays no role in determining the minimum number of modules required to guarantee conflict-free access. This is due to the overlapping of the occurences of template instances in the host. On the other hand, let us see how the scenario changes when we consider templates whose children must satisfy some propoerties with respect to the children of the host nodes. This will be called an *oriented subtree template* $O_v^{t,\xi}$, which is a complete t-ary subtree of height v and rooted at the node ξ such that the indices i_j of the t children of an entire internal node (in the host) cover the interval $[0, t-1]$ when the function $f(i_j) = (i_j \bmod t)$ is applied. Formally,

Definition 3. The *oriented subtree template* $O_v^{t,\xi}$ is a complete t-ary subtree of height v, with root node ξ at level $l(\xi) \leq w - v$, and is recursively extracted from the host H_w^k as follows: (1) ξ belongs to $O_v^{t,\xi}$, and (2) the children of any node $\eta \in O_v^{t,\xi}$, are those t nodes $(\eta_{i_0}, \ldots, \eta_{i_{t-1}})$ such that $\forall z \in [0, \ldots, t-1], \exists ! \; i_j : (i_j \bmod t) = z$, where $0 \leq j \leq t - 1$.
The family of oriented templates of arity t and height v will be denoted as O_v^t.

The bold lines in Figure 2 depict an oriented template $O_2^{3,p}$ in the host H_2^4.

The procedure below guarantees an optimal conflict-free assignment and also balances the memory load.

Procedure Subtree-Coloring-Oriented /* Conflict-free access to O_v^t in H_w^k */

$M[(0,0)] = 0;$

$$M[(i,j)] = \sum_{r=0}^{i-1} [(j_r \bmod t) + 1] \; t^{i-1-r} \bmod \frac{t^{v+1} - 1}{t - 1}, \quad \text{for } (i,j) \in H_w^k, \; j = \sum_{r=0}^{i-1} j_r k^r, \text{ and } i \geq 1.$$

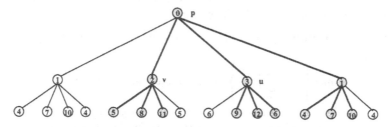

Fig. 2. Mapping of H_2^4 to access conflict-free the O_2^3.

3 Templates in Binary Hypercubes

An n-dimensional binary hypercube, $Q_n = (V, E)$, is a graph of $|V| = 2^n$ nodes, each specified by a binary address $(b_{n-1} b_{n-2} \ldots b_0)$ where the bit b_i corresponds to the ith dimension of Q_n, for $0 \leq i \leq n - 1$. Two nodes X and Y of Q_n are at *Hamming distance* $d(X, Y) = s$ if their addresses differ in exactly s dimensions. There exists an edge between X and Y in the ith dimension of Q_n if and only if $d(X, Y) = 1$ and their addresses differ exactly in the ith bits. Let $\mathcal{D} = \{i_1, \ldots, i_{n-s}\}$, for $0 \leq i_j \leq n - 1$, denote a set of $n - s$ dimensions of Q_n where $0 \leq s \leq n - 1$. A *configuration* is defined by assigning specific values to each dimension in $\mathcal{D}$. Now, let $Q_s(\mathcal{D}, \mathcal{A}) = (V', E')$ be an s-dimensional subcube of Q_n such that $V' \subset V$ consists of all 2^s nodes in V which assume the configuration $\mathcal{A}$, and $E' \subset E$ consists of edges linking the nodes in V' at unit Hamming distance.

Given the host and template definitions, we now show how to distribute the nodes of Q_n into several memory modules such that any occurrence of a subcube Q_s, for $s \leq n$, can be accessed conflict-free. Our first observation is:

Theorem 4. *Any occurrence of $Q_s \subset Q_n$ can be accessed conflict-free if and only if the Hamming distance $d(X, Y) \geq s + 1$ for any pair of nodes X and Y assigned to the same module.*

Letting $\mathcal{M}(n, s)$ denote the total number of modules required for conflict-free access to any occurrence of $Q_s \subseteq Q_n$, for $s \leq n$, the more challenging problem is to find the lower bound on $\mathcal{M}(n, s)$ and a strategy to distribute the hypercube nodes among the available memory modules in such way that the above condition is satisfied. A trivial bound is $\mathcal{M}(n, s) \geq 2^s$, the size of Q_s.

Theorem 5. *Any occurrence of $Q_s \subseteq Q_n$, for $s \leq n$, can be accessed conflict-free using exactly $\mathcal{M}(n, s) = 2^s$ memory modules if and only if there exists a subset $S \subseteq Q_n$ such that every $n - s$ dimensions of the node addresses in Q_n span all possible configurations. Also, exactly 2^{t-s} memory conflicts occur while accessing any occurrence of Q_t, where $s + 1 \leq t \leq n$.*

However, the lower bound on $\mathcal{M}(n, s)$ could be larger than 2^s as argued below. Let $I_t(X)$ be the subset of nodes in Q_n which are at a distance $\leq t$ from a given node X. The Hamming distance satisfies the *triangle inequality*, so each pair of nodes Y_1, $Y_2 \in I_t(X)$ are at a pairwise distance $d(Y_1, Y_2) \leq d(Y_1, X) + d(X, Y_2) \leq 2t$. Then, by Theorem 4, two nodes in $I_t(X)$ cannot be mapped onto the same module if $t \leq \lfloor \frac{s}{2} \rfloor$. Therefore,

Theorem 6. (Hamming Bound) *The conflict-free access to any occurrence of $Q_s \subseteq Q_n$ requires*

$$\mathcal{M}(n, s) \geq max \left\{ 2^s, \sum_{i=0}^{t} \binom{n}{i} \right\} \text{ memory modules where } t = \lfloor s/2 \rfloor.$$

Now, in order to find a suitable strategy for assigning the hypercube nodes, let us highlight the similarities between this problem and the coding theory.

3.1   A Coding Theoretic Approach

Let us first recall that an (n, M, d) code C is a set of M binary words of length n such that any two codewords are at a Hamming distance $\geq d$. The *minimum distance*, d, of the code C plays an important role in the theory of *error correcting codes*, since a code with distance d detects at least all non-zero error patterns of weight $\leq d - 1$ and also corrects all error patterns of weight $\leq \lfloor (d-1)/2 \rfloor$. Thus the most basic question in coding theory is to find the largest (n, M, d) code.

This is identical with the subcube access problem which aims to find in the hypercube Q_n, the largest subset S of nodes at a pairwise distance $\geq d$ that can be assigned to the same memory module. By Theorem 4, the codewords of C form such subset S for conflict-free access to any occurrence of the subcube $Q_{d-1} \subset Q_n$. Among various kinds of existing codes, let us review the *linear codes*. A code C is *linear* if $v + w$ is a codeword in C whenever v and w are in C and $+$ is the bitwise addition modulo 2 operation. Thus, a linear binary $[n, k, d]$ code consists of a set of 2^k words of length n, spanning a subspace $\{0, 1\}^k$ such that any pair of words is at distance $\geq d$. Linear codes are also *systematic codes* since they have distinguishable information digits. The zero word belongs to each linear code since given a codeword v, the sum $v + v = 0 \in [n, k, d]$. The minimum distance, d, of a linear code $[n, k, d]$ coincides with the minimum weight of the non-zero codewords. The parameter k related to the code size 2^k is the number of information digits, that is the bit positions which span all their possible configurations in the code C.

A linear code is said to be in *standard form* if the k information bits are the most significant ones. If C is not in standard form, by a suitable permutation of the n digits, we can obtain a new block code $C' = [n, k, d]$ equivalent to C and also in standard form. Finally, the *coset of C determined by u* (any binary vector of length n) is defined as the set of words $v + u$ for all codewords $v \in C$. Note that there are exactly 2^{n-k} distinct cosets of C, each containing 2^k words.

Subcube Access Strategy Applying the theory of linear codes and their subsets, we are now ready to describe a strategy to decluster the nodes of a hypercube host so that appropriate subcubes can be accessed without conflicts.

Definition 7. Let the *seed set* $S(n, k, d)$ of *dimension* k and *distance* d be a subset of 2^k nodes of Q_n such that the k most significant bits of their addresses span all possible configurations, and any two nodes in S are at a pairwise Hamming distance $\geq d$.

Given $S(n, k, d)$, we show how to distribute the nodes of Q_n into $\mathcal{M}(n, d-1) = 2^{n-k}$ memory modules such that any occurrence of the subcube $Q_{d-1} \subset Q_n$ is accessed conflict-free.

Definition 8. Given the binary representations of the integers X and j as $(x_{n-1}x_{n-2}\ldots x_0)$ and $(j_{n-1}j_{n-2}\ldots j_0)$ respectively, let the bitwise XOR be $X'_j = X + j = (x'_{n-1}, \ldots, x'_0)$ where $x'_i = (x_i + j_i) \bmod 2$. For $0 \leq j \leq 2^{n-k} - 1$, the *shifted set* $S[j]$ is defined as $S[j] = \{X'_j = X + j, \forall X \in S(n, k, d)\}$.

By definition, $S[0]$ and the seed set S are the same. Moreover,

Lemma 9. *The subsets* $S[j] \subset Q_n$, *where* $0 \leq j \leq 2^{n-k} - 1$, *of a seed set* $S(n, k, d)$ *preserve the pairwise distance* d *and collectively form a partition of the nodes in* Q_n.

The correlation with the coding theory is stated as:

Proposition 10. : *The codewords of a linear code* $C = [n, k, d]$ *in standard form constitute a seed set* $S(n, k, d)$. *The coset of* C *determined by* j *is the* shifted set $S[j]$, *where* $0 \leq j \leq 2^{n-k} - 1$. ∎

Our strategy for declustering the hypercube nodes is explicitly stated below.

Declustering Algorithm /* Conflict-free access to $Q_{d-1} \subset Q_n$, for $d \leq n$ */

1. Find a seed set, $S(n, k, d)$, or equivalently, choose $C[n, k, d]$.
2. Construct the corresponding shifted sets, i.e., the cosets of $C[n, k, d]$.
3. Assign the nodes in each shifted set (coset) to a distinct memory module.

Theorem 11. : *The Declustering algorithm guarantees conflict-free access to* $Q_{d-1} \subset Q_n$, *where* $d \leq n$, *requiring* $\mathcal{M}(n, d-1) = 2^{n-k}$ *modules, each having a perfectly balanced load of* 2^k *nodes.* ∎

¿From a practical viewpoint, a linear code $C = [n, k, d]$ and the corresponding seed set can be generated by a suitable matrix G of rank k consisting of k rows and n columns. In fact, any codeword $v \in C$ corresponds to a binary row vector u of length k such that $v = uG$. The memory module to which the hypercube node v is assigned is given by vH, where H is the *parity-check* matrix of the linear code C. As an example, consider the code $C[7, 4, 3]$ whose generating and parity-check matrices are given by G and H.

$$G = \begin{bmatrix} 1 & 0 & 0 & 0 & 1 & 1 & 1 \\ 0 & 1 & 0 & 0 & 1 & 1 & 0 \\ 0 & 0 & 1 & 0 & 1 & 0 & 1 \\ 0 & 0 & 0 & 1 & 0 & 1 & 1 \end{bmatrix} \qquad H = \begin{bmatrix} 1 & 1 & 1 \\ 1 & 1 & 0 \\ 1 & 0 & 1 \\ 0 & 1 & 1 \\ 1 & 0 & 0 \\ 0 & 1 & 0 \\ 0 & 0 & 1 \end{bmatrix}.$$

Code Facts and Hypercube Analyses

Definition 12. (Optimality) A conflict-free access to all occurrences of $Q_s \subseteq Q_n$, for $s \leq n$, is *optimal* if the number $M(n, s)$ of required memory modules attains the equality in the lower bound of Theorem 6.

For example, the *perfect codes*, which attain the equality in the **Hamming Bound**, lead to an optimal conflict-free access using $\sum_{i=0}^{t} \binom{n}{i}$ memory modules. Unfortunately, it has been proved in [8] that:

Code-Fact 1 . *For any n, the trivial perfect codes $[n, n, 1]$ and $[n, 1, n]$ exist. A non-trivial perfect code of length n and distance $d = 2t + 1$ is either $[23, 12, 7]$ (Golay code), or $[2^r - 1, 2^r - r - 1, 3]$ for $r \geq 2$ (Hamming perfect code).*

In terms of our problem, this implies:

Analysis 1 . *The seed set $S(n, n, 1)$ based on the code $[n, n, 1]$ coincides with the nodes of Q_n and allows conflict-free access to Q_0 using a single memory module. Similarly, the seed set $S(n, 1, n)$ allows conflict-free access to any occurrence of Q_{n-1} using $M(n, n-1) = 2^{n-1}$ modules. Based on the non-trivial perfect codes, optimal conflict-free access is guaranteed only for $Q_6 \subset Q_{23}$ and $Q_2 \subset Q_{2^r-1}$, for $r \geq 2$, using respectively 2^{11} and 2^r modules.*

As a consequence of Theorems 4 and 5, and by the seed set definition, we get

Corollary 13. *Any occurrence of $Q_s \subset Q_n$ can be optimally accessed in a conflict-free manner using $M(n, s) = 2^s$ memory modules if and only if the seed set $S(n, n - s, s + 1)$ exists.*

The following result gives us insight on the existence of such seed sets.

Code-Fact 2 .*(Singleton Bound [10])* *For any linear code $[n, k, d]$, it holds $k \leq n - d + 1$. Those codes attaining the equality in this bound are called the maximum distance separable (MDS) codes.*

Analysis 2 . *A seed set $S(n, n - s, s + 1)$ on an MDS code provides an optimal, conflict-free subcube access to any occurrence of $Q_s \subset Q_n$ using 2^s modules.*

Other than the *trivial perfect linear* codes $[n, n, 1]$ and $[n, 1, n]$, the two other binary codes known to be MDS codes are $[3, 2, 2]$ and $[4, 3, 2]$. A very large class of MDS codes exist for non-binary hypercubes, as discussed in [6].

For the binary hypercubes, other results from the coding theory can also be applied. Although they do not always guarantee optimality, yet they offer easy solution to conflict-free subcube access. For example, the linear code $[2^k - 1, k, 2^{k-1}]$ known as the *Simplex* code or the *dual* code of the perfect Hamming Code is the shortest linear code with cardinality 2^k and distance 2^{k-1}. It is a special case of the *Reed-Muller* (RM) codes, defined below.

Code-Fact 3 .*[10]* *For $0 \leq r \leq m$, there exists a binary linear rth order Reed-Muller code $RM(r, m) = [n = 2^m, k = \sum_{i=0}^{r} \binom{m}{i}, d = 2^{m-r}]$.*

Analysis 3 . *Choosing a seed set S based on $RM(r, m)$, any occurrence of $Q_{2^{m-r}-1} \subset Q_{2^m}$ can be accessed conflict-free using $M = 2^{n-k}$ memory modules.*

All previous results can be extended to any subcube $Q_t \subseteq Q_s$ as follows.

Corollary 14. *If any occurrence $Q_s \subset Q_n$ can be accessed conflict-free, then any $Q_t \subset Q_n$, $0 \leq t \leq s$, is also conflict-free. There are at most 2^{t-s} conflicts at each module for accessing $Q_t \subset Q_n$, where $s + 1 \leq t \leq n$.*

Table 1 shows the application of the coding theory approach to a few host and template sizes, while for a given template Q_s, Table 2 shows the number of memory modules required for different host sizes.

In conclusion, most of the codes discussed so far suffer from the constraints on their length n, hence a constraint on the host size. Nevertheless, the coding theory offers methods for combining codes to shorten or lengthen a codeword [10]. The results on how to overcome the size constraints or capturing the scenario when the host size changes dynamically are given in [6], which also extends our approach to the q-ary n-cubes and mixed-radix hypercubes with pairwise-prime moduli as well as non-pairwise-prime moduli.

References

1. V. Auletta, A. De Vivo and V. Scarano, "Multiple Templates Access of Trees in Parallel Memory Systems", *Proc. Int'l Parallel Processing Symp*, Geneva, 1997.

2. S. K. Das and F. Sarkar, "Conflict-Free Data Access of Arrays and Trees in Parallel Memory Systems", *Proc. Sixth IEEE Symp on Parallel and Distributed Processing*, Dallas, TX, Oct 1994, pp. 377-384.

3. S. K. Das, M.C. Pinotti, and F. Sarkar, "Optimal and Load Balanced Mapping of Parallel Priority Queues in Hypercubes," *IEEE Trans. on Parallel and Distributed Systems*, Vol. 7, No. 6, June 1996, pp. 555-564.

4. S. K. Das, F. Sarkar, and M.C. Pinotti, "Conflict-Free Access of Trees in Parallel Memory Systems and Its Generalization with Applications to Distributed Heap Implementation", *Proc. Int'l Conf. on Parallel Processing*, Oconomowoc, Wisconsin, Aug 1995, Vol. III, pp. 164-167.

5. S. K. Das and M.C. Pinotti, "Conflict-Free Access to Templates of Trees and Hypercubes," *Third Annual International Computing and Combinatorics Conference* (COCOON'97), Shanghai, China, Aug 20-22, 1997.

6. S.K. Das and and M.C. Pinotti, "A Coding Theory Approach to the Conflict-Free Subcube Access in Parallel Memory Systems", *Tech. Rep.* CRPDC-96-7, Dept. of Comp. Sci., University of North Texas, Denton, 1996.

7. M. Gössel, and B. Rebel, "Memories for Parallel Subtree Access", *Proc. of Int'l Workshop on Parallel Algorithms and Architectures*, in *Lecture Notes in Computer Science*, Springer-Verlag, Vol. 299, 1987, pp. 120-130.

8. D. G. Hoffmann, D. A. Leonard, C.C. Lindner, K.T. Phelps, C.A. Rodger, J.R. Wall, *Coding Theory: The Essentials*, Marcel Dekker Inc., New York, 1991.

9. K. Kim, and V. K. Prasanna Kumar, "Latin Squares for Parallel Array Access", *IEEE Trans. Parallel and Distributed Systems*, Vol. 4, No. 4, Apr 1993, pp. 361-370.

10. F. J. MacWilliams and N. J. A. Sloane, *The Theory of Error-Correcting Codes, Parts I and II*, North-Holland, New York, 1977.

11. H. A. G. Wijshoff, *Data Organization in Parallel Memories*, Kluwer Academic Publishers, 1989.

12. H. A. G. Wijshoff, "Storing Trees into Parallel Memories", *Parallel Computing 85*, (Ed. Feilmeier), Elsevier Science Pub., 1986, pp. 253-261.

Table 1. Number of modules $\mathcal{M}$ and the code on which the seed set is based to access all the occurrences of a subcube template Q_s in a binary hypercube host Q_n.

Host	Template	$\mathcal{M}$	Seed Set	Code	Comments
Q_4	Q_1	2	$[4,3,2]$	Maximum Distance Separable Code = $RM(1,2)$	Optimal
Q_4	Q_3	8	$[4,1,4]$	Trivial Perfect Code	Optimal
Q_7	Q_2	8	$[7,4,3]$	Hamming Perfect Code	Optimal
Q_7	Q_3	16	$[7,3,4]$	Simplex Code	—
Q_7	Q_6	64	$[7,1,7]$	Trivial Perfect Code	Optimal
Q_8	Q_1	2	$[8,7,2]$	$RM(2,3)$	Optimal
Q_8	Q_2	16	$[8,4,3]$	Concatenated Code	—
Q_8	Q_3	16	$[8,4,4]$	$RM(1,3)$	—
Q_8	Q_7	128	$[8,1,8]$	Trivial Perfect Code	Optimal
Q_{15}	Q_2	16	$[15,11,3]$	Hamming Perfect Code	Optimal
Q_{15}	Q_3	2^9	$[15,6,4]$	Shortened Binary Reed-Solomon Code	—
Q_{15}	Q_7	2^{11}	$[15,4,8]$	Simplex Code	—
Q_{15}	Q_{14}	2^{14}	$[15,1,15]$	Trivial Perfect Code	Optimal

Table 2. For a given template Q_s, the number of memory modules $\mathcal{M}$ required for different host sizes.

Template	Host	$\mathcal{M}$	Seed Set	Code	Comments
Q_2	Q_3	4	$[3,1,3]$	Trivial Perfect Code	Optimal
	Q_7	8	$[7,4,3]$	Trivial Perfect Code	Optimal
	Q_8	16	$[8,4,3]$	Concatenated Code	—
	Q_{15}	16	$[15,11,3]$	Hamming Perfect Code	Optimal
	Q_{31}	32	$[31,25,3]$	Hamming Perfect Code	Optimal
Q_3	Q_4	8	$[4,1,4]$	Trivial Perfect Code	Optimal
	Q_7	16	$[7,3,4]$	Simplex Code	—
	Q_8	16	$[8,4,4]$	Reed-Mullar Code, $RM(1,3)$	—
	Q_{16}	32	$[16,11,4]$	$RM(2,4)$	—
	Q_{32}	64	$[32,26,4]$	$RM(3,5)$	—
Q_7	Q_8	2^7	$[8,1,8]$	Trivial Perfect Code	Optimal
	Q_{16}	2^{12}	$[16,4,8]$	Simplex Code	—
	Q_{18}	2^{12}	$[18,6,8]$	Shortened Binary Reed-Solomon Code	—
	Q_{32}	2^{16}	$[32,16,8]$	$RM(2,5)$	—
Q_4	Q_{14}	2^{12}	$[14,2,5]$	Concatenated Code	—
	Q_{20}	2^{16}	$[20,4,5]$	Shortened Binary RS Code	—
	Q_{24}	2^{16}	$[24,8,5]$	Shortened Binary RS Code	—
	Q_{36}	2^{16}	$[36,20,5]$	Shortened Binary RS Code	—

Parallel vs. Parametric Complexity

Ketan Mulmuley

The University of Chicago and Indian Institute of Technology, Bombay
email: mulmuley@cs.uchicago.edu

Abstract. In [5] (also see [4]) we defined a natural and realistic model of parallel computation called *the PRAM model without bit operations*. It is like the usual PRAM model, the main difference being that no bit operations are provided. It encompasses virtually all known parallel algorithms for (weighted) combinatorial optimization and algebraic problems. In this model we proved that, for some large enough constant b, the mincost-flow for graphs with n vertices cannot be solved deterministically (or with randomization) in $\sqrt{n}/b$ (expected) time using $2^{\sqrt{n}/b}$ processors; this is so even if we restrict every cost and capacity to be an integer (nonnegative if it is a capacity) of bitlength at most an for some large enough constant a. A similar lower bound is also proved for the max flow problem. It follows that these problems cannot be solved in this model deterministically (or with randomization) in $\Omega(N^c)$ (expected) time with $2^{\Omega(N^c)}$ processors, where c is an appropriate positive constant and N is the total bitlength of the input. Since these problems were known to be P-complete, this provides concrete support for the belief that P-completeness implies high parallel complexity, and for the $P \neq NC$ conjecture itself.

These lower bounds actually follow from a general lower bound which roughly states that if the so-called parametric complexity of a problem is high then it is hard to parallelize in the PRAM model without (or with limited) bit operations. In this lecture I shall explore this relationship between parametric and parallel complexities in detail–it may open a way for investigating parallel complexities of many weighted optimization problems. Main motivation behind the lower bounds for the mincost flow and max flow problems–which are P-complete–was to prove unconditionally a weaker implication of the $P \neq NC$ conjecture in model that is restricted but realistic. But now one can ask if similar lower bounds can also be proved for other weighted optimization problems whose parallel complexities are open because they are neither known to be P-complete nor have fast parallel algorithms. One prime example is the minimum-weight perfect matching problem. If the edge-weights are in unary, the problem has fast parallel algorithms [2, 6]. In general, its parallel complexity is open. We conjecture that the parametric complexity of this problem is $2^{\Omega(n^\epsilon)}$ for some small enough positive ϵ. It would then follow from the general lower bound that it cannot be solved in the PRAM model without bit operations in $o(N^{\epsilon'})$ time using $2^{o(N^{\epsilon'})}$ processors, where N is the input bitlength and ϵ' is a small enough positive constant. Several other problems may be investigated in this way: e.g. matroid intersection problems [3], matroid parity problems [3], construction of blocking flows [1], and several problems in computational geometry, robot motion planning, and so forth.

References

1. A. Goldberg, R. Tarjan, A new approach to the maximum-flow problem, JACM, 35: 921–940, 1988.
2. R. Karp, E. Upfal, A. Wigderson, Constructing a perfect matching is in random NC, Combinatorica, 6:35–48, 1986.
3. L. Lovász, M. Plummer, Matching theory, Akadémiai Kiadó, Budapest, 1986.
4. K. Mulmuley, Is there an algebraic proof for $P \neq NC$?, Proceedings of the ACM Symposium on Theory of Computing, 1997.
5. K. Mulmuley, Lower bounds in a parallel model without bit operations, to appear in the SIAM Journal of Computing, available at the author's web address.
6. K. Mulmuley, U. Vazirani, V. Vazirani, Matching is as easy as matrix inversion, Combinatorica, 7(1):105–113, 1987.

Position-Independent Near Optimal Searching and On-line Recognition in Star Polygons*

Alejandro López-Ortiz[1] and Sven Schuierer[2]

[1] Research Scientist, Open Text Corp., 180 Columbia St. W., Waterloo, Ontario CANADA N2L 5Z5, e-mail: alopez-o@daisy.UWaterloo.ca
[2] Institut für Informatik, Universität Freiburg, Am Flughafen 17, Geb. 051, D-79110 Freiburg, FRG, e-mail: schuiere@informatik.uni-freiburg.de

Abstract. We study the problem of on-line searching for a target inside a polygon. In particular we propose a strategy for finding a target of unknown location in a star polygon with a competitive ratio of 14.5, and we further refine it to 12.72. This makes star polygons the first non-trivial class of polygons known to admit constant competitive searches independent of the position of the target. We also provide a lower bound of 9 for the competitive ratio of searching in a star polygon—which is close to the upper bound.

A similar task consists of the problem of on-line recognition of star polygons for which we also present a strategy with a constant competitive ratio *including* negative instances.

1 Introduction

In the past years on-line searching has been an active area of research in Computer Science (e.g. [1, 2, 4, 7, 8, 11]). In its full generality, an on-line search problem consists of an agent or robot searching for a target on an unknown terrain. In the worst case a search by a robot on a general domain can be arbitrarily inefficient as compared to the shortest path from the initial position to the target. However, as it is to be expected, strategies can be improved depending on the type of terrain and the searching capabilities of the robot.

The robot is assumed to be equipped, as it is standard in the field, with an on-board vision system that allows it to see its local environment. Since the robot has to make decisions about the search based only on the part of its environment that it has seen before, the search of the robot can be viewed as an *on-line* problem. The performance of an on-line search strategy is measured by comparing the distance traveled by the robot with the length of the shortest path from the starting point s to the target location t. The ratio of the distance traveled by the robot to the optimal distance from s to t is called the *competitive ratio* of the search strategy.

There are several known classes of polygons that admit search strategies for some targets with a constant competitive ratio, most notably *streets* [7], $\mathcal{G}$-*streets* [4, 10], *HV-streets* [3] and θ-*streets* [3]. However, the existence of a constant competitive searching strategy for these classes of polygons is strongly dependent on the position of the target.

A natural question is to find a class of polygons which the robot may search at a constant competitive ratio independently of the position of the target. Since the target might be hiding anywhere inside the polygon, a natural choice is to explore the class of polygons where one polygon can be seen in its entirety from a single point, known as star polygons.

Icking and Klein studied the problem of on-line *kernel* searching in a star polygon. In this case, the competitive ratio is given by the ratio of the length traversed by the robot from the starting point to a kernel point and the optimal distance, which is the

---

* This research is partially supported by the DFG-Project "Diskrete Probleme", No. Ot 64/8-1.

the distance from the starting point to the kernel set. In [4] Icking and Klein presented a ~ 5.81 competitive strategy for walking into the kernel of a star polygon.

In this paper we present the first non-trivial class of polygons which admits constant competitive-ratio position-independent target searching. In section 2 we introduce some concepts and definitions of use in searching polygons. In section 3 we present a 14.5-competitive algorithm for target searching in star polygons and prove a lower bound of 9 for the competitive ratio of any search strategy for star polygons, we further refine this strategy to achieve a competitive-ratio of 12.72. In section 4 we use this strategy to construct the first constant competitive algorithm for recognition of star polygons. That is, given a polygon, the robot follows a path that proves or disproves that the polygon is a star where the path is no more than a constant times longer than a shortest path with the same property. Furthermore, such path leads into the kernel in a constant competitive ratio as well. We also improve the $\sqrt{2}$ lower bound for walking into the kernel of a star polygon to ~ 1.48.

2 Definitions

We say two points p_1 and p_2 in a polygon P are visible to each other if the line segment $\overline{p_1 p_2}$ is contained in P.

Definition 1. Let p be a point in P. The visibility polygon of p is the subset of P visible to p and denoted by $V_P(p)$.

We assume that the robot has access to its local visibility polygon by a range sensing device, e.g. a ladar.

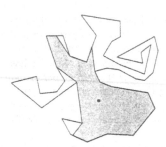

Fig. 1. Visibility polygon.

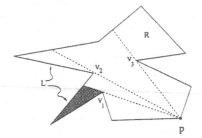

Fig. 2. Left and right pockets.

Definition 2. [12] A simple polygon P is a star polygon if there exists a point z in P such that $V_P(z) = P$. The set of all points z inside P with $V_P(z) = P$ is the kernel of P.

Star polygons are often referred to as star-*shaped* polygons [12], in this paper we use the equally common but shorter name of star polygons.

If the robot does not start in the kernel of P, then there are regions in P that cannot be seen by it. The connected components of $P \setminus V_P(p)$ are called pockets. The boundary of a pocket consists of some polygon edges and a single line segment not belonging to the boundary of P. The edge of the pocket which is not a polygon edge is called the window of the pocket. Note that a window intersects the boundary of P only in its end points. More generally, a line segment that intersects the boundary of P only in its end points is called a chord.

A pocket edge of p is a ray emanating from p which contains a window. Each pocket edge passes through at least one reflex vertex of the polygon, which is also an end point of the window associated with the pocket edge. This reflex vertex is called the entrance point of the pocket.

A pocket is said to be a left pocket if it lies locally to the left of the pocket ray that contains its window. A pocket edge is said to be a left pocket edge if it defines a left pocket. Right pocket and right pocket edge are defined analogously.

Since a point in the kernel of P sees all the points in P, in particular p, a pocket of $V_P(p)$ does not intersect the kernel of P which implies the following observation.

Observation 1 *The kernel lies to the right of all left pocket edges and to the left of all right pocket edges.*

For example, in the polygon of Figure 2, the kernel, if it exists, lies to the right of $\overrightarrow{pv_1^i}$ and $\overrightarrow{pv_2^i}$ and to the left of $\overrightarrow{pv_3^i}$.

This also implies that, for star polygons, starting from a left pocket and moving clockwise, all left pocket edges appear consecutively; at some point the first right pocket edge is seen and from then onwards all pocket edges are right pocket edges, until the full circle back to the sequence of left pocket edges is completed. This is so as the extension of each pocket defines a half plane which contains the kernel of P, if the pockets were to alternate between left and right, the intersection of these halfplanes would be empty which is a contradiction.

If the robot is initially located on a point s on the boundary of the polygon, the robot can scan all left pocket edges by starting from the edge on which s lies, and proceeding on the clockwise direction the interior of the polygon. At some point, a right pocket edge is seen and from then onwards all pocket edges are right pocket edges until the robot reaches the edge containing s again, which completes the scanning process.

3 Target Searching in Star Polygons

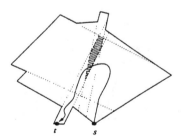
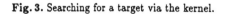

Fig. 3. Searching for a target via the kernel.

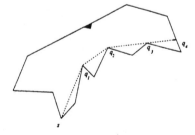

Fig. 4. An extended pocket edge.

There are many similarities between searching for the target and searching for the kernel. However, note that in general, when searching for a target, it is not an efficient strategy to first go to the kernel or towards the center and from there move to the target as illustrated in Figure 3. As illustrated in this case, a path advancing towards the kernel can be made arbitrarily larger than the distance from s to t.

Searching for a target of unknown location inside a star polygon is a provably harder problem than searching for the kernel, as we shall see in the second part of this section.

First we present a strategy to search for a target in a star polygon.

Consider the set of pocket edges seen by the robot from the starting position. We extend this set as follows.

Definition 3. Given a polygon P, an extended pocket edge from a point s is a polygonal chain $q_0, q_1, q_2, \ldots, q_k$ such that $q_0 = s$, and each of q_i is a reflex vertex of P, save possibly for q_k. Furthermore q_{k-2}, q_{k-1} and q_k are collinear and form a pocket edge with $\overline{q_{k-1}q_k}$ as associated window. If $\overline{q_{k-2}q_k}$ is a left (right) pocket edge, then each of $\angle q_{i-1}q_iq_{i+1}$ is a counterclockwise (clockwise) reflex angle (see Figure 4).

If A and B are two sets, then A is weakly visible from B if every point in A is visible from some point in B.

Lemma 4. *If c is a chord in star polygon P that splits P into two parts P_1 and P_2, then one of P_1 and P_2 is weakly visible from c and the other contains at least one point of the kernel of P.*

Proof. Let q be a point in the kernel of P. q is contained in one of the two parts, say in P_1. As q is in the kernel, all of P_2 can be seen from it. But any line contained in the polygon and joining a point in P_1 with a point in P_2 intersects the chord c. This implies that the chord weakly sees all points on the opposite side as well. □

Theorem 5. *There exists a strategy for searching for a target inside a star polygon with a competitive ratio of at most 14.5.*

Proof. Let $\mathcal{F}$ denote the set of all extended pocket edges starting from s. From the definition it follows that, in general, the robot may not see all of $\mathcal{F}$ from s (see for example the star polygon of Figure 5). The robot thus uses a strategy that starts with a subset $\mathcal{F}_0$ of $\mathcal{F}$. This set is enlarged as the robot sees new pocket edges. Given an extended pocket edge E, let l_E denote the last point in the chain, and p_E denote the second to last point of E.

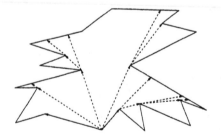

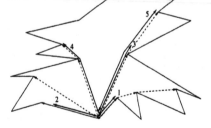

Fig. 5. The extended pocket edges of a polygon.

Fig. 6. Searching on the extended pocket edges.

Let $side \in \{left, right\}$ and if $side = right$, then $\neg side = left$ and vice versa.

Algorithm Star Search
Input: A star polygon P and a starting point s;
Output: The location of the target point t;

1   let $\mathcal{F}$ be the set of extended pocket edges currently seen but not explored;
    (∗ Initially $\mathcal{F}$ contains only simple pocket edges; ∗)
2   let p_E be the closest entrance point to s and $d = d(p_E, s)$
3   if E is a left pocket edge then let $side \leftarrow left$
4   else let $side \leftarrow right$;
5   while $\mathcal{F}$ is non-empty do
6   traverse d units on E starting from s;
7   if t is seen then exit;
8   add the new pocket edges seen in this trajectory to $\mathcal{F}$ as extended pocket
            edges starting from s;
9   remove from $\mathcal{F}$ all extended $side$ pocket edges to the $side$ side of the extended
            pocket edge $\overline{sp_E}$, including E if p_E is reached;
10   move back to s;
11      $side \leftarrow \neg side; \, d \leftarrow c \cdot d$;
12   if $side = left$
13   then let p_E be the rightmost entrance point on a left pocket edge such
                that the length of the extended pocket edge from s to p_E is less than
                d
14   if there is no such edge
15   then let E be the leftmost edge in $\mathcal{F}$
16   if $side = right$
17   then select E analogously to the case $side = left$;
    end while;

In the following we show that when the algorithm terminates, it has seen the target, and it traveled no more than 14.5 times the distance from s to t.

Note that after the first two iterations the while-loop has the following invariant:

Invariant: All pockets at a distance of d/c^2 or less on the $side$ side have been explored.

The correctness of the algorithm follows from Observation 1 and Lemma 4 as follows. As the robot visits extended pocket edges, it eventually visits the leftmost right pocket edge and the rightmost left pocket edge if t is not found before.

Once the robot has visited the extreme leftmost and rightmost pocket edges, it has explored the part to the left of the extreme left-pocket edge, and to the right of the extreme right-pocket edge. Furthermore, the part of the polygon contained in between the two extreme pocket edges has no hidden regions as it contains no pockets. Thus, the entire polygon is seen, and the target must have been found.

We claim that Algorithm *Star Search* has a competitive ratio of 14.5. At the end of Step 17, the invariant holds because if there was a, say, left pocket at a distance of less than d/c^2 it means it was part of the set $\mathcal{F}$ two steps before. Thus, if it was unexplored then, it either was traversed, or another left pocket of length at most d/c^2 which is to the right of it was traversed. But exploring this second edge entails exploring the earlier edge as shown in Lemma 4.

A consequence of the invariant is that if the current distance to be traversed by the robot is d, then the target cannot be at a distance of less than d/c^2. The worst case occurs when the robot sees the target at a distance of $d/c^2 + \epsilon$, at the very end of a search of length d (see Figure 7). This means that the ratio of the distance traversed by the robot according to Algorithm *star search* to the distance from s to t is at most

$$2\frac{\sum_{i=0}^{n} c^i}{c^{n-2}} + 1 = \frac{2c^3}{c-1} + 1 - O(1/c^{n-1}).$$

Substituting the value $3/2$ which minimizes $2c^3/(c-1)$ gives a competitive ratio of $1 + 27/2 = 14.5$. In fact, it can be shown that there is no choice of the step lengths that yields a better competitive ratio for the above algorithm [1, 5]. ◻

We observe that the worst case configuration occurs when the angle $\angle L_{i-2} s L_i$ is relatively flat. In this case the competitive ratio can be improved if the robot does not follow the straight line segment $\overline{sL_i}$ but follows a curve that allows it to detect the target earlier (see Figure 7).

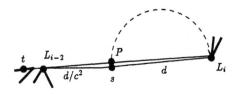

Fig. 7. The worst case to discover the target. If the robot follows the dashed path, then t is detected at P instead of L_i.

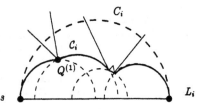

Fig. 8. The new strategy of the robot.

So instead of traveling along the line segment $\overline{sL_i}$ the robot now travels along the semi-circle C_i that is spanned by $\overline{sL_i}$. More precisely, the robot computes a curve C_i that connects s and L_i and that consists of parts of circles $C^{(1)}, \ldots, C^{(k_i)}$ as follows. The center $c^{(j)}$ of each circle $C^{(j)}$ is contained in $\overline{sL_i}$ with $c^{(j)}$ to the left of $c^{(j+1)}$, for $1 \leq j \leq k_i - 1$. The curve C_i is defined inductively. The circle $C^{(1)}$ is the first circle with its center to the right of s that contains s and intersects either L_i or the boundary of P in a point $Q^{(1)}$ above $\overline{sL_i}$. The part of $C^{(1)}$ between s and $Q^{(1)}$ is the first part of C_i. Now assume that C_i is already constructed up to circle $C^{(j)}$ with $1 \leq j \leq k_i - 1$. There is a point $Q^{(j)}$ such that $C^{(j)}$ intersects the boundary of P in $Q^{(j)}$. The circle $C^{(j+1)}$ is the first circle with its center to the right of $c^{(j)}$ that contains $Q^{(j)}$ and intersects either L_i or the boundary of P in a point $Q^{(j+1)}$ different from $Q^{(j)}$ above $\overline{sL_i}$. For illustration refer to Figure 8. Because of the limited space and the fairly involved analysis of the above strategy, we just mention that the competitive ratio can be improved to 12.72 in this way.

3.1 A Lower Bound on the Competitive Ratio

In this section we prove a lower bound of nine for the competitive ratio for searching in star polygons. Our proof is based on the following theorem about on-line searching on the line.

Theorem 6. [9, Theorem 2.2] *Any on-line search strategy on the line for a target at a distance of at most D is at least $(9 - f(D))$-competitive, where $f(n) \leq 24/\log_4 n$, for sufficiently large D.*

Theorem 7. *Any strategy for searching for a target inside a star polygon is at least 9-competitive.*

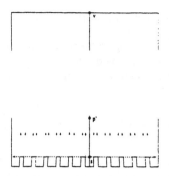

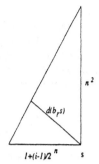

Fig. 9. Lower bound for searching for a target.   **Fig. 10.** Distance to a beam.

Proof. Consider the polygon of Figure 9. Let s be located at the origin. This polygon is made of $(n-1)2^{n-1}+1$ teeth with $(n-1)/2^{n-3}+4$ vertices attached to a rectangle of height n^2 and width $2n$. Teeth are equally spaced at a distance $1/2^n$, and of width $1/2^{n+1}$ save for the tooth containing s which is of width $2-1/2^n$. Each tooth defines a beam (see Definition 9). All beams intersect at the point $v=(0,n^2)$ which sees the entire interior of the polygon.

We claim that the robot must essentially do a doubling search on the teeth, in which case Theorem 6 gives a lower bound of 9. However, in this case there are several differences that must be considered. First, the movement of the robot is not restricted to a line; second, the lower bound is for searches on any point of the interval rather than on discrete positions. Thus the proof proceeds as follows: first we argue that any search strategy is sufficiently close to a search on the real line, and secondly we show that the bound for the continuous case implies a similar bound for the discrete case.

For the robot to explore a tooth it must reach the beam above it. We number the beams symmetrically, and consecutively starting from the origin; thus beam b_i is at the same distance from the origin as beam $-b_i$. The distance from s to the the base of the ith beam on either side is $d_i = 1 + (i-1)/2^n$. The distance from s to the closest point in the beam is (see Figure 10)

$$d(b_i, s) = \frac{d_i}{\sqrt{1+(1+(d_i)^2/n^4}} \geq \frac{d_i}{\sqrt{1+1/n^2}}$$

However the robot is not forced to move back to s after each search. Since the robot cannot reach a height past $9n$ as that alone would imply a competitive ratio above 9 we consider the point p' located at $(0,9n)$ and it follows that

$$d(b_i, p') \geq d_i \frac{n^2 - 9n}{\sqrt{n^4 + d_i^2}} \geq d_i \frac{n^2 - 9n}{\sqrt{n^4 + n^2}} = d_i \frac{n-9}{\sqrt{n^2+1}}$$

The order in which beams are visited can be denoted by the sequence $S = \{s_j\}_{1 \leq j \leq N}$ of the distances d_i from the origin to the base of those beams in which the robot changed direction (turn points).

Consider the beams associated to two consecutive terms in the sequence S above, say b_{k_i} and $b_{k_{i+1}}$. Without loss of generality, let us assume that b_{k_i} is on the left side and $b_{k_{i+1}}$ on the right side. Then, the distance traversed by the robot from beam b_{k_i} to beam $b_{k_{i+1}}$ is at least $d(q_i, q_{i+1}) \geq d(q_i, p_i) + d(p_i, q_{i+1})$, where q_j denotes the position of the robot in b_{k_j} for $j = \{i, i+1\}$, and p_i is the intersection of $\overline{q_i q_{i+1}}$ with the y-axis. Furthermore, $d(q_j, p_j) \geq d(b_{k_j}, p_j) \geq d(b_{k_j}, p')$.

Let $\mathcal{C}_S$ denote the competitive strategy of the strategy S on the real line. Analogously, let C_S^D denote the competitive ratio of a strategy S in the discrete case.

Now we will show that the search strategy S, applied to a target hiding at a point at distance d_i on the real line has competitive ratio:

$$C_S^D \geq \sup_{1 \leq j \leq N} \left\{ 1 + 2 \frac{\sum_{i=1}^{j} |s_i|}{|s_{j-1}| + 1/2^n} \right\} \geq 9 - 25/\log_4 n.$$

Assume that, to the contrary, $C_S^D < 9 - 25/\log_4 n$. We know from Theorem 7 that any sequence S visiting the interval $[-n, n]$ and searching for a target located in any interior point has a competitive ratio greater or equal to $9 - 24/\log_4 n$. Since S is such a sequence, we have then that

$$C_S = 1 + 2 \frac{\sum_{i=1}^{k} |s_i|}{|s_{k-1}|} \geq 9 - 24/\log_4 n$$

for some k such that $1 \leq k \leq N$. Let $C_S(s_k)$ and $C_S^D(s_k)$ denote the competitive ratio of strategy S to find a target hiding at the point s_k for the real and discrete case respectively. Note that $C_S = C_S(s_k) \geq 9 - 24/\log_4 n$ and that

$$1 + 2 \frac{\sum_{i=1}^{k} |s_i|}{|s_{k-1}| + 1/2^n} \leq C_S^D(s_k) < 9 - \frac{25}{\log_4 n} \implies \sum_{i=1}^{k} |s_i| < \left(8 - \frac{25}{\log_4 n} \right) \left(|s_{k-1}| + \frac{1}{2^n} \right).$$

The additive factor of $1/2^n$ in the denominator accounts for the next possible position of the target on that side. We claim that $0 \leq C_S(s_k) - C_S^D(s_k) < 1/\log_4 n$. Indeed,

$$C_S(s_k) - C_S^D(s_k) = 2 \sum_{i=1}^{k} |s_i| \left[\frac{1}{|s_{k-1}|} - \frac{1}{|s_{k-1}| + 1/2^n} \right]$$

$$< \left(8 - \frac{25}{\log_4 n} \right) \frac{1}{2^{n-1}|s_{k-1}|} \leq \frac{1}{2^{n-4}} \leq \frac{1}{\log_4 n} \quad \text{for } n > 3$$

as claimed. Thus, from $C_S(s_k) \geq 9 - 24/\log_4 n$ it follows that $C_S^D \geq C_S^D(s_k) > 9 - 25/\log_4 n$, which is a contradiction.

Thus it follows that $C_S^D \geq 9 - 25/\log_4 n$. Now, we know that the robot traversed, for each s_j a distance $d(q_j, p_j)$ which is at least $|s_j| \frac{n-9}{\sqrt{n^2+1}}$, for a total competitive ratio of at least

$$1 + \sup_{k \in Z} \left\{ \frac{n-9}{\sqrt{n^2+1}} \left(C_S^D(s_k) - 1 \right) \right\} \geq 1 + \sup_{k \in Z} \left\{ \left(\frac{n-9}{\sqrt{n^2+1}} \right) \left(8 - \frac{25}{\log_4 n} \right) \right\}$$

The value above is a lower bound for the competitive ratio of the robot searching a polygon. As the construction of the polygon of Figure 9 is valid for any n, we have that, in the limit, the competitive ratio is bounded by

$$\lim_{n \to \infty} 1 + \left(\frac{n-9}{\sqrt{n^2+1}} \right) \left(8 - \frac{25}{\log_4 n} \right) = 9$$

as claimed. $\qquad \square$

4 Recognition of Star Polygons

For the on-line star recognition problem, we assume that given a polygon, the robot aims to determine if it is star shaped. Similarly to target searches, the competitive ratio is given by the quotient between the shortest path that proves or disproves that a given polygon is a star and the distance traversed by the robot.

As Figure 11 shows, the problem of on-line search for the kernel of a polygon is at least $\sqrt{2}$ competitive [6].

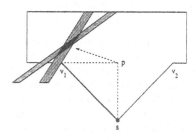

Fig. 11. A lower bound of $\sqrt{2}$ for the competitive ratio of searching for the kernel of a star polygon. Any on-line strategy with a competitive ratio of $\leq \sqrt{2}$ has to follow the dashed path.

Fig. 12. Polygon with two beams.

The next theorem shows that kernel searches are strictly worse than $\sqrt{2}$-competitive. This result stands out against several other lower bounds for searching in simple domains, for which it seems that a robot can find an optimal path on-line for the L_1 metric [8].

Definition 8. The visibility region of a subset B of a polygon is the set of all points in the polygon which see all points in B.

Definition 9. Given the current position of the robot p and a pocket B with respect to that point, the beam of the pocket is the visibility region of B.

Notice that if the pocket is a trapezoid, the visibility region resembles a search light beam (see Figure 12).

Observation 2 *The kernel lies in the intersection of all beams.*

Theorem 10. *Searching for the kernel of a polygon is at least $1/2 + 3/8\sqrt{2} + (2 + \sqrt{2})/8\sqrt{10\sqrt{2} - 13} \sim 1.48$-competitive.*

Proof. Consider the polygon of Figure 12. Notice that the robot must reach the line segment $\overline{v_1 v_2}$ before it reaches the kernel. As well, the robot must reach $\overline{v_1 v_2}$ at its midpoint p, as otherwise the following construction can be made on the opposite side and it follows from the triangle inequality that the competitive ratio would only worsen. Again, from p it is not yet clear where the kernel is located. In fact, depending upon the specific angle and location of the pockets, the beams might specify a small kernel located anywhere in the visibility polygon region of s which is above $\overline{v_1 v_2}$.

We now use an adversary argument. After the robot reaches p the adversary closes one side, and selects two candidate kernels, illustrated by the large dots in Figure 13, such that one is next to v_1 the other right above the midpoint, and the line joining them is at a $\pi/4$ angle to the horizontal. This can be achieved by locating a beam A along the line

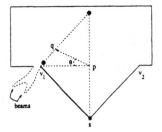

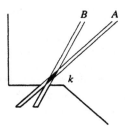

Fig. 13. Lower bound configuration.　　　　**Fig. 14.** Progressively thinner beams.

joining the two candidate regions, and a second one, B, nearly parallel and to the right of A (see Figure 14). The intersection of both beams defines the kernel of visibility.

At this point, we assume that the robot learns of this decision and thus can restrict itself, to its benefit, to determining which of the two regions is the kernel.

In this case, the robot cannot decide which of the candidates is the kernel before it reaches at least one of A or B. As the beams become progressively thinner, the robot reaches either beam at an ϵ distance of the $\pi/4$ line joining the two candidate regions (that is, the right edge of the A beam).

Assume this happens at a point q located, as indicated in the previous paragraph, arbitrarily close to the $\pi/4$ line. Let θ be the angle given by $\angle v_1 pq$. Without loss of generality, let the distance $d(s, v_1) = \sqrt{2}$. We compute the competitive ratio on the left. Let C_1 and C_2 be the two candidate regions. To compute the competitive ratio we first notice that $d(C_1, C_2) = \sqrt{2} - \epsilon$. Then from elementary trigonometry we obtain $d(q, C_1) = d(p, q) \sin(\theta)/\sin(\pi/4)$ and $1 - d(q, C - 2) = d(p, q) \cos(\theta)$, from which follows

$$d(p, q) = \frac{\sin(\pi/4)}{\sin(\pi/4 + \theta)} \quad \text{and} \quad d(q, C_1) = \frac{\sin(\theta)}{\sin(\theta + \pi/4)}.$$

Similarly, $d(q, C_2) = \sin(\pi/2 - \theta)/\sin(\pi/2 - \theta + \pi/4)$. Thus the competitive ratio for the kernel on the left side is given by

$$\frac{\sqrt{2}\sin(\theta) + \sqrt{2}\cos(\theta) + \sqrt{2} + 2\sin(\theta)}{2\sin(\theta) + 2\cos(\theta)}$$

and on the right side

$$\frac{\sin(\theta) + \cos(\theta) + 1 + \sqrt{2}\cos(\theta)}{2\sin(\theta) + 2\cos(\theta)}$$

As the competitive ratio is the maximum of both quantities above, the robot selects θ such that the competitive ratio on either side is the same. Solving the equation we obtain,

$$\theta = \arctan\left(1/4 + 1/8\sqrt{2}\left(1 - \sqrt{10\sqrt{2} - 13}\right)\right).$$

For this value, the competitive ratio is

$$(2 + \sqrt{2})/8\sqrt{10\sqrt{2} - 13} + 1/2 + 3/8\sqrt{2} \approx 1.48642$$

as required.　　　　　　　　　　　　　　　　　　　　　　　　　　　　　　□

The best known search strategy for finding the kernel of a given star polygon, is by Icking and Klein [6] and results in a no worse than $\sqrt{4 + (2 + \pi)^2} \sim 5.5$-competitive strategy. However, it is unclear if the same algorithm applied to a general polygon would terminate at a constant competitive ratio for negative instances. A modification of the target searching strategy of Theorem 5 can be used for this purpose. Furthermore, if the polygon is a star the proposed modified strategy reaches the kernel, if it exists, at a constant competitive ratio as well.

Definition 11. Let s be the starting position of the robot inside a polygon P. Let $V(\Gamma)$ denote the visibility region of a continuous path Γ inside P. Then we denote by OPT the length of the shortest path such that a computational agent (such as a Turing machine) can determine from $V(\Gamma)$ that P is or is not a star.

Theorem 12. *There exists a 46.35-competitive strategy that identifies if a polygon is or is not a star.*

Proof. The algorithm is somewhat similar to the one proposed for target searching in Theorem 5. However there are some key differences. Let $side \in \{left, right\}$ as before. For this theorem we say that a straight chord is a local $side$ pocket edge if it joins two points which are in between two consecutive $side$ extended pocket edges with one endpoint lying on the $side$-most of the two pocket edges. Similarly, local pocket edges together with the extended pocket edge in which they are anchored will be considered as extended pocket edges themselves.

The new strategy *Circle-Swipe* replaces Steps 7 and 10–11 from strategy *Star-Search* of Theorem 5.

Step 7 If the intersection of the half planes defined by the extension of the rectilinear segments of explored pocket edges becomes empty the strategy rejects. Otherwise continue until all pockets have been explored and accept.

Steps 10–11 The robot changes side $side \leftarrow \neg side$. Let $E_i = \langle q_0^i = s, q_1^i, \ldots, q_{k_i}^i \rangle$. The robot moves on a circular arc centered on q_{k-2}^i, of radius $d(q_{k-2}^i, q_{k-1}^i)$ to side $side$ until it sees q_{k-3}^i. The robot then updates the radius to be $d(q_{k-3}^i, q_{k-1}^i)$ and continues describing a (new) circular arc centered at q_{k-3}^i. Eventually the robot sees s and continues describing a circle of radius d until it starts reaching the extension of edges of the next pocket edge $E_{i+1} = \langle q_0^{i+1} = s, q_1^{i+1}, \ldots, q_{k_{i+1}}^{i+1} \rangle$ to be visited on the $side$ side. In each of this cases, the robot does the reverse process, reducing the radius by the length of the edge seen $(d(q_{j-1}^{i+1}, q_j^{i+1}))$ and centering the arc on q_j^{i+1} where j takes the values $0, 1, 2 \ldots k_{i+1}$ successively (see Figure 15). When the robot reaches the boundary of the polygon it returns to s moving over E_{i+1} and sets d to cd.

The invariant is now as follows.

Invariant: The visibility region of the path explored thus far by the robot contains the visibility region of any path of length d/c or less.

Again we must show correctness and analyze the competitiveness of the strategy. In Steps 10–11, while walking on the circular arc if the robot cannot reach a $side$ pocket edge, it means that the robot was blocked by the boundary of P. This boundary point must necessarily lie between the leftmost right pocket and the rightmost left pocket, as otherwise it would have been considered an extended pocket edge of the $\neg side$ side and, chosen as target pocket edge (since it is at a distance of at most d from s).

Now if we have reached a boundary point that is to the right (left) of the rightmost (leftmost) left (right) pocket edge but before a right (left) pocket edge then this point must be visible from s. In this case the robot traverses to s and has completed searching the left (right) side and continues searching on circular arcs on the remaining side to be explored.

Since after the ith iteration of Steps 10–11 the robot has enclosed by a simple connected curve all points at distance at most dc^{i-1} it follows from Observation 4 that the invariant is correct. Step 7 accepts or rejects when either an impossibility for a star polygon has been found or the whole polygon has been explored, which is trivially correct, concluding the proof of correctness.

In turn, the analysis has two components. First we must determine the length of the worst case longest path that may be traversed up to and including the ith iteration of Steps 10–11. Secondly, we shall show that all correct algorithms must accept according to Step 7.

In iteration i the robot traverses a distance no greater than dc^i to reach the circular arc, at most $2\gamma_i dc^i$ on the arc itself for some angle γ_i and then at most dc^i back to the point s. In the worst case, in step k we surround a point at distance $dc^{k-1} + \epsilon$ as given by the invariant of Steps 10–11. Clearly, if the polygon is a star $\gamma_i \leq 2\pi$ as otherwise the intersection of the half planes determined by the edges forming and extended pocket edge would be empty.

The distance traversed before surrounding the point depends on whether or not the pocket edges on one side were exhausted. If they were, the competitive ratio is:

$$\frac{[\sum_{i=0}^{k} c^i(2\pi + 2)] + c^{k+1}(2 + \alpha) + c^{k+2}(2 + 2\pi - \alpha)}{c^k}$$
$$\leq \frac{c}{c-1}(2\pi + 2) + c(2 + \alpha) + c^2(2 + 2\pi - \alpha)$$

where $0 \leq \alpha \leq 2\pi$. Differentiation shows that the maximum is attained when $\alpha = 0$, for all c. Minimizing with respect to c we obtain a cubic equation –which can be solved symbolically– and is minimized when $c \sim 1.547$ with a competitive ratio of 46.35.

If the pocket edges were not exhausted the competitive ratio is,

$$\frac{\sum_{i=0}^{k+1} c^i(2\pi + 2)}{c^k} \leq \frac{c^2}{c-1}(2\pi + 2).$$

This ratio is smaller than 46.35 for $c = 1.547$ which gives a maximum between the two expressions of 46.35 as required.

Secondly, it is easy to see using an adversarial argument, that if the robot accepts a star polygon without having looked into all of its pockets or rejects without having found non-intersecting half planes the adversary can suitably modify the pockets and make the robot fail. That is, if the robot accepts P without exploring a pocket the adversary creates a spiral in that pocket, and the polygon is not a star. On the other hand if the robot rejects without having found non-intersecting half-planes the adversary "empties" all the pockets by means of inserting an almost flat two edge chain closing the pocket (the chain is ϵ dented by a vertex on its midpoint). Because at any time there are only a finite number of pockets and the interior of the intersection of non-degenerate set of half planes is an open set, it follows that there exists small enough ϵ such that the intersection of all the half-planes of this modified polygon is not-empty and thus the polygon is a star, contradicting the robot.

Thus we have established that an agent optimally recognizing a star-polygon traverses the shortest path Γ that satisfies the visibility conditions of step 7 for the given polygon. The invariant then states that a robot using the Circle-Swipe strategy swipes a region that is at most c times farther than the given path, for a total competitive ratio of 46.35 as computed. □

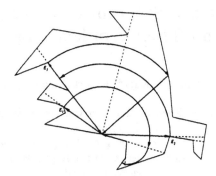

Fig. 15. Recognizing a polygon.

5 Conclusions

We have presented a strategy for on-line searching in a star polygon and for on-line recognition of a star polygon. Our strategies have constant competitive ratios independent of the starting position of the robot and the position of the target. This is in contrast to on-line searching in other classes of polygons where both the position of the target and the starting position are heavily limited.

We have also presented a lower bound for on-line searching in a star polygon which is close to the upper bound obtained by our strategy. Finally, we show that no strategy which searches for the kernel of a star polygon can achieve a competitive-ratio better than 1.48 which improves on the best previously known lower bound of $\sqrt{2}$.

References

1. R. Baeza-Yates, J. Culberson and G. Rawlins. "Searching in the plane", *Information and Computation*, Vol. **106**, (1993), pp. 234-252.
2. K.-F. Chan and T. W. Lam. "An on-line algorithm for navigating in an unknown environment", *International Journal of Computational Geometry & Applications*, Vol. 3, (1993), pp. 227-244.
3. A. Datta, Ch. Hipke, and S. Schuierer. "Competitive searching in polygons—beyond generalized streets", in *Proc. Sixth Annual Int. Symp. on Algorithms and Computation*, pages 32–41. LNCS 1004, 1995.
4. A. Datta and Ch. Icking. "Competitive searching in a generalized street", *Proceedings 10th ACM Symposium on Computational Geometry*, (1994), pp. 175-182.
5. S. Gal. *Search Games*, Academic Press, 1980.
6. Ch. Icking and R. Klein. "Searching for the kernel of a polygon. A competitive strategy", *Proceedings 11th ACM Symposium on Computational Geometry*, (1995).
7. R. Klein. "Walking an unknown street with bounded detour", *Computational Geometry: Theory and Applications*, Vol. 1, (1992), pp. 325-351.
8. J. Kleinberg. "On-line search in a simple polygon", *Proc. 5th ACM-SIAM Symp. on Discrete Algorithms*, (1994), pp. 8-15.
9. A. Lopez-Ortiz. "On-line target searching in bounded and unbounded domains", Ph.D. thesis, University of Waterloo, 1996.
10. A. López-Ortiz and S. Schuierer. "Generalized streets revisited", In J. Diaz and M. Serna, editors, *Proc. 4th European Symposium on Algorithms*, LNCS 1136, pages 546–558. Springer Verlag, 1996.
11. A. Lopez-Ortiz and S. Schuierer. "Walking streets faster", *Proceedings of 5th Scandinavian Workshop in Algorithmic Theory Algorithms*, 1996. LNCS 1097, Springer-Verlag, pp. 345-356.
12. F. P. Preparata, M. I. Shamos. *Computational Geometry*, Springer-Verlag, New York, 1985.

Dynamic Data Structures for Fat Objects and Their Applications[*]

Alon Efrat[1], Matthew J. Katz[2], Franck Nielsen[3], Micha Sharir[4]

[1] School of Mathematical Sciences, Tel Aviv University, Tel-Aviv 69982, Israel.
alone@math.tau.ac.il
[2] Departments of Industrial Engineering & Management and Mathematics &
Computer Science, Ben-Gurion University of the Negev, Beer-Sheva 84105, Israel.
matya@cs.bgu.ac.il
[3] INRIA, BP93, 06902 Sophia-Antipolis cedex (France), and École Polytechnique –
LIX, Paris. *nielsen@sophia.inria.fr* or *nielsen@lix.polytechnique.fr*
[4] School of Mathematical Sciences, Tel Aviv University, and Courant Institute of
Mathematical Sciences, New York University. *sharir@math.tau.ac.il*

Abstract. We present several efficient dynamic data structures for point-
enclosure queries involving convex fat objects in $\mathbb{R}^2$ or $\mathbb{R}^3$. These structures are more efficient than alternative known structures because they
exploit the fatness of the objects. We then apply these structures to obtain efficient solutions to three problems: (i) Finding a perfect matching
between a set of points and a set of convex fat objects. (ii) Finding a
piercing set for a collection of convex fat objects, whose size is within a
constant factor off the optimum size. (iii) Constructing a data structure
for answering *bounded-length segment-shooting* queries: Given a set of fat
objects in the plane as above, and a query oriented segment $\vec{r}$, whose
length is relatively short, find the first object hit by $\vec{r}$.

1 Introduction

A convex object c in $\mathbb{R}^d$ is α-fat, for some parameter $\alpha > 1$, if there exists an
axis-parallel cube s^+ containing c and an axis-parallel cube s^- that is contained
in c, such that the ratio between the edge lengths of s^+ and s^- is at most α.
Practical instances of many geometric problems tend to have fat objects as input.
Fat objects have several desirable properties, which were used by many authors
to obtain more efficient solutions to a variety of algorithmic problems, when the
underlying objects are fat. See [3, 5, 7, 15, 16, 21, 23, 26, 27, 28] for a sample of
these results.

In a recent paper [15], Katz has designed a data structure of nearly linear size
for certain kinds of queries involving a set of convex α-fat objects in the plane.
By augmenting the data structure in various ways he obtains efficient and simple
solutions to several query-type problems, including the *point enclosure* problem,
where we wish to determine whether a query point q lies in the union of the
input set, and, if so, to report a witness object containing q, or, alternatively,
report all k objects containing q. The cost of such a query is $O(\text{polylog } n)$ (or
$O(\text{polylog } n + k \cdot \text{polylog } n)$), as opposed to roughly $O(\sqrt{n})$ (or $O(\sqrt{n} + k)$),
which is the cost of a query when the fatness assumption is dropped and only

---

[*] A full version of this paper appears in *http://www.math.tau.ac.il/~alone/dfat.ps.gz*

nearly linear storage is allowed [18, 19]. Although not noted in [15], this structure can be maintained dynamically, using standard techniques, as will be described below.

In this paper we continue the work of [15]. We assume that our set C of objects consists of convex α-fat objects, for some fixed small constant $\alpha > 1$, and we present compact dynamic data structures for C that enable us to answer a point enclosure query efficiently. The specific bounds depend on the dimension (2D or 3D) and on the type of objects that are stored in the data structure (e.g., convex α-fat polygons or polyhedra or convex α-fat general objects). In general, these bounds are significantly better than the corresponding known bounds where fatness is not assumed. Consider for example the case where the objects are (not necessarily axis-parallel) cubes in $\mathbb{R}^3$. The standard storage/query tradeoff in this case lets s (the size of the data structure) vary in the range n to n^3, and the query cost, expressed as a function of both n and s, is close to $n/s^{1/3}$. Since cubes are fat objects, we may use our data structure in this case. The effect of using our structure is equivalent to a reduction by one in the dimension, in the sense that the storage/query tradeoff that we obtain is roughly the same as for triangles in the plane. That is, s varies in the range n to n^2, and the query cost is only about $n/\sqrt{s}$. Moreover, the structure can be maintained dynamically, when inserting or deleting objects, at a cost of about s/n per update.

In Section 3 we present three applications of our data structures as stated in the abstract.

2   Dynamic Data Structures for Fat Objects

In this section we present efficient dynamic data structures for point-enclosure queries involving a collection C of (possibly intersecting) convex *fat* objects in $\mathbb{R}^2$ or $\mathbb{R}^3$. Specifically, we want such a structure to support queries in which we are given a point q and wish to determine whether q lies in the union of the objects of C, and, if so, report an object containing q (or, alternatively, report all objects containing q). We also want to maintain this structure under insertions and deletions of objects into/from C. We present several data structures for this problem, depending on the dimension and on the type of objects in C.

2.1   Fat polytopes in three dimensions
Let C be a set of n convex polytopes in $\mathbb{R}^3$. We assume that each polytope is α-fat, for some fixed constant parameter $\alpha > 1$, and has a constant number of facets. We further assume that each facet of each polytope in C is triangulated.

We first use a straightforward extension to three dimensions of the planar data structure of Katz [15], to obtain a 3-level tree $\mathcal{T}$, so that given a query point q, we can return, in $O(\log^3 n)$ time, $O(\log^3 n)$ disjoint canonical subsets of C, so that any polytope of C containing q belongs to one of these subsets, and so that each canonical subset has a nonempty intersection.(The constant of proportionality in these bounds depends on α.) This structure can be maintained dynamically, using standard techniques, as follows: Using the same ideas as in [6], we partition C into at most $\log n$ subsets, each containing 2^k objects, for distinct integers $k \leq \log n$. As shown in [6], if constructing a static version of the data structure takes time $T(n)$, then inserting an object takes time $O((T(n)/n) \log n)$.

Since in our case $T(n) = O(n \log^3 n)$, the cost of an insertion is $O(\log^4 n)$. Deletion of an object C is done as follows: We mark C as being deleted from each pre-stored subset containing C in each of the three levels of $\mathcal{T}$. When the actual number of objects in such a subset becomes less than half its original cardinality, we reconstruct all the substructures associated with this subset. In this manner we only pay (in an amortized sense) an additional logarithmic factor for both insertions and deletions, and it can be shown that this bound can also be obtained in the worst case.

We augment $\mathcal{T}$ as follows. Let $C^* \subseteq C$ be a canonical subset, and let p^* be a (pre-computed) point common to all its elements (the construction enables us to assume that p^* does not lie on the boundary of any of the elements of C^*). Let Q^* be an axis-parallel unit cube centered at p^*. We centrally project the boundary of each $C \in C^*$ from p^* onto ∂Q^*, to obtain a collection of $O(n)$ polygons, each having $O(1)$ edges, on ∂Q^*. We process each facet f of Q^* for efficient point enclosure queries (in the non-fat planar setting). That is, we construct the data structure of Matoušek [18], using $O(s)$ storage, where s varies between n and n^2, so that for a given point $q \in f$, we can report the set of polygons on f that contain q as the disjoint union of $O(n^{1+\epsilon}/\sqrt{s})$ canonical subsets. The cost of a query is $O(n^{1+\epsilon}/\sqrt{s})$, and the structure can be maintained dynamically, when inserting or deleting polygons, at a cost of $O(s/n^{1-\epsilon})$ per update.[5]

We next construct another layer of our data structure, as follows. For each canonical set $\mathcal{P}$ of polygons stored in one of the point-enclosure substructures, we replace each polygon π in $\mathcal{P}$ by the plane containing the polytope facet that has been projected onto π, and store the (boundary of the) intersection of the halfspaces bounded by these planes and not containing p^*. For this we use the data structure of Agarwal and Matoušek [4, Thm. 2.8], which maintains dynamically the upper envelope of these planes (relative to the normal direction of the facet f). This structure enables us to determine in $O(\log n)$ time whether a query point lies above all these planes. Alternatively, it enables us to report in $O(\log n + k)$ time the k planes of this structure lying above a query point. Moreover, a plane can be inserted or deleted in time $O(n^\epsilon)$. This completes the description of our data structure.

Answering a query: Let q be a query point. We wish to determine whether some polytope of C contains q and, if so, produce a witness polytope that contains q (or, alternatively, report all such polytopes). We start by querying the first layer of our structure, and obtain a collection of $O(\log^3 n)$ canonical subsets, each augmented as above. For each subset C^*, with a common point p^*, we compute the intersection q' of the ray emerging from p^* towards q with the boundary of the cube Q^*, and query the corresponding point-enclosure substructure with q'. The answer to this query consists of $O(n^{1+\epsilon}/\sqrt{s})$ disjoint canonical subsets of projected polytope facets, where all members of such a subset contain q'. We finally query each of the corresponding third-layer upper-envelope substructures with q. It is easy to verify that q lies below the upper envelope of at least one

---

[5] Throughout the paper, ϵ stands for an arbitrarily small positive constant parameter.

such substructure if and only if q lies in the union of the polytopes of C. If this is the case, we can either report all polytopes containing q, by reporting all the planes that lie above q in each of the corresponding substructures, or stop after reporting just one such plane. The overall cost of a query is thus $O(n^{1+\epsilon}/\sqrt{s})$, or $O(n^{1+\epsilon}/\sqrt{s} + k)$ in the reporting version, where k is the output size.

Updating the structure: Each of the three layers of our structure is dynamic, and the updating of the whole structure is easy to do layer-by-layer. We omit the straightforward details. The overall cost of an update operation is $O(s/n^{1-\epsilon})$. The results of this subsection are summarized in Theorem 2 below.

2.2 General fat objects in three dimensions

Next consider the case where C is a collection of general convex α-fat objects in $\mathbb{R}^3$. We assume here that each object in C has *constant description complexity*, in the sense that its boundary is a semialgebraic set defined in terms of a constant number of polynomial equalities and inequalities of constant maximum degree.

In this case we use the first layer of the data structure described in Section 2.1. For each canonical set C^*, with a common point p^*, we represent the boundary of each $C \in C^*$ as a function $r = f_C(\theta, \phi)$ in spherical coordinates about p^*. With an appropriate standard re-parameterization (which we will not detail here), the graphs of these functions are algebraic of constant description complexity, in the above sense. We need to maintain the upper envelope E^* of these functions. Indeed, a query point q lies in the union of C^* if and only if $r_q \leq E^*(\theta_q, \phi_q)$, where (r_q, θ_q, ϕ_q) are the spherical coordinates of q about p^*.

The maintenance of this envelope can be accomplished using the 'shallow-levels' data structure of Agarwal et al. [2]. This structure has size $O((n^*)^{2+\epsilon})$ and can be constructed in $O((n^*)^{2+\epsilon})$ time, where $n^* = |C^*|$. Using this structure, we can determine whether $r_q \leq E^*(\theta_q, \phi_q)$ in $O(\log n)$ time, or report all k objects of C^* that contain q in time $O(\log n + k)$. An insertion or deletion of an object takes $O(n^{1+\epsilon})$ time. It follows that the overall size of the full data structure is also $O(n^{2+\epsilon})$, that a query can be performed in time $O(\log^4 n)$ (or $O(\log^4 n + k)$), and that an update takes $O(n^{1+\epsilon})$ time. These bounds are summarized in Theorem 2 below.

In the case that $C = \{B_1 \ldots B_n\}$ is a set of n (not necessarily congruent) balls in $\mathbb{R}^3$, we use known techniques to obtain exactly the same bounds. These bounds also appear in Theorem 2.

2.3 The planar case

In this subsection we consider the case where C is a collection of general convex α-fat objects in the plane. As before, we assume that each object in C has constant description complexity. In the full version of this paper, we obtain the following result, which is a slight improvement over the data structure of Katz [15]:

Theorem 1. *We can store a (static) set C of n convex α-fat objects in the plane, into a data structure of size $O(n \log n)$, using $O(n \log^2 n)$ preprocessing time, such that one can determine in time $O(\log^2 n)$, whether a query point is contained in some object of C.*

A dynamic data structure Again, our goal is to preprocess C into a dynamic data structure that can support insertions and deletions of objects, and queries where we are given a point q and wish to determine whether q lies in the union of $\mathcal{C}$ and, if so, to report an object of $\mathcal{C}$ containing q, or, alternatively, report all such objects. We use the data structure $\mathcal{T}$ of Theorem 1, so that given a query point q, we can obtain in $O(\log n)$ time a collection of $O(\log n)$ canonical subsets of $\mathcal{C}$, such that each object containing q appears in one of these subsets, and such that each subset C^* has a point p^* common to all its members. For each object $c \in C^*$, we can represent ∂c as a continuous function $r = f_c(\theta)$ in polar coordinates about p^*, and each pair of these functions intersect at most a constant number s of times. Then q is contained in an element of C^* if and only if $r_q \leq E^*(\theta_q)$, where E^* is the upper envelope of these functions, and where (r_q, θ_q) are the polar coordinates of q about p^*. We therefore need to maintain the upper envelopes E^* for the canonical sets, so that searches and updates of them can be performed efficiently. For this we can use (a simplified version of) the shallow-level data structure of Agarwal et al. [2] mentioned in the preceding subsection. Recall that the complexity of E^* is $\lambda_s(n^*)$, where $n^* = |C^*|$ and where $\lambda_s(n)$ is the maximum length of (n, s) Davenport-Schinzel sequences [25]. It follows from [2] that we can construct a data structure of size $O(n^{1+\epsilon})$, in time $O(n^{1+\epsilon})$, using which we can answer a query in $O(\log n)$ time (or in $O(\log n + k)$ time, for reporting all k objects of C^* containing q), and perform an insertion or a deletion in time $O(n^\epsilon)$. Combining all these substructures, using the decomposition technique of van Kreveld [17, Corollary 5.2 (ii)], into one overall structure, we obtain the bounds appearing in Theorem 2 below.

The case of fat polygons: We can do somewhat better if the objects in $\mathcal{C}$ are convex α-fat polygons in $\mathbb{R}^2$. Details are omitted due to lack of space, and appear in the full version. The results we obtain are listed in Theorem 2 below.

Theorem 2. *Let $\mathcal{C}$ be a set of n convex α-fat objects in $\mathbb{R}^d$, each having a constant description complexity, for some fixed constant $\alpha > 1$ and for $d = 2, 3$. For any parameter $n \leq s \leq n^2$, we can preprocess $\mathcal{C}$ into a data structure, such that finding an object of $\mathcal{C}$ containing a query point (point-enclosure query), and inserting or deleting an object into/from $\mathcal{C}$ can be done in the time listed in the following table: In all cases below we can also report all objects containing a query point in time $O(Q(n) + k)$, where $Q(n)$ is the time for a point-enclosure query, and k is the number of reported objects.*

Objects:	general objects	polytopes	balls	general objects	polygons
Dimension:	3D	3D	3D	2D	2D
Preprocessing:	$O(n^{2+\epsilon})$	$O(s^{1+\epsilon})$	$O(s^{1+\epsilon})$	$O(n^{1+\epsilon})$	$O(n \log^3 n)$
Storage:	$O(n^{2+\epsilon})$	$O(s)$	$O(s)$	$O(n^{1+\epsilon})$	$O(n \log^3 n)$
point-enc. query:	$O(\log^4 n)$	$O(\frac{n^{1+\epsilon}}{\sqrt{s}})$	$O(\frac{n^{1+\epsilon}}{\sqrt{s}})$	$O(\log n)$	$O(\log^3 n)$
Update:	$O(n^{1+\epsilon})$	$O(s/n^{1-\epsilon})$	$O(s/n^{1-\epsilon})$	$O(n^\epsilon)$	$O(\log^4 n)$

3 Applications of the Data Structures

3.1 Matching points and fat objects

Let $\mathcal{C}$ be a set of n convex α-fat objects in $\mathbb{R}^2$ or $\mathbb{R}^3$, and let P be a set of n points. We want to solve the *matching problem*, which is to match each point of

P to a distinct object that contains it. Please see the full version of this paper for motivation and a list of relevant results. We can solve the matching problem by applying the matching algorithm of Efrat and Itai [11]. This algorithm maintains a dynamic data structure that stores a subset of the objects of C, and supports queries where we specify a point p and wish to find an object in the current subset that contains p, and then delete that object from the structure. The algorithm performs $O(n^{3/2})$ such operations, and its running time is dominated by the cost of these operations.

We use the appropriate data structure from among those developed in the preceding section, depending on the type of objects in C. In the three-dimensional cases, we set the storage parameter s to be $n^{4/3}$, so that both queries and updates take $O(n^{1/3+\epsilon})$ time each. We thus obtain:

Theorem 3. *Let C be a set of n convex α-fat objects, each of a constant description complexity, in $\mathbb{R}^d$ (for $d = 2, 3$), and let P be a set of n points in $\mathbb{R}^d$. Then we can either find a one-to-one matching between P and C, such that each point $p \in P$ is contained in the object of C matched to p, or determine that no such matching exists. The running time of the algorithm is $O(n^{11/6+\epsilon})$ for polytopes in $\mathbb{R}^3$ and for balls in $\mathbb{R}^3$. The running time is close to $O(n^{3/2})$ for general objects and polygons in $\mathbb{R}^2$.*

3.2 Piercing fat objects

Let C be a (static) set of n objects in $\mathbb{R}^d$. A set of points $\mathcal{P}$ in $\mathbb{R}^d$ is a *piercing set* for C if $\mathcal{P}$ intersects every object in C. Finding a minimal piercing set is NP-complete for $d \geq 2$ [13], so it is natural to seek approximate solutions, in which the size of the computed piercing set is not much larger than the optimum size. The problem of finding a minimal piercing set is a special instance of the well known *set cover* problem, so we can apply the greedy algorithm for finding a set cover [9] to obtain, in polynomial time, a piercing set whose size is larger than the optimum size by a factor of $(1 + \log l)$, where $l \leq n$ is the *depth* of the arrangement of C. Brönnimann and Goodrich [8] presented a polynomial-time algorithm for computing a set cover in which the approximation factor depends both on the optimum cover size c and on the VC-dimension of the underlying set system. If the VC-dimension is some constant, then their algorithm finds a cover of size $O(c \log c)$.

In this subsection we present efficient approximate algorithms for the case of fat objects in two or three dimensions. The algorithms produce piercing sets whose size is within a constant factor off the optimum size.

The high-level description of the algorithm is simple: For each object $C \in \mathcal{C}$, let Q_C denote the smallest axis-parallel cube enclosing C. We sort the objects of C in increasing order of the size of Q_C. The algorithm works in stages, where the i-th stage starts with the subset C_i of C consisting of those objects that have not yet been pierced (initially, $C_1 = C$). Let C_i be the smallest object (in the above order) in C_i. Let bQ_{C_i} be the cube Q_{C_i} scaled by some fixed factor $b > 1$ about its center (we can choose, e.g., $b = 2$). The fatness of the objects of C and the fact that C_i is the smallest object in C_i imply that for any object $C \in C_i$ that intersects C_i, the measure of $C \cap bQ_{C_i}$ is at least some fixed fraction of the measure of bQ_{C_i}. Hence, we can place a constant number of points inside bQ_{C_i}, (this number only depends on α and d) so that any $C \in C_i$ that intersects C_i will contain one of these points. We add these points to the output piercing set, and delete from C_i all the objects that are pierced by any of them. The subset C_{i+1} of the remaining objects is then passed to the next stage. The algorithm terminates when this set becomes empty.

The termination of the algorithm, and the fact that its output is a piercing set are both obvious. Moreover, the objects $C_1, C_2, \ldots$ are pairwise disjoint, so if the algorithm terminates after j stages, then the size of the optimum piercing set is at least j, whereas the size of the output is $O(j)$, so the output size is indeed within a constant factor off the optimum. To implement the algorithm, we use the appropriate data structure developed in the preceding section, to obtain the following result:

Theorem 4. *Let C be a set of n convex α-fat objects in $\mathbb{R}^d$, for some fixed constant $\alpha > 1$ and for $d = 2, 3$. Then we can compute a piercing set for C of size $O(j)$, with the constant of proportionality depending on α and d, where j, is the size of a minimal-cardinality piercing set for C. The running time of the algorithm depends on d, and on the type of objects in C, as follows:*

Objects:	polytopes	balls	general	polygons
dimension	3D	3D	2D	2D
Running time	$O(n^{4/3+\varepsilon})$	$O(n^{4/3+\varepsilon})$	$O(n^{1+\varepsilon})$	$O(n \log^4 n)$

3.3 Segment shooting among fat objects

Let C be a set of n convex α-fat objects in the plane, and let δ be the smallest diameter of any object in C. In the bounded-size arc shooting problem, we wish to preprocess C, so that, for a given oriented query arc $\mathbf{r}$ of length at most $h\delta$, for some constant h, the first object of C hit by r (if such an object exists) can be found efficiently. More precisely, we search for an object $c \in C$ for which there is a point $z \in \mathbf{r}$ such that $z \in c$ and the initial portion of $\mathbf{r}$ preceding z does not intersect any object of C. An efficient solution to this problem is presented in [15] for the special case where C consists of either (constant-complexity) polygons or disks, and the query arc is either a segment or a circular arc. Here we present a solution for the general case, assuming the query arcs are segments. The data structure we describe is based on the data structure for point enclosure, its size is nearly linear in n, and the query cost is polylogarithmic, as opposed to roughly $O(\sqrt{n})$ in the non-fat setting (see [1]). Given an oriented query segment $\mathbf{r} = \overrightarrow{pq}$, we first check whether its first endpoint p lies in one (or more) of the objects of C, using the data structure of Theorem 1. Assuming it does not, we proceed as follows. We obtain in $O(\log^2 n)$ time a collection of $O(\log^2 n)$ canonical subsets of C, such that each object that is intersected by $\mathbf{r}$ appears in at least one of these subsets, and such that each subset C^* has a point p^* common to all its members. This is done by performing a point enclosure query for each of the points in a point set of constant size that is constructed as a function of $\mathbf{r}$, in its vicinity; see [23, 24].

We now perform a segment shooting query in each of these subsets C^* to obtain a collection of $O(\log^2 n)$ candidate objects from which we choose the one that is hit first by **r**. We next show how to perform efficiently a segment shooting query in a subset C^* with a common point.

Assume for simplicity of exposition that the common point of C^* is the origin O, and let U denote the union of the objects in C^*. Let ℓ be the oriented line containing **r** and oriented in the same direction. We will find v, the intersection point of ℓ with U, that lies ahead of p and is closest to p, if a such a point exists. By checking the relative position of p, q and v along ℓ, we discover whether v lies on **r**. We divide U into two regions, U^u, the region above the x-axis, and U^d, the region below the x-axis. We find the first intersection point between ℓ and each of them, and select the one that is closer to p. Let us focus on U^u and let $\xi = \partial U^u$. We consider ξ as the graph of the upper envelope of the boundaries of the sets of C^*, expressed as functions in polar coordinations about O. If we assume that each pair of the boundaries of the objects of C intersect in at most a constant number s of points, then the number of vertices of ξ (i.e. intersection pairs of boundaries that lie on ξ) is at most $(\lambda_s|C^*|)$. We construct a binary tree $\mathcal{T}$ as follows. We compute the convex hull $CH(\xi)$, using the optimal algorithm of Nielsen and Yvinec [20]. Clearly, ℓ intersects U^u if and only if the straight line containing ℓ intersects $CH(\xi)$. We associate $root(\mathcal{T})$ with $CH(\xi)$. Next, we find a vertex $m \in \xi$ which divides ξ into two parts ξ_1 and ξ_2 having approximately the same number of vertices. Let the endpoints of these parts be t_1, m and m, t_2. We add edges that connect these endpoints to O, so that ξ_1 (resp. ξ_2) together with these edges form the boundary of a star-shaped region $\mathcal{K}_1$ (resp. $\mathcal{K}_2$), that contains the origin O. We compute $CH(\mathcal{K}_1)$ and $CH(\mathcal{K}_2)$, and associate the two children of $root(\mathcal{T})$ with $CH(\mathcal{K}_1)$ and $CH(\mathcal{K}_2)$. We continue to construct $\mathcal{T}$ recursively, and stop when $\mathcal{K}_v$, for a leaf v, consists of the boundary of a single object.

Answering a query. Assume none of the endpoints of **r** is in U. We check whether ℓ intersects $CH(\xi)$, the convex hull of U^u. If it does not then we deduce that **r** does not intersect U^u. If it does, we move on to the two children w_1, w_2 of the root of $\mathcal{T}$, storing the regions $CH(\mathcal{K}_1)$ and $CH(\mathcal{K}_2)$, respectively. Clearly ℓ intersects at least one of these regions. If it intersects only one of them, we continue recursively in the appropriate subtree. The interesting case is when ℓ intersects both regions at two respective openly-disjoint intervals I_1 and I_2. suppose, with no loss of generality, that I_1 precedes I_2 along ℓ. We distinguish between the following three cases:

(i) p precedes I_1 on ℓ: In this case, we recurse only at w_1,

(ii) p succeeds I_1 on ℓ: In this case, we recurse only at w_2. (If p also succeeds I_2, we can stop the recursion and conclude that **r** does not meet U^u.)

(iii) $p \in I_1$: In this case it is not clear whether the first hitting point of r with U^u occurs within $\mathcal{K}_1$ or within $\mathcal{K}_2$, since it is possible that **r** emerges from $CH(\mathcal{K}_1)$ without hitting $\mathcal{K}_1$ itself. (In the special case where q also lies in I_1, or precedes I_2 in ℓ, we know that **r** does not hit $\mathcal{K}_2$, so we recurse only at w_1. To resolve this problem, let p' denote the endpoint of I_1 lying ahead of p (between p and I_2). If $p' \in U$ then it suffices to recurse only on w_1. Otherwise, we perform a new shooting query with $\mathbf{r}' = \mathbf{p'p}$ (we shoot in the opposite direction, from

p' towards p). However, we start the recursion from the node w_1 of $\mathcal{T}$. The important observation is that case (iii) will never arise in any successive step of the query, so the query will follow a simple path in $\mathcal{T}$ from w to some leaf, or else it stops earlier with the conclusion that $\mathbf{r}'$ does not hit U^u. If we reach a leaf Z, we check in constant time whether $\mathbf{r}'$ hits the single object associated with Z, to determine whether $\mathbf{r}'$ hits U^u.

(iii.a) If $\mathbf{r}'$ does not hit U^u, we recurse (with the original query) only at w_2. Note that case (iii) will not arise again, at any future recursive step.

(iii.b) If $\mathbf{r}'$ hits U^u, we recurse only at w_1. (Now it is possible for case (iii) to arise again.)

If we have not terminated the query with the conclusion that $\mathbf{r}$ does not hit U^u, we will end at some leaf Z of $\mathcal{T}$. As in case (iii) above, we find the first hitting point of $\mathbf{r}$ with the single object o whose boundary is stored at Z, and return this point (and o) as output. The cost of this process is $O(\log^3 n)$; The cost of intersecting ℓ with any convex hull is $O(\log n)$; the query follows a path of length $O(\log n)$ in $\mathcal{T}$, and at each node of the path we may need to execute a sub-query (if we are in case (iii)) that traces another path of length $O(\log n)$. We repeat this process on U^d, and apply it to each of the $O(\log^2 n)$ canonical sets C^*. Thus we have:

Theorem 5. *Let C be a collection of convex α-fat objects in $\mathbb{R}^2$, each with constant description complexity. We can preprocess C in time $O(n^{1+\epsilon})$, into a data structure of size $O(n^{1+\epsilon})$, so that given an oriented segment $\mathbf{r}$ of length $\leq h\delta$, where h is a constant and δ is the smallest diameter of any object in C, we can find the first object of C hit by $\mathbf{r}$ (if such an object exists) in time $O(\log^5 n)$.*

Dynamization and Generalization. In time $O(\log n)$ we can easily add objects to the data structure, while increasing the query time by $O(\log n)$, using the standard technique of [6]. Concerning deletions, we suspect that one can maintain in $O(\log^2 n)$ the convex hull of a set of convex objects, provided they all share a point. We believe that this is possible using extensions of the ideas of [14]. However, we are not aware of any work on this problem. The special case where the objects are fat polygons, can be handled trivially using the technique of [22], so that each insertion or deletions takes $O(\log^2 n)$, and the time for a query is the same.

References

[1] P. K. Agarwal, Ray shooting and other applications of spanning trees with low stabbing number, *SIAM J. Comput.* 21 (1992), 540–570.

[2] P.K. Agarwal, A. Efrat and M. Sharir, Vertical decomposition of shallow levels in 3-dimensional arrangements and its applications, *Proc. 11th ACM Symp. Comput. Geom.*, 1995, 39–50.

[3] P.K. Agarwal, M.J. Katz and M. Sharir, Computing depth orders and related problems, *Computational Geometry: Theory and Applications* 5 (1995), 187–206.

[4] P.K. Agarwal and J. Matoušek, Dynamic half-space range reporting and its applications, *Algorithmica* 14 (1995), 325–345.

[5] H. Alt, R. Fleischer, M. Kaufmann, K. Mehlhorn, S. Näher, S. Schirra and C. Uhrig, Approximate motion planning and the complexity of the boundary of the union of simple geometric figures, *Algorithmica* 8 (1992), 391–406.

[6] J. Bentley and J. Saxe, Decomposable searching problems I: Static-to-dynamic transformation, *J. Algorithms* 1 (1980), 301–358.

[7] M. de Berg, M. de Groot and M. Overmars, New Results on Binary Space Partitions in the Plane, *Proc. 4th Scandinavian Workshop on Algorithm Theory*, 1994, 61–72.

[8] H. Brönnimann and M.T. Goodrich, Almost optimal set covers in finite VC-Dimension, *Discrete and Computational Geometry* 14 (1995), 263–279.

[9] V. Chvatal, A greedy heuristic for the set-covering problem, *Math. Oper. Res.*, 4 (1979), 233–235.

[10] K. L. Clarkson, K. Mehlhorn, and R. Seidel, Four results on randomized incremental constructions, *Computational Geometry: Theory and Applications* 3 (1993), 185–212.

[11] A. Efrat and A. Itai, Improvements on bottleneck matching and related problems, using geometry, *Proc. 12th ACM Symp. Comput. Geom.*, 1996, 301–310. See also: A. Efrat, M.J. Katz and A. Itai, Improvements on bottleneck matching and related problems, using geometry, in preparation.

[12] A. Efrat and M. Sharir, On the Complexity of the Union of Fat Objects in the Plane, *Proceedings 13 Annual Symposium on Computational Geometry*, 1997, to appear.

[13] R.J. Fowler, M.S. Paterson, and S.L. Tanimoto, Optimal packing and covering in the plane are NP-complete, *Information Processing Letters* 12 (3) (1981), 133–137.

[14] J. Hershberger and S. Suri, Applications of a semi-dynamic convex hull algorithm, *Proceedings 2 Scand. Workshop on Algorithms Theory, Lecture Notes in Computer Science*, 1990, vol. 447, Springer-Verlag, New York–Berlin–Heidelberg, 380–392.

[15] M.J. Katz, 3-D vertical ray shooting and 2-D point enclosure, range searching, and arc shooting amidst convex fat objects, tech. Report 2583, INRIA, BP93, 06902 Sophia-Antipolis, France, 1995.

[16] M.J. Katz, M.H. Overmars, M. Sharir, Efficient hidden surface removal for objects with small union size, *Computational Geometry: Theory and Applications* 2 (1992), 223–234.

[17] M. J. van Kreveld, *New Results on Data Structures in Computational Geometry*, Ph.D. Dissertation, Dept. Comput. Sci., Utrecht Univ. , Utrecht, Netherlands, 1992.

[18] J. Matoušek, Efficient partition trees, *Discrete and Computational Geometry* 8 (1992), 315–334.

[19] J. Matoušek, Range searching with efficient hierarchical cuttings, *Discrete and Computational Geometry* 10 (1993), 157–182.

[20] F. Nielsen and M. Yvinec, Output-Sensitive Convex Hull Algorithms of Planar Convex Objects, *Research Report* 2575 INRIA, BP93, 06902 Sophia-Antipolis, France.

[21] M.H. Overmars, Point location in fat subdivisions, *Information Processing Letters* 44 (1992), 261–265.

[22] H. Overmars and J. van Leeuwen, Maintenance of configurations in the plane, *J. Comput. Syst. Sci.* 23 (1981), 166–204.

[23] M.H. Overmars and A.F. van der Stappen, Range searching and point location among fat objects, *Proc. 2nd Annual European Symp. on Algorithms*, 1994, 240–253. (See also: Tech. Report UU-CS-1994-30, Dept. of Computer Science, Utrecht University.)

[24] O. Schwarzkopf and J. Vleugels, Range Searching in Low-Density Environments, Tech. Report UU-CS-1996-26, Dept. of Computer Science, Utrecht University.

[25] M. Sharir and P.K. Agarwal, *Davenport Schinzel Sequences and Their Geometric Applications*, Cambridge University Press, New York, 1995.

[26] A.F. van der Stappen, *Motion Planning amidst Fat Obstacles*, Ph.D. thesis, Utrecht University, 1994.

[27] A.F. van der Stappen, D. Halperin and M.H. Overmars, The complexity of the free space for a robot moving amidst fat obstacles, *Computational Geometry: Theory and Applications* 3 (1993), 353–373.

[28] A.F. van der Stappen and M.H. Overmars, Motion planning amidst fat obstacles, *Proc. 10th ACM Symp. Comput. Geom.*, 1994, 31–40.

Intractability of Assembly Sequencing: Unit Disks in the Plane

Michael Goldwasser* and Rajeev Motwani**

Department of Computer Science
Stanford University
Stanford, CA 94305-9045
email: {wass,rajeev}@cs.stanford.edu

Abstract. We consider the problem of removing a given disk from a collection of unit disks in the plane. At each step, we allow a disk to be removed by a collision-free translation to infinity, and the goal is to access a given disk using as few steps as possible. This DISKS problem is a version of a common task in assembly sequencing, namely removing a given part from a fully assembled product. Recently there has been a focus on optimizing assembly sequences over various cost measures, however with very limited algorithmic success. We explain this lack of success, proving strong inapproximability results in this simple geometric setting. Namely, we show that approximating the number of steps required to within a factor of $2^{\log^{1-\gamma} n}$ for any $\gamma > 0$ is quasi-NP-hard. This provides the first inapproximability results for assembly sequencing, realized in a geometric setting.

As a stepping stone, we study the approximability of scheduling with AND/OR precedence constraints. The DISKS problem can be formulated as a scheduling problem where the order of removals is to be scheduled. Before scheduling a disk to be removed, a path must be cleared, and so we get precedence constraints on the tasks; however, the form of such constraints differs from traditional scheduling in that there is a choice of which path to clear. We prove our main result by first showing the similar inapproximability of this scheduling problem, and then by showing that a sufficiently hard subproblem can be realized geometrically using unit disks in the plane. Furthermore, our construction is fairly robust, in that is can be placed on a polynomially sized integer grid, it remains valid even when we consider only horizontal and vertical translations, and it also applies to axis-aligned unit squares and higher dimensions.

* Supported by ARO MURI Grant DAAH04-96-1-0007 and by NSF Award CCR-9357849, with matching funds from IBM, Mitsubishi, Schlumberger Foundation, Shell Foundation, and Xerox Corporation.
** Supported by an Alfred P. Sloan Research Fellowship, an IBM Faculty Partnership Award, an ARO MURI Grant DAAH04-96-1-0007, and NSF Young Investigator Award CCR-9357849, with matching funds from IBM, Mitsubishi, Schlumberger Foundation, Shell Foundation, and Xerox Corporation.

1 Introduction

The *assembly sequencing* problem consists of taking a geometric description of a product, consisting of a set of parts, and producing a sequence of collision-free operations which results in the (dis)assembly of the product. Many complexity measures for judging the quality of a sequence were discussed in a precursor to this work [17], where we prove strong inapproximability for a non-geometric generalization of the sequencing problem. In this paper, we show that strong lower bounds can be realized geometrically, for the problem of planning for the removal of a given part from a fully assembled product. This problem is motivated by the issues of maintenance and of recycling. The classic maintenance example is the need to replace a spark plug without taking the entire car apart. A classic recycling example is to strip down an old computer for a valuable part with minimal effort. Unfortunately, there has been little success algorithmically for optimizing (dis)assembly sequences over complexity measures, and several current assembly sequencing packages must rely on either a brute force search through all possibilities, or on heuristic searches with no performance guarantees. Our work proves the difficulty of finding optimal or near-optimal cost assembly sequences, under much simpler geometric conditions than are needed by industrial assembly sequences.

Specifically, we examine what we term the DISKS problem. The goal is to remove a key disk from a collection of unit disks in the plane, where each step allows for a collision-free translation of a single disk to infinity. The cost of a solution is equal to the number of steps required to remove the key disk. We prove that it is quasi-NP-hard to approximate this DISKS problem to within a factor of $2^{\log^{1-\gamma} n}$ for any $\gamma > 0$. These hardness results are the first such inapproximability results involving the cost of assembly sequences, to be realized geometrically.

To prove our results, we study the problem of scheduling with AND/OR precedence constraints, a generalization of the DISKS problem. We prove the inapproximability of this AND/OR scheduling problem, previously shown only to be NP-hard, using a reduction from the LABELCOVER$_{min}$ problem [3,4]. Moreover, we show that the hardness of AND/OR scheduling applies to a more restricted version which can then be realized geometrically as our DISKS problem. Our construction can be placed on a polynomially sized grid, and our results are valid even when translations are limited to two directions. (Limited to one direction, the problem is trivially solvable.) These results also generalize to axis-aligned unit squares as well as to higher dimensions. Full details of this paper are included in a more extensive study of the approximability of cost measures for assembly sequencing [16].

2 Previous Work

Assembly Sequencing: The use of automation in assembly sequencing has increased rapidly over the years [6,11,24–26,35,43,44,46]. Theoretical results show that assembly sequencing, in its most general form, is intractable [27,31,32,37,46], and thus many researchers began considering restricted, yet still interesting, versions of the problem. For many of these restricted settings, polynomial algorithms have been designed which find an assembly sequence if one exists [1,19,22,43,45]. There are also algorithms which enumerate all possible assembly sequences [11], however there may be exponentially many such sequences for a product.

A logical continuation to this success is to use automated reasoning to find the "best" assembly sequence under certain complexity measures. In fact, the IEEE Technical Committee on Assembly and Task Planning summarized the current state of assembly sequencing by explaining [18], "... after years of work in this field, a basic planning methodology has emerged that is capable of producing a feasible plan ... The challenges still facing the field are to develop efficient and robust analysis tools and to develop planners capable of finding optimal or near–optimal sequences rather than just feasible sequences." Unfortunately, meeting this challenge has been difficult. Several complexity measures for assembly sequencing have been suggested, motivated strongly by industrial applications [7,17,45,46]. In fact, some current software systems offer the user the option of optimizing the sequence over a choice of complexity measures [30,40], however these systems currently must rely either on heuristics with no performance guarantees on the cost of the sequence, or else on possibly exponential search techniques to find the true optimal. The NP-completeness of exactly optimizing several measures is shown in a symbolic framework [46], and the inapproximability of several measures, including the one considered here, are shown in a graph-theoretic generalization of assembly sequencing [17]. The strongest of these lower bounds, however, are not realized geometrically. For a restricted class of inputs which have a so-called "total ordering" property, a greedy algorithm is given which claims to produce the minimal length sequence to remove any given part [47], however the required property does not have a clean definition, and as our results will show, this problem is quite difficult in general.

Approximability Theory. As our optimization problems turn out to be NP-hard, we approach the problems using techniques common to the theory of approximability [4,13,28,36]. Since we cannot expect to find the optimal sequence in polynomial time, we look for a polynomial time *approximation algorithm* which returns a solution whose cost can be bounded by some function of the true optimal cost. A standard measure for the quality of an approximation algorithm is the *approximation ratio* between the cost of the solution returned by the algorithm versus the cost of the optimal solution. Although all NP-complete decision problems can be reduced to one another, the approximability of such problems can be quite different, ranging from problems which can be approximated arbitrarily close to optimal, to problems where getting even a very rough approximation is already NP-hard. Many researchers have worked towards classifying the approximability of different NP-hard problems. We consider four broad classes defined in [4], which group problems based on the strength of the inapproximability results which have been proven. Class I includes all problems for which approximating the optimal solution to within a factor of $(1 + \epsilon)$ is NP-hard for some $\epsilon > 0$ (e.g., MAX-3SAT [38]). Class II groups those problems for which it is quasi-NP-hard[1] to achieve an approximation ratio of $c \cdot \log n$ for some $c > 0$ (e.g., SET COVER [12]). For problems in Class III, it is quasi-NP-hard to achieve a $2^{\log^{1-\gamma} n}$ factor[2] approximation for any $\gamma > 0$ (e.g., LABELCOVER [3]). Finally, Class IV consists of the hardest problems, namely those for which it is NP-hard to achieve an n^ϵ approximation factor for some $\epsilon > 0$ (e.g., CLIQUE [23]). Our work places both the AND/OR scheduling problem and the DISKS problem into Class III of this hierarchy.

Computational Geometry. Assembly sequencing is an intriguing combination of a combinatorial and geometric problem. Quite naturally, research from computational geometry relates very closely to assembly sequencing. The separability of objects has been well studied in the geometric community [8–10,20,41,42]. In a single direction, a *depth order* of a set of parts is an ordering of the parts which allows for collision-free translations of the individual parts to infinity. A classic result states that given a collection of convex shapes in two dimensions, for any translational direction there exists some ordering, such that the parts can be translated away one at a time [20]. Similar separability issues are studied in two dimensions for more general classes of shapes, such as monotone or star-shaped polygons [42]. For a collection of balls in $\mathcal{R}^d$, there exist at least $d+1$ balls, each of which can be translated to infinity in some direction [8]. Unfortunately, there exist collections of unit disks in the plane for which the removal of a certain disk may require the prior removal of $\Omega(n)$ other disks [21]. Also, there exists a set of convex parts in three dimensions which cannot be disassembled using two hands, with only translations, or even generalized rotations and translations [41].

A surprising element of our problem is that the general lower bounds can be realized for a simple geometric setting consisting of a collection of unit disks in the plane. More often than not, an optimization problem becomes significantly easier when its input is restricted to a geometric setting. For example, there exists some $c > 0$ for which achieving a $(1 + c)$-approximation for the METRIC TRAVELING SALESMAN problem is NP-hard [39], however in the Euclidean plane, TSP can be approximated to within $(1 + \epsilon)$ for all $\epsilon > 0$ [2]. Similarly, achieving an n^ϵ-approximation for MINIMUM INDEPENDENT SET is NP-hard [23], however for planar graphs, MIS can be approximated to within $(1 + \epsilon)$ [5]. Similar results hold for most optimization problem when restricted to planar graphs [33]. There exists a $\ln n$ lower bound for approximating the SET COVER problem [12], however the RECTANGLE COVER problem, covering a set of axis-aligned rectangles with minimum number of points, has no such inapproximability results [36].

---

[1] That is, this would imply $NP \subseteq DTIME(n^{\text{poly}(\log n)})$. "A proof of quasi-NP-hardness is good evidence that the problem has no polynomial-time algorithm." [4]

[2] This factor, $2^{\log^{1-\gamma} n}$, lies between polynomial and polylogarithmic in that $2^{\log^{1-\gamma} n} = o(n^\epsilon)$ for any $\epsilon > 0$, and $2^{\log^{1-\gamma} n} = \omega(\log^c n)$ for any constant c.

3 Scheduling with AND/OR Precedence Constraints

We can view an instance of the DISKS problem quite naturally as a scheduling problem. The removal of each part is thought of as a task which can be scheduled, and the goal is to schedule a key task as early as possible. The cost of a solution is equal to the total number of tasks scheduled.

Of course, our tasks have certain precedence constraints relating their order of removal. That is, it may be the case that a certain part cannot be removed until after some other parts are removed. What distinguishes this setting from more traditional scheduling is the form of the precedence constraints. Commonly, a task may have what we term an AND-precedence constraint, in that it has an associated set of tasks, all of which must be scheduled before that task [13]. Unfortunately, this is not the case in our assembly sequencing problem. When considering a single direction, if a part is to be removed, then indeed there is a clear set of associated parts which block the removal, and thus all of these parts must be removed prior to removing our part in that direction. However, we may choose to remove that same part in some other direction, in which case a different set of parts may block the removal.

Instead, our problem can be viewed as an instance of scheduling with AND/OR precedence constraints, where the OR's allow us to offer a choice of directions. We will consider scheduling where every task has a set of direct predecessors, and each task is either an AND-task, in which case it cannot be scheduled until after all of its predecessors, or the task is an OR-task, in which case it cannot be scheduled until after at least one of its predecessors. For our setting, we consider a single processor and unit processing time for all tasks. It is worth nothing that with classical AND precedence constraints, this problem of minimizing the number of scheduled tasks can be solved exactly, in polynomially time by computing a depth order.

Notation and Definitions. We define the problem of scheduling with AND/OR precedence constraints as follows. The input contains a set of tasks, $\mathcal{T}$. Each task, $t_i \in \mathcal{T}$, is labeled as either an AND-*task* or an OR-*task*. Each task, $t_i \in \mathcal{T}$, has an associated set of tasks, P_i, as direct *predecessors*; we refer to $|P_i|$ as the *degree* of the task. An AND-task, t_i, cannot be scheduled until after *all* tasks in P_i. An OR-task, t_j, cannot be scheduled until after *at least one* task of P_j. The *max* AND-*degree* of an instance is the maximum size $|P_i|$ over all AND-tasks t_i. The *max* OR-*degree* of an instance is the maximum size $|P_j|$ over all OR-tasks t_j.

The constraints can be represented by a *precedence graph*, with a node for each task, t_i, and a directed edge from t_i to t_j whenever $t_j \in P_i$, is a direct predecessor of t_i. A *leaf-task* is one with $P_i = \emptyset$, and thus no outgoing edges. Such a task can be scheduled at any time. We say that an instance of AND/OR scheduling has *partial-order* precedence constraints if there are no cycles in the precedence graph. We say an instance of AND/OR scheduling has *internal-tree* precedence constraints if there are no cycles, and if all non-leaf nodes have at most one incoming edge.

The goal for this problem is to successfully schedule a specific task, and the cost is equal to the total number of tasks which must be scheduled. We consider a single processor and unit processing time for all tasks. Additionally, we will consider the problem of minimizing the number of scheduled *leaves*, when constrained to internal-tree precedence constraints. Notice that internal-tree precedence constraints define a monotone, boolean formula on the leaf nodes, in which setting a leaf's variable to "one" signifies that the leaf will be scheduled. Minimizing the number of scheduled leaves is equivalent to satisfying a monotone, boolean formula with the minimum number of ones. We are unaware of any previous results for this exact approximation problem. Minimizing the number of ones in satisfying a 3CNF formula is known to be $n^{0.5-\epsilon}$-hard to approximate [29], and related minimization problems are studied in [34].

Previous Work. A model for scheduling with AND/OR precedence constraints was introduced in [14,15], however with one key difference. In this previous work, only the case of partial order precedence constraints is considered. Notice that in traditional scheduling, with AND-precedence constraints, the existence of a cycle in the precedence constraints makes the scheduling problem infeasible, and thus a partial order is a standard assumption. With AND/OR constraints, this absence of cycles is no longer a necessary condition for the existence of a valid solution. In fact, cycles will often exist in instances drawn from assembly sequencing, as it may be the case that part A blocks part B in one direction, part B blocks part C in another, and part C blocks part A in a third direction. For this reason, we make no apriori assumptions about the structure of the precedence constraints.

The work of [14,15] studies a larger variety of settings, including multiple processors, deadlines, and individual processing times. They prove the NP-hardness of finding feasible schedules in many setting which were polynomially solvable with more traditional AND-precedence constraints, however they do not consider the approximability of the corresponding optimization problems.

3.1 Inapproximability of AND/OR Scheduling

Theorem 1. *It is quasi-NP-hard to find a solution to any of the following problems which is within a factor of $2^{\log^{1-\gamma} n}$ of the optimal solution, for any $\gamma > 0$.*

- *Minimizing the number of* leaves *scheduled, for* AND/OR *scheduling with internal-tree precedence constraints.*
- *Minimizing the number of* leaves *scheduled, for* AND/OR *scheduling with internal-tree precedence constraints, and max-degree bounded by two.*
- *Minimizing the number of* tasks *scheduled, for general* AND/OR *scheduling.*
- *Minimizing the number of* tasks *scheduled, for general* AND/OR *scheduling with max-degree bounded by two.*

Proof. We begin by considering the AND/OR scheduling problem when restricted to *internal-tree* precedence constraints, when charged for the number of *leaves* scheduled. We show the inapproximability of this problem by showing that the

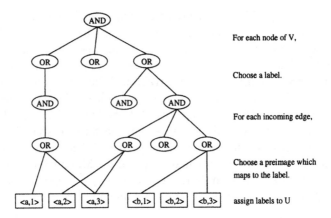

For each node of V,

Choose a label.

For each incoming edge,

Choose a preimage which maps to the label.

assign labels to U

Fig. 1. LABELCOVER as AND/OR scheduling with internal-tree precedence

LABELCOVER$_{\text{min}}$ problem is a special case. The LABELCOVER$_{\text{min}}$ problem, defined in [4], is an artificial generalization of the SET COVER problem introduced in approximability theory. The input is a regular bipartite graph, $G = (U, V, E)$, a set of labels $\{1, 2, \ldots, N\}$, and a partial function $\Pi_e : \{1, 2, \ldots, N\} \longrightarrow \{1, 2, \ldots, N\}$ for each edge $e \in E$. A labeling associates a non-empty set of labels with every vertex in $U \cup V$. It is said to *cover* an edge $e = (u, v)$, if for every label b assigned to v, there is some label a assigned to u such that $\Pi_e(a) = b$. The goal of LABELCOVER$_{\text{min}}$ is to give a labeling which covers all edges, while minimizing the total number of labels assigned to nodes of U.

Given an instance of LABELCOVER$_{\text{min}}$, we express it as an instance of AND/OR scheduling with internal-tree precedence constraints. The AND/OR instance has five levels, which alternate between AND-nodes and OR-nodes. The highest level contains solely the root of the internal-tree, and the lowest level contains exactly the leaves. The tasks at the five levels are as follows:

- The first level has a single AND-node, which is the root of the internal-tree. This task enforces that for a valid labeling, every node in V must have a non-empty set of labels.
- The second level has an OR-node for each vertex in V. This nodes requires that for a given node v to have a non-empty label set, at least one label must be assigned to it.
- The third level has an AND-node for each pair $\langle v, l' \rangle$, where $v \in V$, and $l' \in \{1, \ldots, N\}$. This node signifies that for label l to be assigned to vertex v, it must be the case that for each edge $e = (u, v)$ incident to v, the mapping Π_e on that edge, must respect the labeling.
- The fourth level has an OR-node for each pair $\langle e, l' \rangle$, where $e = (u, v)$ is an edge, and l' is a label. If l' is to be assigned to v, then edge e cannot be covered unless one of the pre-images of l' from mapping Π_e is assigned to u.
- The fifth level has a leaf for each pair $\langle u, l \rangle$, and corresponds to label l being assigned to vertex u.

This completes the construction. It can be seen that there is a one-to-one correspondence between valid labeling in the LABELCOVER$_{min}$ instance and valid solutions to the AND/OR scheduling instance. It is easy to verify that the AND/OR instance has internal-tree precedence constraints. Notice that the number of non-leaf tasks in this construction is polynomially bounded in the size of the LABELCOVER$_{min}$ instance (namely, in $|U|$, $|V|$ and N). Combining this with the result of [4] which proves a similar lower bound for the approximability of LABELCOVER$_{min}$, we get that it is quasi-NP-hard to achieve an $2^{\log^{1-\gamma} n}$ approximation for any $\gamma 0$, when minimizing the number of leaves with internal-tree precedence constraints.

Given an instance with unbounded degree, we can bound the in-degree in the obvious way, by replacing each internal node with a tree of bounded degree nodes. Assume there were originally I internal nodes and L leaves, and that I is polynomially bounded in L. The maximum fan-in for any node is at most $(I + L)$, and thus that node must be replaced by a tree of at most $(I + L)$ nodes, each with fan-in two. The new instance has $I(I + L)$ internal nodes, which is still polynomially-bounded in L.

The only difficulty in switching from the measure of counting scheduled leaves to counting all scheduled tasks is that the overhead of the internal nodes may have a significant cost, changing our approximation ratio. This can be remedied quite easily. Assume we have a hard instance of AND/OR scheduling with internal-tree precedence constraints, containing I internal nodes and L leaves. We convert this to a general instance of AND/OR scheduling by hanging from each leaf a chain of I new nodes. Notice that the OR-degree is not increased, although this new instance no longer has internal-tree precedence constraints. In this way, the problematic additive cost can be made arbitrarily small, and so the only issue in completing the proof is to address the input size. Originally, the value n was assumed to be the number of leaves. In our new instance, the total number of nodes is n', however we can rely on the fact that $n' = \text{poly}(n)$, and thus an approximation ratio of $2^{\log^{1-\gamma} n'}$ is less than a ratio of $2^{\log^{1-\gamma'} n}$ for some $\gamma' > 0$.

4   The DISKS Problem

Theorem 2. *It is quasi-NP-hard to approximate the DISKS problem to within a factor of $2^{\log^{1-\gamma} n}$ for any $\gamma > 0$. This bound also applies if we consider only translations along the positive X-axis and Y-axis. Additionally, this construction generalizes to axis-aligned unit squares, and to higher dimensions.*

Proof. Assume we are given a hard instance from Theorem 1, of AND/OR scheduling with internal-tree precedence constraints, and OR-degree bounded by two (we do not require such a bound on the AND-degree). Without loss of generality, we assume that OR-nodes rely only on internal nodes.

We construct the following instance of the DISKS problem. Our scene consists entirely of disks with radius one, whose centers lie on a polynomially-sized, integer grid. We prove this result directly for the case where only two directions

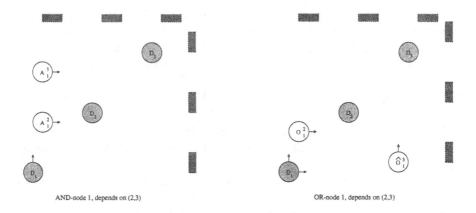

AND-node 1, depends on (2,3)   OR-node 1, depends on (2,3)

Fig. 2. Internal node mechanisms

of translations are allowed, namely North and East. We place a wall of width $2W$ around the perimeter of our working area which we consider immovable. We will place some holes in the wall, described later, which allow a clear path out for some disks. We consider our main working area to have two sections, one for the mechanisms involving the interior nodes, and the second section for the leaf node mechanisms.

First we describe the mechanism involving the internal nodes. Since the internal-tree defines a partial order on these nodes, we can number the internal nodes, $T_1, \ldots, T_I$ so that if an internal node depends on another internal node, it will have a higher index. For each internal node, T_i, we create a disk, D_i, centered at $(6i, 6i)$. We define the wall to the North by placing a column of W disks with x-coordinates centered at $6i + 2$ for each disk D_i, assuring us that the disk itself has an "escape route" to the North. For the East wall, we place a row of disks centered at y-coordinate $6i + 2$ in the case that disk D_i is an OR-disk, or else at $6i + 1$ in the case that disk D_i is an AND-disk. In this way, we assure an additional escape passage to the East for an OR-disk, but not for an AND-disk.

Next, we add in additional disks to enforce the precedence constraints. For AND-node, T_i, blocked by node $T_k \in P_i$ (and thus $i < k$), we add a disk A_i^k centered at $(6i + 1, 6k - 1)$, which will be forced to the East by our previous placement of the walls. For an OR-node, T_i, which depends on 2 nodes, T_k and T_l, we create two new disks, O_i^k located at $(6i + 1, 6k - 1)$, which will be forced East by our walls, and $\hat{O}_i^l$ located at $(6l - 1, 6i + 1)$, which will be forced North. The entire internal node mechanisms are contained in a $(6I+1) \times (6I+1)$ square. Examples are given in Figure 2.

The section for the leaf mechanisms begins at height $6(I + 1)$ so as to be higher than the internal mechanisms. We can number the leaf nodes in any order, and we create a separate mechanism for each leaf in a strip of height $2I$. For a given leaf, L_a, we create what we term a *blockade*, to the right of this strip. The blockade consists first of a diagonal chain of to the Northeast of height $2I$, followed by a horizontal chain of B disks to the East of the end of the first chain

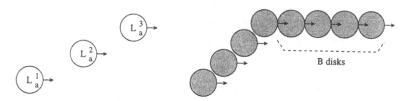

Internal nodes 1,2,3 depend on Leaf a

Fig. 3. Leaf node mechanism

(where B is determined later). The disk beginning the blockade is centered at $(6(I+1), 6(I+1)+Ia)$. The wall to the East of the blockade is removed, allowing the disks of the blockade an escape. For any disk located in the horizontal strip associated with L_a, escaping to the East will require an additional cost of at least B to break through the blockade. However this cost is only charged once per blockade, after which any disks in the horizontal strip may escape. Now, for every internal node T_i which depends on leaf L_a, we create a disk L_a^i, located at $(6i+1, 6(I+1)+Ia+2i)$, which is forced East by the walls. Figure 3 shows an example of a leaf mechanism.

To complete the construction, we set the blockade value, $B = 4I(L+I)$, to be greater than the total number of disks in the remainder of the internal and leaf mechanisms combined. In this way, the number of blockades removed dominates any additive costs in the rest of the construction. Finally, we assign $W = B(L+1)$, so that the cost of removing all non-wall disks is less than the cost of digging a single new hole through any part of the wall. For this reason, we may assume without loss of generality that any solution to this DISKS instance has cost at most W. Finally, we note that the wall has perimeter which is $O(BL)$, and hence the total number of disks in our construction is polynomially bounded. An example of the final construction is given in Figure 4.

It is not hard to verify that for this DISKS instance, a solution for removing the root disk with cost at most kB can be translated to an AND/OR solution of cost at most k. Similarly, an AND/OR solution of cost k can be translated to a DISKS solution with cost less than $(k+1)B$. Therefore, approximating the DISKS problem to within a factor of $2^{\log^{1-\gamma} n}$ for any $\gamma > 0$ is quasi-NP-hard, as the additive error and the polynomial increase of the input size disappear by adjusting γ.

Our proof shows the hardness of the DISKS problem when translations are limited to the North and East. In fact, if we allow translations in arbitrary directions, the theorem holds using this same construction. Furthermore, even if we are not restricted to linear moves, we could prove the same lower bound for minimizing the number of disks removed.

It is also easy to see that the disks can be replaced by axis-aligned, 2×2 squares and the construction still holds. For higher dimensions, the walls can be extended to block any useful motions in other dimensions, while still using polynomially many disks.

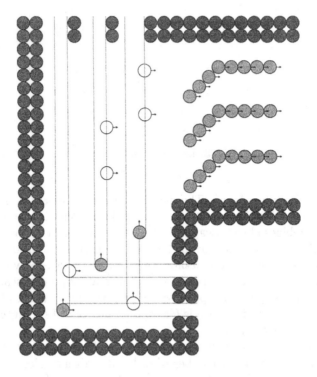

Fig. 4. A complete DISKS construction

5 Future Directions

Open directions for future research fall into three quite different directions. First, we offer no non-trivial (i.e. $o(n)$) approximation algorithms for either the DISKS problem or any variants of AND/OR scheduling. It so happens that there is no known, non-trivial approximation algorithm for LABELCOVER, the "easiest" of these problems. However our reductions are a bit misleading in this respect, because the polynomial blowup in the input size. It may be possible to achieve a non-trivial upper bound for many of our variants of AND/OR scheduling, without providing such an upper bound on LABELCOVER or the other variants. For example, the existence of an $n^{1/2}$-approximation for the DISKS problem would guarantee through our reductions, an n^{β}-approximation for LABELCOVER$_{min}$, however it may be that $\beta > 1$, in which case the approximation would be trivial.

Secondly, it is open to strengthen the lower bounds for any of these problems. The LABELCOVER$_{min}$ results provide our strongest results even for the most general AND/OR scheduling problem, yet there is reason to believe this may be an even more difficult problem. It is already conjectured that LABELCOVER is truly n^{ϵ}-hard to approximate for some $\epsilon > 0$ [4], a result which would carry over through all of our reductions. However, we feel that the study of AND/OR scheduling may provide more meaningful implications in the approximation complexity hierarchy. It may be possible to strengthen the lower bounds for AND/OR scheduling without necessarily settling the LABELCOVER conjecture.

We examined a very structured class of instances of AND/OR scheduling which had what we termed *internal-tree* precedence constraints, and we considered charging only for the *leaves* that are scheduled. Without a bound on the out-degree, we can collapse internal nodes so that our tree has alternating levels of AND-nodes followed by OR-nodes, with a final level of leaves. If we restrict the number of alternating levels to one, this problem becomes *exactly* SET COVER, and thus lies in Class II of the approximation hierarchy. When the number of alternating levels is equal to two, we saw in Figure 1, that this problem already captures LABELCOVER, and thus lies in Class III. However it is not at all clear that this problem is equivalent to LABELCOVER, as it may capture more instances. Furthermore, what happens when we go to three full alternations, or to an arbitrary depth internal tree? Does this hierarchy collapse at some point, and if so when? Can the inapproximability bounds be strengthened for any of these versions? What about when no constraints at all are placed on the structure of the precedence graph?

Finally, the DISKS problem is just one version of a collection of optimization problems for assembly sequencing. A more complete framework for analyzing the approximability of many other cost measures for assembly sequencing problems is presented in [16]. In that work, many strong inapproximability results are shown in a non-geometric generalization, however with the exception of the DISKS setting, the geometric lower bounds shown for other cost measures are far weaker than their non-geometric counterparts. A complete list of related problems and open questions is given there.

Acknowledgments

The authors wish to thank Danny Halperin for suggesting that we consider the scenario with unit disks. Also, thanks go to Chandra Chekuri and Sanjeev Khanna for their involvement in the study of the AND/OR scheduling model. Finally, to Cyprien Godard, Jean-Claude Latombe, G. Ramkumar, Bruce Romney, and Randy Wilson, for their involvement in earlier parts of this work as well as for their many helpful discussions and suggestions.

References

1. P. Agarwal, M. de Berg, D. Halperin, and M. Sharir. Efficient generation of k-directional assembly sequences. In *Proc. 7th ACM Symp. on Discrete Algorithms*, pages 122–131, 1996.
2. S. Arora. Polynomial-time approximation schemes for Euclidean TSP and other geometric problems. In *Proc. 37th Symp. on Found. Comput. Sci.*, pages 1–11, 1996.
3. S. Arora, L. Babai, J. Stern, and Z. Sweedyk. The hardness of approximate optima in lattices, codes and linear equations. In *Proc. 34th Symp. on Found. Comput. Sci.*, pages 724–733, 1993.

4. S. Arora and C. Lund. Hardness of approximations. In D. Hochbaum, editor, *Approximation Algorithms for NP-Hard Problems*. PWS Publishing Company, Boston, MA, 1996.

5. B. Baker. Approximation algorithms for NP-complete problems on planar graphs. *J. ACM*, 41(1):153–180, 1994.

6. D. Baldwin. Algorithmic methods and software tools for the generation of mechanical assembly sequences. M.Sc. thesis, MIT, Cambridge, MA, 1990.

7. G. Boothroyd. *Assembly Automation and Product Design*. Marcel Dekker, Inc., New York, NY, 1991.

8. R. Dawson. On removing a ball without disturbing the others. *Mathematics Magazine*, 57(1):27–30, 1984.

9. M. de Berg, M. Overmars, and O. Schwarzkopf. Computing and verifying depth orders. *SIAM J. Comput.*, 23(2):432–446, 1994.

10. F. Dehne and J.-R. Sack. Translation separability of polygons. *Visual Computer*, 3(4):227–235, 1987.

11. T. D. Fazio and D. Whitney. Simplified generation of all mechanical assembly sequences. *IEEE Trans. on Robotics and Automation*, 3(6):640–658, 1987.

12. U. Feige. A threshold of $\ln n$ for approximating set cover. In *Proc. 28th ACM Symp. Theory Comput.*, pages 314–318, 1996.

13. M. Garey and D. Johnson. *Computers and Intractability: A Guide to the Theory of NP-Completeness*. W. H. Freeman, New York, NY, 1979.

14. D. Gillies. *Algorithms to schedule tasks with AND/OR precedence constraints*. Ph.D. thesis, University of Illinois, Urbana, IL, 1993.

15. D. Gillies and J. Liu. Scheduling tasks with AND/OR precedence constraints. *SIAM J. Comput.*, 24(4):797–810, 1995.

16. M. Goldwasser. *Complexity Measures for Assembly Sequencing*. Ph.D. thesis, Dept. Comput. Sci., Stanford Univ., Stanford, CA, 1997.

17. M. Goldwasser, J.-C. Latombe, and R. Motwani. Complexity measures for assembly sequences. In *Proc IEEE Int. Conf. on Robotics and Automation*, pages 1581–1587, 1996.

18. S. Gottschlich, C. Ramos, and D. Lyons. Assembly and task planning: A taxonomy. *IEEE Robotics and Automation Magazine*, 1(3):4–12, 1994.

19. L. Guibas, D. Halperin, H. Hirukawa, and J.-C. L. R. Wilson. A simple and efficient procedure for polyhedral assembly partitioning under infinitesimal motions. In *Proc IEEE Int. Conf. on Robotics and Automation*, pages 2553–2560, 1995.

20. L. Guibas and F. Yao. On translating a set of rectangles. In F. Preparata, editor, *Computational Geometry*, Advances in Computing Research, pages 61–77. JAI Press Inc., 1983.

21. D. Halperin. Personal communication. 1995.

22. D. Halperin and R. Wilson. Assembly partitioning along simple paths: the case of multiple translations. In *Proc. IEEE Int. Conf. on Robotics and Automation*, pages 1585–1592, 1995.

23. J. Hstad. Clique is hard to approximate within $n^{1-\epsilon}$. In *Proc. 37th Symp. on Found. Comput. Sci.*, pages 627–636, 1996.

24. R. Hoffman. A common sense approach to assembly sequence planning. In *Computer-Aided Mechanical Assembly Planning*, pages 289–314. Kluwer Academic Publishers, Boston, 1991.

25. L. Homem de Mello and A. Sanderson. *Computer-Aided Mechanical Assembly Planning*. Kluwer Academic Publishers, Boston, 1991.

26. L. Homem de Mello and A. Sanderson. A correct and complete algorithms for the generation of mechanical assembly sequences. *IEEE Trans. on Robotics and Automation*, 7(2):228–240, 1991.

27. J. Hopcroft, J. Schwartz, and M. Sharir. On the complexity of motion planning for multiple independent objects: P-space hardness of the "Warehouseman's Problem". *Int. J. Robotics Research*, 3(4):76–88, 1984.

28. D. Johnson. Approximation algorithms for combinatorial problems. *J. Comput. Systems Sci.*, 9:256–278, 1974.

29. V. Kann. Polynomially bounded minimization problems which are hard to approximate. In *Automata, Languages and Programming (Proc. 20th ICALP)*, volume 700 of *Lecture Notes in Computer Science*, pages 52–63. Springer-Verlag, 1993.

30. S. Kaufman, R. Wilson, R. Jones, T. Calton, and A. Ames. The Archimedes 2 mechanical assembly planning system. In *Proc IEEE Int. Conf. on Robotics and Automation*, pages 3361–3368, 1996.

31. L. Kavraki and M. Kolountzakis. Partitioning a planar assembly into two connected parts is NP-complete. *Information Processing Letters*, 55(3):159–165, 1995.

32. L. Kavraki, J.-C. Latombe, and R. Wilson. Complexity of partitioning an assembly. In *Proc. 5th Canad. Conf. Comput. Geom.*, pages 12–17, Waterloo, Canada, 1993.

33. S. Khanna and R. Motwani. Towards a syntactic characterization of PTAS. In *Proc. 28th ACM Symp. Theory Comput.*, pages 329–337, 1996.

34. S. Khanna, M. Sudan, and L. Trevisan. Constraint satisfaction: The approximability of minimization problems. In *Proc. 12th IEEE Comput. Complexity Conference*, 1997.

35. S. Lee and Y. Shin. Assembly planning based on geometric reasoning. *Computers and Graphics*, 14(2):237–250, 1990.

36. R. Motwani. Approximation algorithms. Stanford Technical Report STAN-CS-92-1435, 1992.

37. B. Natarajan. On planning assemblies. In *Proc. 4th ACM Symp. on Computational Geometry*, pages 299–308, 1988.

38. C. Papadimitriou and M. Yannakakis. Optimization, approximation, and complexity classes. *J. Comput. Systems Sci.*, 43(3):425–440, 1991.

39. C. Papadimitriou and M. Yannakakis. The traveling salesman problem with distances one and two. *Mathematics of Operations Research*, 18(1):1–11, 1993.

40. B. Romney, C. Godard, M. Goldwasser, and G. Ramkumar. An efficient system for geometric assembly sequence generation and evaluation. In *Proc. ASME Int. Computers in Engineering Conference*, pages 699–712, 1995.

41. J. Snoeyink and J. Stolfi. Objects that cannot be taken apart with two hands. In *Proc. ACM Symp. on Computational Geometry*, pages 247–256, 1993.

42. G. Toussaint. Movable separability of sets. In G. Toussaint, editor, *Computational Geometry*, pages 335–375. North-Holland, Amsterdam, Netherlands, 1985.

43. R. Wilson. *On Geometric Assembly Planning*. Ph.D. thesis, Dept. Comput. Sci., Stanford Univ., Stanford, CA, 1992. Stanford Technical Report STAN-CS-92-1416.

44. R. Wilson, L. Kavraki, J.-C. Latombe, and T. Lozano-Pérez. Two-handed assembly sequencing. *Int. J. of Robotics Research*, 14(4):335–350, 1995.

45. R. Wilson and J.-C. Latombe. Geometric reasoning about mechanical assembly. *Artificial Intelligence*, 71(2):371–396, 1994.

46. J. Wolter. *On the Automatic Generation of Plans for Mechanical Assembly*. Ph.D. thesis, University of Michigan, 1988.

47. T. Woo and D. Dutta. Automatic disassembly and total ordering in three dimensions. *J. Engineering for Industry*, 113(2):207–213, 1991.

On Hamiltonian Triangulations in Simple Polygons

(Extended Abstract)

Giri NARASIMHAN *

Department of Mathematical Sciences
The University of Memphis
Memphis TN 38152.
e-mail: giri@next1.msci.memphis.edu

Abstract. An n-vertex simple polygon P is said to have a *Hamiltonian Triangulation* if it has a triangulation whose dual graph contains a Hamiltonian path. Such triangulations are useful in fast rendering engines in Computer Graphics. We give a new characterization of polygons with Hamiltonian triangulations. and use it to devise $O(n \log n)$-time algorithms to recognize such polygons.

1 Introduction

A simple polygon P is said to have a *Hamiltonian Triangulation* if it has a triangulation whose dual graph contains a Hamiltonian path. Such triangulations are useful in fast rendering engines in Computer Graphics, since most visualization of surfaces is currently done via triangles [AHMS, ESV]. In order to transmit triangulation data for a simple polygon to a rendering engine, one needs to send information about each of its triangular faces. For a polygon with a hamiltonian triangulation, its triangulation data can be sent at a faster rate since consecutive triangles share a face, and hence only incremental changes need to be specified per triangle. Benefits include potential reduction in rendering time by a factor of three, and compression for storing models. *OpenGL*, a triangular-mesh rendering product from Silicon Graphics uses such a specification to render hamiltonian triangulations. In a related work, Evans et. al [ESV] report a program for generating triangular strips from polygonal models. Since interactive display rates are becoming increasingly important in virtual reality and visualization applications, many rendering engines use special-purpose hardware to speed up the rendering process by maintaining a buffer to store k previously transmitted vertices. For polygons with hamiltonian triangulations, the size of the buffer needed is minimum since only the last 2 transmitted vertices need to be stored.

Hence it would be useful to recognize such polygons, especially since the recognition need be done only once, while the rendering may be performed many times. The first step towards the recognition of polygons with hamiltonian triangulations is a characterization of such polygons. Arkin et al. [AHMS] observed

* Supported in part by NSF Grant CCR-940-9752

that a polygon has a hamiltonian triangulation if and only if it is *discretely straight walkable*.

They introduced the concept of *Discretely Straight Walkability*, which is a discrete version of the *Straight Walkability* concept introduced by Icking and Klein [IK]. The original *walkability* concepts grew out of research on the *two-guards problem*, which in turn arose from the study of different versions of *art-gallery problems* (for example see [O87]). Briefly, the two-guards problem involves determining whether two guards can move along the boundary of a simple polygonal room from a common starting point s to a common destination point t while staying mutually visible at all times. Straight walkability deals with a version of the two-guards problem where the guards do not backtrack during their walk. Discrete straight walkability deals with a variant of straight walkability where only one guard moves at any one time while the other guard is stationary at a vertex. Guard problems are important because they contribute to the study of visibility in polygons. It is interesting to note its relationship to hamiltonian triangulations in polygons as well.

Using the connections to discrete straight walkability, Arkin et al. showed an algorithm to recognize polygons with hamiltonian triangulations in time that is linear in the size of the visibility graph of the given polygon P. The size of the visibility graph of P could be quadratic in the size of P and hence their algorithm could be very inefficient even for nearly convex polygons.

In section 4 we show a new characterization of discretely straight walkable polygons. Using this characterization we show more efficient algorithms for many problems related to such polygons. In particular we show (in section 5) the following results:

1. An $O(n \log n)$-time algorithm to recognize polygons with a Hamiltonian triangulation.
2. An $O(n \log n)$-time algorithm to construct such a triangulation.
3. Given vertices p and q on a simple polygon, an $O(n)$-time algorithm to determine whether the polygon is discretely straight walkable with respect to the two vertices.
4. An $O(n \log n)$-time algorithm to list out all pairs of points on a simple polygon with respect to which the polygon is discretely straight walkable.

The algorithms in [AHMS] for the first two problems had a time complexity of $O(|E|)$, where E is the set of edges in a visibility graph of the polygon. The first two results improve on the best-known time complexity for the problems. The last result subsumes the first result quoted above. The reason three of the four algorithms run in $O(n \log n)$ time instead of linear time is that the algorithms compute all the ray shots obtained by extending the edges. There does not seem to be any way of avoiding the computation of possibly redundant ray shots.

As observed in [AHMS], even if a triangulation is hamiltonian, the topology of the triangulation is not necessarily specified by an encoding sequence of vertices. A variant of the concept of hamiltonian triangulations called *Sequential Triangulations* was introduced in [AHMS]. It turns out that even in *OpenGL*, sequential triangulations can be more efficiently specified than hamiltonian triangulations. We show that any hamiltonian triangulation can be converted into a sequential triangulation by the introduction of at most $n/2$ steiner points.

2 Preliminaries

We define notation for this paper; much of our notation is borrowed from [IK, H]. A *polygonal chain* in the plane is a concatenation of line segments. The endpoints of the segments are called *vertices*, and the segments themselves are *edges*. If the segments intersect only at the endpoints of adjacent segments, then the chain is *simple*, and if a polygonal chain is closed we call it a *polygon*. In this paper, we deal only with simple polygons. Two points in the polygon are (mutually) *visible* if the line joining them is contained in the polygon. We assume that the input is in general position, which means that no three vertices are collinear, and no three lines defined by edges intersect in a common point. If x and y are points of P, then $P_{CW}(x, y)$ ($P_{CCW}(x, y)$) is the subchain obtained by traversing P clockwise (counterclockwise) from x to y.

In most of the problems we consider, two vertices of P are specially designated as the start vertex, s, and the goal vertex, t. We refer to the subchains $P_{CW}(s, t)$ and $P_{CCW}(s, t)$ as L and R, respectively; both L and R are oriented from s to t. Points on L (R) are denoted by p, p', p_1, etc. (q, q', q_1, etc.). If p is a vertex of a polygonal chain, $Succ(p)$ represents the vertex of the chain immediately succeeding p, and $Pred(p)$ the vertex preceding p. For $p, p' \in L$, we say that p *precedes* p' (and p' *succeeds* p) if we encounter p before p' when traversing L from s to t. We write $p < p'$.

A *(general) walk* is a pair of continuous functions

$$f : [0, 1] \to L, \quad g : [0, 1] \to R,$$

where $f(0) = g(0) = s$, $f(1) = g(1) = t$, and $f(t)$ and $g(t)$ are co-visible for all t. A walk is *straight* if f and g are monotonic functions; that is, $f(t_1) \le f(t_2)$ and $g(t_1) \le g(t_2)$ whenever $t_1 < t_2$.

For points x and y, $d(x, y)$ is the direction of the directed segment from x to y. The (undirected) segment between x and y is $\overline{xy}$. The ray $\mathbf{r}(x, y)$ is the ray with terminus x in direction $d(x, y)$. The line containing x and y is denoted $\ell(x, y)$, and the directed line from x through y is $\vec{\ell}(x, y)$. An important definition is that of *(ray) shots*: the *backward (ray) shot* (or *hit point*) from a reflex vertex $p \in L$, denoted $Backw(p)$, is the first point of P encountered by a "bullet" shot from p in direction $d(Succ(p), p)$, and the *forward (ray) shot* $Forw(p)$ is the point encountered by the bullet shot from p in direction $d(Pred(p), p)$. We let $d(Backw(p))$ represent the direction of the backward shot from p, i.e. $d(Succ(p), p)$. We define the backward shot $Backw(q)$ and forward shot $Forw(q)$ for a reflex vertex $q \in R$ in similar fashion.

The *visibility graph* of is the graph obtained by representing each of the vertices of the polygon by vertices in the graph and having an edge between two vertices in the graph if the corresponding two vertices in the polygon are visible.

3 Previous Work: Straight Walkability

In order to explain our results, we repeat the two-guards problem studied by Icking and Klein [IK]. Given a simple polygon, P with n vertices and two distinguished vertices s and t, determine if two guards can move from s to t along the two boundary chains while remaining mutually visible at all times. If such a movement is possible then the polygon P is said to be *walkable* with respect to s and t. Furthermore, if it is possible for the two guards to reach the destination t without ever backtracking along the way, then the polygon P is said to be *straight walkable* with respect to s and t. Additionally, if it is possible for the two guards to move from s to t in such a way that when one guard is moving the other guard is stationary at a vertex, then the polygon is said to be *discretely straight walkable* with respect to s and t. We will simply use the terms walkable polygon, straight walkable polygon, or discretely straight walkable polygon to mean that there exists a pair of vertices on the polygon with respect to which it is walkable, straight walkable, or discretely straight walkable, respectively.

A simple polygon is said to be LR-visible with respect to two points on its boundary if the two chains determined by the two points are weakly visible from each other, i.e., every point on one chain is visible from some point on the other. As before, we will simply use the term LR-visible polygon to refer to a simple polygon that is LR-visible with respect to some two points on the polygon.

The main result of Icking and Klein's paper [IK] deals with a precise characterization of straight walkable polygons. They consider two type of forbidden configurations, known as *deadlocks* and *wedges*, which are shown in Fig. 3. Each of the cases is shown consisting of the two polygonal chains L and R and the ray shots going from one chain to the other that causes the forbidden configuration. The first two cases show a deadlock with respect to s and t respectively. The next two cases show a wedge configuration on L and R respectively. Note that in all the four configurations the two ray shots (dotted) must intersect in the interior of the polygon). It is easy to see that an instance of each of the configurations prevents a polygon from being straight walkable; [IK] show that the absence of these configurations is also sufficient to ensure straight walkability of a LR-visible polygon. This result is restated in theorem 1.

Theorem 1 *[IK] A simple polygon P is* straight walkable *with respect to two of its vertices s and t if and only if:*

1. *P is LR-visible with respect to s and t,*
2. *P does not have any deadlocks with respect to either s or t,*
3. *P does not have any wedges on the two boundary chains L and R.*

In the process of developing an algorithm to recognize straight walkable polygons, [IK] developed a considerable collection of definitions, functions and theorems, some of which are summarized here since these are necessary for our work. They define functions lo and hi on the vertices of P. Specifically, the following definition is given for L, with a symmetric definition applying to R. (The operations min and max are defined with respect to the ordering on the chain L, so that min of a set of points is a point of the set not preceded by any other.)

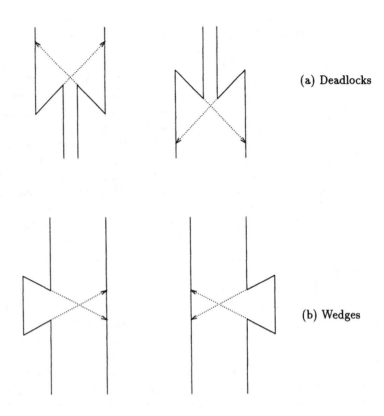

(a) Deadlocks

(b) Wedges

Fig. 1. Forbidden configurations for straight walkable polygons

Definition 1 *[IK] for a vertex $p \in L$, we define:*

$$hiP(p) = \min\{q \mid q \text{ vertex of } R \text{ and } L \ni Backw(q) > p\}$$
$$hiS(p) = \min\{Forw(p') \in R \mid p' \text{ vertex of } L_{>p}\}$$
$$hi(p) = \min\{hiP(p), hiS(p), g\}$$
$$loP(p) = \max\{q \mid q \text{ vertex of } R \text{ and } L \ni Forw(q) < p\}$$
$$loS(p) = \max\{Backw(p') \in R \mid p' \text{ vertex of } L_{<p}\}$$
$$lo(p) = \max\{loP(p), loS(p), s\}$$

We can think of hi and lo as functions from the vertices of L to the points of R, and also from the vertices of R to the points of L. It is important to note that these are monotonic functions. [IK] are able to show that a polygon P is straight walkable if and only if $[lo(v), hi(v)]$ is a non-empty interval for every vertex v. Furthermore, if P is straight walkable, then the set of possible *walk partners* for a vertex v is precisely the points of $[lo(v), hi(v)]$, where two points $p \in L$, $q \in R$ are walk partners in a given straight walk if the two guards are at points p and q at some moment of the walk.

4 Characterization

Theorem 2 gives a precise characterization of discretely straight walkable polygons, i.e., polygons with hamiltonian triangulations. For the characterization, we consider a variant of the *wedge* configuration, which we call as the *semi-wedge* configuration. Fig. 4 shows a semi-wedge formed on the chain L. A similar one for R is not shown. Again, note that the two ray shots (dotted) must either intersect in the interior of the polygon or they must terminate in the interior of the same edge of the polygon. We show that this is a forbidden configuration for discretely straight walkable polygons. We further go on to show that this condition is also sufficient for a straight walkable polygon to be discretely straight walkable.

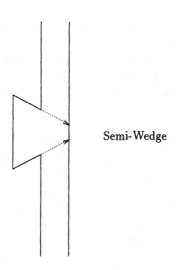

Semi-Wedge

Fig. 2. Forbidden configuration for discretely straight walkable polygons

Hence a stricter version of the wedge condition is enough to ensure discrete straight walkability. An alternate characterization of discretely straight walkable polygons is the following. A polygon P is discretely straight walkable if and only if $[lo(v), hi(v)]$ contains a vertex of the polygon for each vertex v. The proof of theorem 2 also proves this alternate characterization. Compare this characterization to the lemma in [IK], which shows that the polygon P is straight walkable if and only if $[lo(v), hi(v)]$ is nonempty for each vertex v. The proof of theorem 2 also provides ideas for constructing a discrete straight walk (and consequently a hamiltonian triangulation), if one exists. This is because the vertices in the interval $[lo(v), hi(v)]$ provide a set of possible walk partners for the vertex v.

Theorem 2 *A simple polygon P is* discretely straight walkable *with respect to two of its vertices s and t if and only if:*

1. *P is straight walkable with respect to s and t,*
2. *P does not have any* semi-wedges *on the two boundary chains L and R.*

Proof. It is clear that an instance of a semi-wedge prevents a polygon from being discretely straight walkable, since a vertex inside the semi-wedge would not have a walk partner.

As a first step to prove sufficiency, assume that vertex v is such that $[lo(v), hi(v)]$ is non-empty, but devoid of vertices. This implies that $lo(v)$ and $hi(v)$ are not vertices either, and that they lie on a common edge of the polygon. Furthermore, this also implies $lo(v) = loS(v)$ and $hi(v) = hiS(v)$, since loP and hiP take only vertex values. But this is precisely a semi-wedge configuration, and hence the contradiction.

The proof will be complete if we can show that discrete straight walkability is ensured if for every vertex v, $[lo(v), hi(v)]$ has at least one vertex of P. Assume that it does not. This can only happen if for adjacent vertices v_i, v_{i+1} on L (the argument is symmetric for R), $[lo(v_i), hi(v_i)] \cap [lo(v_{i+1}), hi(v_{i+1})]$ is devoid of vertices, since in this case, the two guards cannot complete the discrete straight walk.

We define the following two functions on the vertices of L and R:
$$loV(v) = \min\{q \mid q \text{ vertex from } [lo(v), hi(v)]\}$$
$$hiV(v) = \max\{q \mid q \text{ vertex from } [lo(v), hi(v)]\}$$
Note that the values of loV and hiV must exist for both the vertices v_i and v_{i+1} by our assumption. Furthermore, since lo and hi are monotonic functions, so are loV and hiV. It was also shown in [IK] that $x \in [lo(y), hi(y)]$ if and only if $y \in [lo(x), hi(x)]$. We now argue that if $hiV(v_i) = x_j$ then $loV(v_{i+1}) = x_k$, then $k = j + 1$, since otherwise there would be a vertex between x_j and x_k, say x_l whose interval $[lo(x_l), hi(x_l)]$ would be devoid of vertices. Since every vertex is visible from its interval $[lo(v), hi(v)]$, we infer the following facts: v_i is visible from x_j, v_{i+1} is visible from x_{j+1}. Since v_i must be visible from v_{i+1} just as x_j is from x_{j+1}, either v_i or v_{i+1} must be visible from the entire edge joining x_j and x_{j+1}. This contradicts the assumption for the values of either $hiV(v_i)$ or $loV(v_{i+1})$.

The proof of the above theorem gives us a way for constructing the discrete straight walk (and consequently the hamiltonian triangulation), if one exists. The construction is described in the next section.

5 Algorithmic Applications

The previous section gives us a characterization for discretely straight walkable polygons. In this section we present efficient algorithms for several variants of the discretely straight walkability problem. Due to space constraints, we only give brief sketches in each case.

5.1 Walkability with respect to two given points

We first consider the problem of determining whether a given polygon P is discretely straight walkable with respect to two given points. Heffernan [H] presents a surprising linear-time algorithm for finding hi and lo values by ignoring redundant ray shots and computing only relevant information. The search structure used in his algorithm makes it more efficient. We use Heffernan's algorithm and compute the hi and lo values for each vertex. The only work left to check for discrete straight walkability with respect to the given s and t is to check whether any of the output intervals is devoid of vertices. Clearly this can be achieved in optimal linear time.

5.2 Recognizing Polygons with hamiltonian triangulations

This algorithm is similar to the one by Tseng and Lee [TL] to determine whether a polygon is straight walkable. The basic idea is to enumerate all deadlocks and all semi-wedges and all LR-visible pairs of the polygon P. All three can be computed in $O(n \log n)$ by simply computing all ray shots and scanning all the information after sorting. The scanning will help determine regions where the the points s and/or t cannot lie. If there is a portion that is not eliminated in the scanning, then clearly there exists a pair of points with respect to which the polygon has a hamiltonian triangulation. Even though the scanning is non-trivial it is straight-forward after sorting all the vertices and the ray shots.

5.3 Constructing a hamiltonian triangulation

The strategy for doing this is as follows. Theorem 2 shows that if P is discretely straight walkable (i.e., it has a hamiltonian triangulation), then loV and hiV values exist for each vertex of the chains L and R and that the intervals $[loV(v), hiV(v)]$ intersect for adjacent vertices on each of the two chains. Now the simple strategy for constructing the discrete straight walk is going to be: the guards move alternately (with guard one moving first) in such a way that when the stationary guard is at vertex v, the other guard moves until the vertex $hiV(v)$ is reached.

It is an easy step to now generate the triangulation. Every time the mobile guard moves to the next vertex, the edge connecting the stationary guard to the new position of the mobile guard is an edge of the triangulation. It also gives the next triangle visited by a hamiltonian path in the dual of the triangulation.

Clearly, this algorithm runs in optimal linear time, once the points s and t are known.

5.4 All-pairs discrete straight walkability

Several papers deal with problems of this nature [DHN, DHN2, TL]. Given a simple polygon this algorithm will output all pairs of points with respect to which the polygon is discrete straight walkable. As in the references mentioned, the output will be in the form of pairs of intervals. This algorithm follows the all-pairs straight walkability problem; the modification needed is to generate all semi-wedges, instead of all wedges. The extra processing that is needed is to inspect all ray shots that end on an edge and to identify all the semi-wedges that are not wedges. Details are omitted once again. Like the algorithm in [TL], this algorithm takes $O(n \log n)$ time.

6 Sequential Triangulations and Steiner Points

A hamiltonian triangulation is not a sequential triangulation when the hamiltonian path in its dual involves two consecutive *left* (or *right*) turns. Informally, a left (resp. right) turn in a hamiltonian path is easily understood by imagining the path as stepping on the triangles and then the next step must involve either a left or a right turn. A sequential triangulation is constrained to use a sequence of alternate left and right turns.

Given a triangulation, introducing a steiner point and retriangulating the triangle in which it is contained, changes the sequence of left and right turns in the associated hamiltonian. We state without proof that a hamiltonian triangulation can be transformed into a sequential triangulation by introducing at most $n/2$ steiner points and retriangulating the triangles. This is optimal in the sense that there are polygons with hamiltonian triangulations which need $n/2$ steiner points.

7 Open Problems

It is known [AHMS] that given any set of points in the plane there exists a hamiltonian triangulation of these points and that such a triangulation can be computed efficiently. The most interesting related problem that remains open is the following [AHMS]: given a set of points in 3-dimensional space, does there exist a *hamiltonian tetrahedralization* of the points? This problem is open even if the set of points are in convex, general position. Another variant of this problem is to consider hamiltonian triangulations in the skeleton of a tetrahedralization of the set of points. We can show that when the points are in convex, general position in 3-d, both the variants are equivalent. This problem seems to be related to variants of the *unfolding* problem for polytopes.

Other interesting problems relate to discrete versions of LR-visibility and their possible connections to discrete straight walkability.

References

[AHMS] E.M. Arkin, M. Held, J.S.B. Mitchell, S.S. Skienna, "Hamiltonian Triangulations for Fast Rendering," *Proc. of the 2nd ESA*, 1994.

[DHN] G. Das, P. J. Heffernan and G. Narasimhan, "LR-visibility in polygons," *Computational Geometry - Theory and Appln.*, 5(1-2):37-57, (1997).

[DHN2] G. Das, P. J. Heffernan and G. Narasimhan, "Finding all weakly-visible chords of a polygon in linear time," *Nordic Journal of Computing,* **1** (1994), 433-457.

[ESV] F. Evans, S.S. Skienna, A. Varshney, "Optimizing Triangle Strips for Fast Rendering," *Proc. of VISUALIZATION,* 1996.

[H] P. J. Heffernan, "An optimal algorithm for the two-guard problem," *Proc. 9th Annual ACM Symp. on Computational Geometry,* (1993), pp. 348-358.

[IK] C. Icking and R. Klein, "The two guards problem," *Proc. 7th Annual ACM Symp. on Computational Geometry,* 1991, pp. 166-175.

[O87] J. O'Rourke, "Art Gallery Theorems and Algorithms," Oxford University Press, 1987.

[TL] L. H. Tseng and D. T. Lee, "Two-guard walkability of simple polygons," manuscript, 1993.

Computing Orthogonal Drawings with the Minimum Number of Bends*

Paola Bertolazzi[†]   *Giuseppe Di Battista*[‡]

bertola@iasi.rm.cnr.it   dibattista@iasi.rm.cnr.it

Walter Didimo[§]

didimo@inf.uniroma3.it

Abstract

We describe a branch-and-bound algorithm for computing an orthogonal grid drawing with the minimum number of bends of a biconnected planar graph. Such algorithm is based on an efficient enumeration schema of the embeddings of a planar graph and on several new methods for computing lower bounds of the number of bends. We experiment such algorithm on a large test suite and compare the results with the state-of-the-art. The experiments show how minimizing the number of bends strongly improves several quality measures of the effectiveness of the drawing. We also present a graphic tool with animation that embodies the algorithm and allows interacting with all the phases of the computation.

1 Introduction

Various graphic standards have been proposed to draw graphs, each one devoted to a specific class of applications. An extensive literature on the subject can be found in [4, 23, 2]. In particular, an *orthogonal drawing* maps each edge into a chain of horizontal and vertical segments and an *orthogonal grid drawing* is an orthogonal drawing such that vertices and bends along the edges have integer coordinates. Orthogonal grid drawings are widely used for graph visualization in many applications including database systems (entity-relationship diagrams), software engineering (data-flow diagrams), and circuit design (circuit schematics).

---

*Research supported in part by the ESPRIT LTR Project no. 20244 - ALCOM-IT.

[†]IASI, CNR, viale Manzoni 30, 00185 Roma Italy.

[‡]Dipartimento di Informatica e Automazione, Università di Roma Tre, via della Vasca Navale 84, 00146 Roma, Italy.

[§]Dipartimento di Informatica e Automazione, Università di Roma Tre, via della Vasca Navale 84, 00146 Roma, Italy.

Many algorithms for constructing orthogonal grid drawings have been proposed in the literature and implemented into industrial tools. They can be roughly classified according to two main approaches: the *topology-shape-metrics* approach determines the final drawing through an intermediate step in which a planar embedding of the graph is constructed (see, e.g., [21, 19, 22, 20]); the *draw-and-adjust* approach reaches the final drawing by working directly on its geometry (see, e.g., [1, 18]). Since a planar graph has an orthogonal grid drawing iff its vertices have degree at most 4, both the approaches assume that vertices have degree at most 4. Such a limitation is usually removed by "expanding" higher degree vertices into two or more vertices. Examples of expansion techniques can be found in [20].

In the topology-shape-metrics approach, the drawing is incrementally specified in three phases. The first phase, *planarization*, determines a planar embedding of the graph, by possibly adding ficticious vertices that represent crossings. The second phase, *orthogonalization*, receives as input a planar embedding and computes an orthogonal drawing. The third phase, *compaction*, produces the final orthogonal grid drawing trying to minimize the area.

The orthogonalization step is crucial for the effectiveness of the drawing and has been extensively investigated. A very elegant $O(n^2 \log n)$ time algorithm for constructing an orthogonal drawing with the minimum number of bends of an n-vertex embedded planar graph has been presented by Tamassia in [19]. It is based on a minimum cost network flow problem that considers bends along edges as units of flow.

However, the algorithm in [19] minimizes the number of bends only within the given planar embedding. Observe that a planar graph can have an exponential number of planar embeddings and it has been shown [6] that the choice of the embedding can deeply affect the number of bends of the drawing. Namely, there exist graphs that, for a certain embedding have a linear number of bends, and for another embedding have only a constant number. Unfortunately, the problem of minimizing the number of bends in a variable embedding setting is NP-complete [10, 11]. Optimal polynomial-time algorithms for subclasses of graphs are shown in [6].

Because of the tight interaction between the graph drawing area and applications, the attention to experimental work on graph drawing is rapidly increasing. In [5] it is presented an experimental study comparing topology-shape-metrics and draw-and-adjust algorithms for orthogonal grid drawings. The test graphs were generated from a core set of 112 graphs used in "real-life" software engineering and database applications with number of vertices ranging from 10 to 100. The experiments provide a detailed quantitative evaluation of the performance of seven algorithms, and show that they exhibit trade-offs between "aesthetic" properties (e.g., crossings, bends, edge length) and running time. The algorithm GIOTTO (topology-shape-metrics with the Tamassia's algorithm in the orthogonalization step) performs better than the others at the expenses of a worst time performance.

Other examples of experimental work in graph drawing follow. The performance of four planar straight-line drawing algorithms is compared in [14]. Himsolt [12] presents a comparative study of twelve graph drawings algorithms; the algorithms selected are based on various approaches (e.g., force-directed, layering, and planarization) and use a variety of graphic standards (e.g., orthogonal, straight-line, polyline). The experiments are conducted with the graph drawing system GraphEd [12]. The test suite consists of about 100 graphs. Brandenburg and Rohrer [3] compare five "force-directed" methods for constructing straight-line drawings of general undirected graphs. Juenger and Mutzel [15] investigate crossing minization strategies for straight-line drawings of 2-layer graphs, and compare the performance of popular heuristics for this problem.

In this paper, we present the following results. Let G be a biconnected planar graph such that each vertex has degree at most 4 (4-*planar graph*).

- We describe a branch-and-bound algorithm called BB-ORTH that computes an orthogonal grid drawing of G with the minimum number of bends in the variable embedding setting. The algorithm is based on: several new methods for computing lower bounds on the number of bends of a planar graph (Section 3). Such methods give new insights on the relationships between the structure of the triconnected components and the number of bends; a new enumeration schema that allows to enumerate without repetitions all the planar embeddings of G (Section 4). Such enumeration schema exploits the capability of SPQR-trees [7, 8] in implicitly representing the embeddings of a planar graph.

- We present a system that implements BB-ORTH. Such a system is provided with a graphical interface that animates all the phases of the algorithm and displays partial results, exploiting existing graph drawing algorithms to represent all the graphs involved in the computation. The interaction with the system allows to stop the computation once a sufficiently good orthogonal drawing is displayed. (Section 5).

- We test BB-ORTH against a large test suite of randomly generated graphs with up to 60 vertices and compare the experimental results with the best state-of-the-art results (algoritm GIOTTO) (Section 6). Our experiments show: an improvement in the number of bends of 20-30%; an improvement of several other quality measures of the drawing that are affected from the number of bends. For example the length of the longest edge of the drawings obtained with BB-ORTH is about 50% smaller than the length of the longest edge of the drawings obtained with GIOTTO; a sensible increasing of the CPU-time that however is perfectly affordable within the typical size of graphs of real-life applications [5].

- Also, BB-ORTH can be easily applied on all biconnected components of graphs. This yields a powerful heuristic for reducing the number of bends in connected graphs. Further, the limitation on the degree of the vertices can be easily removed by using the expansion techniques cited above.

2 Preliminaries

We assume familiarity with planarity and connectivity of graphs [9, 17]. Since we consider only planar graphs, we use the term *embedding* instead of *planar embedding*.

An orthogonal drawing of a 4-planar graph G is *optimal* if it has the minimum number of bends among all the possible orthogonal drawings of G. When this is not ambiguous, given an embedding ϕ of G, we also call *optimal* the porthogonal drawing of G with the minimum number of bends that preserves the embedding ϕ.

Let G be a biconnected graph. A *split pair* of G is either a separation-pair or a pair of adjacent vertices. A *split component* of a split pair $\{u, v\}$ is either an edge (u, v) or a maximal subgraph C of G such that C contains u and v, and $\{u, v\}$ is not a split pair of C. A vertex w distinct from u and v belongs to exactly one split component of $\{u, v\}$.

Suppose $G_1, \ldots, G_k$ are some pairwise edge disjoint split components of G with split pairs $u_1, v_1 \ldots u_k, v_k$, respectively. The graph G' obtained by substituting each of $G_1, \ldots, G_k$ with *virtual edges* $e_1 = (u_1, v_1), \ldots, e_k = (u_k, v_k)$ is a *partial graph* of G. We denote E^{virt} ($E^{nonvirt}$) the set of (non-)virtual edges of G'. We say that G_i is the *pertinent graph* of $e_i = (u_i, v_i)$ and that $e_i = (u_i, v_i)$ is the *representative edge* of G_i.

Let ϕ be an embedding of G and let ϕ' be an embedding of G'. We say that ϕ *preserves* ϕ' if $G'_{\phi'}$ can be obtained from G_ϕ by substituting each component G_i with its representative edge.

In the following we summarize SPQR-trees. For more details, see [7, 8]. SPQR-trees are closely related to the classical decomposition of biconnected graphs into triconnected components [13].

Let $\{s, t\}$ be a split pair of G. A *maximal split pair* $\{u, v\}$ of G with respect to $\{s, t\}$ is a split pair of G distinct from $\{s, t\}$ such that for any other split pair $\{u', v'\}$ of G, there exists a split component of $\{u', v'\}$ containing vertices u, v, s, and t.

Let $e = (s, t)$ be an edge of G, called *reference edge*. The *SPQR-tree* $\mathcal{T}$ of G with respect to e describes a recursive decomposition of G induced by its split pairs. Tree $\mathcal{T}$ is a rooted ordered tree whose nodes are of four types: S, P, Q, and R. Each node μ of $\mathcal{T}$ has an associated biconnected multigraph, called the *skeleton* of μ, and denoted by *skeleton*(μ). Also, it is associated with an edge of the skeleton of the parent ν of μ, called the *virtual edge* of μ in *skeleton*(ν). Tree $\mathcal{T}$ is recursively defined as follows.

If G consists of exactly two parallel edges between s and t, then $\mathcal{T}$ consists of a single Q-node whose skeleton is G itself.

If the split pair $\{s, t\}$ has at least three split components $G_1, \cdots, G_k$ ($k \geq 3$), the root of $\mathcal{T}$ is a P-node μ. Graph *skeleton*(μ) consists of k parallel edges between s and t, denoted $e_1, \cdots, e_k$, with $e_1 = e$.

Otherwise, the split pair $\{s, t\}$ has exactly two split components, one of them is the reference edge e, and we denote with G' the other split component. If G' has cutvertices $c_1, \cdots, c_{k-1}$ ($k \geq 2$) that partition G into its blocks $G_1, \cdots, G_k$, in this order from s to t, the root of $\mathcal{T}$ is an S-node μ. Graph $skeleton(\mu)$ is the cycle $e_0, e_1, \cdots, e_k$, where $e_0 = e$, $c_0 = s$, $c_k = t$, and e_i connects c_{i-1} with c_i ($i = 1 \cdots k$).

If none of the above cases applies, let $\{s_1, t_1\}, \cdots, \{s_k, t_k\}$ be the maximal split pairs of G with respect to $\{s, t\}$ ($k \geq 1$), and for $i = 1, \cdots, k$, let G_i be the union of all the split components of $\{s_i, t_i\}$ but the one containing the reference edge e. The root of $\mathcal{T}$ is an R-node μ. Graph $skeleton(\mu)$ is obtained from G by replacing each subgraph G_i with the edge e_i between s_i and t_i.

Except for the trivial case, μ has children $\mu_1, \cdots, \mu_k$ in this order, such that μ_i is the root of the SPQR-tree of graph $G_i \cup e_i$ with respect to reference edge e_i ($i = 1, \cdots, k$). Edge e_i is said to be the *virtual edge* of node μ_i in $skeleton(\mu)$ and of node μ in $skeleton(\mu_i)$. Graph G_i is called the *pertinent graph* of node μ_i, and of edge e_i.

The tree $\mathcal{T}$ so obtained has a Q-node associated with each edge of G, except the reference edge e. We complete the SPQR-tree by adding another Q-node, representing the reference edge e, and making it the parent of μ so that it becomes the root.

Let μ be a node of $\mathcal{T}$. We have: if μ is an R-node, then $skeleton(\mu)$ is a triconnected graph; if μ is an S-node, then $skeleton(\mu)$ is a cycle; if μ is a P-node, then $skeleton(\mu)$ is a triconnected multigraph consisting of a bundle of multiple edges; and if μ is a Q-node, then $skeleton(\mu)$ is a biconnected multigraph consisting of two multiple edges.

The skeletons of the nodes of $\mathcal{T}$ are homeomorphic to subgraphs of G. The SPQR-trees of G with respect to different reference edges are isomorphic and are obtained one from the other by selecting a different Q-node as the root. Hence, we can define the *unrooted SPQR-tree* of G without ambiguity.

The SPQR-tree $\mathcal{T}$ of a graph G with n vertices and m edges has m Q-nodes and $O(n)$ S-, P-, and R-nodes. Also, the total number of vertices of the skeletons stored at the nodes of $\mathcal{T}$ is $O(n)$.

A graph G is planar if and only if the skeletons of all the nodes of the SPQR-tree $\mathcal{T}$ of G are planar. An SPQR-tree $\mathcal{T}$ rooted at a given Q-node represents all the planar embeddings of G having the reference edge (associated to the Q-node at the root) on the external face.

3 Lower Bounds for Orthogonal Drawings

In this section we propose some new lower bounds on the number of bends of a biconnected planar graph. The proofs of the theorems are omitted due to space limitation.

Let $G = (V, E)$ be a biconnected 4-planar graph and H be an orthogonal drawing of G, we denote by $b(H)$ the total number of bends of H and by $b_{E'}(H)$ the number of bends along the edges of E' ($E' \subseteq E$).

Property 1 *Let* $G_i = (V_i, E_i)$, $i = 1, \ldots, k$ *be* k *subgraphs of* G *such that* $E_i \cap E_j = \emptyset$ $i \neq j$, *and* $\cup_{i=1,\ldots,k} E_i = E$. *Let* H_i *be an optimal orthogonal drawing of* G_i *and let* H *be an orthogonal drawing of* G. *Then,* $b(H) \geq \sum_{i=1,\ldots,k} b(H_i)$.

Let $G'_{\phi'}$ be an embedded partial graph of G and let $H'_{\phi'}$ be an orthogonal drawing of $G'_{\phi'}$. Suppose $H'_{\phi'}$ is such that $b_{E^{nonvirt}}(H'_{\phi'})$ is minimum. Consider an embedding ϕ of G that preserves ϕ' and an optimal orthogonal drawing H_ϕ of G_ϕ.

Lemma 1

$$b_{E^{nonvirt}}(H'_{\phi'}) \le b_{E^{nonvirt}}(H_\phi)$$

From Property 1 and Lemma 1 it follows a first lower bound.

Theorem 1 *Let* $G = (V, E)$ *be a biconnected 4-planar graph and* $G'_{\phi'}$ *an embedded partial graph of* G. *For each virtual edge* e_i *of* G', $i = 1, \ldots, k$, *let* b_i *be a lower bound on the number of bends of the pertinent graph* G_i *of* e_i. *Consider an orthogonal drawing* $H'_{\phi'}$ *of* $G'_{\phi'}$ *such that* $b_{E^{nonvirt}}(H'_{\phi'})$ *is minimum. Let* ϕ *be an embedding of* G *that preserves* ϕ', *consider any orthogonal drawing* H_ϕ *of* G_ϕ. *We have that:*

$$b(H_\phi) \ge b_{E^{nonvirt}}(H'_{\phi'}) + \sum_{i=1,\ldots,k} b_i.$$

Remark 1 *An orthogonal drawing* $H'_{\phi'}$ *of* $G'_{\phi'}$ *such that* $b_{E^{nonvirt}}(H'_{\phi'})$ *is minimum can be easily obtained by using the Tamassia's algorithm [19]. Namely, when two faces* f *and* g *share a virtual edge, the corresponding edge of the dual graph in the minimum cost flow problem of [19] is set to zero.*

A second lower bound is described in the following theorem.

Theorem 2 *Let* G_ϕ *be an embedded biconnected 4-planar graph and* $G'_{\phi'}$ *an embedded partial graph of* G, *such that* ϕ *preserves* ϕ'. *Derive from* $G'_{\phi'}$ *the embedded graph* $\overline{G}_{\phi'}$ *by substituting each virtual edge* e_i *of* G', $i = 1, \ldots, k$, *with any simple path from* u_i *to* v_i *in* G_i. *Consider an optimal orthogonal drawing* $\overline{H}_{\phi'}$ *of* $\overline{G}_{\phi'}$ *and an orthogonal drawing* H_ϕ *of* G_ϕ. *Then we have that:*

$$b(H_\phi) \ge b(\overline{H}_{\phi'})$$

The next corollary allows us to combine the above lower bounds into a hybrid technique.

Corollary 1 *Let* G_ϕ *be an embedded biconnected 4-planar graph and* $G'_{\phi'}$ *an embedded partial graph of* G. *Consider a subset* F^{virt} *of the set of the virtual edges of* $G'_{\phi'}$ *and derive from* $G'_{\phi'}$ *the graph* $G''_{\phi'}$ *by substituting each edge* $e_i \in F^{virt}$ *with any path from* u_i *to* v_i *in* G_i. *Denote by* E_p *the set of edges of* G *introduced by such substitution. For each* $e_j \in E^{virt} - F^{virt}$ *let* b_j *be a lower bound on the number of bends of the pertinent graph* G_j *of* e_j. *Consider an orthogonal drawing* $H''_{\phi'}$ *of* $G''_{\phi'}$, *such that* $b_{E^{nonvirt}}(H''_{\phi'}) + b_{E_p}(H''_{\phi'})$ *is minimum. Let* H_ϕ *be an orthogonal drawing of* G_ϕ, *we have that:*

$$b(H_\phi) \ge b_{E^{nonvirt}}(H''_{\phi'}) + b_{E_p}(H''_{\phi'}) + \sum_{e_j \in E^{virt} - F^{virt}} b_j$$

4 A Branch and Bound Strategy

Let G be a biconnected 4-planar graph. In this section we describe a technique for enumerating all the possible orthogonal drawings of G and rules to avoid examining all of them in the computation of the optimum.

The enumeration exploits the SPQR-tree $\mathcal{T}$ of G. Namely, we enumerate all the orthogonal drawings of G with edge e on the external face by rooting $\mathcal{T}$ at e and exploiting the capacity of $\mathcal{T}$ rooted at e in representing all the embeddings having e on the external face. A complete enumeration is done by rooting $\mathcal{T}$ at all the possible edges.

We encode all the possible embeddings of G, implicitly represented by $\mathcal{T}$ rooted at e, as follows.

We visit the SPQR-tree such that a node is visited after its parent, e.g. depth first or breadth first. This induces a numbering $1, \ldots, r$ of the P- and R-nodes of $\mathcal{T}$. We define a r-uple of variables $X = x_1, \ldots, x_r$ that are in one-to-one correspondence with the P- and R-nodes $\mu_1, \ldots, \mu_r$ of $\mathcal{T}$. Each variable x_i of X corresponding to a R-node μ_i can be set to three values corresponding to two swaps of the pertinent graph of μ_i plus one unknown value. Each variable x_j of X corresponding to a P-node μ_j can be set to up to seven values corresponding to the possible permutations of the pertinent graphs of the children of μ_j plus one unknown value. Unknown values represent portions of the embedding that are not yet specified.

A *search tree* $\mathcal{B}$ is defined as follows. Each node β of $\mathcal{B}$ corresponds to a different setting X_β of X. Such setting is partitioned into two contiguous (one of them possibly empty) subsets $x_1, \ldots, x_h$ and $x_{h+1}, \ldots, x_r$. Elements of the first subset contain values specifying embeddings while elements in the second subset contain unknown values. Leaves of $\mathcal{B}$ are in correspondence with settings of X without unknown values. Internal nodes of $\mathcal{B}$ are in correspondence with settings of X with at least one unknown value. The setting of the root of $\mathcal{B}$ consists of unknown values only. The children of β (with subsets $x_1, \ldots, x_h$ and $x_{h+1}, \ldots, x_r$) have subsets $x_1, \ldots, x_{h+1}$ and $x_{h+2}, \ldots, x_r$, a child for each possible value of x_{h+1}.

Observe that there is a mapping between embedded partial graphs of G and nodes of $\mathcal{B}$. Namely, the embedded partial graph G_β of G associated to node β of $\mathcal{B}$ with subsets $x_1, \ldots, x_h$ and $x_{h+1}, \ldots, x_r$ is obtained as follows. First, set G_β to $skeleton(\mu_1)$ embedded according to x_1. Second, substitute each virtual edge e_i of μ_1 with the skeleton of the child μ_i of μ_1, embedded according to x_i, only for $2 \leq i \leq h$. Then, recursively substitute virtual edges with embedded skeletons until all the skeletons in $\{skeleton(\mu_1), \ldots, skeleton(\mu_h)\}$ are used.

We visit $\mathcal{B}$ breadth-first starting from the root. At each node β of $\mathcal{B}$ with setting X_β we compute a lower bound and an upper bound of the number of bends of any orthogonal drawing of G such that its embedding is (partially) specified by X_β. The current optimal solution is updated accordingly. Children are not visited if the lower bound is greater than the current optimum.

For each β, lower bounds and upper bounds are computed as follows by using the results presented in Section 3.

1. We construct G_β using an array of pointers to the nodes of $\mathcal{T}$. Observe that at each substitution of a virtual edge with a skeleton a reverse of the adjacency lists might be needed.

2. We compute lower bounds. G_β is a partial graph of G, with embedding derived from X_β. Let E^{virt} be the set of virtual edges of G_β. For each $e_i \in E^{virt}$ consider the pertinent graph G_i and compute a lower bound b_i on the number of bends of G_i using Theorem 1. For each e_i such that the value of such lower bound is zero we substitute in G_β edge e_i with any simple path from u_i to v_i in G_i, so deriving a new graph G'_β. Denote by F^{virt} the set of such edges and by E_p the set of edges of G introduced by the substitution (observe that E_p may be empty). By Remark 1 we apply Tamassia's algorithm on G'_β, assigning zero costs to the edges of the dual graph associated to the virtual edges. We obtain an orthogonal drawing H'_β of G'_β with minimum number of bends on the set $E^{nonvirt} \cup E_p$. Let $b(H'_\beta)$ be such a number of bends. Then, by Corollary 1 we compute the lower bound L_β at node β, as

$$L_\beta = b(H'_\beta) + \sum_{e_i \in E^{virt} - F^{virt}} b_i$$

Lower bounds (b_i) can be pre-computed with a suitable pre-processing by visiting $\mathcal{T}$ bottom-up. The pre-computation consists of two phases. We apply Tamassia's algorithm on the skeleton of each R- and P-node, with zero cost for virtual edges; in this way we associate a lower bound to each R- and P-node of $\mathcal{T}$. Note that these pre-computed bounds do not depend on the choice of the reference edge, so they are computed only once, at the beginning of BB-ORTH. We visit $\mathcal{T}$ bottom-up summing for each R- and P-node μ the lower bounds of the children of μ to the lower bound of μ. Note that these pre-computed bounds depend on the choice of the reference edge, so they are re-computed at any choice of the reference edge.

3. We compute upper bounds. Namely, we consider the embedded partial graph G_β and complete it to a pertinent embedded graph G_ϕ. The embedding of G_ϕ is obtained by substituting the unknown values of X_β with embedding values in a random way. Then we apply Tamassia's algorithm to G_ϕ so obtaining the upper bound. We also avoid multiple generations of the same embedded graph in completing the partial graph.

We compute the optimal solution over all possible choices of the reference edge. To do so, our implementation of SPQR-trees supports efficient evert operations. Also, to expedite the process we: reuse upper bounds computed during computations referring to different choices of the virtual edge; avoid to compute solutions referring to embeddings that have been already explored. If an edge e that has already been reference edge appears on the external face of some G_β, since we have already explored all the embeddings with e on the external face, we cut from the search tree all the descendants of β.

One of the consequences of the above discussion is summarized in the following theorem:

Theorem 3 *Let G be a biconnected planar graph. The enumeration schema adopted by* BB-ORTH *examines each planar embedding of G exactly once.*

5   Implementation

The system has been developed with the C++ language and has full Leda [16] compatibility. Namely, several classes of Leda have been refined into new classes: *Embedded Graphs*, preserve the embedding for each update operation. We have taken care of a new efficient implementation of several basic methods like DFS, BFS, *st*-numbering, topological sorting, etc; *Directed Embedded Graphs*, with several network flow facilities. In particular we implemented the minimum cost flow algorithm that searches for minimum cost augmenting paths with the Dijkstra algorithm, used in the original paper of Tamassia [19]. We implemented the corresponding priority queues with the Fibonacci-heaps of Leda; *Planar Embedded Graphs*, with faces and dual graph; *Orthogonal planar embedded graphs*, to describe orthogonal drawings; *Drawn Graphs*, with several graphical attributes; *SPQR-trees*, with several supporting classes, like split-components, skeletons, etc. As far as we know, this is the first implementation of the SPQR-trees. Further, to support experiments, we have developed a new Leda graph editor, based on the previous Leda editor. The graphical interface shows an animation of the algorithm in which the SPQR-tree is rooted at different reference edges. Also, it displays the relationships between the SPQR-tree and the search tree and the evolution of the search tree. Clicking at SPQR-tree nodes shows skeletons of nodes. Of course, to show the animation, the graphical interface exploits several graph drawing algorithms.

Fig. 1 shows a 17-verticed graph drawn by the GIOTTO, and by BB-ORTH, respectively. Fig. 2 shows a 100-vertices graph drawn by the GIOTTO, and by BB-ORTH, respectively.

Finally, we observe that a careful cases analysis of each P-node μ allows faster computations of preprocessing lower bounds; such analysis is based on the number of virtual edges of the skeleton of μ, on the values of the precomputed lower bounds of the children of μ and on length of the shortest paths in the pertinent graphs of the children of μ.

6   Experimental Setting and Computational Results

We have tested our algorithm against two randomly generated test suites *T1* and *T2* overall consisting of about 200 graphs. Such graphs are available on the web (www.inf.uniroma3.it/people/gdb/wp12/LOG.html) and have been generated as follows. It is well known that any embedded planar biconnected graph can be generated from the triangle graph by means of a sequence of *insert-vertex* and *insert-edge* operations. Insert-vertex subdivides an existing edge into two new edges separated by a new vertex. Insert-edge inserts a new edge between two existing vertices that are on the same face. Thus, we have implemented a generation mechanism that at each step randomly chooses which operation to apply and where to apply it. The two test suites differ in the probability distributions we have given for insert-vertex and insert-edge. Also, to stress the performance of the algorithm we have discarded graphs with "a small number" of triconnected components. Observe that we decided to define a new test suite instead of using the one of [5] because the biconnected components of those graphs do not have a sufficiently large number of triconnected components (and thus a sufficient number of different embeddings) to stress enough BB-ORTH.

For each drawing obtained by the algorithm, we have measured the following quality parameters. *bends*: total number of edge bends. *maxedgebends*: max number of bends on an edge. *unifbends*: standard deviation of the number of edge bends. *area*: area of the smallest rectangle with horizontal and vertical sides covering the drawing. *totaledgelen*: total edge length. *maxedgelen*: maximum length of an edge. *uniflen*: standard deviation of the edge length. *screenratio*: deviation of the aspect ratio of the drawing (width/height or height/width) from the optimal one (4/3). Measured performance parameters are: cpu time, total number of search tree nodes, number of visited search tree nodes.

Due to space limitations, we show the experimental results with respect to *T1* only. They are summarized in the graphics of Fig. 3

On the x-axis we either have the number of vertices or the total number (RP) of R- and P-nodes. All values in the graphics represent average values.

Fig. 3.A, 3.C and 3.D compare respectively the number of bends, the area and the maxedgelen of the drawing of BB-ORTH with the ones of GIOTTO. Graphics with white points represent the behavior of BB-ORTH. Graphics with black points represent the behavior of GIOTTO. The improvement on the number of bends positively affects all the quality measures. In Fig. 3.B, 3.D and 3.F we show the percentages of the improvements.

Concerning performance, Fig. 3.G shows the CPU-time of BB-ORTH on a RISC6000 IBM. The algorithm can take about one hour on graphs with 60 vertices. Finally, Fig. 3.H compares the number of nodes of the search tree and the number of nodes that are actually visited. It shows that for high values of RP the percentage of visited nodes is dramatically low (below 10 %).

7 Conclusions and Open Problems

Most graph drawing problems involve the solution of optimization problems that are computationally hard. Thus, for several years the resarch in graph drawing has focused on the development of efficient heuristics. However, the graphs to be drawn are frequently of small size (it is unusual to draw an Entity-Relationship or Data Flow diagrams with more than 70-80 vertices) and the available workstations allow faster and faster computation. Also, the requirements of the users in terms of aesthetic features of the interfaces is always growing.

Hence, recently there has been an increasing attention towards the development of graph drawing tools that privilege the effectiveness of the drawing against the efficiency of the computation (see e.g., [15]).

In this paper we presented an algorithm for computing an orthogonal drawing with the minimum number of bends of a graph. The computational results that we obtained are encouraging both in terms of the quality of the drawings and in terms of the time performance.

Several problems are still open. For example: is it possible to apply variations of the techniques presented in this paper to solve the *upward drawability testing* problem? A directed graph is upward planar if it is planar and can be drawn with all the edges following the same direction. The problem to test a directed graph for upward drawability is NP-complete [10]; how difficult is it to find an enumeration schema and lower bounds for the problem of computing the orthogonal drawing in 3-space with the minimum number of bends?

Acknowledgements

We are grateful to Antonio Leonforte for implementing part of the system. We are also grateful to Sandra Follaro, Armando Parise, and Maurizio Patrignani for their support.

References

[1] T. Biedl and G. Kant. A better heuristic for orthogonal graph drawings. In *Proc. 2nd Annu. European Sympos. Algorithms (ESA '94)*, volume 855 of *Lecture Notes in Computer Science*, pages 24–35. Springer-Verlag, 1994.

[2] F. J. Brandenburg, editor. *Graph Drawing (Proc. GD '95)*, volume 1027 of *Lecture Notes in Computer Science*. Springer-Verlag, 1996.

[3] F. J. Brandenburg, M. Himsolt, and C. Rohrer. An experimental comparison of force-directed and randomized graph drawing algorithms. In F. J. Brandenburg, editor, *Graph Drawing (Proc. GD '95)*, volume 1027 of *Lecture Notes in Computer Science*, pages 76–87. Springer-Verlag, 1996.

[4] G. Di Battista, P. Eades, R. Tamassia, and I. G. Tollis. Algorithms for drawing graphs: an annotated bibliography. *Comput. Geom. Theory Appl.*, 4:235–282, 1994.

[5] G. Di Battista, A. Garg, G. Liotta, R. Tamassia, E. Tassinari, and F. Vargiu. An experimental comparison of three graph drawing algorithms. In *Proc. 11th Annu. ACM Sympos. Comput. Geom.*, pages 306–315, 1995.

[6] G. Di Battista, G. Liotta, and F. Vargiu. Spirality of orthogonal representations and optimal drawings of series-parallel graphs and 3-planar graphs.

In *Proc. Workshop Algorithms Data Struct.*, volume 709 of *Lecture Notes in Computer Science*, pages 151–162. Springer-Verlag, 1993.

[7] G. Di Battista and R. Tamassia. On-line maintenance of triconnected components with SPQR-tree s. *Algorithmica*, 15:302–318, 1996. Preprint: Technical Report CS-92-40, Comput. Sci. Dept., Brown Univ. (1992).

[8] G. Di Battista and R. Tamassia. On-line planarity testing. *SIAM J. Comput.*, to appear. Preprint: Technical Report CS-92-39, Comput. Sci. Dept., Brown Univ. (1992).

[9] S. Even. *Graph Algorithms*. Computer Science Press, Potomac, Maryland, 1979.

[10] A. Garg and R. Tamassia. On the computational complexity of upward and rectilinear planarity testing. Report CS-94-10, Comput. Sci. Dept., Brown Univ., Providence, RI, 1994.

[11] A. Garg and R. Tamassia. On the computational complexity of upward and rectilinear planarity testing. Submitted to *SIAM Journal on Computing*, 1995.

[12] M. Himsolt. Comparing and evaluating layout algorithms within GraphEd. *J. Visual Lang. Comput.*, 6(3), 1995. (special issue on Graph Visualization, edited by I. F. Cruz and P. Eades).

[13] J. Hopcroft and R. E. Tarjan. Dividing a graph into triconnected components. *SIAM J. Comput.*, 2:135–158, 1973.

[14] S. Jones, P. Eades, A. Moran, N. Ward, G. Delott, and R. Tamassia. A note on planar graph drawing algorithms. Technical Report 216, Department of Computer Science, University of Queensland, 1991.

[15] M. Jünger and P. Mutzel. Exact and heuristic algorithms for 2-layer straightline crossing minimization. In F. J. Brandenburg, editor, *Graph Drawing (Proc. GD '95)*, volume 1027 of *Lecture Notes in Computer Science*, pages 337–348. Springer-Verlag, 1996.

[16] K. Mehlhorn and S. Näher. LEDA: a platform for combinatorial and geometric computing. *Commun. ACM*, 38:96–102, 1995.

[17] T. Nishizeki and N. Chiba. Planar graphs: Theory and algorithms. *Ann. Discrete Math.*, 32, 1988.

[18] A. Papakostas and I. G. Tollis. Improved algorithms and bounds for orthogonal drawings. In R. Tamassia and I. G. Tollis, editors, *Graph Drawing (Proc. GD '94)*, volume 894 of *Lecture Notes in Computer Science*, pages 40–51. Springer-Verlag, 1995.

[19] R. Tamassia. On embedding a graph in the grid with the minimum number of bends. *SIAM J. Comput.*, 16(3):421–444, 1987.

[20] R. Tamassia, G. Di Battista, and C. Batini. Automatic graph drawing and readability of diagrams. *IEEE Trans. Syst. Man Cybern.*, SMC-18(1):61–79, 1988.

[21] R. Tamassia and I. G. Tollis. A unified approach to visibility representations of planar graphs. *Discrete Comput. Geom.*, 1(4):321–341, 1986.

[22] R. Tamassia and I. G. Tollis. Efficient embedding of planar graphs in linear time. In *Proc. IEEE Internat. Sympos. on Circuits and Systems*, pages 495–498, 1987.

[23] R. Tamassia and I. G. Tollis, editors. *Graph Drawing (Proc. GD '94)*, volume 894 of *Lecture Notes in Computer Science*. Springer-Verlag, 1995.

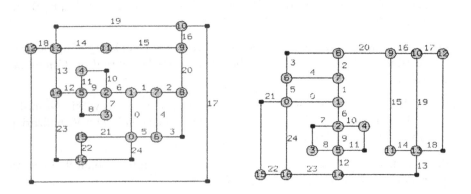

Figure 1: A 17-vertices graph drawn by GIOTTO, and by BB-ORTH

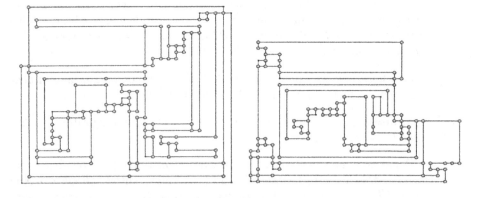

Figure 2: A 100-vertices graph drawn by GIOTTO, and by BB-ORTH

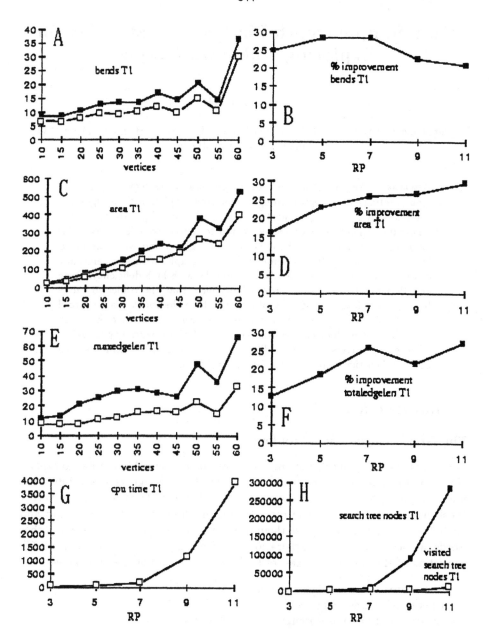

Figure 3: Results of the experiments.

On a Simple Depth-First Search Strategy for Exploring Unknown Graphs

Stephen Kwek

Department of Computer Science, Washington University, St. Louis, MO 63130
Email: kwek@cs.wustl.edu

Abstract. We present a simple depth-first search strategy for exploring (constructing) an unknown strongly connected graph G with m edges and n vertices by traversing at most $\min(mn, dn^2 + m)$ edges. Here, d is the minimum number of edges needed to add to G to make it Eulerian. This parameter d is known as the deficiency of a graph and was introduced by Kutten [Kut88]. It was conjectured that graphs with low deficiency can be traversed more efficiently than graphs with high deficiency. Deng and Papadimitriou [DP90] provided evidence that the conjecture may be true by exhibiting a family of graphs where the robot can be forced to traverse $\Omega\left(d^2 m\right)$ edges in the worst case. Since then, there has been some interest in determining whether a graph with deficiency d can be explored by traversing $O(poly(d)m)$ edges. Our algorithm achieves such bound when the graph is dense, say $m = \Omega(n^2)$.

1 Introduction

The Graph Exploration Problem

We look at the following classic *graph exploration* problem: A robot has to *explore* (*i.e.*, construct) an unknown terrain represented by a strongly connected directed multi-graph G, by repeatedly select an outgoing edge from its current vertex (position) and traverse it. The robot knows all the traversed (*seen*) edges and visited nodes, and can recognize them when they are encountered again. In other words, at each stage of the exploration, the robot knows a proper subgraph G', consisting of the traversed edges, of G. However, the robot does not know:

- how many nodes or edges are there in the graph G,
- where each unseen edge leads to, and
- the origin of any unseen edge of a visited node.

The robot is said to have *learned* the graph G if all the edges have been traversed (and thus, the graph G' is the same as G). Since the graph is strongly connected, it is trivial that the robot can learn any graph and knows that its exploration task is complete when all of the edges in G' have been traversed. The goal is to learn the graph by traversing as few edges as possible.

It is easy to show that if the graph is Eulerian, then the robot only needs to traverse at most $4m$ edges [DP90] where m is the number of edges. To measure "how far is a graph from being Eulerian", Kutten [Kut88] introduced the *deficiency* d of a graph which is the number of edges that are needed to make a graph Eulerian. It has been conjectured that graphs with low deficiency can be traversed more efficiently than graphs with high deficiency. Deng and Papadimitriou [DP90] provided evidence that the conjecture may be true by exhibiting a family of graphs where the robot can be forced to traverse $\Omega\left(d^2 m\right)$ edges in the worst case. Thus, ideally, one would like to construct an algorithm that traverses $O(f(d)m)$ edges for some polynomial f. Deng and Papadimitriou [DP90] gave an algorithm that traverses $O(d^{O(d)}m)$ edges. This has recently been improved by Albers and Henzinger [AH97] which gave an algorithm that achieves an upper bound of $O\left(d^6 d^{2\log d}m\right)$ edge traversals.

In this paper, we present a simple depth-first search algorithm that traverses at most $\min(mn, dn^2 + m)$ edges where n is the number of vertices. Note that if the unknown graph is dense, say $m = \Omega(n^2)$, then our bound becomes $O(dm)$. It can be shown easily that any reasonable[1] algorithm traverses at most mn edges. Conversely, kutten showed in [Kut88] that there exist graphs for every n and m for which these mn edge traversals are optimal.

Algorithm that acheives the mistake bound $\min(mn, dn^2 + m)$ had also been independently discovered by Bar-Noy et. al. [BNKSP97], and Albers and Henzinger [AH97]. The former employed a simple nearest-neighbor approach, while the latter adopted a more sophisticated approach. In [AH97], Albers and Henzinger argued that for every d, there is a family of graphs that can force both approaches to traverse $2^{\Omega(d)}m$ edges. However, since both approaches traverse at most mn edges, the deficiency of the family of graphs has to be bounded by $O(\log n)$. In other words, it is still possible that for $d = \Omega(\log n)$, these simple approaches could yield an edge traversal bound of $O(f(d)m)$ for some polynomial or sub-exponential function f.

This problem has also been investigated by Koenig and Smirnov [KS96] where they considered directed graphs obtained from undirected graphs by replacing each edge with a pair of edges, one for each direction. Notice that such digraphs are always Eulerian and hence can be explored by traversing each edge at most four times [DP90].

Other Related Work

Motivated by robotics, various related graph exploration (navigation) problems have been investigated by both theoreticians and AI practitioners. Frequently, geometric constraints of the environment are ignored and the environment is assumed to be represented by a graph. The research conducted by AI practitioners is mostly empirical in their approach and we refer the reader to [Kor90, PK92, Ste95, SKVS96].

Theoretical studies of graph exploration problems were first initiated by Papadimitriou and Yannakakis [PY91] where they were interested in the problem

---

[1] When the robot reaches a node where all the out-going edges have been traversed, any reasonable algorithm should traverse at most n seen edges to reach an unexhausted vertex from which it traverses an unseen edge.

of finding a shortest path between a pair of points under various graph theoretic and geometric settings. Motivated by this earlier work, Blum *et. al.* considered the exploration problem in the Euclidean plane with convex obstacles [BRS91] and other variations [BRS91, BC93].

The exploration problem examined in this paper is due to Deng and Papadimitriou [DP90]. Subsequently, they together with Kameda [DKP91] studied a geometric version of this problem where the robot's task is to explore a room with polygonal obstacles and the goal is to minimize the total distance traveled. Recently, Berman *et. al.* [BBF$^+$96] investigated the case where the obstacles are oriented rectangles. Other work relating to exploring geometric domain are be found in [LS86, LS87, EBEY92, LT94, TK94, AWZ96].

Betke *et. al.* [BRS95] and Awerbuch *et. al.* [ABRS95] investigated the problem with the additional constraints that the robot is required to return to its starting point for 'refueling' periodically. Bender and Slonim [BS94] illustrated how two robots can collaborate in exploring directed graph with regular degree and indistinguishable vertices.

Preliminary

Let $in(v)$ and $out(v)$ denote the indegree and outdegree of the vertex v. We say a vertex is *deficient* if its indegree is greater than its outdegree. The *deficiency* of a vertex v, $d(v)$, is

$$d(v) = \begin{cases} in(v) - out(v) & \text{if } v \text{ is deficient} \\ 0 & \text{if } v \text{ is not deficient} \end{cases}$$

The minimum number of edges needed to add to a graph G to make it Eulerian is called the *deficiency* of G and is denoted here by d. Clearly,

$$d = \sum_{v \in V} d(v).$$

We say that a vertex is *exhausted* if the robot has traversed all its outgoing edges. If the robot reaches an exhausted vertex v then we say the robot is *stuck* at v.

2 A Simple Depth-First Search Strategy

Suppose a is the vertex where the robot begins its exploration. To simplify our discussion, we assume the graph the robot is exploring is a 'modified' graph obtained by adding a new vertex s with an edge (s, a) and the robot begins at s. Note that although the modified graph is no longer strongly connected, it is still possible for the robot to traverse all the edges. With this assumption, the vertices along any path that begins with s have even degree except the last vertex and s.

Our strategy EXPLORE is shown Figure 1. It simply traverses an unexplored edge when there is one (Step 6 to 13) until the robot is stuck. As the robot traverses the edges in this greedily manner, we push the current vertex into a stack (Step 11).

```
EXPLORE(s)
1      stack := ∅
2      V(G') = {s}
3      E(G') = ∅
4      u := s
5      do
6            do
7                  if u is not exhausted then
8                        traverse an unexplored edge (u, v)
9                        V(G') := V(G') ∪ {v}
10                       E(G') = E(G') ∪ {(u, v)}
11                       push(v)
12                       u := v
13                 until stuck at u (i.e., u is exhausted)
14                 do
15                       y := pop(stack)
16                 until stack = ∅ or y is not exhausted
17                 if (stack ≠ ∅) then
18                       find a shortest path from u to y in G'
19                       traverse the path so that the robot is at y
20                       u := y
21          until stack = ∅
22    return G'
```

Fig. 1. An algorithm for exploring an unknown graph that traverses at most $\min(mn, dn^2 + m)$ edges.

When the robot is stuck, we simply pop the stack until either the stack is empty or the poped vertex y is not exhausted. In the latter case, we claim (see Claim 2) that there is a path from u that leads to y in the graph G' obtained by the robot's exploration so far. The robot then traverses this path to y and repeat the greedy traversal of the unexplored edges (by letting u be y and repeat Steps 6 to 13). In the former case, we show below that the robot has traversed all the edges of G.

Since we pop a vertex out of the stack only if it is exhausted, we have the following observation about the vertices in G' that have not been exhausted.

Observation 1 *The vertices in G' that are not exhausted are in the stack.*

Thus, if we can show that (see Claim 2) there is a path connecting u to the unexhausted vertex y in G' at Step 18, then we are sure that the robot is able to traverse an unseen edge whenever there is an unexhausted vertex in G'. Thus, by Observation 1, when the stack is empty, all the vertices in G' are exhausted. Since G is strongly connected, the latter implies that the robot has completed it's exploration of G.

Claim 2 *In Step 18, there is a path from u to y in G'.*

Proof: The stack obtained just after the execution of the loop in Steps 6 to 13, induces a path P from s to u which passes through y (maybe several times). Observe that the vertices from the last occurrence (on P) of y to u in the path P are all exhausted except y. Consider the graph G_P induced by P. It is clear that (see Figure 2) G_P is a graph with strongly connected components C_1, ..., C_k such that there is exactly one edge going from C_i to C_{i+1} and u is in C_k. If y is also in C_k as u then there is a path from u to y.

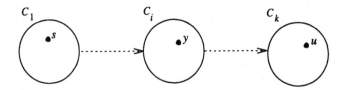

Fig. 2. The graph G_P induced by P

Thus, suppose y is in C_i such that $i \neq k$. Notice that by our choice of y, the vertices in $C_{i+1}, \cdots, C_k$ must be exhausted. Consider the graph G'. If there is no path in G' from u to any vertex in $C_j, j \leq i$, then the vertices in G' reachable by u cannot reach the vertices in $C_1, \cdots, C_i$. By Observation 1, only the vertices in $C_1, \cdots, C_i$ can be unexhausted, thus, there is no path from u to y in the unknown graph G. This contradicts the strongly connectedness of G. □

In the following, we derived an upper bound on the number of edge traversals by EXPLORE. The proof involves assignment of labels to the edges as we traverse them. This label assignment is to facilitate the analysis of the algorithm and has nothing to do with the correctness of the algorithm.

Clearly, the first vertex u where the robot is stuck in Step 13 has to be a deficient vertex. The sequence of vertices which we poped from the stack in Steps 14 to 16 induces a path P' in G from u to y. Give the edges in P' excluding those lying on a cycle a label '1' and set a counter D to 2. Subsequently, each execution of Step 14 to 16 induces a path P' from u to y. We label the edges in P' according to the following rules:

Rule 1: Suppose the robot is stuck at a vertex u where there is a label i such that there are more[2] outgoing edges with label i than incoming edges with label i. Then we add to each edge in P' excluding those lying on a cycle (in P') a label i. If there are more than one such i, EXPLORE arbitrarily selects one such i.

Rule 2: Suppose the condition in Rule 1 does not hold. It is clear that for the vertex u, the number of incoming traversed edges exceeds the number of traversed outgoing edges (and all outgoing edges have been traversed). That is, u must be a deficient vertex. In this case, we label the edges in P' excluding those lying on a cycle (in P') the value of D and increment D by 1.

Notice that a new label is created only when Rule 2 is applied. This implies that u is deficient and we can 'blame' the activation of Rule 2 on 1 unit of G's deficiency. Thus, we have the following observation.

Observation 3 *There are at most d labels and Rule 2 is invoked at most d times.*

Lemma 4. *For each vertex u and each label i, Rule 1 is applied at most once. In other words, each edge is assigned at most d (distinct) labels.*

Proof: Consider the following two cases when an outgoing edge of u is labeled i for the first time:

Case 1: u is exhausted. In this case, We assign an incoming edge of u a label of i and continue popping the stack. Subsequently, whenever a new outgoing edge of u is labeled i, an unlabeled incoming edge of u is labeled i. Thus, for the vertex u, the number of incoming edges that are labeled i is the same as the number of outgoing edges. That is, in the future, Rule 1 will never be applied to u with label i.

Case 2: u is not exhausted. In this case, we apply Rule 2. We stop popping the stack and unlike Case 1, we do not assign a label of i to any of u's incoming edges. The robot then continue to explore an unseen outgoing edge of u. Notice that for all the other vertices, the number of incoming edges with label i is the same as the number of outgoing edges with label i. Subsequently, no new edge is labeled i until Rule 1 is applied to u with label i. When this happens, there is exactly one incoming edge and one outgoing edge of u that are labeled i. Note that u is now exhausted and we can apply similar argument as in Case 1.

□

Observation 3 and Lemma 4, imply that the robot is stuck at most dn times. Each time the robot is stuck, it traverses at most n traversed edges to reach an unexhaust vertex in Step 18. The robot traverses seen edges only in Step 18 when it is stuck.

---

[2] In fact, it is easy to see that in this case, there is exactly one more outgoing edge than incoming edges with label i.

We can view the popping of a vertex from the stack in step 14 as poping an edge and each edge can be shown to poped exactly once. Moreover, by Observation 3, Step 18 is executed at most dn time and each execution of Step 18 takes $O(n)$. Thus we have the following theorem.

Theorem 5. EXPLORE *traverses at most* $\min(mn, dn^2 + m)$ *number of edges and has time complexity* $O(dn^2 + m)$. $\square$

3 Discussion

The algorithm EXPLORE presented in this paper is based on a simple depth first search strategy. Our analysis has been quite lax and it indicates a possibility that a variant or tighter analysis of EXPLORE might give rise to a better bound. In particular, it could be possible that for $d = \Omega(\log n)$, the number of edge traversals of EXPLORE or the nearest neighbor approach presented in [BNKSP97] are bounded by $O(f(d)m)$ for some polynomial function f? We emphasize again that such result, if true, does not contradict the lower bound on the edge traversals for these simple approaches presented in [AH97]. This is because the latter has an implicit assumption[3] that $d = O(\log n)$ (see discussion in Section 1).

Notice that EXPLORE does not assume that the robot knows how many unseen edges are leading into or out of the visited vertex. Such information may be available in practice and may result in more efficient exploration algorithm. Another research direction is to consider using multiple robots to explore the graph. It is interesting to know if the total number of edge traversals can be made smaller than using a single robot.

Although our simple depth-first search approach is unlikely that EXPLORE outperforms known algorithms (see [Kor90, PK92, Ste95, SKVS96]) that are used in practice, we hope that our analysis will provide new insight to this classical problem and give us a more efficient strategy.

Acknowledgements

The author thanks Shay Kutten, Baruch Scieber, Sally Goldman, Monika Henzinger, Subhash Suri and Stephen Scott for helpful discussions regarding this work. This work is supported by NSF NYI Grant CCR-9357707 (of S. Goldman) with matching funds provided by Xerox PARC and WUTA.

References

[ABRS95] B. Awerbuch, M. Betke, R. Rivest, and M. Singh. Piecemeal graph exploration by a mobile robot. In *Proc. 8th Annu. Conf. on Comput. Learning Theory*, pages 321–328. ACM Press, New York, NY, 1995.

---

[3] In fact, the deficiency of the family of graphs constructed in [AH97] seems to be bounded by $o(\log n)$.

[AH97]   S. Albers and M. Henzinger. Exploring unknown environments. In *Proc. 29th Annu. ACM Sympos. Theory Comput.*, pages 416–425, 1997.

[AWZ96]   D. Angluin, J. Westbrook, and W. Zhu. Robot navigation with range queries. In *Proc. 28th Annu. ACM Sympos. Theory Comput.*, 1996. 469–478.

[BBF+96]   P. Berman, A. Blum, A. Fiat, H. Karloff, A. Rosen, and M. Saks. Randomized robot navigation algorithms. In *Proceedings SODA 96*, 1996. to appear.

[BC93]   A. Blum and P. Chalasani. An on-line algorithm for improving performance in navigation. In *Proc. 34th Annu. IEEE Sympos. Found. Comput. Sci.*, pages 2–11. IEEE Computer Society Press, Los Alamitos, CA, 1993.

[BNKSP97] A. Bar-Noy, S. Kutten, B. Scieber, and D. Peleg. Competitive unidirectional learning. Unpublished Manuscript, 1997.

[BRS91]   A. Blum, P. Raghavan, and B. Schieber. Navigating in unfamiliar geometric terrain. In *Proc. 23th Annu. ACM Sympos. Theory Comput.*, pages 494–504. ACM, 1991.

[BRS95]   M. Betke, R. Rivest, and M. Singh. Piecemeal learning of an unknown environment. *Machine Learning*, 18(2/3):231–254, 1995.

[BS94]   M. Bender and D. Slonim. The power of team exploration: two robots can learn unlabeled directed graphs. In *Proceedings of the 35rd Annual Symposium on Foundations of Computer Science*, pages 75–85. IEEE Computer Society Press, Los Alamitos, CA, 1994.

[DKP91]   X. Deng, T. Kameda, and C. Papadimitriou. How to learn in an unknown environment. In *Proc. of the 32nd Symposium on the Foundations of Comp. Sci.*, pages 298–303. IEEE Computer Society Press, Los Alamitos, CA, 1991.

[DP90]   X. Deng and C. H. Papadimitriou. Exploring an unknown graph. In *Proc. 31th Annu. IEEE Sympos. Found. Comput. Sci.*, volume I, pages 355–361, 1990.

[EBEY92]   A. Fiat E. Bar-Eli, P. Berman and P. Yan. On-line navigation in a room. In *Proc. 3rd SODA*, pages 75–84, 1992.

[Kor90]   R.E. Korf. Real-time heuristic search. *Artificial Intelligence*, 42(3):189–211, 1990.

[KS96]   S. Keonig and Y. Smirnov. Graph learning with a nearest neighbor approach. In *Proceedings of the 9th Conference on Computaitonal Learning Theory*, pages 19–28, 1996.

[Kut88]   S. Kutten. Stepwise construction of an efficient distributed traversing algorithm for general strongly connected directed networks. In *9th International Conference on Computer Communication, Tel Aviv, Israel*, pages 446–452, 1988.

[LS86]   V. Lumelsky and A. Stepanov. Dynamic path planning for a mobile automaton with limited information on the environment. *IEEE Trans. on Automatic Control*, AC-31:1059–1063, 1986.

[LS87]   V. Lumelsky and A. Stepanov. Path planning strategies for a point mobile automaton moving amidst unknown obstacles of arbitrary shapes. *Algorithmica*, 2:403–430, 1987.

[LT94]   V. Lumelsky and S. Tiwari. An algorithm for maze searching with azimuth input. In *IEEE Conference on Robotics and Automation*, pages 111–116, 1994.

[PK92]   J.C. Pemberton and R.E. Korf. Incremental path planning on graphs with cycles. In *Proceedings of the AI Planning Systems Conference*, pages 179–188, 1992.

[PY91]   C. Papadimitriou and C. Yannakakis. Shortest path wothout a map. *Theoretical Computer Science*, 84:127–150, 1991.

[SKVS96]   Y. Smirnov, S. Keonig, M. Veloso, and R. Simmons. Efficient goal-directed exploration. In *Proceedings of the National Conference on AI*, 1996. 292-297.

[Ste95]   Stentz. The focussed d^* algorithm for real-time replanning. In *Proceedings of the International Joint Conf. on AI*, pages 1652–1659, 1995.

[TK94]   C. J. Taylor and D. J. Kriegman. Vision-based motion planning and exploration algorithms for mobile robots. In *Proc. of the Workshop on Algorithmic Foundation of Robotics*, 1994.

Orthogonal Drawing of High Degree Graphs with Small Area and Few Bends*

Achilleas Papakostas   and   Ioannis G. Tollis

Dept. of Computer Science
The University of Texas at Dallas
Richardson, TX 75083-0688
email: papakost@utdallas.edu, tollis@utdallas.edu

Abstract. Most of the work that appears in the orthogonal graph drawing literature deals with graphs whose maximum degree is four. In this paper we present an algorithm for orthogonal drawings of simple graphs with degree higher than four. Vertices are represented by rectangular boxes of perimeter less than twice the degree of the vertex. Our algorithm is based on creating groups/pairs of vertices of the graph both ahead of time and in real drawing time. The orthogonal drawings produced by our algorithm have area at most $(m-1) \times (\frac{m}{2}+2)$. Two important properties of our algorithm are that the drawings exhibit small total number of bends (less than m), and that there is at most one bend per edge.

1   Introduction

Most of the work that appears in the orthogonal graph drawing literature deals with graphs whose maximum degree is four [3, 9, 13, 15, 17, 18, 19, 20]. The drawings produced by these algorithms require at least two bends per edge. This is a big restriction since in most applications graphs generally have degree higher than four. Orthogonal drawings of graphs of high degree are useful for visualizing database schemas or the internal structure of large software systems. More specifically, vertices are now boxes called *tables* containing fields, which are placed vertically one below the other. These fields can be attributes of some entity (in the case of a database) or a specification list (in the case of a software module). For a survey of graph drawing algorithms and other related results, see [4]. Our graphs have n vertices and m edges.

Foessmaier and Kaufmann [8] presented an extension of Tamassia's algorithm [19] for minimizing the total number of bends of planar embedded graphs of maximum degree four to planar graphs of arbitrary degree. The vertices are represented by squares of size depending on the degree of each vertex. No discussion is made on the area or the number of bends per edge of the resulting orthogonal drawing. Their algorithm runs in $O(n^2 \log n)$ time.

GIOTTO [1, 19] is another algorithm for orthogonal drawings of graphs of any degree. It is also based on Tamassia's algorithm [19] for minimizing the total

---

* Research supported in part by NIST, Advanced Technology Program grant number 70NANB5H11.

number of bends for planar graphs of maximum degree four. Dummy vertices are used to represent crossings, and all vertices of the drawing are represented by boxes. The disadvantage of GIOTTO is that these boxes may grow arbitrarily in size, regardless of the degree of the vertex they represent. Experimental results of GIOTTO's performance with respect to various aesthetic measures on a large database of about 11,000 graphs (arising from real applications) can be found in [5]. In what follows, the area of a drawing is given in the format $width \times height$.

Even and Granot [6] presented two algorithms for placing rectangular modules and connections between them in an orthogonal fashion in the plane. The size of the modules and the positions of the terminals around the modules are part of the input. The edges of the graph are attached to the terminals of the modules. Their first algorithm produces planar orthogonal drawings starting from a visibility representation of the given graph. Their second algorithm places the modules diagonally in the plane and routes the edges around them. In both cases, the final orthogonal drawing has area at most $(W + m) \times (H + m)$ where W (H) is the total width (height) of all modules, and m is the number of edges of the graph. No edge has more than four bends and both algorithms run in $O(m)$.

There is a short discussion in [3] about producing a drawing of a general (i.e., not necessarily planar) graph with degree higher than four. The authors extended their algorithm for maximum degree four graphs and they proposed an approach in which each vertex is represented by a vertical line segment consisting of multiple grid points. The area of the resulting drawing was at most $(m - n + 1) \times (m - \frac{n}{2} + \frac{n_2}{2})$, where n_2 is the number of vertices of degree two. If we change the vertex representation to that of a box of the same height as the line segment and width one, we obtain an orthogonal drawing with area at most $(m + 1) \times (m - \frac{n}{2} + \frac{n_2}{2})$. Each edge has at most two bends. Drawings resulting from this approach turn out to be very tall, skinny, they have at most $2m - 2n + 4$ bends, and many crossings.

In this paper we present a different approach for dealing with orthogonal drawings of simple graphs with degree higher than four. Vertices are represented by rectangular boxes of perimeter less than twice the degree of the vertex. Our algorithm is based on creating groups/pairs of vertices of the graph both ahead of time and in real drawing time. The orthogonal drawings produced by our algorithm have area at most $(m - 1) \times (\frac{m}{2} + 2)$. Two important properties of our algorithm are that there is at most one bend per edge, and the drawings have a small total number of bends. For more details on this technique and for results on 3D orthogonal drawing of high degree graphs see [12].

It is interesting to note that our algorithm provides a new, and significantly better than the previously known, upper bound on the area for drawings of graphs of arbitrary degree. If χ denotes the minimum number of crossings over all drawings of a given graph G, then a straight-forward upper bound on the area of any drawing of G is $O((n + \chi)^2)$. We know [11] that for any graph G, $\chi > 0.029 \frac{m^3}{n^2} - 0.9n$. Therefore, in the worst case, the above area upper bound becomes $O(n^8)$, whereas our algorithm guarantees a bound of $O(n^4)$.

Independently, Biedl [2] recently presented a technique that produces orthogonal drawings of high degree graphs with at most m bends and $\left(\frac{m+2n}{2}\right) \times \left(\frac{m+2n}{2}\right)$ area. For sparse graphs, that have only a few nodes of high degree, our area bounds are better than the ones of [2]. Notice that in practice, graphs that are involved in visualization applications are typically rather sparse. For example, in the large experimental study reported in [5] the average degree of all 11,000 graphs considered was about 2.7 (in other words, the total number of edges was about $1.35n$).

If s and t are two distinct vertices of G, then a numbering of its vertices is an *st-numbering* for G if and only if both the following conditions are satisfied:

1. Number 1 is assigned to vertex s and number n is assigned to vertex t.
2. Each vertex j $(2 \leq j \leq n-1)$ has at least one neighbor (predecessor) i and and one neighbor (successor), such that that $i < j < k$.

Lempel et al. [10] showed that for every biconnected graph and every edge (s, t), there exists an *st*-numbering. An algorithm for computing an *st*-numbering in $O(m)$ time has been presented by Even and Tarjan [7].

2 A Box Representation for Vertices and a Simple Algorithm

Clearly, a point representation for vertices of a graph does not suffice if we want to remove any restriction about the degree of its vertices. In this paper, we use a *rectangular box* to represent vertices of the graph. From an aesthetic point of view, using boxes reduces significantly the number of bends in the drawing, both total and per edge. We call the boundary edges of a box *sides*. Figure 1a shows a box with its four sides, *top, left, bottom,* and *right*. Each side has a number of *connectors* where the edges that are incident to this box attach to. The sides of a box always lie along lines of the underlying integer grid. Therefore, connectors have integer coordinates.

We present a Simple Algorithm for producing an orthogonal drawing. The Simple Algorithm inserts the vertices in the drawing, one vertex at a time. For simplicity, we assume that the given graph is biconnected, and that an st-numbering has been computed on the graph. Note that the edges of the graph are directed from lower to higher numbered vertices, as a result of the st-numbering. The size of each box to be inserted (say v) is decided when v is the next vertex to be inserted in the drawing.

All outgoing edges of vertex v are attached to the top side connectors (see Fig. 2a). This implies that the width of the box is at least equal to the number of outgoing edges of the vertex. If the box has only one outgoing edge, we still use two columns for the box (see Fig. 2b). The incoming edges of v are split between the right and left side connectors. More specifically, if v has $indeg(v)$ incoming edges, then $\lfloor \frac{indeg(v)}{2} \rfloor$ incoming edges are attached to the right side and the remaining $\lceil \frac{indeg(v)}{2} \rceil$ incoming edges are attached to the left side of box

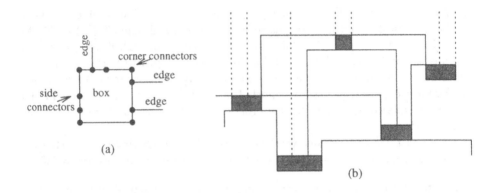

Fig. 1. (a) A box with its sides and connectors (b) a sample orthogonal drawing produced by the Simple Algorithm.

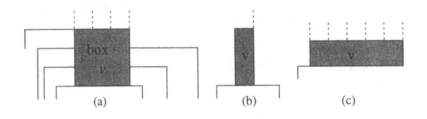

Fig. 2. Various types of box v: (a) 7 incoming and 4 outgoing edges, (b) only one outgoing edge, (c) only one incoming edge.

v (see Fig. 2a). If v has only one incoming edge, we still use two rows for the box (i.e., a box with height one, see Fig. 2c).

When the Simple Algorithm places vertex v, it first creates the box to represent v and then it opens up the appropriate number of new columns between v's median incoming edges and new rows above the current drawing, and places v there. Figure 1b shows a few vertices placed by the Simple Algorithm.

Theorem 1. *Let G be a biconnected graph with n vertices and m edges. Consider each edge as being oriented from its lower to its higher numbered vertex, for a given st-numbering of G. The Simple Algorithm produces an orthogonal drawing of G in $O(m)$ time, so that the following hold:*

1. *each vertex v with degree $deg(v)$ is represented by a box whose perimeter is less than $2 \times deg(v)$,*
2. *the width of the resulting orthogonal drawing is at most $m + n_{out1}$,*
3. *the height of the drawing is at most $\frac{m}{2} + \frac{n_{odd}}{2} + n_{in1} + n_{in2}$,*
4. *the drawing has at most m bends, and*
5. *no edge bends more than once.*

Note that n_{out1} is the number of vertices with one outgoing edge, n_{odd} is the number of vertices with an odd number of incoming edges, and n_{in1} (n_{in2}) is the number of vertices with one (two) incoming edge(s).

Sketch of the Proof. The width of the drawing depends on the total number of outgoing edges, m, and the fact that boxes with one outgoing edge require two columns. The height involves half the total number of incoming edges, $\frac{m}{2}$, and the following two facts: Boxes with one or two incoming edges take up two rows; a box v with an odd number of incoming edges requires $\frac{indeg(v)}{2} + \frac{1}{2}$ rows. The complete proof can be found in [14]. □

3 Algorithm BOX_ORTHOGONAL: Numbering and Grouping of Vertices

In this section we describe a new algorithm which produces orthogonal drawings for any given graph, with better bounds than the Simple Algorithm of the previous section. BOX_ORTHOGONAL is the name of the new algorithm, and it has many similarities to the Simple Algorithm. It uses boxes to represent vertices of the graph, except for vertices of degree one, two, and some vertices of degree three, which are represented by points. We use the point representation for some small degree vertices mainly for aesthetic reasons.

First we compute an *initial numbering* for the given graph; this numbering may be modified (if necessary) resulting to the *final numbering*. This numbering will provide the order according to which the vertices are considered. Let us assume that we are given a connected n-vertex graph G. If G is biconnected, then the initial numbering is an st-numbering [7] of the vertices of G. If G is not biconnected, then we break G into its biconnected components, and compute a numbering for G so that there is a unique source and multiple sinks (one per biconnected component). The initial numbering of the vertices implies an orientation of the edges of G, from the lower numbered vertex to the higher numbered vertex. A vertex v with b incoming and a outgoing edges is called a b-a vertex.

The purpose of modifying the initial numbering to the final numbering is to eliminate specific patterns of vertex types that result in drawings with large area under the Simple Algorithm. More specifically, we consider each $b - 1$ ($b > 3$ is odd) vertex whose outgoing edge enters a vertex w which is not a sink, a $1 - a$ ($a > 1$), or another $b - 1$ ($b \geq 1$) vertex. We reverse the direction of the edge between the $b - 1$ vertex and w. As a result, the $b - 1$ vertex becomes a $(b+1) - 0$ sink, and $b + 1$ is an even number. Notice that no edge reversal creates a new $b - 1$ vertex. Also, no directed cycle is formed since one of the vertices affected by the edge reversal is now a sink.

Let G' be the directed graph resulting after all possible edge reversals. Before we actually draw G' in the plane, we perform a *grouping/pairing* of its vertices. As we will see later, grouping/pairing will contribute to orthogonal drawings with better bounds in terms of area. Roughly speaking, we try to group/pair vertices of types $1 - a$, $b - 1$, and $2 - a$. Vertex grouping/pairing takes place in four passes.

In the first pass, we scan the vertices of G' looking for:

1. $b - 1$ $(b \geq 1)$ vertices whose outgoing edge enters a $1 - a$ $(a > 1)$ vertex; we pair the two vertices.
2. $c - d$ vertices such that $c > 1$ is odd and $d \geq 2$, which are predecessors of at least one $1 - a$ $(a > 1)$ vertex; we pair each such vertex with the $1 - a$ vertex.
3. $c - d$ vertices such that $c > 1$ is odd and $d \geq 2$, which have at least one $b - 1$ $(b \geq 1)$ predecessor; we pair each such vertex with its $b - 1$ predecessor.

In the second pass, we look for $2 - a$ $(a > 1)$ vertices which have at least one $b - 1$ (b even) predecessor, or two $b - 1$ predecessors so that $b = 1$ or $b = 3$. In the first case, we pair the $2 - a$ vertex with its $b - 1$ predecessor, and in the second case we group the $2 - a$ vertex with both its $b - 1$ predecessors. In the third pass, we scan the unpaired $1 - a$ vertices in decreasing order of their assigned number. For as long as we encounter a $1 - a$ vertex whose predecessor is also a $1 - a$ vertex, we put them both in the same group, which is called *chain*. In the fourth and final pass, we scan the remaining $1 - a$ vertices and we pair $1 - a$ vertices that have the same predecessor. Vertex grouping/pairing can be completed in $O(m)$ time.

Lemma 2. *After vertex grouping/pairing is complete, any $1 - a$ vertex, v, which does not belong to a group/pair has a predecessor, u, which is exactly one of the following:*

1. *u is a source,*
2. *u is a $1 - a$ vertex that participates in some group/pair,*
3. *u is a $c - d$ vertex where $c, d \geq 2$.*

Also, there cannot be another $1 - a$ vertex, v', which does not belong to any group/pair and has the same predecessor as v.

Sketch of the Proof. Consider a $1 - a$ vertex v which does not belong to any group/pair, and its predecessor u. Vertex u cannot be a $b - 1$ vertex, because it would have formed a pair with v. If vertex u is a $1 - a$ vertex, it has to belong to some group/pair, because otherwise, v and u would have formed a chain. The complete proof can be found in [14]. □

4   Algorithm BOX_ORTHOGONAL: Placing Vertices

A disadvantage of the Simple Algorithm is that vertices with one, two, or an odd number of incoming edges contribute extra to the height of the drawings, and vertices with one outgoing edge contribute extra to the width of the drawings (see Theorem 1). We call these vertices the *problematic* vertices of the drawing process. In Algorithm BOX_ORTHOGONAL we follow the general rules of the Simple Algorithm for creating and placing boxes of vertices in the orthogonal drawing. However, we introduce a special placement technique for the problematic vertices, so that the resulting drawing has at most m columns and at most $\frac{m}{2} + c$ rows (c is a small constant). In what follows, we assume that v is the next vertex to be considered by the drawing algorithm.

4.1 Vertex v Has Only One Outgoing Edge

We use a box to represent such a vertex, as shown in Fig. 2b, when v's degree is at least four; otherwise, v is represented by a point. Vertices with only one outgoing edge are always placed directly on top of one of their predecessor vertices (boxes). Our goal is to ensure that all boxes placed on top of a single common predecessor vertex share each other's columns. Let v be a vertex with three or more incoming edges and one outgoing edge; let u be the vertex on top of which v is going to be placed. We use the next available column of u's outgoing edges from right to left, and we attach box v along that column as shown in Fig. 3. Let v' be a vertex with only one or two incoming edges and only one outgoing edge. We represent v' with a point, and we attach it along the next available column of u's outgoing edges from left to right, as shown in Fig. 3a.

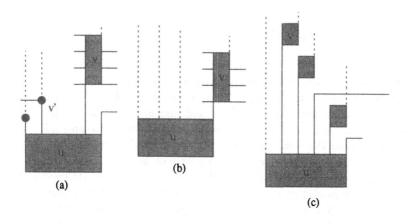

Fig. 3. Placing v with one outgoing edge on top ot its predecessor u.

4.2 Vertex v Is a Source Or a Sink

The initial numbering of the vertices of the given yields one source and at least one sink. However, more sources/sinks may be created after modifying the initial numbering. A source is drawn as a box of height one, and can be placed right off the end of the right margin of the current drawing. No new rows open up this way. Vertex number 1 (i.e., the original source) is the only exception since it opens up two new rows. Placing a sink follows the same rules as placing $b - 1$ vertices discussed above. Placing sinks does not require opening up any new columns.

4.3 Vertex v Has Only One Incoming Edge

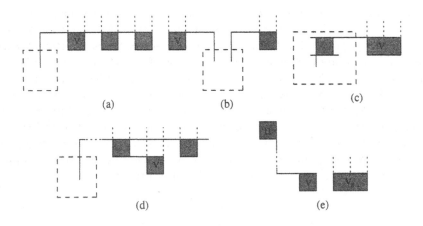

Fig. 4. Placing next vertex v when it is $1 - a$.

If the next vertex to be inserted v is a $1 - a$ vertex ($a > 1$), then we have the following cases for placing v: (a) v is the first vertex of a chain of at least two $1 - a$ vertices (see Fig. 4a). (b) v and another $1 - a$ vertex have the same predecessor and belong to the same pair (see Fig. 4b). (c) v and a $b - 1$ ($b \geq 1$) vertex belong to the same pair (see Fig. 4c). If v does not belong to any pair or chain and its predecessor is either a source or another $1 - a$ vertex belonging to some pair or chain, v is placed right next to the box of its predecessor (see Fig. 4d). If v's predecessor, u, is a $c - d$ vertex where $c, d \geq 2$, then we look for the most recently placed source in the drawing, say v_s. We draw an edge from the bottom of u towards the row of v_s's top side and place v next to v_s without opening any new rows, as shown in Fig. 4e.

4.4 Vertex v Has an Odd Number of Incoming Edges

Note that v has more than one incoming edges and more than one outgoing edges, otherwise we have one of the previous cases. If v belongs to a pair, the two vertices of the pair are placed as follows: (a) v is the predecessor of a $1 - a$ vertex and the two vertices form a pair (see Fig. 5a). (b) v belongs to the same

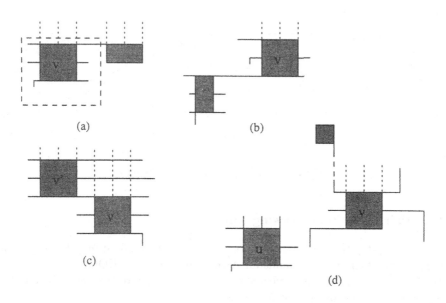

Fig. 5. Placing v when it has an odd number of incoming edges.

pair with a $b - 1$ ($b \geq 1$) vertex which is one of v's predecessors (see Fig. 5b). If v is not paired, we try to pair it with the first vertex v' to be inserted after v which is neither a source nor a $2 - a$ vertex. We place v and v' so that one row is reused (see Fig. 5c). If we are unable to pair v with v', then we have two different options: (a) If there is another already placed vertex u with an odd number of incoming edges that we were unable to pair in the past, then we pair v with u (see Fig. 5d). (b) Otherwise, we simply place v in the drawing as described in the Simple Algorithm.

4.5 Vertex v Has Two Incoming Edges

If v belongs to a group/pair, then v is grouped/paired with either one (see Fig. 6a), or both its predecessors (see Fig. 6b). Recall that the predecessors are $b - 1$ vertices. If v does not belong to any group/pair, then we try to group/pair it now, in one of the following ways: (a) We consider the first vertex to be inserted after v which is not a source, say v'. We pair v with v' and we place them in the way shown in Fig. 6c. (b) If there is an already placed $2 - a$ vertex u that we were unable to pair in the past, then we place v so that v shares one or two rows with u (the placement is done similarly to Fig. 5d of the previous subsection). (c) If none of the above is possible, then we simply place v according to the Simple Algorithm.

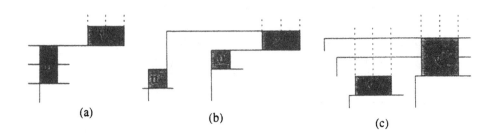

Fig. 6. Grouping/pairing $2 - a$ vertices.

4.6 Pairing in Real Drawing Time

The pairing step that we describe here is not actually required in order to achieve the area and bend upper bounds for Algorithm BOX_ORTHOGONAL (see next section). However, if it is performed, it can only improve the area and number of bends of the resulting drawing.

Let us consider a vertex v with an even number of incoming edges greater than two and at least two outgoing edges, which does not belong to any group/pair. Let us also assume that v is the next vertex to be inserted into the drawing by Algorithm BOX_ORTHOGONAL. We try to pair v with the first vertex to be inserted after v which is not a source, say v'. If v' already belongs to a group/pair, then we do not pair v and we simply insert it in a way similar to the Simple Algorithm.

If, on the other hand, v' does not participate in any group/pair, we check whether we can place the two vertices in the drawing so that one row is reused. Two such cases are illustrated in Fig. 7. Observe that the top side of v and the bottom side of v' share the same row. If we have none of the above two cases, then no pairing is possible; in this case, we simply insert v and continue with the next vertex.

5 The Drawing Algorithm and Its Analysis

After the grouping/pairing procedure is complete, Algorithm BOX_ORTHOGONAL considers the vertices in the order of the final numbering, places them in the plane following the special instructions discussed above for groups, pairs and individual vertices, and draws the edges. The vertices are represented as described above (point or box), depending on the number of their incoming edges. Here is an outline of our algorithm:

Algorithm: *BOX_ORTHOGONAL*
Input: A graph G.
Output: An orthogonal drawing of G.

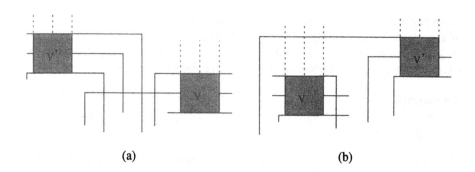

Fig. 7. Pairing two vertices in real drawing time so that one row is shared.

1. Compute an initial numbering of G having the properties specified in Section 3. If G is biconnected, the initial numbering is an st-numbering.
2. Modify the initial numbering to obtain the final numbering.
3. Apply the grouping/pairing procedure on the vertices of G.
4. Place vertex v_1 using a box of height one and width $outdeg(v_1) - 1$.
5. REPEAT
   (a) Consider next vertex v_i according to G's final numbering.
   (b) If v_i has already been placed, then continue with the next vertex.
   (c) If v_i does not belong to a group/pair, then try to pair it in real drawing time as discussed in the last subsection of the previous section. If this is not possible, place v_i in a way similar to the Simple Algorithm.
   (d) If v_i belongs to a group/pair, then place all members of this group/pair in the way described in the previous section for the specific kind of group/pair.
6. UNTIL there are no vertices left.
7. End.

As far as the total area of the drawing is concerned, observe the following: For every box we balance its incoming edges. This means that we attach half of them to the left and the other half to the right side of the box. We also introduce row reuse when placing problematic vertices. The row reuse is a result of grouping/pairing these vertices with other vertices. Every box u of the drawing with $outdeg(u) \geq 2$ occupies $outdeg(u)$ new columns. Sinks and vertices with only one outgoing edge are always placed on top of boxes with at least two outgoing edges (see Section 4.1). Sinks do not open up any new columns; $b - 1$ vertices placed along the rightmost outgoing edge of their predecessor box (see Fig. 3b) require one extra column.

An important advantage of this algorithm is that it introduces very few bends. More specifically, every edge of the drawing has at most one bend, no matter whether it enters a box through its right, left or bottom side. Also, all boxes/points representing vertices with only one outgoing edge have one incoming edge with no bends (see Fig. 3). Hence, the total number of bends of any drawing under Algorithm BOX_ORTHOGONAL is at most $m - n_{b-1}$ ($b \geq 1$), where n_{b-1} is the number of vertices with one outgoing edge. The complete (and lengthy) proof of the following theorem can be found in [14].

Theorem 3. *Algorithm BOX_ORTHOGONAL produces an orthogonal drawing of any graph G with m edges in $O(m)$ time. The produced drawing has the following properties:*

1. *each vertex v with degree $deg(v) > 3$ is represented by a box whose perimeter is less than $2 \times deg(v)$,*
2. *the width is at most $m - 1$,*
3. *the height is at most $\frac{m}{2} + 2$,*
4. *the total number of bends is less than m, and*
5. *each edge has at most one bend.*

In practice, we expect that Algorithm BOX_ORTHOGONAL will produce drawings that are better than the above bounds in terms of height, width, and total number of bends. More specifically, if more row reuse can take place when a new vertex is inserted, we take it. As discussed above, additional row reuse (and therefore smaller height for the drawing) may result from pairing that is done in real drawing time for vertices with an even number of more than two incoming and at least two outgoing edges. Finally, a *refining* step following the conclusion of the algorithm may reduce the area and bends further.

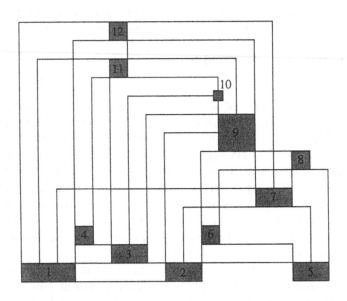

Fig. 8. An example drawing using Algorithm BOX_ORTHOGONAL.

In Fig. 8, we show a drawing produced by Algorithm BOX_ORTHOGONAL for an example nonplanar graph G with 12 vertices and 31 edges. Notice that G has one vertex with degree eight (vertex 9) and five vertices with degree six each (vertices 1, 2, 3, 7 and 11). In the drawing of Fig. 8 we have already applied the refining step which provided column reuse, thus saving more bends. The width of the final drawing is 17, the height is 14, and we have a total of 18 bends. Finally, note that the aspect ratio of the drawing is $\frac{17}{14} = 1.21$.

6 Contributions and Future Work

In this paper we presented algorithms for orthogonal drawing of graphs with degree higher than four. The major contributions of this work can be summarized in the following:

- Vertices are represented by boxes of perimeter less than twice the degree of the vertex,
- the area is low when compared to previous algorithms,
- our algorithms provide the first upper bound on the area of orthogonal drawings of high degree graphs,
- the width is larger than the height of the resulting orthogonal drawing (by approximately $\frac{m}{2}$) thus resulting in a good aspect ratio,
- the edges have either one bend or no bends; this increases the clarity of the drawing,
- our algorithm is geared towards high degree vertices; the performance in terms of area increases when the graph has many vertices of high degree,
- the input graph does not have to be biconnected; our algorithm avoids the extra overhead of dealing with non-biconnected graphs, and
- if the graph is not biconnected, the various components can be clearly identified in the drawing since all the vertices of the same component are drawn together.

It is an interesting open problem to find other representation schemes for vertices when their degree is higher than four. For the technique that we proposed, we would like to improve the upper bounds in the area and the number of bends. Results on interactive orthogonal graph drawing for graphs of maximum degree four can be found in [16]. Extending the interactive techniques to work for high degree graphs seems to be a natural direction for future research.

References

1. C. Batini, E. Nardelli, and R. Tamassia, *A Layout Algorithm for Data-Flow Diagrams*, IEEE Trans. on Software Engineering SE-12(4) (1986), pp. 538-546.
2. T. Biedl, *Orthogonal Graph Visualization: The Three-Phase Method With Applications*, Ph.D. Thesis, RUTCOR, Rutgers University, May 1997.

3. T. Biedl and G. Kant, *A Better Heuristic for Orthogonal Graph Drawings*, Proc. 2nd Ann. European Symposium on Algorithms (ESA '94), Lecture Notes in Computer Science, vol. 855, pp. 24-35, Springer-Verlag, 1994.

4. G. Di Battista, P. Eades, R. Tamassia and I. Tollis, *Algorithms for Drawing Graphs: An Annotated Bibliography*, Computational Geometry: Theory and Applications, vol. 4, no 5, 1994, pp. 235-282. Also available via anonymous `ftp` from `ftp.cs.brown.edu`, `gdbiblio.tex.Z` and `gdbiblio.ps.Z` in `/pub/papers/compgeo`.

5. G. Di Battista, A. Garg, G. Liotta, R. Tamassia, E. Tassinari and F. Vargiu, *An Experimental Comparison of Three Graph Drawing Algorithms*, Proc. of ACM Symp. on Computational Geometry, 1995, pp. 306-315. The version of the paper with the four algorithms can be obtained from http://www.cs.brown/people/rt.

6. S. Even and G. Granot, *Grid Layouts of Block Diagrams - Bounding the Number of Bends in Each Connection*, Proc. DIMACS Workshop GD '94, Lecture Notes in Comp. Sci. 894, Springer-Verlag, 1994, pp. 64-75.

7. S. Even and R.E. Tarjan, *Computing an st-numbering*, Theor. Comp. Sci. 2 (1976), pp. 339-344.

8. U. Foessmaier and M. Kaufmann, *Drawing High Degree Graphs with Low Bend Numbers*, Proc. Symposium on Graph Drawing (GD '95), Lecture Notes in Comp. Sci. 1027, Springer-Verlag, Sept. 1995, pp. 254-266.

9. Goos Kant, *Drawing Planar Graphs Using the Canonical Ordering*, Algorithmica, vol. 16, no. 1, 1996, pp. 4-32.

10. A. Lempel, S. Even and I. Cederbaum, *An algorithm for planarity testing in graphs*, Theory of Graphs, 215-232, Gordon and Breach, New York, 1967.

11. J. Pach and G. Toth, *Graphs Drawn with Few Crossings per Edge*, Proc. of Symposium on Graph Drawing (GD'96), Lecture Notes in Comp. Sci. 1190, Springer-Verlag, pp. 345-354.

12. A. Papakostas, *Information Visualization: Orthogonal Drawings of Graphs*, Ph.D. Thesis, Comp. Science Dept., The University of Texas at Dallas, November 1996.

13. A. Papakostas and I. G. Tollis, *Algorithms for Area-Efficient Orthogonal Drawings*, Technical Report UTDCS-06-95, The University of Texas at Dallas, 1995.

14. A. Papakostas and I. G. Tollis, *Orthogonal Drawing of High Degree Graphs with Small Area and Few Bends*, Technical Report UTDCS-04-96, The University of Texas at Dallas, 1996.

15. A. Papakostas and I. G. Tollis, *A Pairing Technique for Area-Efficient Orthogonal Drawings*, Proc. of Symposium on Graph Drawing (GD'96), Lecture Notes in Comp. Sci. 1190, Springer-Verlag, pp. 355-370.

16. A. Papakostas, J.M. Six, and I. G. Tollis, *Experimental and Theoretical Results in Interactive Orthogonal Graph Drawing*, Proc. of Symposium on Graph Drawing (GD'96), Lecture Notes in Comp. Sci. 1190, Springer-Verlag, pp. 371-386.

17. M. Schäffter, *Drawing Graphs on Rectangular Grids*, Discr. Appl. Math. 63 (1995), pp. 75-89.

18. J. Storer, *On minimal node-cost planar embeddings*, Networks 14 (1984), pp. 181-212.

19. R. Tamassia, *On embedding a graph in the grid with the minimum number of bends*, SIAM J. Computing 16 (1987), pp. 421-444.

20. R. Tamassia and I. Tollis, *Planar Grid Embeddings in Linear Time*, IEEE Trans. on Circuits and Systems CAS-36 (1989), pp. 1230-1234.

A Centroid Labelling Technique and Its Application to Path Selection in Trees
(Extended Abstract)

Sarnath Ramnath[1] and Hong Shen[2]

[1] Department of Computer Science, St Cloud State University
St Cloud, MN 56301, USA
[2] School of Computing and Information Technology, Griffith University
Nathan, QLD 4111, Australia

1 Introduction

Finding the kth longest path in a tree of n vertices with weighted edges is an interesting problem with both theoretical significance and practical applications [5, 8]. The known results require $O(n \log n)$ sequential time [5], and $O(\log^2 n)$ parallel time and $O(n \log^2 n)$ operations [6, 9], respectively.

In this paper we present a centroid labelling technique for weighted lists that is then used to compute all the centroids needed to decompose a tree in $O(\log n)$ time using $n/\log n$ processors on an EREW PRAM. Applying this technique, we obtain a parallel algorithm for selecting the k-th longest path from a tree in $O(\log n \log \log n \log^* n)$ time and $O(n \log n)$ operations on a EREW PRAM. To the best of our knowledge, our algorithm is the first work-optimal NC algorithm for solving this problem.

Due to the space limitations, we will omit the proofs to all lemmas and theorems. Interested reader may obtain them from [?].

2 Centroid labelling of a tree

We start with a few definitions and lemmas:

Let $T = (V, E)$ be a rooted binary tree with n, $2^p < n \le 2^{p+1}$, vertices. A *2-prune* of T at vertex v produces two trees: A tree $T_2^{2,v} = (V_1, E_1)$ which is the subtree rooted at v; A tree $T_1^{2,v}$, which is the subtree induced by the remaining vertices in T.

A *3-prune* of T at vertex v produces three trees: $T_1^{3,v} = (V_1, E_1)$, which is the subtree rooted at the left child of v; $T_2^{3,v} = (V_2, E_2)$, which is the subtree rooted at the right child of v; $T_3^{3,v}$, which is the subtree induced by the vertex set $V - V_1 - V_2$. (Note. The two upper subscripts specify the kind of prune and the vertex of pruning; the lower subscript specifies one of the resulting subtrees.)

Definition 1. A tree T, with root r, is said to be *centroid labelled* if we have assigned a pair of labels, *(first(T), last(T))*, such that exactly one of the following conditions hold:

1. *Type 1.* In this case, $first(T) = \alpha$ must be the index of a vertex in T, $second(T)$ is unassigned. All three trees $T_1^{3,\alpha}$, $T_2^{3,\alpha}$ and $T_3^{3,\alpha}$ must also be centroid labelled and each of these trees must contain no more than 2^p nodes. We then say that T is a Type 1 tree.

2. *Type 2.* In this case, both $first(T)$ and $second(T)$ are unassigned, r has only one child, and the subtree rooted at r's child is of Type 1.

3. *Type 3.* In this case, $first(T) = \alpha$ must be the index of a vertex in T and $second(T) = \beta$ is either *null*, or the index of a vertex in $T_2^{2,\alpha}$, such that

   (a) If $second(T)$ is null, then both $T_1^{2,\alpha}$ and $T_2^{2,\alpha}$ must be centroid labelled and each of these must *either* (i) contain no more than 2^p nodes *or* (ii) be a Type 2 tree.

   (b) If $second(T)$ is not null, then all three trees $T_1^{2,\alpha}$, $(T_2^{2,\alpha})_1^{2,\beta}$ and $(T_2^{2,\alpha})_2^{2,\beta}$, resulting from a 2-prune at α followed by a 2-prune at β must be centroid labelled and each of these must *either* (i) contain no more than 2^p nodes *or* (ii) be a Type 2 tree.

   We then say that T is a Type 3 tree.

Lemma 2. *Let T be any tree at level l of the decomposition, T containing n, $n \le 2^{p+1}$ nodes. Then, all trees at level $l + 3$ of the decomposition will have at most 2^p nodes.*

Proof: [11].

This centroid decomposition using this labelling thus reduces the the tree to a single vertex in at most $3 \log n$ levels.

We therefore need to solve the following problem:

Problem: *Given a binary tree T, compute a centroid labelling of T.*

Since our algorithm for centroid labelling uses the centroid path decomposition of Cole and Vishkin, we shall first define the centroid-path decomposition problem.

Let $dist(x, y)$ denote the distance from vertex x to vertex y in the tree, and $SIZE(v)$ be the number of nodes in the subtree of T rooted at v.

Following the definitions in [4], we have that the centroid level of v is $\lceil \log SIZE(v) \rceil$, and the *centroid path* of v is the longest path passing through v such that all the nodes on the path have the same centroid level as v. Starting at the root of T, we can partition T into classes of centroid paths $C - PATH_{\lceil \log n \rceil}$, $C - PATH_{\lceil \log n \rceil - 1}, \ldots, C - PATH_1$, where $C - PATH_i$ contains all centroid paths at level i for $1 \le i \le \lceil \log n \rceil$. We call this partition the *centroid-path decomposition*.

Lemma 3. *In the centroid-path decomposition of binary tree T, each vertex in T belongs to exactly one centroid path.*

Proof: [11].

The above lemma implies that the total number of centroid paths generated by centroid-path decomposition is at most $|T| = n$.

Lemma 4. *Let path $P(u \to v) \in C - PATH_i$ starting from node u and ending at node v, v_l and v_r be v's left and right children respectively. Then we have*

$$\max\{SIZE(u) - SIZE(v) + 1, SIZE(v_l), SIZE(v_r)\} \leq 2^{i-1}.$$

Proof: [11].

Returning to the centroid labelling problem, consider the situation where the root r' of the subtree $T' = (V', E')$, $2^k <| V' | \leq 2^{k+1}$, that we are currently interested in also happens to be the first node (head) in a centroid path of the original tree T. In that situation, a 3-prune operation at the last node (tail) of this centroid path will produce three subtrees each of which contains no more than 2^k nodes. Clearly assigning the index of the *tail* of this centroid path to first(T') and leaving second(T') unassigned will suffice.

Now consider the problem of recursively decomposing the subtree $T_3'^{3,tail}$ created by this 3-prune. A moments reflection reveals that the centroid-path decomposition does not give us enough infomation to decide how this tree is to be further decomposed, because, by Lemma 2, it excludes each centroid path, once generated, from further decomposition, whereas we need to decompose all the resulting subtrees in the next stage. What we shall do here is to identify one/two nodes on this path such that carrying out a 2-prune on these node(s) yields two/three subtrees satisfying the constraints in Clause 3 of Definition 1.

Our algorithm for centroid labelling therefore runs in two phases:

Phase 1: Find the centroid-path decomposition for T.

Phase 2: Carry out the centroid labelling for every centroid path of T.

We define some more notation. Let $v_1, \ldots, v_c$ be the vertices on the centroid path $p + 1$. For every vertex v_i, $1 \leq i \leq c - 1$, let $orphan(v_i)$ be the child of v_i that is not on the centroid path; $size(v_i)$ be the number of nodes in the subtree rooted at v. For all v_i, define $weight(v_i) = 1 + size(orphan(v_i))$.

We can now view the problem as follows: *we have a list of c nodes each with some positive integer weight. We have to compute a centroid labelling for the nodes on this list.* We shall now define the centroid labelling of a weighted list in a manner analogous to the one for trees.

Definition 5. Let S be a list of nodes $v_1, \ldots v_c$, $c > 1$, with positive integer weights, such that the sum of all weights of nodes in S is N, $2^p < N \leq 2^{p+1}$, for some p. Let S_{ij} denote the contiguous segment $v_i, \ldots v_j$ of S, and $| S_{ij} |$ denote the *weight* of S_{ij}, i.e., $\Sigma_{m=i}^{j}(weight(v_m))$. S is said to be *centroid labelled* if we have assigned a pair of indices $(first(S), second(S))$ to S, such that:

1. $1 < first(S) = \alpha \leq c$.
2. Exactly one of the following conditions hold:
   (a) $second(S)$ is null. In that case we have *either*
       (i) $| S_{1(\alpha-1)} |$ and $| S_{\alpha c} |$ are both $\leq 2^p$
       or (ii) $| S_{1(\alpha-1)} | \leq 2^p$ and $\alpha = c$,
       or (iii) $| S_{\alpha c} | \leq 2^p$ and $\alpha = 2$.

(b) $\alpha < second(S) = \beta \le c$. In that case we have $\mid S_{1(\alpha-1)} \mid + \mid S_{\alpha(\beta-1)} \mid >$
2^p, $\mid S_{\alpha(\beta-1)} \mid + \mid S_{\beta n} \mid > 2^p$ and *either*
   (i) $\mid S_{1(\alpha-1)} \mid$, $\mid S_{\alpha(\beta-1)} \mid$ and $\mid S_{\beta n} \mid$ are all $\le 2^p$;
or (ii) $\alpha = 2$ and $\mid S_{\alpha(\beta-1)} \mid$, $\mid S_{\beta n} \mid \le 2^p$;
or (iii) $\beta = \alpha + 1$ and $\mid S_{1(\alpha-1)} \mid$, $\mid S_{\beta n} \mid \le 2^p$;
or (iv) $\mid S_{1(\alpha-1)} \mid$, $\mid S_{\alpha(\beta-1)} \mid \le 2^p$ and $\beta = c$.

If $c = 1$, we shall say that the list is centroid labelled by definition. The motivation behind this (somewhat contrived) definition is captured in the following lemma.

Lemma 6. *Given a tree T, a centroid-path decomposition of T followed by a centroid labelling of all the centroid paths of T yields a centroid labelling of T.*

Proof: [11].

We shall now describe an algorithm for computing centroid labels for a list. We shall associate a doubly linked chain, $Chain(v)$, with each node v. $\mid Chain(v) \mid$ denotes the number of items in $Chain(v)$; initially all the chains are empty, i.e., $\mid Chain(v) \mid = 0$. (*Note: These chains are simply doubly linked lists; we call them chains so as not to confuse the terminology being used to describe the list S of nodes that has to be centroid labelled.*) $Top(v)$ is the pointer to the first item in $Chain(v)$. Each item on the chain has four fields: *first* and *second* that are to be used for the centroid labelling, and *next* and *prev* that serve as pointers. Let ptr_i be a pointer to some item in $Chain(v_i)$. The operation *Insert(ptr_i, first, second)* inserts a new item containing the values *first* and *second* just above the item pointed to by ptr_i, as described by the following code segment:

```
newitem := create(first, second);
If Top(vi) = ptri then Top(vi) := newitem; end if;
newitem.next := ptri;
newitem.prev := ptri.prev;
newitem.prev.next := newitem;
ptri.prev := newitem.
```

To begin with, we have c singleton lists, each of which is centroid labelled by definition. In each step we merge pairs of adjacent lists to create larger centroid labelled lists. Before we describe the algorithm, we shall explain how this data structure is being used. Let it be that at some stage of the algorithm, we have computed a centroid labelling for S_{ij}. At this stage, the first item in $Chain(v_i)$ will store the values $(first(S_{ij}), second(S_{ij}))$ $(= (\alpha, \beta)$, say) in its first and second fields. The next item in $Chain(v_i)$ will store $(first(S_{i(\alpha-1)}), second(S_{i(\alpha-1)}))$; the first item in $Chain(v_\alpha)$ will store $(first(S_{\alpha(\beta-1)}), second(S_{\alpha(\beta-1)}))$ and so on. Using the chains, we can thus recursively decompose the list into smaller sublists.

Algorithm *CLabel* describes how to compute a centroid labelling for a list.

Algorithm CLabel
1. Initialize all the Chains, i.e., set Top to null.
2. For $t := 1, \ldots, \log n$
    $\forall i$, such that $i = 1 \pmod{2^t}$, *in parallel* do
    LMerge $(S_{i(i+2^{t-1}-1)}, S_{(i+2^{t-1})(i+2^t-1)}, Top(v_i), Top(v_j))$.
  end for;
end CLabel

The following recursive algorithm *LMerge* explains how two adjacent lists are to be merged. (Note. The labels given to the various cases in the algorithm description correspond to the cases in the following proof of correctness.)

Algorithm LMerge $(S_{ij}, S_{(j+1)k}, ptr_i, ptr_j)$
  /* ptr_i and ptr_{j+1} are pointers to items in Chain(v_i) and Chain(v_{j+1}) respectively. */
  If $i = j$ and $j + 1 = k$, then $Insert(ptr_i, j + 1, null)$; Exit.   /* case i*/
  Let m be the integer such that $2^m <| S_{ij} |\leq 2^{m+1}$, and l be such that $2^l <| S_{(j+1)k} |\leq 2^{l+1}$.
    If $l = m$, then $Insert(ptr_i, j + 1, null)$; Exit.   /* case ii*/
    If $l < m$ then
        if $i = j$ then $Insert(ptr_i, j + 1, null)$; Exit.   /* case iii(a)*/
        if $| S_{ij} | + | S_{(j+1)k} |\leq 2^{m+1}$ then
          if $ptr_i.second = \beta \neq null$ then
            LMerge$(S_{\beta j}, S_{(j+1)k}, ptr_\beta, ptr_{j+1})$;   /* case v(a)*/
          else
            $\alpha := ptr_i.first$;
            if $| S_{\alpha j} | + | S_{(j+1)k} |\leq 2^m$ then
              LMerge$(S_{\alpha j}, S_{(j+1)k}, ptr_\beta, ptr_{j+1})$;   /* case v(b)*/
            else
              $ptr_i.second := j + 1$; Exit.   /* case iii(c)*/
            end if;
          end if;
        else
          $Insert(ptr_i, j + 1, null)$; Exit.   /* case iii(b)*/
        end if;
    end if;
    If $l > m$ then
        if $(j + 1 = k)$ then $Insert(ptr_i, j + 1, null)$; Exit.   /* case iv(a)*/
        if $| S_{ij} | + | S_{(j+1)k} |\leq 2^{l+1}$ then
          $\alpha := ptr_{j+1}.first$;
          if $ptr_{j+1}.second = \beta \neq null$ then
            $Insert(ptr_i, \alpha, \beta)$;
            $Top(v_{j+1}) := ptr_{j+1}.next$; $Delete(ptr_{j+1})$;
            LMerge$(S_{ij}, S_{(j+1)(\alpha-1)}, ptr_i, ptr_{j+1}.next)$;   /* case vi(a)*/
          else
            if $| S_{ij} | + | S_{(j+1)(\alpha-1)} |> 2^l$ then

$Insert(ptr_i, j + 1, \alpha)$;
$\text{Top}(v_{j+1}) := ptr_{j+1}.next$; $Delete(ptr_{j+1})$; Exit. /* case iv(c)*/
else
$Insert(ptr_i, \alpha, null)$;
$\text{Top}(v_{j+1}) := ptr_{j+1}.next$; $Delete(ptr_{j+1})$;
$LMerge(S_{ij}, S_{(j+1)(\alpha-1)}, ptr_i, ptr_{j+1}.next)$;   /* case vi(b)*/
end if;
end if;
else
$Insert(ptr_i, j + 1, null)$; Exit.   /* case iv(b)*/
end if;
end if;
end LMerge

We use an Exit statement rather than Return, because the control need never return to the calling program. Each recursive call therefore causes the remaining task to be "pipelined" into the next recursive call. As we shall see, this is crucial in establishing the $O(\log N)$ running time.

Before we prove correctness, we establish the following property of the centroid labeling produced by CLabel.

Property 1: Let S_{ij}, $2^p <| S_{ij} |\leq 2^{p+1}$, be a sublist that has been centroid labelled by CLabel, such that $first(S_{ij}) = \alpha$ and $second(S_{ij}) = \beta \neq null$. The following inequalities must hold

$| S_{i(\alpha-1)} | + | S_{\alpha(\beta-1)} |> 2^p$ and $| S_{\alpha(\beta-1)} | + | S_{\beta j} |> 2^p$.

Proof: [11].

Lemma 7. *Given two centroid labelled sublists S_{ij} and $S_{(j+1)k}$, the algorithm LMerge correctly computes the centroid labelling for S_{ik}.*

Proof: [11].

We denote the levels in the recursive structure of algorithm LMerge as follows: when a call is made from CLabel, we say this call is at level 0; the next recursive call (from LMerge) is at level 1, and so on.

Lemma 8. *The number of levels of recursion needed to merge S_{ij} and $S_{(j+1)k}$ (i.e, the number of stages in the round that merges S_{ij} and $S_{(j+1)k}$) is at most $min(\lceil\log(max(| S_{ij} |,| S_{(j+1)k} |))\rceil, k-i)$. Consequently, the number of levels of recursion in the entire algorithm is bounded above by $min(c - 1, \lceil\log N\rceil)$.*

Proof: [11].

Theorem 9. *Algorithm CLabel computes the centroid labelling of a given weighted list in $O(min(c, \log N))$ time using c processors on an EREW PRAM.*

Proof: [11].

Theorem 10. *The centroid labelling of a given binary tree T can be completed in $O(\log n)$ time and n processors on an EREW PRAM.*

Proof: [11].

2.1 An optimal algorithm

We shall now show how to obtain an optimal parallel algorithm for the centroid labelling problem of a weighted list. Although this is not central to the main result of the paper, we include this for the sake of completeness.

Proposition 11. *The centroid labelling of a weighted list of c nodes can be carried out in $O(c)$ sequential steps.*

This is done using a "greedy" approach. In the first step, pick a pair of neighbors (v_i, v_{i+1}) such that $weight(v_i) + weight(v_{i+1})$ is the least. These are then combined into one "supernode", and the process is repeated until we are left with one node. Some doubly linked data-structures are needed for the book-keeping. We omit the details.

Theorem 12. *The centroid labelling of a weighted list of c nodes can be carried out in $O(min(c, \log N))$ parallel steps using $\frac{c}{\log N}$ processors on an EREW PRAM.*

Proof: [11].

3 Selecting the k-th longest path

We shall now outline the complete algorithm for selecting the k-th longest path.

Algorithm Tree-select (T)
1. Convert T into a binary tree T', as explained in Section 2.
2. Compute a centroid labelling for T' using CLabel.
3. Construct a set of sorted matrices that capture all the intervertex distances in T.
4. Select the k-th element from the set of sorted matrices.
end Tree-select

Steps 1 and 2 have already been described in detail. Step 4 can be carried out using the algorithm in [10]. We shall now expand on Step 3.

Our approach is as follows: given a tree T, we use $first(T)$ and $second(T)$ to decompose T into subtrees. We then solve the subproblems recursively until we are down to trees of a small size. Using the solutions to the subproblems, we can construct the solution for the original problem. The subtree corresponding to each subproblem is characterized by two vertices - *head* and *tail* - which connect the subtree to the other vertices in the tree. The *head* is the root of the subtree;

the *tail* is the vertex at which the 2-*prune* (or 3-*prune*) that generated this subtree was carried out. Note that *tail* would be undefined (or *nil*) unless *head* was the root of some "supertree" whose decomposition produced this subtree.

If X is any matrix, and e is an element, let $X + e$ denote the matrix obtained by adding e to every element in X. Let $dist(x, y)$ denote the distance from vertex x to vertex y in the tree, and $size(v)$ be the number of nodes in the subtree of T rooted at v. We use a variable $MSet$ to store the set of sorted matrices. Each sorted matrix is either of the form $X + Y$ and represented by the pair of sorted arrays (X, Y), or it is a sorted one-dimensional array, represented by X.

Algorithm Construct-Matrix(T, r, $tail$, X, Y).
/* T is a tree with root r; $tail$ is a specially designated vertex, which may be nil; X (resp. Y) is an output parameter used to return the distance of each *real* node in T from the root r (resp *tail*). */

3.1 If $|T| \leq c$ (for some suitable constant c) then $MSet := MSet \cup \{$all intervertex distances between real vertices of T arranged in a sorted array$\}$;
$X := \{$ distance from r of all real nodes of T arranged in a sorted array$\}$;
If $tail \neq null$ then $Y := \{$ distance from $tail$ of all real nodes of T arranged in a sorted array$\}$;
Return.

3.2 If T is a tree of Type 2, then
$r_1 :=$ child of r that is a vertex in T; $T_1 :=$ subtree rooted at r_1.
Construct Matrix($T_1, r_1, null, X_1, Y_1$);
$X := X_1 + dist(r_1, parent(r_1))$; $Y := nil$;
If r is a *real* node then
$MSet := MSet \cup \{X\}$; Return.

3.3 $\alpha := first(T)$;

3.4 If T is a tree of Type 1, then
Construct-Matrix($T_1^{3,\alpha}$, $lchild(\alpha)$, nil, X_1, Y_1);
Construct-Matrix($T_2^{3,\alpha}$, $rchild(\alpha)$, nil, X_2, Y_2);
Construct-Matrix($T_3^{3,\alpha}$, r, α, X_3, Y_3);
$MSet := MSet \cup \{(X_2 + dist(\alpha, rchild(\alpha)), X_1 + dist(\alpha, lchild(\alpha))),$
$(Y_3, X_1 + dist(\alpha, lchild(\alpha))), (Y_3, X_2 + dist(\alpha, rchild(\alpha)))\}$;
$X := Merge(X_1 + dist(r, lchild(\alpha)), X_2 + dist(r, rchild(\alpha)), X_3,)$;
$Y := nil$; Return;

3.5 $\beta := second(T)$;

3.6 If $\beta = null$ then
Construct-Matrix($T_1^{2,\alpha}$, r, $parent(\alpha)$, X_1, Y_1);
Construct-Matrix($T_2^{2,\alpha}$, α, $tail$, X_2, Y_2);
$MSet := MSet \cup \{(Y_1, X_2 + dist(\alpha, parent(\alpha)))\}$
$X := Merge(X_1, X_2 + dist(r, \alpha))$;
$Y := Merge(Y_2, Y_1 + dist(tail, parent(\alpha)))$; Return;

3.7 Construct-Matrix($T_1^{2,\alpha}$, r, $parent(\alpha)$, X_1, Y_1);
Construct-Matrix($((T_2^{2,\alpha})_1^{2,\beta}$, α, $parent(\beta)$, X_2, Y_2);
Construct-Matrix($((T_2^{2,\alpha})_2^{2,\beta}$, β, $tail$, X_3, Y_3);

$MSet := MSet \cup \{(Y_1, X_2+dist(\alpha, parent(\alpha))), (Y_1, X_3+dist(\beta, parent(\alpha))), (Y_2, X_3+$
$dist(\beta, parent(\beta)))\};$
$X := Merge(X_1, X_2 + dist(r, \alpha), X_3 + dist(r, \beta));$
$Y := Merge(Y_1 + dist(tail, parent(\alpha)), Y_2 + dist(tail, parent(\beta)), Y_3);$ Return;
end Construct-Matrix

The initial call to Construct-Matrix is Construct-Matrix (T', r', nil, X, Y)
where T' is the result of binarizing T, and r' is the root of T'. Note that all the
dummy vertices are filtered out by excluding them from the sorted matrices.

Lemma 13. *Algorithm Construct-Matrix runs in at most $3 \log n$ levels of recursion and computes all intervertex distances in T.*

Proof: [11].

Since X and Y computed in each level recursion are each sorted, $MSet$ contains a set of sorted matrices. Moreover, since at any level of recursion, the sum
of the sizes of all subtrees after decomposition is always bounded by n (the size
of the original tree), the sum of dimensions of sorted matrices representing the
intervertex distances among these subtrees is $O(n)$. By Lemma 8 the following
lemma is immediate:

Lemma 14. *The sum of one-side dimensions of all sorted matrices in $MSet$ is $O(n \log n)$.*

The number of levels of recursion is $3 \log n$ by Lemma 8. We need to carry out
a merge operation in each recursive call (return) that can be either 2-way merge
(3.6) or 3-way merge (3.4 and 3.7). Each 3-way merge operation can be broken
up as two 2-way merge. This results in a 2-way merge tree T_M^2 of height at most
$6 \log n$ (corresponding to the worst-case 3-way merge tree, T_M^3, of height $3 \log n$)
composed of all the merges to be performed in all recursive calls to compute
all sorted arrays X's and Y's. The total number of nodes in each level of T_M^2 is
clearly n. Employing Cole's cascaded mergesort [2], we can have all the merges in
T_M^2 (and hence in T_M^3) computed, in bottom-up (pipelining) fashion in $O(\log n)$
time using n processors on an EREW PRAM.

We can easily show that generating the sorted matrices is equivalent to the
generalized merging problem [11], and hence have the following lemma:

Lemma 15. *Algorithm Construct-Matrix can be completed in $O(\log n)$ time using n processors on an EREW PRAM.*

Let the set $MSet$ produced by algorithm Construct-matrix consist of sorted
matrices $S_1, S_2, \ldots, S_t$. Clearly S_i is a square matrix and let $|S_i| = n_i \times n_i$. We
now need to select the kth largest element in S.

We construct a sorted matrix S' as follows: place $S_1, \ldots, S_t$ along the main
diagonal of S', where each S_i occupies disjoint rows and columns from the others;
set all the elements below S_i to $-\infty$ and all those above S_i to ∞. In this setting,

S' has dimensions $n' \times n'$, where $n' = \sum_{i=1}^{t} n_i$. Now selecting the kth largest element in S is equivalent to selecting the k'th largest in S', where $k' = k + \sum_{i=1}^{t}(n_i \sum_{j=i+1}^{t} n_j)$, which is equivalent to selecting the $(n'^2 - k' + 1)$th smallest element in S'. By Lemma 9 we know that n' is $O(n \log n)$. Using the optimal selection algorithm for sorted matrices [10], we have the following lemma:

Lemma 16. *Step 4 in Algorithm Tree-Select for selecting the kth largest element in MSet can be done in $O(\log n \log \log n \log^* n)$ time and $O(n \log n)$ operations on an EREW PRAM.*

We have thus shown that all four steps of algorithm Tree-select can be carried out in $O(\log n \log \log n \log^* n)$ time and $O(n \log n)$ operations on an EREW PRAM, giving us the following result.

Theorem 17. *Given a tree with n vertices, we can find the kth longest path in T in $O(\log n \log \log n \log^* n)$ time and $O(n \log n)$ work on an EREW PRAM.*

References

1. M. Atallah, R. Cole and M. Goodrich. Cascading divide-and-conquer: A technique for designing parallel algorithms. *SIAM J. Computing*, 499-532, 1989
2. R. Cole. Parallel merge sort. *SIAM Journal on Computing*, 17:770–785, 1988.
3. R. Cole. Personal Communication.
4. R. Cole and U. Vishkin. The accelerated centroid decomposition technique for optimal parallel tree evaluation in logarithmic time. *Algorithmica*, 3:329–346, 1988.
5. G. N. Frederickson and D. B. Johnson. Finding kth paths and p-centers by generating and searching good data structures. *J. Algorithms*, 4:61–80, 1983.
6. He and Yesha. Efficient parallel algorithms for r-dominating set and p-center problems on trees. *Algorithmica*, 5, 1990.
7. O. Kariv and S. L. Hakimi. An algorithmic approach to network location problems i: The p-centers. *SIAM J. Appl. Math.*, 37:539–560, 1979.
8. N. Megiddo, A. Tamir, E. Zemel, and R. Chandrasekaran. An $O(n \log^2 n)$ algorithm for the kth longest path in a tree with applications to location problems. *SIAM Journal on Computing*, 1981.
9. H. Shen. Fast parallel algorithm for finding the kth longest path in a tree. Proc. International Conference on Advances in Parallel and Distributed Computing (APDC'97), Shanghai, 1997.
10. H. Shen and S. Ramnath. Optimal parallel selection in sorted matrices. *Information Processing Letters*, 59:117–122, 1996.
11. S. Ramnath and H. Shen. Optimal Parallel Algorithm for Selecting the kth Longest Path in A Tree. Technical Report, St Cloud State University, 1997.

Offset-Polygon Annulus Placement Problems*

Gill Barequet[1], Amy J. Briggs[2], Matthew T. Dickerson[2], and Michael T. Goodrich[1]

[1] Center for Geometric Computing, Dept. of Computer Science, Johns Hopkins University, Baltimore, MD 21218. E-mail: [barequet|goodrich]@cs.jhu.edu
[2] Department of Mathematics and Computer Science, Middlebury College, Middlebury, VT 05753. E-mail: [briggs|dickerso]@middlebury.edu

Abstract. In this paper we address several variants of the polygon annulus placement problem: given an input polygon P and a set S of points, find an *optimal* placement of P that maximizes the number of points in S that fall in a certain annulus region defined by P and some offset distance $\delta > 0$. We address the following variants of the problem: placement of a convex polygon as well as a simple polygon; placement by translation only, or by a translation and a rotation; off-line and on-line versions of the corresponding decision problems; and decision as well as optimization versions of the problems. We present efficient algorithms in each case.

Keywords: optimal polygon placement, tolerancing, robot localization, offsetting.

1 Introduction

1.1 Background and Applications

In this paper we address several variants of the problem of placing an annulus defined by a given polygon such that it covers all (or a maximum number of) points of a given set of points. This problem is motivated by several applications. For example, in the *robot localization* problem (see, e.g., [GMR]), a robot should determine its current location in some environment map from a set of points obtained by a distance range sensor. Due to the inherent errors in range finding, the points usually do not define an exact match. Most points, however, fall within some distance $\delta > 0$ of the environment boundary. Thus the localization problem can be viewed as finding some *optimal* placement of the environment model (typically a polygon) with respect to the set of points and a distance $\delta > 0$. A second application is a pattern matching problem arising in computer vision (see, e.g., [HU]), where the input consists of a set of points taken from some image and a pattern (polygon) that one would like to locate in this image. A good match can be found by determining a placement of the polygon that maximizes the number of points within some distance $\delta > 0$ of the image points. Yet another application arises in geometric tolerancing. Chang and Yap [CY]

---

* Work on this paper by the first and the fourth authors has been supported in part by the U.S. ARO under Grant DAAH04-96-1-0013. Work by the third author has been supported in part by the National Science Foundation under Grant CCR-93-1714. Work by the fourth author has been supported also by NSF grant CCR-96-25289.

describe geometric tolerancing as being concerned with the specification of geometric shapes for use in manufacturing of mechanical parts, and they note that, since manufacturing processes are inherently imprecise, it is imperative that such geometric designs be accompanied by tolerance specifications. An instance of the tolerancing problem is to take a set of points representing an actual measurement of a manufactured object (using a coordinate-measuring machine, laser range-finder, or scanning electron microscope [DMSS]) and determine whether the manufactured object matches a polygon (the design) within some tolerance $\delta > 0$. This corresponds, for example, to the *tolerance zone* semantics described by Requicha [Re], Srinivasan [Sr], and Yap [Ya].

1.2 Previous Related Work

The notion of polygon annulus placement relative to a set of points appears to be new in the computational geometry literature. There are nevertheless several related problems that have been studied before, including variants directed at placing an entire polygon (not an annulus) to cover a set or subset of points (see, e.g., [ESZ, EE, BDP, DS]). These problems do not model important aspects for optimizing polygon placement (as mentioned in the applications above). Previous work directed at annulus problems, on the other hand, have dealt exclusively with circular annuli (see, e.g. [HT, LL, AST, AS, SJ, SLW, DGR]). These characterizations capture well the notion of "roundness" present in a set of points, but they do not easily extend to polygonal shape matching.

1.3 Definitions and Problems

We start with definitions for convex polygons to simplify the presentation. Extensions to simple polygons are made in Section 5.

Definition 1 (Offset Annulus) *The δ-annulus of a convex polygon P is the closed region defined by all points in the plane at distance at most δ from the boundary of P.*

Definition 2 (Offset Polygons) *Given a convex polygon P and a distance $\delta > 0$, the δ-offset polygons are defined as follows: The inner δ-offset polygon $I_{P,\delta}$ is the boundary portions of the δ-annulus of P that are properly contained by P. Similarly, the outer δ-offset polygon $O_{P,\delta}$ is the boundary portions of the δ-annulus of P outside of (i.e., properly containing) P.*

Note that $I_{P,\delta}$ is made up of edges that are parallel to edges of P (although there may be some edges of P that are not parallel to any in $I_{P,\delta}$). The offset polygon $O_{P,\delta}$, on the other hand, is made up of alternating line segments and circular arcs, and every edge of P is parallel to some edge of $O_{P,\delta}$. One can also imagine a fully *linearized* version of the outer offset polygon, where one extends each of the linear edges until they meet the extensions of neighboring linear edges. (For simplicity, we will first discuss algorithms for solving polygon-annulus problems adopting this linearized view, and we will then show how to

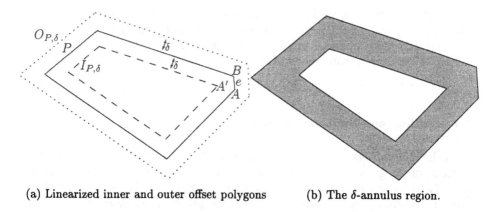

(a) Linearized inner and outer offset polygons   (b) The δ-annulus region.

Fig. 1. Offsetting a polygon

extend these to the more-natural standard notion of a δ-offset without affecting the running times by more than a constant factor.)

Figure 1(a) shows a convex polygon P (with solid edges) and its inner and linearized outer offset polygons $I_{P,\delta}$ and $O_{P,\delta}$ (with dashed and dotted edges, respectively) for some value of δ. Note that for any convex polygon P and for any value of δ the outer offset polygon $O_{P,\delta}$ always has the same number of edges as P, but the inner offset polygon $I_{P,\delta}$ may have fewer edges. In this example the edge $e \in P$ does not have a counterpart in $I_{P,\delta}$. More specifically, the point A, edge e, and point B, all in P, collapse into a single point A' in $I_{P,\delta}$. Also, unless P is a regular polygon, the offset polygons $I_{P,\delta}$ and $O_{P,\delta}$ are *not* scaled versions of P. The δ-annulus region of a polygon is shown shaded in Figure 1(b). Note that the annulus region is defined to include the boundary edges. Although these definitions are stated for convex polygons, we show that in many cases they can easily be extended to simple polygons (see Section 5). In any case, the definition of δ-annulus regions naturally gives rise to the following problems:

- **Offset-Polygon Max Cover:** Given a set S of n points in the plane, a convex polygon P, and a distance δ, find a placement τ of P that maximizes the number of points of S contained in the δ-annulus region of $\tau(P)$. Report the placement τ and the set of contained points.
- **Offset-Polygon Containment (Decision Version):** Given a set S of n points in the plane, a convex polygon P, and a distance δ, determine if there exists a placement τ of P such that *all* n points of S are contained in the δ-annulus region of $\tau(P)$. Report such a placement τ if one exists.
- **Offset-Polygon Containment (Optimization Version):** Given a set S of n points in the plane, and a convex polygon P, find the smallest value of $\delta > 0$ such that there exists a placement τ of P with *all* n points of S being contained in the δ-annulus region of $\tau(P)$. Report such a placement τ if one exists, together with this optimal value of $\delta > 0$.

Note that we can use an algorithm for either the offset-polygon max-cover problem or for the width-optimization problem to solve the offset-polygon containment decision problem. In particular, the answer for the decision problem is "yes" if and only if for the former problem the value of k—the maximum number of points contained in the δ-annulus for P—is n, or for the latter problem the value of δ'—the minimum width of an annulus that contains all the points—is δ.

1.4   Outline and Summary of Results

Let n be the number of input points and let m be the number of edges (and vertices) of the given polygon P. In this paper we give several results for solving the offset-polygon max-cover and containment problems. We show that if we restrict the offset-polygon containment decision problem to convex polygons under translations only, then we can determine a containing placement of P, if one exists, in $O(n \log n \log m + m)$ time. Our method involves a non-trivial extension of the roundness method of Duncan et al. [DGR] to offset polygons using the polygon-offset nearest-neighbor and furthest-neighbor diagrams [BDG]. Moreover, we show how to solve the optimization version of this problem in the same time bound, by using the simplest (and most practical) version of parametric searching.

We also study the offset-polygon max-cover problem for convex polygons under translations, showing that this more-general problem can be solved in $O(n^2 \log(nm) + m)$ time and $O(n + m)$ space. Our methods involve a non-trivial extension of the techniques of Barequet et al. [BDP]. In addition, we show how to solve this problem under translations and rotations by combining this approach with extensions of the rotation-diagram techniques of Dickerson and Scharstein [DS]. The resulting time bound in this case is $O(n^3 \log(nm) + m)$ using $O(n + m)$ space in the worst case. Under some very reasonable "fatness" conditions (which we make precise in Section 5), we show that our techniques can be generalized for simple polygons under translations to result in an algorithm running in $O(n^2 m^2 \log(nm))$ time and $O(nm^2)$ space.

In addition to the off-line results discussed above, we also describe a method, based upon an interesting dynamic data structure, that solves an on-line version of the offset-polygon containment decision problem under translations. The algorithm reads points one at a time, halting and answering "no" when a placement containing all points read so far is no longer possible, or, alternatively, running to completion on n points and answering "yes." In the worst case, this on-line algorithm runs in $O(n^2 m^2 \log(nm))$ time and $O(n^2 m^2)$ space for simple polygons. For many distributions of points, however, it performs significantly better. In particular, for convex polygons our on-line algorithm runs in $O(nh \log(nm) + m)$ time and requires only $O(nh + m)$ space, where h depends on the distribution (see Section 4.3). (In the worst case $h = \Theta(n)$, but for many distributions h is substantially smaller.)

The outline of the paper is as follows. We begin in Section 2 with some important geometric properties and primitives. In Section 3 we present the algorithms for convex polygons, and in Section 4 we give our on-line solution to the offset-polygon max-cover problem. In Section 5 we extend our solutions to the offset-polygon max-cover problem to simple polygons. We conclude with Section 6.

2 Key Geometric Properties

An important step of our algorithms is the computation of the intersections between translated copies of offset polygons. For simplicity of expression, let us assume we are dealing with linearized offset polygons; we show later how to remove this restriction to deal with the more-standard definition of δ-annulus region with only a constant-factor increase in the running times of our algorithms. Let us therefore consider an upper bound on the number of intersections between translated copies of linearized offset polygons, and a description of how to compute them. It is well-known that two translated homothetic copies of the same convex polygon can intersect at most twice (where in the degenerate case an intersection may be a segment rather than a point). The following theorem states that translations of an inner and outer offset convex polygon can also intersect at most twice.

Theorem 1. *Given a polygon P, a distance δ, and a translation τ, the offset polygons $\tau(I_{P,\delta})$ and $O_{P,\delta}$ intersect at most twice, where each intersection may be a point or (in the degenerate case) a segment.*

Proof Omitted in this version of the paper. $\square$

The technique used in this proof also provides the necessary framework for the proof of the following lemma (using the *tentative prune-and-search* technique of Kirkpatrick and Snoeyink [KS]).

Lemma 2. *The intersections between offset polygons $\tau(I_{P,\delta})$ and $O_{P,\delta}$ can be found in worst case $O(\log m)$ time, where m is the number of vertices of P.*

We compute these intersections because they correspond to placements of the annulus region such that two (or more) points of S are in contact with the boundary of the annulus region.

The following lemmas are generalizations of lemmas from [BDP, DS] that deal with intersections between two copies of the same polygon.

Lemma 3. *Let P be a convex polygon, q_1, q_2 points, and τ_1 and τ_2 the translations mapping the origin to points q_1 and q_2, respectively. For any point x, let $\tau_x = q_2 - x$ be the translation that maps x to q_2. Then both q_1 and q_2 are contained in the δ-annulus region of $\tau_x(\tau_1(P))$ if and only if x is contained in the intersection of the δ-annulus regions of $\tau_1(P)$ and $\tau_2(P)$.*

Lemma 4. *Let P be a convex polygon and S be a non-empty set of points contained in the δ-annulus of P. Then there exists a translation τ such that S is contained in the δ-annulus of $\tau(P)$ and at least one point of S is on the boundary of the annulus region.*

3 Algorithms for Convex Polygons

3.1 Offset-Polygon Containment under Translation

We first briefly describe a deterministic $O(n \log n \log m + m)$-time algorithm for solving the annulus-width optimization problem: Given a set S of n points and a convex polygon P with m vertices, find the minimum-width annulus of P that covers S. For this purpose we define the *convex polygon-offset* distance-function $\mathcal{D}_P$ that corresponds to P and compute the nearest- and furthest-site Voronoi diagrams of S with respect to $\mathcal{D}_P$ (see [BDG]). This can be performed in $O(n(\log n + \log m) + m)$ time. Next we use the method of [DGR] (where the authors minimize the width of a *circular* annulus) and consider the overlay of the two diagrams. As is well-known, the center of the minimum-width annulus that contains S is either a vertex of one of the two diagrams (possibly a vertex at infinity in the furthest-site diagram) or a point of intersection between the two diagrams. Given a specific value of δ, we place δ-annuli centered at all the points of S and observe (like in [DGR]) the overlay for determining whether the intersection of all annuli is nonempty. (The intersection contains the loci of all feasible placements of the annulus so that it covers S.) This step takes $O(n \log m)$ time. Finally, a parametric-searching algorithm is applied for optimizing (minimizing) the value of δ for which the intersection of all the annuli is nonempty. Over all, the whole procedure requires $O(n \log n \log m + m)$ time.

3.2 Offset-Polygon Max-Cover under Translation

In this section we consider offset-polygon max-cover under translation. Our algorithm extends the techniques of Barequet et al. [BDP] to allow for containment within the annulus region rather than containment by the entire polygon. The idea is to do an anchored sweep of *both* the inner and outer offset polygons around each point of S. The critical events of the sweep occur when some point of S either enters or exits the δ-annulus. The full algorithm is given in Figure 2. The correctness of this algorithm follows from Lemmas 3 and 4. There exists at least one optimal placement with a point in contact with the annulus boundary, and this placement will be found by the sweep. The only additional detail regards the processing of degenerate intersections, where the intersection between two offset polygons is a segment (along a connected portion of an edge) rather than a discrete point. In this case only one of the two endpoints of the segment corresponds to an event. If the point q_i is currently marked "in" then it is at the second endpoint of the intersection segment where it changes to "not in." Conversely for points marked "not in," it is at the first endpoint of the segment where it changes to "in." This follows from the fact that the entire segment corresponds to a translation in which both points q_i and q_j are on the boundary of the translated polygon and so points that are "in" remain so until the *end* of the segment, whereas points that are "not in" become "in" at the *start* of the segment.

We measure the complexity of our max-cover algorithm under translations as a function of two variables: m, the number of vertices of P, and n, the number

I. Preprocessing:
1. Preprocess offset polygons $I = I_{P,\delta}$ and $O = O_{P,\delta}$ for intersection computation.
2. Initialize a priority queue Q which will store points in clockwise order around the boundaries of the offset polygons I and O.

II. Iteration:
1. Set $max := 0$.   {# of points so far}
2. **FOR** each point $q_i \in S$ **DO BEGIN**   {Anchored sweep around q_i}
3.   Let P' be I.   {First sweep I}
4.   Set $c := 1$.   {Points contained}
5.   **FOR** each $j \neq i$ and $q_j \in S$ **DO BEGIN**   {Examine nearby points}
6.   Set $X := \{x | x \in \partial \tau_i(I) \cap \partial \tau_j(P')\} \bigcup \{x | x \in \partial \tau_i(O) \cap \partial \tau_j(P')\}$.
7.   **FOR** all $x \in X$ **DO**
8.   Add (x, j) to Q.   {Add intersections to queue}
           END FOR
9.   **IF** q_j is contained in the δ-annulus of $\tau_i(P)$ **THEN**
10.   Mark q_j 'In'; Set $c := c + 1$;   {Mark and count points}
           ELSE
11.   Mark q_j 'Not In'.
           END IF
       END FOR
12.   **WHILE** $Q \neq \emptyset$ **DO BEGIN**   {Sweep with intersections}
13.   Delete (x, j) from front of Q.   {Update structures}
14.   **IF** q_j is 'Not In' **THEN**   {See comments in text}
15.   Set $c := c + 1$; Mark q_j 'In'.
           ELSE
16.   Set $c := c - 1$; Mark q_j 'Not In'.
           END IF
17.   **IF** $c > max$ **THEN**
18.   Set $max := c$; Store translation.
           END IF
       END WHILE
19.   **REPEAT** steps 4 through 18 with $P' = O$. {Now sweep O}
   END FOR

Fig. 2. Max-cover algorithm under translations, for convex polygons

of points in the set S. The preprocessing step requires $O(m)$ time and space for computing and storing the offset polygons $I = I_{P,\delta}$ and $O = O_{P,\delta}$. The offset polygons are stored such that later intersection tests can be performed in $O(\log m)$ time and space (see [KS] and Section 2). The steps inside the inner nested loop execute $O(n^2)$ times. Since each pair of points has two offset polygons, each of which has at most two intersections with the polygon being swept, the total size of the queue is $O(n)$ and queue operations can be performed in $O(\log n)$ time. Polygon intersections in Step 6 can be computed in $O(\log m)$ time (by Lemma 3). The total running time is therefore $O(n^2 \log(nm) + m)$ in the worst case. The algorithm requires $O(n + m)$ space. (In the full version of the paper we show how to significantly improve the running time of the algorithm in many instances by the use of bucketing.)

3.3 Offset-Polygon Max-Cover under Translation and Rotation

We now describe how the offset-polygon max-cover problem can be solved for convex polygons when we allow for translations and rotations. To solve this problem we extend the results of Dickerson and Scharstein [DS] and make use of their *rotation diagram* technique. We refer the reader to [DS] for details on this method; here we describe only the necessary modifications in the approach and in the complexity analysis. This method creates a rotation diagram R_{q_i} for each point q_i. The diagram R_{q_i} is a description of the configuration space of all placements of the polygon P that keep the boundary of P in contact with q_i. The horizontal axis of this diagram represents the angle of rotation (from 0 to 2π). The vertical axis represents the arclength along ∂P (from 0 to the circumference of P). For each other point q_j, the rotation diagram for q_i includes the region of all such placements that contain q_j. It is shown in [DS] that this containing region for q_j can be decomposed into $O(m^2)$ subregions of constant complexity. The left and right boundaries of these subregions are certain critical angles of rotation, where vertices of one polygon pass through edges of another. The upper and lower boundaries are shown to be sine curves. To solve the optimal placement problem, the algorithm performs a plane sweep of each rotation diagram R_{q_i} to find the region of greatest depth. This gives the optimal placement of P that is in contact with q_i. The main difference for the annulus placement problem is that we need two rotation diagrams for each point q_i: one for the inner offset polygon $I_{P,\delta}$ and one for the outer offset polygon $O_{P,\delta}$. Furthermore, each of these two rotation diagrams for q_i has regions for each $q_j \neq q_i$ that represent containment in the annulus region rather than in the entire polygon. The following lemma states that these modified rotation diagrams have the same complexities.

Lemma 5. *For convex polygons, the polygon annulus containing regions for a given point is decomposed into $O(m^2)$ subregions each of which have constant complexity: vertical left and right boundaries and a sine curve for the top and bottom boundaries.*

The proofs of [DS] suffice to show that the upper and lower boundaries are still sine curves. The $O(m^2)$ is a trivial upper bound which is attainable. There is however a constant factor increase in the complexity of the diagrams. The number of critical angles are doubled because we now count intersections of both the inner and outer polygon placed at point q_i and either the inner or outer polygon at q_j (depending on which rotation diagram we are computing). Therefore, since the number of subregions can double, the number of intersection points can increase by a factor of four. To solve the offset-polygon max-cover problem we use the same idea of the rotation diagram and perform plane sweeps of each of the $2n$ diagrams. Lemma 4 tells us that this suffices because even with a restriction to translation only there is at least one optimal placement that has a point on an inner or an outer boundary of the annulus region. Thus we can state the following theorem:

Theorem 6. *The convex offset-polygon maximum-cover problem can be solved in $O(n^3 \log(nm) + m)$ time and $O(n + m)$ space in the worst case for translation and rotation.*

3.4 True δ-Tolerancing

As mentioned earlier, our algorithms assume a linearized outer polygon boundary. For adapting the linearized versions of the offset-polygon max-cover and offset-polygon containment problems to their (standard) non-linear forms, we need only show that the framework for the offset-annulus translation variant works also for the true δ-tolerancing case. The key to this adaptation lies in the fact that for every convex polygon P, tolerance δ, and a translation τ, the number of intersections of $\tau(I_{P,\delta})$ and the true outer boundary $O_{P,\delta}$ is still at most two. The proof of this claim is almost identical to that of Theorem 1 (see Section 2). Indeed, the weak monotonicity of the curves is preserved (we do not need the curves to be piecewise-linear). Furthermore, we can still apply the prune-and-search technique, since the simplicity of the pieces of the curves is also maintained: it takes a constant amount of time to evaluate the intersection of a circular arc with a line segment or with another circular arc. Therefore we are able to apply the same algorithm (for the translation-only variant) as in Section 3.2 and obtain the same asymptotic running time and space. In the full version of the paper we explain how to extend also the translation and rotation version of the problem for the true δ-Tolerancing case.

4 An On-Line Decision of the Containment Problem

In the previous section we provided solutions to several variants of the offset-polygon max-cover and containment problems, under various rigid transformations. In this section we present an alternate "on-line" approach to offset-polygon containment decision problems for the translation-only case. As before, we assume convex polygons and deal with simple polygons in a later section. The idea of this on-line approach is that instead of being given the entire set S at once, the points are read one at a time, and for each new point we decide whether there is a placement of the annulus region of P that contains all the points *seen so far*. There are several motivations for the on-line approach. One is that for the decision problem we need not necessarily process the entire point set; if after a certain number of points there is no longer a placement containing them all then we can halt immediately and answer 'No' (thus offering some savings in running time over unnecessarily processing all the points). This may be particularly useful for the tolerancing problem. A second advantage is the ability to process incoming points *as they arrive* while simultaneously reading subsequent points (a form of pipelining). This is an advantage in the cases of the proposed applications where the points are not stored in a file but are read one-at-a-time by an external device. A third possible advantage is that as more points are read we can slowly *refine* the space of possible placements of P. This can be helpful for both the robot localization and geometric tolerancing problems where we might direct the input device for further measurements. Finally, the on-line approach allows for the *pruning* of the data structures providing a more efficient approach for most practical applications.

4.1 Basic Algorithm Approach

We begin with the basic ideas of the on-line approach. We want to read input points one at a time. For each point q_i we construct and store a data structure (similar to that of Algorithm 1) that maintains optimal placements of the annulus region around P in contact with q_i. We also update the data structures for the existing points q_j for $j < i$. That is, for each $j < i$ we: (1) Compute the translations that keep the annulus region of P in contact with q_i and contain q_j and add this information to the new data structure of q_i; and (2) Compute translations that keep the annulus region in contact with q_j and contain q_i and update the data structure of q_j. Remember that for each point q_j these translations are computed from the intersections of the translated offset polygons in $O(\log m)$ time by Lemma 2. However our use of data structures for the on-line algorithm differs in two ways from Algorithm 1. The first difference is that (unfortunately) we need to store several data structures simultaneously, rather than computing the optimal placement for one and then discarding it. This is because each data structure is continually being updated as new points are added. The second difference is more advantageous: since we are concerned only with the decision problem of whether there is a placement containing *all* n points, we need keep track of only those placements containing all points seen so far. Any placement that does not contain all points can be discarded. That is, we want the *intersections* of all the pairwise containing regions, where each region is given by a pair of segments (possibly empty) on the inner offset polygon and another pair of segments (also possibly empty) on the outer offset polygon. If at any point in the algorithm there are no such remaining placements, then we can halt and output 'No'.

4.2 Analysis and Details of Data Structure

How do we store the set of placements containing all points? Recall that the region of placements containing q_i and with q_j on the boundary corresponds to a pair of segments along the inner and outer boundaries of the annulus region. For each point, we store these placements in two balanced binary search trees (one for the inner polygon and one for the outer polygon) ordered clockwise around the boundary of the polygon. Unfortunately, it is possible to construct a case where the complexity of the set of placements containing all points is $\Theta(n)$. (Each new pair of segments increases the complexity of the arrangement by 2.) Thus the space required *per point* may be as high as $\Theta(n)$ for a total of $\Theta(n^2)$ space. The searches, inserts, and deletes can all be performed in $O(\log n)$ time. In particular, for each new point q_i added to the structure of point q_j, there are at most two segments to be added to both the inner and outer offset polygons. Since we want only placements containing all points, we store the *intersections* of these two new segments with all existing segments. We find the endpoints in $O(\log m)$ time and delete all regions not inside the endpoints. Deleting one segment and rebalancing the tree requires $O(\log m)$ time. The total number of insertions and deletions to each tree is $O(n)$. Hence we need a total of $O(n^2 \log n)$ time for updating all the trees and $O(n^2 \log m)$ time for computing all the intersections.

The overall complexity is thus $O(n^2 \log(nm) + m)$ time and $O(n^2 + m)$ space in the worst case (when no pruning is done). The algorithm may terminate early with a 'No' answer.

4.3 Improvement by Pruning

Both the space and time complexity of the algorithm can be improved considerably by an on-line pruning. Recall from the previous section that for each point q_i we need to store only the placements containing *all* the points: that is, the intersections of all $i - 1$ intersection regions. We can discard the data structure of a point q_i when this intersection region becomes empty. This happens when there are no longer any placements containing all other points with q_i on the boundary. We define $H_1 = \{p_1\}$ and H_i (for $2 \leq i \leq n$) to contain all the points $x \in H_{i-1} \cup \{p_i\}$ such that there exists a placement $\tau(P)$ which contains $H_{i-1} \cup \{p_i\}$ with x on the boundary of $\tau(P)$. Let $h_i = |H_i|$. (In case $h_i = 0$ for some i the algorithm terminates with a 'No' answer.) Also, let h be the maximum value of h_i for $1 \leq i \leq n$. Then the total number of data structures after the ith step of the algorithm is h_i and the total at any time is $O(h)$. The total time required to update existing data structures for a new point q_{i+1} is $O(h_i \log n) = O(h \log n)$. The pruning step can be implemented efficiently. The main idea, which we explore in the full version of the paper, is marking points for deletion (and temporarily ignoring them), but using them at a later stage of the algorithm. We omit in this version of the paper the full analysis of the pruning version of the algorithm. The total running time of the algorithm is $O(nh \log(nm) + m)$ time. The algorithm requires only $O(nh + m)$ space in the worst case.

5 Simple Polygons

In this section we extend our results to the case of simple polygons.[3] We restrict the class of simple polygons to a natural set of "fat" polygons, which are more-natural candidates for offset-polygon placement problems. Specifically, we disallow polygons with narrow corridors, as specified in the following definition.

Definition 3 (δ-wide Polygons) *A δ-wide polygon P is a simple polygon with the property that if $p, q \in \partial P$ with $dist(p, q) \leq 2\delta$ then there is a path connecting p and q along the boundary of P such that every point on the path is at most 2δ away from p or every point is at most 2δ away from q.*

This restriction is reasonable for the proposed applications since the offset polygons are meant to capture measurements that are close to the input polygon and in actual production the allowed tolerance has to be twice the minimum feature-size. It eliminates cases in which the inner and outer δ-offsets of P become disconnected or non-simply-connected. Figure 3(a) shows a simple polygon

---

[3] See [AAAG] for a discussion of the *straight skeleton* of a simple polygon which is closely related to the notion of the inner offset polygon. This discussion is however not in the context of the problems discussed in this paper.

 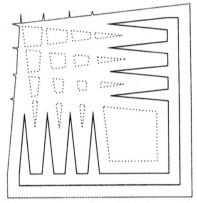

(a) Non-simple inner and outer offsets (b) Intersecting offsets of narrow spikes

Fig. 3. Simple polygons with non-simple offsets

(with solid edges) whose outer offset is *perforated*: it contains a boundary (dotted densely) and a hole (dotted sparsely). The inner offset of the same polygon consists of three distinct polygons (with dashed edges). Figure 3(b) shows another simple polygon whose outer offset is perforated.

We can solve simple-polygon variants of the offset-polygon max-cover problem for δ-wide simple polygons with a slightly modified version of the algorithm given in Figure 2. Let us use $\overline{O}_{P,\delta}$ to denote the *true* (non-linear) outer δ-offset of P. Similarly, we call the inner curve formed by straight segments and circular arcs at distance δ the *true* inner δ-offset of P and denote it by $\overline{I}_{P,\delta}$. Note that $\overline{I}_{P,\delta}$ and $\overline{O}_{P,\delta}$ are each of complexity $O(m)$ for a δ-wide simple polygon P of size m. Instead of having at most two intersections between $\tau(\overline{I}_{P,\delta})$ and $\overline{O}_{P,\delta}$, we can have $\Theta(m^2)$ pairwise intersections in the worst case requiring $\Theta(m^2)$ time to compute. Each pair of points has two offset polygons, each of which has $O(m^2)$ intersections with the polygon being swept. So the size of the queue is $O(nm^2)$ and queue operations can be performed in $O(\log(nm))$ time. The overall complexity of the algorithm becomes $O(n^2m^2\log(nm))$ time and $O(nm^2)$ space. The above running times also hold for linearized versions of the offset-polygon max-cover problem if we disallow polygons with narrow features that cause the outer or inner offset polygon to intersect itself. Note that for a general simple polygon P, both linearized outer and inner boundaries, $O_{P,\delta}$ and $I_{P,\delta}$, can contain some points further than δ from ∂P. As a result, $I_{P,\delta}$ and $O_{P,\delta}$ can each be of complexity $\Theta(m^2)$ for a polygon P with m vertices. Figure 3(b) gives an illustration of this. Thus there can be $\Theta(m^4)$ intersections between $\tau(I_{P,\delta})$ and $O_{P,\delta}$. The simple polygons to which our algorithm applies must therefore be δ-wide without narrow spikes (as shown in Figure 3(b)).

The on-line algorithm can also be modified for simple polygons. If we make the assumption that the features of the polygon are such that the annulus region has $O(m)$ complexity, then in the worst case the number of intersections between two translated copies of the annulus is $O(m^2)$ and the complexity of the arrangement of containing regions for a given point is $O(nm^2)$. The on-line algorithm therefore requires $O(n^2m^2 \log(nm))$ time and $O(n^2m^2)$ space.

6   Conclusion

In this paper we provide efficient algorithms for polygon offset placement problems. We handle the convex and non-convex cases, the translation-only variant as well as the translation and rotation variant, the static and dynamic modes of input points, and decision and optimization versions of the problems. There are several possible further research directions, which include the following:

1. Minimizing δ such that the placement of the given polygon annulus contains some given value $k < n$ of them (offset-polygon partial containment).
2. Generalizing from a polygon to a collection of polygonal *chains*. (This variant often occurs in applications to robot localization.)
3. Generalizing from polygons to smooth shapes.
4. Computing *approximate* solutions to all of these problems.
5. Proving lower bounds for the problems.
6. Solving similar problems in higher dimensions.
7. Analizing the *expected* value of h in the pruning version of the on-line algorithm (Section 4.3).

Acknowledgement. We would like to thank Vincent Mirelli for several stimulating conversations related to the topics of this paper.

References

[AAAG] O. Aichholzer, D. Alberts, F. Aurenhammer, and B. Gärtner, A novel type of skeleton for polygons, *J. of Universal Computer Science* (an electronic journal), 1 (1995), 752–761

[AS] P.K. Agarwal and M. Sharir, Efficient randomized algorithms for some geometric optimization problems, *Proc. 11th Ann. ACM Symp. on Computational Geometry*, Vancouver, Canada, 1995, 326–335.

[AST] P.K. Agarwal, M. Sharir, and S. Toledo, Applications of parametric searching in geometric optimization, *J. Algorithms*, 17 (1994), 292–318.

[BDG] G. Barequet, M. Dickerson, and M.T. Goodrich, Voronoi diagrams for medial-axis distance functions, these proceedings, 1997.

[BDP] G. Barequet, M. Dickerson, and P. Pau, Translating a convex polygon to contain a maximum number of points, *Proc. 7th Canadian Conf. on Computational Geometry*, Québec City, Québec, Canada, 1995, 61–66; full version to appear in: *Computational Geometry: Theory and Applications*.

[Ch] B. Chazelle, The polygon placement problem, *Advances in Computing Research: volume 1* (F. Preparata, ed.), JAI Press, 1983, 1–34.

[CK] I.J. Cox and J.B. Kruskal, Determining the 2- or 3-dimensional similarity transformation between a point set and a model made of lines and arcs, *Proc. 28th Conf. on Decision and Control*, 1989, 1167–1172.

[CY] E.-C. CHANG AND C.-K. YAP, Issues in the Metrology of Geometric Tolerancing, Courant Institute of Mathematical Sciences, New York University, Unpublished manuscript.

[DGR] C.A. DUNCAN, M.T. GOODRICH, AND E.A. RAMOS, Efficient approximation and optimization algorithms for computational metrology, Proc. 8th Ann. ACM-SIAM Symp. on Discrete Algorithms, New Orleans, LA, 1997, to appear.

[DS] M. DICKERSON AND D. SCHARSTEIN, Optimal placement of convex polygons to maximize point containment, Proc. 7th Ann. ACM-SIAM Symp. on Discrete Algorithms, Atlanta, GA, 1996, 114–121.

[DMSS] W.P. DONG, E. MAINSAH, P.F. SULLIVAN, AND K.F. STOUT, Instruments and Measurement Techniques of 3-Dimensional Surface Topography, Three-Dimensional Surface Topography: Measurement, Interpretation and Applications (K.F. Stout, Ed.), Penton Press, Bristol, PA, 1994.

[EE] D. EPPSTEIN AND J. ERICKSON, Iterated nearest neighbors and finding minimal polytopes, Discrete & Computational Geometry, 11 (1994) 321–350.

[ESZ] A. EFRAT, M. SHARIR, AND A. ZIV, Computing the smallest k-enclosing circle and related problems, Computational Geometry: Theory and Applications, 4 (1994), 119–136.

[GMR] L. GUIBAS, R. MOTWANI, AND P. RAGHAVAN, The robot localization problem, in: Algorithmic Foundations of Robotics, A K Peters, Ltd., 1995, 269–282.

[HT] M.E. HOULE AND G.T. TOUSSAINT, Computing the width of a set, Proc. 1st Ann. ACM Symp. on Computational Geometry, 1985, 1–7.

[HU] D.P. HUTTENLOCHER AND S. ULLMAN, Recognizing solid objects by alignment with an image, Int. J. of Computer Vision, 5 (1990), 195–212.

[KS] D. KIRKPATRICK AND J. SNOEYINK, Tentative prune-and-search for computing fixed-points with applications to geometric computation, Fundamental Informaticæ, 22 (1995), 353–370.

[LL] V.B. LE AND D.T. LEE, Out-of-roundness problem revisited, IEEE Trans. on Pattern Analysis and Machine Intelligence, 13 (1991), 217–223.

[LS] H.P. LENHOF AND M. SMID, Sequential and parallel algorithms for the k closest pairs problem, Int. J. Computational Geometry and Applications, 5 (1995), 273–288.

[Re] A.A.G. REQUICHA, Mathematical meaning and computational representation of tolerance specifications, Proc. Int. Forum on Dimensional Tolerancing and Metrology, 1993, 61–68.

[SJ] M. SMID AND R. JANARDAN, On the width and roundness of a set of points in the plane, Proc. 7th Canadian Conf. on Computational Geometry, Québec City, Québec, Canada, 1995, 193–198.

[Sr] V. SRINIVASAN, Role of sweeps in tolerance semantics, Proc. Int. Forum on Dimensional Tolerancing and Metrology, 1993, 69–78.

[SLW] K. SWANSON, D.T. LEE, AND V.L. WU, An optimal algorithm for roundness determination on convex polygons, Computational Geometry: Theory and Applications, 5 (1995), 225–235.

[Ya] C.-K. YAP, Exact computational geometry and tolerancing metrology, in: Snapshots of Computational and Discrete Geometry, Vol. 3, Technical Report SOCS-94.50 (D. Avis and J. Bose, Eds.), McGill School of Computer Science, 1995.

Computing Constrained Minimum-Width Annuli of Point Sets*

Mark de Berg[1], Prosenjit Bose[2], David Bremner[3], Suneeta Ramaswami[3] and Gordon Wilfong[4]

[1] Department of Computer Science, Utrecht University, PO Box 80.089, 3508 TB Utrecht, The Netherlands
[2] Département de Mathematique et Informatique, Université du Québec à Trois-Rivières, Trois-Rivières, Québec, Canada
[3] School of Computer Science, 3480 University Street, McGill University, Montréal, Québec H3A 2A7, Canada
[4] Bell Laboratories, Lucent Technologies, Murray Hill, New Jersey, USA

Abstract. We study the problem of determining whether a manufactured disc of certain radius r is within tolerance. More precisely, we present algorithms that, given a set of n probe points on the surface of the manufactured object, compute the thinnest annulus whose outer (or inner, or median) radius is r and that contains all the probe points. Our algorithms run in $O(n \log n)$ time.

1 Introduction

One of the fundamental aspects of the manufacturing sciences is quality control for various objects that are built by some physical process, i.e. the measurement of the "shape" and "size" of such manufactured objects in order to determine whether they conform sufficiently to the specified ideal measures. Metrology is the science of measurement. One of its subareas is *tolerancing* (or *dimensional tolerancing*), which addresses questions pertaining to specifications of the tolerance of shape characteristics and the methodology used to measure it. Coordinate Measurement Machines (CMMs) are machines that take measurements on a physical object by probing it and record such information (for example, this could be the coordinate values of various points on the object). This information may then be used to compute various characteristics of the object in order to determine whether they lie within the specified tolerance. Such problems fall in the domain of *computational metrology*. Many of these problems have a rich geometric flavor and the advent of CMMs has led to much interaction between computational geometry and metrology.

A fundamental and well-studied problem in tolerancing metrology is the problem of measuring the *roundness* of an object. In this case, the manufactured objects are round (circular) and the CMM gives the coordinate values of n points sampled on the boundary of the manufactured object. The goal is to

---

* This research was partially supported by the Dutch Organization for Scientific Research N.W.O, NSERC Canada and FCAR Quebec. We also thank the Bellairs Research Institute, McGill University for the use of their facilities.

determine whether the object is close to circular or not. This is the problem of computing the *minimum width annulus* containing the given set of points. If the thickness of such an annulus is less than the specified tolerance, the manufactured object is acceptable. This is, in fact, the measure used by the American National Standards Institute (see Foster [11], pp. 40-42) and the International Standards Office for testing roundness.

1.1 Previous Work

The minimum width annulus problem has been well-studied. Rivlin [17] first showed that the minimum width annulus of n points is either the width of the set or must have two points on the inner circle and two points on the outer circle of the annulus. In the case of the former, the radius of the annulus is infinite. In the case of the latter, it is finite and its center must lie on an intersection point of the furthest-point Voronoi diagram of S and nearest-point Voronoi diagram of S. Smid and Janardan [20] give an alternate proof of Rivlin's characterization of the minimum width annulus of a set of planar points.

An algorithm for this problem was given by Ebara et al. [9] in 1989, where they give an $O(n^2)$ time algorithm for the problem by computing the union of the nearest and furthest-point Voronoi diagrams of the input set of n points. This algorithm can be improved to run in $O(n \log n + k)$ time by using the planar map overlay algorithm due to Guibas and Seidel [12], where k represents the number of intersection points between the nearest and furthest-point Voronoi diagrams (in the worst case k can be $\Theta(n^2)$) This algorithm was re-discovered in the manufacturing community by Roy and Zhang in 1992 [18].

Recently, Bose and Devroye [6] have shown that the expected number of intersections between the furthest-point and nearest-point Voronoi diagram of a random point set is linear (the set of points is drawn from a density bounded from 0 and ∞ on a convex compact set). This implies that the above algorithm has $O(n \log n)$ expected running time in this case.

In the case where the points are not random, the above characterization for the center of the minimum-width annulus has also been used to obtain algorithms with improved running times, by using Megiddo's *parametric search* [14] and other techniques. Agarwal et al. [1, 2, 3] obtain randomized algorithms with expected running times that are $o(n^2)$. The best-known algorithm to date appears in [2] and has an expected running time of $O(n^{3/2+\epsilon})$, where $\epsilon > 0$ is an arbitrarily small constant.

Recently, Duncan, Goodrich and Ramos [8] have presented a number of efficient algorithms for various roundness problems. One such problem, that they refer to as the *optimal referenced roundness* problem, deals with finding a minimum width annulus covering a point set when the circle equidistant from the inner and outer circles defining the annulus has a fixed radius. (This is referred to as the *fixed median radius* problem in this paper.) One of their results for this problem is a deterministic $O(n \log n)$ time algorithm based on parametric search. Independently, we have derived an alternate $O(n \log n)$ time algorithm for this problem based instead on computing a generalized Voronoi diagram.

1.2 Our Results

Suppose we would like to manufacture a disc of radius r. There are two aspects of the produced object that we can test, namely whether it is round enough and whether its size (that is, radius) is appropriate. Computing the smallest width annulus amounts to testing only the roundness of the object. In this paper, we shall test both aspects simultaneously. Thus we shall look for the smallest width annulus of radius r. We consider three versions of this problem in which the radius of an annulus is defined to be the radius of its outer circle, its inner circle, or its median circle (that is, the circle in between and equidistant from the outer and inner circle).

Note that the number of degrees of freedom of the annulus in our restricted settings is one less than in the general setting, where one wishes to compute the smallest width annulus without constraints on either its inner or outer radius. We show how to use this fact to improve upon the $O(n^{3/2+\epsilon})$ bound given by Agarwal et al. [2] for the general case. Our algorithms for the three restricted settings each have only $O(n \log n)$ running time. Moreover, they are fairly simple, and should present no real problems to an implementor. In this extended abstract, some proofs have been omitted due to space constraints. Full details may be found in [5].

2 Notation and Definitions

In this section, we establish notation that will be used throughout the paper. The input set of points will always be denoted by $S = \{s_1, s_2, \ldots, s_n\}$. The Euclidean distance between two points p_1 and p_2 in the plane is denoted by $d(p_1, p_2)$. Given a point p in the plane and a radius r, we denote the circle with center p and radius r by $C(p, r)$, the closed disc with center p and radius r by $D(p, r)$ and the open disc with center p and radius r by $\widetilde{D}(p, r)$. Given a closed region R, we use $\mathrm{bd}(R)$ to represent the boundary of R and $\mathrm{int}(R)$ to represent the interior of R, i.e. the set $R - \mathrm{bd}(R)$.

We say a closed, connected region in the plane is a *circle-gon* if it is a convex region bounded by circular arcs. A *feasible annulus* for S is a concentric pair of circles $C(p, r_i)$ and $C(p, r_o)$ $(r_i < r_o)$ such that all points of S lie in the region $D(p, r_o) \cap (E^2 - \widetilde{D}(p, r_i))$, where E^2 is the Euclidean plane. A *diametral pair* of S is a pair of points in S whose distance is exactly the diameter of the set of points. The straight line bisector of two points p_1 and p_2 in the plane is denoted by $\mathrm{bis}(p_1, p_2)$. *Voronoi diagrams* are used extensively in our algorithms.

- The *nearest point Voronoi diagram* of a set of points S is denoted by $\mathrm{NPVD}(S)$. The *Voronoi region* associated with an element s_i from S is denoted by $V(s_i)$.
- The *furthest point Voronoi diagram* of a set of points S is denoted by $\mathrm{FPVD}(S)$. The furthest point Voronoi region of $s_i \in S$ is denoted by $V_F(s_i)$.
- The *generalized Voronoi diagram* is simply the generalization of the Voronoi diagram, where the input set may now consist of objects other than points. In Section 5, we use the generalized nearest point Voronoi diagram of a set consisting of points and circular arcs, where the Voronoi regions are bounded by straight line segments and curves of degree 2.

3 Fixed Outer Radius

Input A radius r and a set of points S;

Question What is a minimum width feasible annulus with outer radius r?

Let F be the intersection of the n discs $D(s_i, r)$, $1 \leq i \leq n$. Any annulus with outer radius r containing every point in S must be centered at some point in F. Furthermore, to find the *minimum width* annulus with outer radius r, it is sufficient to find the *largest* empty circle with center in F. In the remainder of this section, we describe a simple $O(n \log n)$ time algorithm to compute the minimum width annulus with outer radius r that contains S. Before proceeding, we state the following frequently used fact about closest point Voronoi regions: Given a point $p \in V(s_i)$, the largest empty circle centered at p has radius $d(p, s_i)$. The two lemmas stated below follow from this fact.

Lemma 1. *Given a planar set of points S and a closed bounded region R contained entirely in the Voronoi cell of some $s_i \in S$, the largest empty circle with center in R is centered at a point on $\mathsf{bd}(R)$.*

Lemma 2. *Let $\alpha \subseteq C(s_j, r)$ be a circular arc contained entirely within a Voronoi region $V(s_i)$. The largest empty circle constrained to have its center on α will have one of the endpoints of α as its center.*

Since F is the intersection of a set of discs, it is a convex circle-gon with at most n vertices. Our goal is to compute the intersection of $\mathsf{NPVD}(S)$ and F. Note that the number of intersection points on $\mathsf{bd}(F)$ is $O(n)$, since an edge of $\mathsf{NPVD}(S)$ intersects $\mathsf{bd}(F)$ at most twice.

Let $F(s_i)$ denote the convex region obtained by intersecting F and $V(s_i)$. For each non-empty $F(s_i)$, we find the largest empty circle $C(s_i)$ with center in $F(s_i)$. From Lemma 1, it follows that $C(s_i)$ will have its center c on $\mathsf{bd}(F(s_i))$. Also, $C(s_i)$ will have radius $d(c, s_i)$. The minimum width annulus with outer radius r will be given by the best of these candidates, i.e. by a circle from the set $\{C(s_i) \mid 1 \leq i \leq n\}$ that has the maximum radius. Observe that $\mathsf{bd}(F(s_i))$ is composed of circular arcs and/or straight line segments (which come from the Voronoi edges of $V(s_i)$). Thus, using Lemma 2, we have the following characterization of the candidate centers for the minimum width annulus with outer radius r.

Theorem 3. *The largest empty circle with center in F must be centered at a Voronoi vertex lying in F, at a vertex of $\mathsf{bd}(F)$ or at an intersection point between $\mathsf{bd}(F)$ and a Voronoi edge.*

We describe below a simple $O(n \log n)$ time algorithm to determine the intersection points between $\mathsf{NPVD}(S)$ and $\mathsf{bd}(F)$. We would like to point out that it is possible to use Guibas and Seidel's output-sensitive planar-map overlay algorithm [12] to solve our problem within the same time bounds. However, we present here an extremely simple algorithm that is more practical than that implied by the more general algorithm in [12].

Let $\ell(c, \theta)$ be an oriented line, where c is a point on the line and $0 \leq \theta < 2\pi$ is the angle the line makes with the positive x-axis. We say that an oriented line is a *supporting line* of a region R if it is incident on $\mathsf{bd}(R)$ and R lies entirely on or to the left of the line. We use $L(R, \theta)$ to denote the supporting line of R that is parallel to $\ell(c, \theta)$. Let $v_1, v_2, \ldots, v_n$ be the vertices of a convex circle-gon F in counter-clockwise order, and let $a_i = (v_i, v_{i+1})$ be its circular-arc edges $(v_{n+1} \equiv v_1)$. The convex polygon P obtained from this vertex set is said to be the polygon defined by F. We then have the following lemmas (see [5] for proofs).

Lemma 4. *Given an oriented line $\ell(c, \theta)$, let $L(P, \theta)$ be incident on vertex v_i. Then either $L(F, \theta) = L(P, \theta)$, or $L(F, \theta)$ is tangent to one of the arcs a_i or a_{i-1} $(a_0 \equiv a_n)$.*

Lemma 5. *Given a set of k line segments and a convex circle-gon F with n vertices, whether each segment intersects F and the points of intersection on $\mathsf{bd}(F)$, if any, can be determined in $O(k \log n)$ total time with an $O(n)$ time pre-processing step.*

By carrying out the above procedure for every Voronoi edge of $\mathsf{NPVD}(S)$, we can identify all Voronoi vertices that lie inside F and also compute all the intersection points between $\mathsf{NPVD}(S)$ and $\mathsf{bd}(F)$. This takes $O(n \log n)$ time, since there are $O(n)$ Voronoi edges.

For each vertex of F, it is necessary to identify the Voronoi region in which it lies. This can be done in $O(\log n)$ time for each vertex, after a linear-time pre-processing step, by using a point location algorithm for planar subdivisions [13]. However, a simple sort step, followed by a linear scan, will give us the desired information within the same time bounds: Sort the vertices and intersection points on $\mathsf{bd}(F)$ in counter-clockwise order with respect to some point in the interior of F (say the centroid of three vertices of F). Starting at an intersection point, step through this sorted list. (i) If we step from an intersection point p to a vertex v of F, then let $e \subset \mathsf{bis}(s_i, s_j)$ be the Voronoi edge that caused the intersection p. If v lies in the half-plane closer to s_i, then v must lie in $V(s_i)$. If not, it must lie in $V(s_j)$. (ii) If we step from a vertex to another vertex, the latter must lie in the same Voronoi region as the former. Note that if there are no intersection points on $\mathsf{bd}(F)$, it means that F lies entirely within some $V(s_i)$, and such an s_i can be identified in linear time.

In summary, the outline of the algorithm to compute the minimum width annulus with outer radius r is as follows:

1. Compute $\mathsf{NPVD}(S)$ in $O(n \log n)$ time (*e.g.* [10, 19]).
2. Compute the intersection F of the discs $D(s_i, r), s_i \in S$ in $O(n \log n)$ time [7].
3. Identify the Voronoi vertices that lie in F and find the intersection points between $\mathsf{NPVD}(S)$ and F (refer Lemma 5). Find the Voronoi region in which each vertex of $\mathsf{bd}(F)$ lies, as described above. This takes $O(n \log n)$ time.
4. Find the largest empty circle with center in F in $O(n)$ time (refer Theorem 3).

Theorem 6. *Given a set S of n points in the plane, a minimum width annulus containing S and having outer radius r can be computed in $O(n \log n)$ time.*

4   Fixed Inner Radius

Input: A radius r and a set of points S;
Question: What is a minimum width feasible annulus with inside radius r?

Let U be the union of open discs $\tilde{D}(s_i, r), 1 \leq i \leq n$. Note that the center of an annulus with fixed inner radius r must be outside the union of these open discs. Therefore, the center of the minimum spanning disc of S whose center lies outside U is the center of the thinnest annulus with fixed inner radius r. In the following two lemmas, we first determine the different configurations of points of S that must be on the boundary of the thinnest annulus and then use these configurations to show that there exists a linear size set C of locations that contains the center of the thinnest annulus with fixed inner radius r.

Let F be the complement of U (i.e. F is the complement of the union of n open discs, each of radius r centered at each $s \in S$). The *centroid* of a pair of points is defined as the midpoint of the line segment between them.

Lemma 7. *The minimum width feasible annulus with fixed inner radius r must have (at least)*

1. *three points or a diametral pair on the outer circle,*
2. *two points on the inner circle and one on the outer, or*
3. *two points on the outer and one point on the inner.*

Proof. Omitted. See [5] for details.

Lemma 8. *There are $O(n)$ positions where the the center of the thinnest annulus for S constrained to have center in F can be located. They are the following:*

1. *the midpoint of the diameter of S,*
2. *a vertex of FPVD(S),*
3. *the intersection point of an edge e of FPVD(S) and bd(F) closest to the centroid of the the two generators of e,*
4. *a vertex v of bd(F).*

Proof. We use the configurations outlined in Lemma 7 to prove that the center can be located only in one of the four places enumerated in the statement of the lemma. See [5] for details. □

The approach used to solve the problem is to verify each of the candidate positions for the center as shown in Lemma 8 and select the position yielding the smallest spanning disc. The complement of the union of n open discs has linear boundary complexity [7]. The edges of the boundary are arcs of circles of radius r and the vertices are the intersections of two or more arcs. Given a set of n open discs, the complement of their union can be computed in $O(n \log n)$

time [4, 7]. Therefore, the set F can be computed in $O(n \log n)$ time. We show below that the total time taken to enumerate and test the candidate centers is $O(n \log n)$.

First, the centers d_i of the diameters of S can be computed in $O(n \log n)$ time [16]. Second, we compute all of the vertices of the FPVD(S). We need to determine if a center of a diameter of S or a vertex of FPVD(S) is in F. To do this, we first compute NPVD(S) in $O(n \log n)$ time [10, 15, 16]. We preprocess NPVD(S) for point location [13] (i.e. with $O(n)$ preprocessing, we can locate the cell of NPVD(S) containing a query point in $O(\log n)$ time). For each center d_i of the diameter of S and each vertex v_i of FPVD(S), we determine which cell of NPVD(S) it lies in. If the distance between d_i (resp. v_i) and the generator of the cell it is in is at least r, then d_i (resp. v_i) lies inside F. Therefore, since there are $O(n)$ centers of the diameter of S and $O(n)$ vertices in FPVD(S), generating all these candidate centers can be accomplished in $O(n \log n)$ time.

Third, we need to compute every intersection point of an edge e of FPVD(S) and bd(F) closest to the centroid of the two generators of e. Let p and q be the two points that generate edge e. Let p^* be the intersection of the line containing e and segment $\overline{pq}$. We first test if p^* is in e and in F. If it is, then it is a candidate center and we generate it. Since there are only a linear number of edges in FPVD(S) there can only be a linear number of such candidate centers. If p^* is not in e and F, then conceptually we perform two ray shooting queries along the line containing e in the two directions away from p^* to find the two closest feasible centers to p^*. If these intersection points are on e, then we generate them. Once again, for each p^* we generate at most two intersection points, therefore there are at most $O(n)$ such candidate centers.

Preprocessing F for ray shooting queries is too costly, so to generate the latter set of candidates, we perform two plane sweeps. The boundary of F is broken into y-monotone pieces and we perform one sweep for left rays away from p^* and one for right rays. After the first intersection for a given ray is found, we remove it from the event queue. Each ray is therefore processed at most twice, once when it is placed in the queue and once for its first intersection. Since there are only $O(n)$ rays and the boundary of F has $O(n)$ size, each of the two sweeps takes $O(n \log n)$ time to generate the $O(n)$ candidate centers.

Finally, each vertex of F is a candidate center. Thus, there are only $O(n)$ such candidate centers and they can be generated in $O(n \log n)$ time [4, 7].

Once all of the $O(n)$ candidates have been generated, computing the location of the center and the radius of the minimum spanning disc is simple. For each candidate center c_i, we determine in which cell of the FPVD(S) it lies. This can be accomplished in $O(\log n)$ time per candidate with $O(n)$ preprocessing time [13]. The distance between c_i and the generator of the cell containing c_i is the length of the radius of the spanning disc. We output the smallest spanning disc. Since none of the steps outlined above take more than $O(n \log n)$ time, we conclude with the following theorem.

Theorem 9. *Computing a thinnest annulus with inner radius fixed at r of n points in the plane can be accomplished in $O(n \log n)$ time.*

5 Fixed Median Radius

Input A radius r and a set of points S;
Question What is a minimum width feasible annulus with median radius r?

Lemma 10. *The minimum width feasible annulus of given median radius r contains either a diametral pair on the outer circle or at least 3 elements of S on its boundary.*

Proof. Omitted. See [5] for details.

Place a disc of radius $2r$ centered at each point in S; this is the maximum possible outside radius, since it drives the inside radius to zero. In the rest of this section let F denote the intersection of these discs. Any feasible annulus with median radius r must have a center in F; conversely for every point in F, there is a feasible annulus of median radius r centered at that point. The region F is a circle-gon, a convex region bounded by circular arcs. For a point p and a set X, let $d_{\min}(p, X)$ denote $\min_{x \in X} d(p, x)$. We say that $C(s, 2r)$ is a *closest circle* to p, if for all $s' \in S$ $d_{\min}(p, C(s', 2r)) \geq d_{\min}(p, C(s, 2r))$. For any point p in F, $d_{\min}(p, C(s, 2r)) = 2r - d(p, s)$. Thus we have the following:

Lemma 11. *For any $p \in F$, if s is a farthest element of S from p, then $C(s, 2r)$ is a closest circle to p.*

Rather than concerning ourselves with the intersection of two Voronoi diagrams as in the characterization of candidate centers given in Section 1, the previous lemma allows us to compute the generalized (nearest point) Voronoi diagram of the points s and certain arcs of the circles $C(s, 2r)$. In general we could break each circle $C(s, 2r)$ into some constant number circular arcs and use all of these arcs as sites in the Voronoi diagram. In order to ensure that the arcs intersect only at their endpoints (which is necessary to ensure a linear bound on the size of the generalized Voronoi diagram), we show that we need only concern ourselves with the arcs bounding F.

Lemma 12. *For any $p \in F$, if $C(s, 2r)$ is a closest circle to p, then an arc of $C(s, 2r)$ containing the point on $C(s, 2r)$ closest to p is part of the boundary of F.*

Lemma 10 characterized candidate centers for the minimum width annulus of median radius r as being either midpoints of diameters of the input set S, or as those points c such that the minimum width feasible annulus of median radius r contains at least three points of S on its boundary. Notice that the second type of candidate are exactly the vertices of the generalized Voronoi diagram G of S and the boundary of F. Each vertex of G is closest to three sites; in the case where a site is some $s \in S$, this corresponds to s being on the inner boundary of the annulus with median radius r and centered at the vertex. In the case where a closest site is an arc of $C(s, 2r)$ (part of the boundary of F), by Lemma 11 and Lemma 12 this corresponds to s being on the outer boundary of the annulus with median radius r. We summarize in a theorem.

Theorem 13. *Let G be the generalized Voronoi diagram of S and the arcs bounding F. The center of the minimum enclosing annulus of median radius r is either a vertex of G contained in F or the center $c_d \in F$ of a diameter of S.*

The outline of our algorithm is as follows.

1. Compute F in $O(n \log n)$ time by divide and conquer [7].
2. As in the previous section, we can compute the centers of the diameter of S in $O(n \log n)$ time.
3. Compute G in $O(n \log n)$ time by the algorithm of Yap [21].
4. Test each vertex of G, and each diameter center, as follows.
   (a) $O(\log n)$ time for point location to determine feasibility.
   (b) $O(1)$ time to compute the width of the candidate annulus.

Testing a candidate vertex v of G involves first determining if $v \in F$ and if so computing the minimum width median radius r annulus centered at v. Testing a point p for inclusion in F proceeds much as if F were a convex polygon. Choose some arbitrary point c interior to F (say the centroid of the vertices of F). Draw rays from c through the vertices of F, dividing the plane into sectors. Determine by binary search in $O(\log n)$ time which sector p lies in. Each sector corresponds to a unique arc A of the boundary of F, which in turn lies on a unique circle $C(s, 2r)$. The point p is inside F if and only if it is inside $C(s, 2r)$.

To compute the width of a minimum width annulus of median radius r centered at a vertex v of G, we need only consider a single generator g (arc or point) of v. If g is an element of S then the radius of the inner circle is $d(v, g)$. If g is an arc of some circle $C(s, 2r)$, then the radius of the outer circle is $d(v, s) = 2r - d(v, g)$. In either case, once we have computed one radius r_g, the radius of the other circle is given by $2r - r_g$. If our candidate center is a center of the diameter of S, then the outer radius is naturally half of the diameter; the inner radius is determined as before.

Theorem 14. *Given a set S of n points in the plane, a minimum width annulus containing S and having median radius r can be computed in $O(n \log n)$ time.*

6 Conclusions

We have presented efficient algorithms to compute the thinnest annulus containing a set of n points whose outer (or inner, or median) radius has a given value r. This is useful for testing whether a manufactured disc of radius r is within tolerance. Our algorithms run in $O(n \log n)$ time. Hence, they are more efficient than the best known algorithm for computing the thinnest annulus containing a set of points whose radius is not given, which runs in $O(n^{3/2+\epsilon})$ time.

Acknowledgments: We would like to thank Godfried Toussaint for organizing and inviting us to the workshop on Computational Metrology, held at the Bellairs Research Institute, McGill University. We would also like to thank all the participants of the workshop for interesting and useful discussions.

References

1. P.K. Agarwal, B. Aronov, and M. Sharir. Computing envelopes in four dimensions with applications. In *Proceedings of the 10th Annual ACM Symposium on Computational Geometry*, pages 348–358, 1994.
2. P.K. Agarwal and M. Sharir. Efficient randomized algorithms for some geometric optimization problems. In *Proceedings of the 11th Annual ACM Symposium on Computational Geometry*, pages 326–335, 1995.
3. P.K. Agarwal, M. Sharir, and S. Toledo. Applications of parametric searching in geometric optimization. *Journal of Algorithms*, 17:292–318, 1994.
4. F. Aurenhammer. Improved algorithms for discs and balls using power diagrams. *Journal of Algorithms*, 9:151–161, 1988.
5. Mark de Berg, Prosenjit Bose, David Bremner, Suneeta Ramaswami, and Gordon Wilfong. Computing constrained minimum-width annuli of point sets. Technical Report SOCS-96-7, McGill University, 1996.
6. P. Bose and L. Devroye. Intersections with random geometric objects. Technical report, McGill University, 1996.
7. K. Q. Brown. *Geometric transforms for fast geometric algorithms*. PhD thesis, Carnegie-Mellon University, Pittsburgh, PA, 1980.
8. C.A. Duncan, M.T. Goodrich, and E.A. Ramos. Efficient approximation and optimization algorithms for computational metrology. In *Proceedings of 8th ACM-SIAM Symp. Discrete Algorithms*, pages 121–130, 1997.
9. H. Ebara, N. Fukuyama, H. Nakano, and Y. Nakanishi. Roundness algorithms using the Voronoi diagrams. In *Proc. of the First CCCG*, page 41, 1989.
10. S. J. Fortune. A sweepline algorithm for Voronoi diagrams. *Algorithmica*, 2:153–174, 1987.
11. L. W. Foster. *GEO-METRICS II: The application of geometric tolerancing techniques*. Addison-Wesley Publishing Co., 1982.
12. L. J. Guibas and R. Seidel. Computing convolutions by reciprocal search. *Discrete Comput. Geom.*, 2:175–193, 1987.
13. D. G. Kirkpatrick. Optimal search in planar subdivisions. *SIAM J. Computing*, 12:28–35, 1983.
14. N. Megiddo. Applying parallel computation algorithms in the design of serial algorithms. *Journal of the ACM*, 30:852–865, 1983.
15. J. O'Rourke. *Computational Geometry in C*. Cambridge University Press, 1994.
16. F. Preparata and M. I. Shamos. *Computational Geometry*. Springer-Verlag, 1985.
17. T. J. Rivlin. Approximation by circles. *Computing*, 21:93–104, 1979.
18. U. Roy and X. Zhang. Establishment of a pair of concentric circles with the minimum radial separation for assessing roundness errors. *Computer Aided Design*, 24(3):161–168, 1992.
19. M.I. Shamos and D. Hoey. Closest-point problems. In *Proc. Sixteenth Annual IEEE FOCS*, pages 151–162, October 1975.
20. M. Smid and R. Janardan. On the width and roundness of a set of points in the plane. In *Proc. Seventh CCCG*, pages 193–198, Quebec City, Quebec, August 1995.
21. C.K. Yap. An $O(n \log n)$ algorithm for the Voronoi diagram of a set of simple curve segments. *Discrete Comput. Geom.*, 2:365–393, 1987.

Geometric Applications of Posets*

Michael Segal[1] and Klara Kedem[1]

Department of Mathematics and Computer Science,
Ben-Gurion University of the Negev, Beer-Sheva 84105, Israel

Abstract. We show the power of posets in computational geometry by solving several problems posed on a set S of n points in the plane: (1) find the k rectilinear nearest neighbors to every point of S (extendable to higher dimensions), (2) enumerate the k largest (smallest) rectilinear distances in decreasing (increasing) order among the points of S, (3) given a distance $\delta > 0$, report all the pairs of points that belong to S and are of rectilinear distance δ or more (less), covering $k \geq \frac{n}{2}$ points of S by rectilinear (4) and circular (5) concentric rings, and (6) given a number $k \geq \frac{n}{2}$ decide whether a query rectangle contains k points or less.

1 Introduction

1.1 Problems

Given a set S of n points in the plane and an integer k we solve the following problems in this paper:

1. Find the k ($k \geq \frac{n}{2}$) nearest rectilinear neighbors (under L_∞ metric) for each point of S (by reporting the $n - k$ farthest rectilinear neighbors).
2. Enumerate the k largest (smallest) rectilinear distances in decreasing (increasing) order.
3. Given a distance $\delta > 0$, report all the pairs of points of S which are of rectilinear distance δ or less (more).
4. Find the smallest "rectangular" axis-aligned (constrained or not constrained) ring that contains k ($k \geq \frac{n}{2}$) points of S. A *rectangular ring* is two concentric rectangles, the inner rectangle fully contained in the external one. As a measure we take the maximum width or area of the ring. By *constrained* we mean that the center of the ring is one of the points of S.
5. Find the smallest constrained circular ring (or a sector of a constrained ring) that contains k ($k \geq \frac{n}{2}$) points of S.
6. Given a number $k \geq \frac{n}{2}$, decide whether a query rectangle contains k points or less.

* Work by Klara Kedem has been supported by a grant from the U.S.-Israeli Binational Science Foundation.

1.2 Background

Most of the problems mentioned above have been considered in previous papers [6, 7, 8, 10, 16]. Dickerson et al. [6] present an algorithm for the first problem which runs in time $O(n \log n + nk \log k)$, and works for any convex distance function. Eppstein and Erickson [10] solve the first problem on a random access machine model in time $O(n \log n + kn)$ and $O(n \log n)$ space. In the algebraic desicion tree model their time bound increases by a factor of $O(\log \log n)$. Flatland and Stewart [11] present an algorithm for the first problem which runs in time $O(n \log n + kn)$ in the algebraic decision tree model. Finally, a recent paper of Dickerson and Eppstein [8] describes an $O(n \log n + kn)$ time and $O(n)$ space algorithm for the first problem, it works for any metric and is extendable to higher dimensions. For our best knowledge only Dickerson and Shugart [7] present an algorithm for the second problem (for the *largest k* distances) for any metric, and their algorithm requires $O(n + k)$ space with expected runtime of $O(n \log n + \frac{k \log k \log n}{\log \log n})$. Dickerson et al. [6] present an algorithm for the problem: enumerate all the *k smallest* distances in S in *increasing* order. Their algorithm works in time $O(n \log n + k \log k)$ and uses $O(n + k)$ space. Lenhof et al. [16], Salowe [17], Dickerson and Eppstein [8] also solved this problem but they just report the k closest pairs of points without sorting the distances, spending $O(n \log n + k)$ time and $O(n + k)$ space. An algorithm for solving the second problem (for the smallest k distances) is also presented in [8], spending $O(n \log n + k \log k)$ time and using $O(n + k)$ space. [8] also considered the third problem: find all the pairs of points of S separated by distance δ or less. They give an $O(n \log n + k)$ time and $O(n)$ space algorithm, where k is the number of distances not greater than δ.

   Problem 6 is a variant of the orthogonal range search where we are given a set S of n points and want to find which points are enclosed by the query rectangle. This problem was efficiently solved by Bentley [3] in $O(\log n + m)$ query time, where m is the number of points contained in the given query rectangle, using the range search tree and with preprocessing time and space $O(n \log n)$.

   Some variations of problems 4 and 5 have been considered in previous papers. Efrat et al.[9] consider the problem of enclosing k points within a minimal area circle and pose an open problem of covering k points by a ring. They gave two solutions for the smallest k-enclosing circle. When using $O(nk)$ storage, the problem can be solved in time $O(nk \log^2 n)$. When only $O(n \log n)$ storage is allowed, the running time is $O(nk \log^2 n \log \frac{n}{k})$. The problem of computing the roundness of a set of points, which is defined as the minimum width concentric annulus that contains all points of the set was solved in [2, 14, 19]. The best known running time is $O(n^{\frac{3}{2}+\epsilon})$, given in [2], where $\epsilon > 0$ is an arbitrary small constant. Segal and Kedem [18] considered the problem of enclosing k $(k \geq \frac{n}{2})$ points in the smallest axis parallel rectangle. Their algorithm runs in time $O(n + k(n - k)^2)$ and uses $O(n)$ space. Their method and algorithm are one of the tools used in this paper, and we review it below. It is based on *posets* (partially ordered sets) [1]. A *poset* being a partially ordered set of elements.

   Segal and Kedem [18] describe how to construct a poset such that a subset R of S contains the $n - k$ elements of S with the largest x coordinates. They represent S as a tournament tree. The tournament tree can be implemented as a heap. The operations of creating R and updating the tournament tree run in times $O(n + (n - k) \log n)$

and $O(\log n)$ respectively. Space requirement is $O(n)$. (For more details see the full version of [18].) With the use of posets they [18] device an algorithm that finds the smallest axis parallel rectangle with $k \geq \frac{n}{2}$ points in $O(n + k(n - k)^2)$ time and $O(n)$ space.

The runtimes achieved in the previously described works for problems 1,4 and 5 are not attractive when the k is close to n. We show that in this case the use of posets can significantly reduce the runtime of the algorithms. Our algorithm for solving the first problem runs in time $O((n - k)n)$ (assuming $k \geq \frac{n}{2}$) and uses linear space. For problem 2 we present two algorithms : for enumerating the largest and smallest distances. The first one runs in time $O(k \log n + n)$, and uses $O(n)$ space. The second algorithm runs in time $O(n \log n + k \log n)$, and uses $O(n)$ space. We solve both cases of problem 3 by a similar technique. For our best knowledge the second case of problem 3 has not been considered before. The runtime and space requirements of both algorithms for Problem 3 are as in [8], namely $O(n \log n + k)$ time and $O(n)$ space. We solve problem 4, rectangular ring containing $k(k \geq \frac{n}{2})$ points for the constrained case in $O(n(n - k) \log (n - k))$ time and $O(n)$ space, while for the non constrained case we present an algorithm with runtime $O(n(n - k)^4 \log n)$ and $O(n)$ space. We find a constrained circular ring that covers k ($k \geq \frac{n}{2}$) points (Problem 5) in $O(n^2 + n(n - k) \log n)$ time and $O(n)$ space, and we find a *sector* of a constrained circular ring that covers k points ($k \geq \frac{n}{2}$) in $O(n^2 + nk(n-k)^2)$ time and $O(n^2)$ space. For the sixth problem we obtain an algorithm with $O(n + (n - k) \log n)$ preprocessing time and space and $O(\log (n - k))$ query time. We also show how to extend the algorithms of all the problems to higher dimensional space.

1.3   Motivation

Another algorithm that runs efficiently for large k values was presented by Matoušek [15]. It finds the smallest circle enclosing all but few of the given n points in the plane. Given a large integer $k \leq n$ his algorithm runs in time $O(n \log n + (n - k)^3 n^\varepsilon)$ for some $\varepsilon > 0$.

A possible motivation to cover all but a small number of points by one or more objects comes from statistics. In the analysis of statistical data one would like to get rid of outlyers in the data. Assuming $n - k$ data points are outlyers, one way to find the k "good" data points is to enclose them in a small given shape (or shapes).

2   Rectilinear nearest neighbors (Problem 1)

Problem: Find the k nearest rectilinear neighbors to all points of S, where $\frac{n}{2} \leq k \leq n - 1$. Instead of finding the k nearest neighbors we will determine the $n - k$ farthest neighbors using the technique of [18]. Thus we implicitly find (but do not report) the nearest neighbors.

We define the nearest x-neighbor of a point $p_i \in S$ as point $q \in S$, such that $|x(p_i) - x(q)| = \min\{|x(p_i) - x(p)|, p \in S, p \neq p_i\}$, where $x(p)$ is the x-coordinate of p. First we find the k nearest x-neighbors for each point of S. To solve this subproblem we find the points with the $n - k - 1$ smallest and the $n - k - 1$ largest x-coordinates by posets [1]. Let A' (respectively A'') be the set of the $n - k - 1$ points of S with the smallest (largest) x-coordinates. Note that from the technique in [1] it follows that A' and A'' are sorted. Let A be the set of points of S with x-coordinates between those of the points of A' and A'' ($A = S - A' - A''$) (see Figure 1).

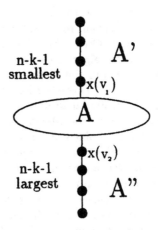

Fig. 1. Poset for $n - k - 1$ largest and $n - k - 1$ smallest values.

The number of points in A is $2k+2-n$. Since $\frac{n}{2} \leq k < n$, for a every point $p_i \in S$ all the points of A are among the k nearest x-neighbors of p_i, and the $n - k - 1$ farthest x-neighbors of p_i can be only in $A' \cup A''$. For the same reason, for a point $p_i \in A$ we will look for the farthest x-neighbors in A'' and among all the points in A' whose x-coordinate is smaller than $x(p_i)$. Symmetrically, if $p_i \in A''$ we will look for the farthest x-neighbors in A' and among all the points in A'' whose x-coordinate is greater than $x(p_i)$. Assume $p_i \in A$. Then by a simple merge on A' and A'' we can find the $n - k - 1$ points farthest from p_i. If $p_i \in A'(A'')$ then we perform a similar merge on $A''(A')$ and the set containing all the points in $A'(A'')$ whose x-coordinate is smaller (greater) than $x(p_i)$.

Returning to the two-dimensional problem, we store all the points of S in an array T. We create separate posets for the x and y axes. We call them the x-poset and the y-poset. Entry i for point p_i in T will contain 2 pointers: one to the leaf in the x-poset containing p_i, and one to the leaf in the y-poset. Our goal is to find for every point $p_i \in S$ all the $n - k - 1$ farthest rectilinear neighbors.

We create a set L of candidate neighbors with their L_∞ distances. For each point $p_i \in S$ it is enough to store the entry i_1 (i_2) in A' (A'') where the search for the $n - k$ farthest x-neighbors halted. Symmetrically for the y-neighbors. There is a possibility that the same point appears in both the set of farthest x-neighbors and the farthest y-neighbors of p_i. We go over all the $n - k - 1$ farthest y-neighbors of p_i and check if their corresponding x-coordinate is in the range $[1, i_1]$ and $[i_2, n]$ in the x-poset. If the answer is "YES" then the same point, say p_j, appears as the farthest neighbor of p_i in both axes, we choose the maximum distance of the two distances. Assume, that the maximum distance was obtained on the x-axis. Then we put into the set L the point p_j with a flag noting it x and skip in the x-poset and y-poset to the next farthest points. At the end of the process L has l points, where $(n - k - 1) \leq l \leq 2(n - k - 1)$. We find the $(n - k - 1)$th point in L using the linear time selection algorithm of [4] and thus solve the problem.

Considering the time complexity. Creating the posets takes $O(n + \sum_{i=k}^{n} \log i) = O(n + (n - k) \log n)$ time. The merge step over A', A'' and the selection take $O(n - k)$ time per point of S. The required storage, $O(n)$, is used for storing the posets, the auxiliary array T, L, and the indices. We conclude by the following theorem:

Theorem 1. *Given a set S of n points in the plane, we can find the k nearest rectilinear neighbors of all the points in S (or, equivalently, the $n - k - 1$ farthest neighbors) in $O(n + (n - k) \log n + n(n - k)) = O((n - k)n)$ time, using linear space.*

Remark 1 This problem can be easily extended to d-dimensional space, $d \geq 3$. Perform, for each axis i, $3 \leq i \leq d$ the same algorithm as for the y axis in the previous algorithm. The set L will have $(n - k - 1) \leq l \leq d(n - k - 1)$ points, and the $(n - k - 1)$th point in L will be determined by the selection algorithm. So the total runtime and space are not changed for a constant dimension d.

Remark 2 The algorithm described above still works when $k < \frac{n}{2}$. First we sort all the points according to their x and y-coordinates. Then for each point we find the $n - k - 1$ farthest neighbors in both axes by the same algorithm as before, create L and use the selection algorithm. In this case we add factor of $O(n \log n)$ to the runtime of the algorithm.

3 Enumerating rectilinear distances (Problem 2)

Problem 2: Given a set S of n distinct points in the plane, let $D = \{d_1, d_2, \ldots, d_N\}$, where $N = \frac{n(n-1)}{2}$ and $d_1 \geq d_2 \geq d_3 \geq \ldots \geq d_N$ denote the rectilinear distances determined by all the pairs of points in S. For a given positive integer $k \leq N$, we want to enumerate all the k pairs of points which realize the k *largest* distances in D. For some values of k we do not need to know the total order of the points (in x or y axis). For example, if $k = 1$ then the maximum and minimum values of the x and y coordinates suffice.

As in the previous section we first show an algorithm that enumerates all the k pairs of points which realize the k largest distances on the x axis.

Assume that the points of S are sorted by their x-coordinate in increasing order and name them by this order, namely points $1, 2, \ldots, n$. For d_1 we know that the points 1 and n (according to the sorting) realize this distance. We denote this pair by $(1, n)$. One can also think about the interval $[1, n]$ containing the n x-consecutive points. We will use the notation (i, j) to denote both the pair of points i and j and the interval $[i, j]$. Clearly, the next distance, d_2, can be realized by one of the *candidate pairs* $(1, n - 1)$ or $(2, n)$. Depending on the pair that realized d_2, the distance d_3 has also two candidate pairs. It is possible that the number of candidate pairs in step i will grow, if, for example, the pair $(1, n - 1)$ realized d_2 and the pair $(2, n)$ realized d_3, then the candidates for realizing d_4 are the pairs $(1, n - 2), (2, n - 1), (3, n)$. We denote the set of candidate pairs for distance i by L_i. This is the set of pairs of points that can potentially realize d_i, after the pair that realized d_{i-1} is known. An interval (ζ, ψ) is *nested* in (ξ, η) if $(\zeta, \psi) \subseteq (\xi, \eta)$. Throughout the algorithm we will make sure that L_i does not contain nested intervals.

We say that the candidate pair (i, j), where $i < j + 1$ *blocks* $(i+1, j)$ and $(i, j-1)$ because the x-distance defined by points i and j is greater than the distances defined by the pairs $(i + 1, j)$ and $(i, j - 1)$.

Claim 1: L_i differs from L_{i-1}, $i \geq 2$ by at most three candidate pairs : one that is deleted from L_{i-1} and at most two new pairs that are inserted into L_i.

Proof. For L_1 we have only candidate pair $(1, n)$. L_2 consists of the pairs $(2, n)$ and $(1, n - 1)$. If, wlog, the pair $(1, n - 1)$ in L_2 realizes d_2, then L_3 will consist of $(2, n)$ and $(1, n - 2)$. This is because $(2, n)$ blocks $(3, n)$ and $(2, n - 1)$. If the distance defined by the pair $(2, n)$ is always smaller than the distances defined by the $(1, n-j), (1 \leq j \leq n - 2)$ then L_i is different from L_{i-1} by deleting $(1, n - j)$ and inserting $(1, n - j - 1)$. If for some $j, 1 \leq j \leq n - 2$, the distance realized by the pair $(2, n)$ is greater than the distance realized by the pair $(1, n - j)$, then the candidate pairs for the next stage are changed by inserting two candidate pairs $(3, n), (2, n-1)$ and deleting $(2, n)$ and $(1, n - j)$ remains as a candidate as well. Thus, we conclude that if at some stage i there is only one pair (ξ, η) in L_i, then at the next stage this pair is deleted, and two new pairs $(\xi + 1, \eta)$ and $(\xi, \eta - 1)$ (if they exist) are inserted into L_{i+1} as candidate pairs. If, at some stage i there are several candidate pairs and one of them, e.g. (ξ, η) realizes d_i, then for the next stage this pair is deleted and $(\xi+1, \eta)$ and $(\xi, \eta-1)$ (if exist) are inserted into L_{i+1} unless there is exists candidate pair in L_i (except for (ξ, η)) that blocks them. Thus, we delete one candidate pair and insert at most two candidate pairs.

We define *left* and *right neighbors* of a pair (ξ, η) as follows: a left neighbor of (ξ, η) is every pair $(\mu, \eta - 1), \mu < \xi$. A right neighbor of (ξ, η) is every pair $(\xi + 1, \mu), \mu > \eta$.

Throughout the updates of L_i we do not re-insert a pair that had been used before to realize a distance $d_j, j < i$. Moreover, we avoid storing nested intervals in L_i. As we reach stage $i - 1$ we find which pair of L_{i-1} realizes d_{i-1}. Assume (ξ, η) realizes d_{i-1}. We update L_{i-1} to get L_i. We delete (ξ, η) from L_{i-1}. If L_{i-1} contained a left (right) neighbor of (ξ, η) then we do not add the pair $(\xi, \eta - 1)$ $((\xi + 1, \eta))$ to L_i. Otherwise we add these pairs to L_i. This ensures that L_i does not contain nested intervals.

Claim 2: If a pair (ξ, η) realizes d_i, then it will not be added as a candidate pair in L_j, for $j > i$.

Proof. We prove by induction. L_1 consists of only one interval $(1, n)$. L_2 contains two candidate pairs $(1, n - 1)$ and $(2, n)$ that define overlapping , but not nested intervals. Clearly, the pair $(1, n)$ will not be inserted to $L_j, j > 1$, because we always decrease the interval. Assume we are at stage i. By the induction hypothesis L_i does not contain nested intervals. Assume that $(\xi, \eta) \in L_i$ realizes d_i. (ξ, η) can donate two new overlapping intervals to L_{i+1}: namely, $(\xi + 1, \eta)$ and $(\xi, \eta - 1)$. We look at the neighbors of (ξ, η) in L_i. If there exists a left neighbor of (ξ, η), then we do not add $(\xi, \eta - 1)$ to L_{i+1} in our algorithm (same for the right neighbor). Clearly, (ξ, η)

will not re-appear in the next stages because we only decrease the range of intervals and since there is no nesting there is no interval that contains (ξ, η).

Corollary 3: $|L_i| \le i$, $i = 1, \ldots, n-1$, and $|L_i| \le n-1$, $i = n, \ldots, \frac{n(n-1)}{2}$.

Following Corollary 3 we can easily solve Problem 2 for one axis. Since the number of candidates for each stage does not exceed $n-1$, it suffices to find the updates to the candidate list L_i at each stage i, and then find which pair realizes d_i. Naively we can carry out one stage in $O(n)$ time, therefore the k largest distances are found in $O(kn)$ time and linear space. This runtime can be improved by using tournament trees ([1, 18]) with $n-1$ leaves, each storing a candidate pair. Initially we store only one candidate pair, namely $(1, n)$, and the other leaves are empty. As we proceed to L_i we make at most three updates to the tree. The pair that realizes d_i is the winner of the tournament. The update of the tournament tree for L_{i+1} proceeds as follows: If we do not need to add anything we just empty the leaf occupied by the winner for d_i and continue to find the second best (the pair for d_{i+1}) in the tournament tree. If we add one pair, we replace the contents of the leaf that contained the winner with the new pair and update the path to the root to find the pair realizing the next distance. If we add two pairs, than we put one pair instead of the winner's leaf, another pair into the current available leaf (we always have one due to Corollary 3) and update two paths to the root to find the next winner. We take care of not inserting a nested interval by maintaining an array U whose i'th entry is either empty or contains a pointer to the leaf containing the pair (i, j) in the tournament tree for some j. (Notice that there can be only one leaf containing i as the first point, since there is no nesting). The leaves of the tournament tree point to their corresponding entries in U, and each non empty entry in U points also to the closest non empty pairs in U, backwards and forward respectively.

An update of the tree takes $O(\log n)$ time, so the runtime of this algorithm is improved from $O(kn)$ to $O(n + k \log n)$.

Returning to the L_∞ metric. We perform the algorithm for the x axis simultaneously with the algorithm for the y axis. We first compute the winner in both trees and compare the two distances: the largest current x-distance and the largest current y-distance. We choose the largest between them. We check whether these two distances are defined by the same pair of points. If they are, then we choose the largest distance, report the pair and proceed with both the algorithms to the next step (namely, updating the tournament trees, and finding the next winners). If they are not, then we check whether the larger of the distances has been reported before (in $O(1)$ time we compute the distance in the other axis and compare it to the distance we have in that axis at this stage of the algorithm). If it has been reported, we move to the next step in this axis, and if not we report this pair of points and proceed to the next stage.

Theorem 2. *Given a set S of n points in the plane and a number k we can enumerate the k largest rectilinear distances in nonincreasing order in $O(n + k \log n)$ time, using only $O(n)$ space.*

Remark 3 If U is implemented as a linked list, and the tournament tree is implemented as a heap (see the full version) then the space is $O(\min(k, n))$.

The second case of problem (2) is: enumerate the k smallest rectilinear distances in increasing order. The idea is similar to the algorithm above. We first show an algorithm that enumerates all the k pairs of points which realize the k smallest distances on the x axis. We assume that the points of S are sorted by their x-coordinate, in increasing order. A candidate pair for realizing d_1 is either one of the neighboring pairs $(\xi, \xi + 1)$, for $\xi = 1, \dots, n - 1$, We choose the pair that realizes the smallest distance by creating a tournament tree of pairs. At the following step we perform similar updates to the tournament tree, namely, delete the pair that realized d_1 and insert at most two new candidate pairs, avoiding nested pairs. The algorithm that we apply here is almost identical to the previous one, except that here the distances increase, and we have to initially sort the coordinates of the points.

Theorem 3. *Given a set S of n points in the plane and a number k we can enumerate the k smallest rectilinear distances in nondecreasing order in $O(n \log n + k \log k)$ time, using only $O(n)$ space.*

Remark 4 These enumerating problems can be easily extended to d-dimension, d fixed, $d \geq 3$ with the same runtime and space.

4 Reporting δ distances (Problem 3)

In a very recent paper Dickerson and Eppstein [8] considered the following problem:
Problem 3.1: Given a set S of n distinct points in d-dimensional space, $d \geq 2$, and a distance δ. For each point p in S report all pairs of points (p, q) with q in S such that the distance from p to q is less than or equal to δ.

This problem and the problem of enumerating the k smallest distances in non-decreasing order are closely related. If δ of this problem is the unique k^{th} largest distance of the enumerating problem, then the two solutions are identical. [8] solve Problem 3.1 in $O(n \log n + k)$ time and $O(n)$ space algorithm, where k is the number of distances not greater than δ, and the distances are not ordered. Our algorithm reports these distances *sorted* in the same time and space complexity for L_∞.

Another variant of this problem, that has not been considered before, is:
Problem 3.2: Find all pairs of points in S separated by a L_∞ distance δ or more.

For both Problems 3.1 and 3.2, if we want the distances sorted, we can use our algorithms from the previous section to get $O(n + k \log n)$ algorithm with linear space for Problem 3.1, where k is the number of distances not greater than δ, and $O(n \log n + k \log k)$ time algorithm with linear space for Problem 3.2. The only change is that we compare the output distances with δ. Notice that if we use the algorithm of [8] for sorting the distances then we would end up spending $O(n + k)$ space.

We want to solve first Problem 3.2. The technique is similar to the one we used in solving Problem 1. We first describe an algorithm for the x axis.

Throughout the algorithm we will maintain a poset (which is initially empty) that will contain the largest and the smallest x values of the points that have been encountered in the algorithm (as will be seen below). Pick an arbitrary point $p_1 \in S$. The farthest x-neighbor of p_1 can be the point with the smallest (or largest) x coordinate. The smallest point is added to the set s_x and the largest to the set g_x. After we find which point is the farthest x-neighbor of p_1 (say it is p_i and assume wlog

$p_i \in s_x$), we check whether $|x(p_1) - x(p_i)| \geq \delta$. If $|x(p_1) - x(p_i)| < \delta$, then we know that there is no point $q \in S$, such that $|x(p_1) - x(q)| \geq \delta$. If $|x(p_1) - x(p_i)| \geq \delta$ we continue to find the next farthest x-neighbor of p_1 and update s_x and g_x accordingly. It can either be a point with x-coordinate adjacent to $x(p_i)$ in s_x or the next farthest point in g_x. The algorithm for p_1 ends when on both ends of s_x and g_x the distance is smaller than δ. We end up with a poset P_x, where s_x and g_x are sorted in x order and the rest of the points in $S - s_x - g_x$ are not sorted. Similarly, we work on the y distance for p_1, and create P_y, s_y and g_y.

In order to find the δ L_∞ distances for p_1 we go over P_x and P_y. If the same point, p_j, appears either in x or in y sets, then we can output the pair (p_1, p_j) and proceed to the next points till we got all the points whose distance from p_1 is not smaller than δ. We repeat the process with $p_2 \in S$. As for p_1 the x-farthest point is the point with the largest or smallest x-coordinate, but this point is already in g_x or s_x. So we go over P_x as was created for p_1. We might add points to g_x, s_x, if all the distances $|p_2, q| > \delta, q \in s_x$ or $q \in g_x$ or not. Now we use the sets s_x, g_x, s_y, g_y computed before and report the appropriate pairs that have the required distance (not smaller than δ). There are two possibilities, (1) no points are added to s_x (or s_y, g_x, g_y), or (2) some are added. The number of elements in $s_x(g_x, s_y, g_y)$ does not decrease.

Considering the time complexity. The worst case is when we have to know the total x-order and y-order of all the points in S. The worst case runtime is $O(n \log n + k)$ and the space is $O(n)$.

The algorithm for Problem 3.1 is very similar to the above algorithm. The main difference is that instead of starting at the farthest neighbors and constructing $P_x(P_y)$ incrementally, we now sort the $x(y)$ coordinates of the points of S (so we do not need the posets). For each point p_i we go over its x (and y) nearest neighbors in left (up) and right (down) directions and report the distances (similar to algorithm 3.2) as long as they are less than δ.

Theorem 4. *Given a set S of n points in the plane and a distance $\delta > 0$ we can report all the pairs of points of S which are of rectilinear distance δ or more (less) in $O(n \log n + k)$ time, using only $O(n)$ space.*

Note that in the theorem above k is the number of L_∞ distances for the case of "more than δ", and k is the number of distances measured along x and y axes for the case of "less than δ".

5 Rectangular rings (problem 4)

In this section we solve problem 4: Given a set S of n points in the plane, find the smallest rectangular axis-aligned ring (constrained or non-constrained) that contains $k, k \geq \frac{n}{2}$ points of S. As a measure we take the width (for constrained ring) or area (for non-constrained ring) of the ring.

5.1 Constrained rectangular ring

This problem can be translated to the following one:
For every point $p_i \in S$ find the $n - k$ nearest and $n - k$ farthest rectilinear (under

L_∞ metric) neighbors. We can use our algorithm from Section 2 to find the $n - k - 1$ farthest rectilinear neighbors for each point of S, and the algorithm of [8] to find the $n - k - 1$ nearest neighbors. Given the set of the $n - k - 1$ nearest neighbors N_i of $p_i \in S$ and the set of the $n - k - 1$ farthest neighbors F_i, we sort N_i and F_i according to their L_∞ distance from p_i. There are exactly $n - k - 1$ rings centered in p_i and containing k points. The rings $j = 1, \ldots, n - k - 1$ are determined by the j'th points in the sorted N_i and F_i respectively, where the j'th point from F_i determines the outer rectangle and the j'th point from N_i determines the inner rectangle.

The runtime of the algorithm in [8], as well as for our algorithm (Section 2) is $O(n - k)$ for one point $p_i \in S$ (after $O(n \log n)$ time for preprocessing). We spend $O((n - k) \log (n - k))$ time for sorting N_i and F_i for each point $p_i \in S$, and then go over the corresponding rectangles. Therefore,

Theorem 5. *Given a set S of n points in the plane, we can find the smallest rectangular axis-aligned constrained ring that contains $k, k \geq \frac{n}{2}$ points of S in $O(n \log n + n(n - k) \log (n - k))$ time, using $O(n)$ space.*

Remark 5 This problem can be easily extended to d-dimensions, $d \geq 3$, with the same runtime.

5.2 Non-constrained rectangular ring

We find the smallest rectangular ring that contains $k, k \geq \frac{n}{2}$ of given n points by first computing all the rectangles which contain $k + p$ points $(p = 1, \ldots, n - k)$. Each such rectangle defines a center c for which we find the largest rectangle centered at c that contains p points. In [18] an algorithm for finding the smallest axis-aligned rectangle that contains $k, k \geq \frac{n}{2}$ points is presented. The outline of algorithm from [18] is as follows: initially fix the leftmost point of the rectangle to be the leftmost point of S. At the next stage the leftmost point of the rectangle is fixed to be the second left point of S, etc. Within one stage, of a fixed leftmost rectangle point, r, we pick the rightmost point of the rectangle to be the q'th x-consecutive point of S, for $q = k + r - 1, \ldots, n$. For fixed r and q the x boundaries of the rectangle are fixed, and we go over a small number of possibilities to choose the upper and lower boundaries of the rectangle so that it will enclose k points. This algorithm runs in time $O(n + (n - k)^3)$. We use it for computing all the rectangles which contain $k + p$ points $(p = 1, \ldots, n - k)$. We denote the external rectangle by $\mathcal{R}$.

We modify the problem of finding the smallest rectangle with a given center, that contains p points, to find the largest rectangle with a given center, that contains p points. Notice that the external rectangle $\mathcal{R}$ defines the range of boundaries for the internal rectangle. Our algorithm goes over all the possible rectangles with the given center that contain p points and chooses the largest among them as follows. Let Q be an inner rectangle that contains p points. We extend its boundaries until it almost meets, but does not contain another point of S, within the boundaries of $\mathcal{R}$.

The naive approach for finding the largest rectangle with a given center that contains p points is to go over all pairs of points that together with the center c define a rectangle, check whether this rectangle contains p points and find the largest rectangle among those that do. The total running time is $O(n^3)$.

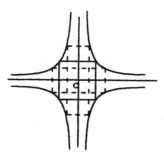

Fig. 2. Hyperbolae define the locus of rectangles with given area

Another approach to this problem is to define the following decision problem: For a given area $\mathcal{A}$ does there exist a rectangle centered at c that covers exactly p points and whose area is $\mathcal{A}$. For the decision algorithm we sort the points of S according to their x and y coordinates respectively. Four hyperbolae define the locus of all rectangles with a given area $\mathcal{A}$, centered at c (see Figure 2). Observe the halfspace defined by the hyperbola H that contains the origin. We consider all the points of S which are inside the intersection of the four halfspaces that correspond to the four hyperbolae. Denote this set by $S' \subseteq S$. Each point $s \in S'$ defines two rectangles with center c and the given area: where s either determines the *width* of the rectangle, or its height. For the time being we look at the rectangle whose width is determined by s. Let s be the point that determines the widest rectangle Q and assume that s is to the left of c.

We shrink the width of the rectangle, keeping its corners in the corresponding hyperbolae until an *event* happens. (Clearly the height of the rectangle grows when the width shrinks.) An event is when a point is added or deleted from the rectangle during the width shrinking. We check if the newly obtained rectangle contains p points. If the obtained rectangle does contain p points, we are done; otherwise we continue to shrink the rectangle until the next event. We perform the same actions for the height as well.

To implement this algorithm efficiently we define four subsets U, D, R, L of S' corresponding to the halfplanes that bound Q. R is the set of all the points of S' contained in the halfplane to the right of the left side of Q and are within the interior of the hyperbolae. L (U, D) is the set of points to the left (up, down) of the right (upper, lower) side of the rectangle Q. We define $p_r(p_l)$ to be the point x-closest to Q in $R(L)$ and $p_u(p_d)$ to be the point y-closest to Q in $U(D)$. Assume that the number of points contained in Q is r and we are shrinking Q in x direction until the next event. Assume that the x-closest neighbor of $p_r(p_l)$ in $R(L)$ is $p_r^h(p_l^h)$ and the y-closest neighbor of $p_u(p_d)$ in $U(D)$ is $p_u^v(p_d^v)$. Thus, our event is when one of p_r^h, p_l^h or p_u^v, p_d^v enters or exists the rectangle Q. If Q contained r points and the next event is a point from R or L, then the new rectangle will contain $r - 1$ points, otherwise $r + 1$. We update p_r, p_l, p_u, p_d (and also the subsets U, D, R, L). When we reach a rectangle with p points we first extend its boundaries with $\mathcal{R}$ until it almost touches the $p+1$'th point and then we move to the next step (with the same center). During the process for this center we keep the largest area inner rectangle encountered so far. The algorithm for solving the decision problem works in time $O(n)$ after preprocessing of $O(n \log n)$, because we can carry each step in constant time, except for the first step where we have to compute the points that lie in the interior of the hyperbolae.

In order to solve the optimization problem, we apply the optimization technique of Frederickson and Johnson [12]. We define the matrix of distances as follows: one dimension of the matrix contains the sorted x-distances from the center (multiplied by 2) , and the other dimension contains the sorted y-distances from the center (multiplied by 2). The matrix values are potential area values of the rectangle. We perform a binary search on the matrix to find the optimal area. Since the rows and columns of the matrix are sorted, we can use the linear time selection algorithm of [12] to find the largest axis-parallel rectangle centered at c and containing p points in $O(n \log n)$ time.

The analysis follows this of [18]: There are $O((n - k)^4)$ external rectangles, and for each of them we apply an $O(n \log n)$ algorithm for finding the largest internal rectangle. So, the total runtime is $O(n(n - k)^4 \log n)$ with linear space. We conclude by the following theorem:

Theorem 6. *Given a set S of n points in the plane, we can find the smallest area rectangular axis-aligned ring that contains $k, k \geq \frac{n}{2}$ points of S in $O(n(n-k)^4 \log n)$ time, using $O(n)$ space.*

Remark 6 This problem can be extended to 3-dimension space. Using the algorithm of [18] and technique of [12] for 3-dimension space we obtain algorithm with runtime $O(n^2(n - k)^6 \log n)$ time.

6 Constrained circular ring (Problem 5)

In this section we solve the following problems: Given a set S of n points, find the smallest *constrained* circular ring (or a sector of a constrained circular ring) that contains k points ($k \geq \frac{n}{2}$) of S. We first describe an algorithm that find the smallest width circular ring containing k points ($k \geq \frac{n}{2}$), and centered at some point $p_i \in S$. We need to know the sorted order of the $n - k$ closest points to p_i and $n - k$ farthest points from p_i and then proceed as in the algorithm for finding a constrained rectangular ring. The time for computing the $n - k$ closest and $n - k$ farthest points for p_i is $O(n + (n - k) \log n)$. Thus we can conclude by

Theorem 7. *Given a set S of n points in the plane, we can find the smallest width constrained ring that contains $k, k \geq \frac{n}{2}$ points of S in $O(n^2 + n(n - k) \log n))$ time, using $O(n)$ space.*

Now we describe how to find minimal **area** sector of a constrained ring that contains $k, k \geq \frac{n}{2}$, points. We first describe an algorithm that finds the smallest area sector of a ring containing k points ($k \geq \frac{n}{2}$) centered at point $O(0, 0)$. We start with finding for $O(0, 0)$ the ordering of S points with respect to the polar angle around the origin. We use the algorithm in [18] to solve our problem in the following way: apply the algorithm in [18] for a smallest axis-aligned rectangle with k points using a polar coordinate system (ρ, θ). This yields the smallest area sector of a ring centered at the origin and containing k points of S. We proceed as in the algorithm of [18]. The running time of this algorithm is $O(n + k(n - k)^2)$. We can use this ring-algorithm as a subroutine to solve the following problem: Find the smallest area sector of a constrained ring (centered on an input point) containing k points. We can perform an angular sort of all the points in $O(n^2)$ time and space [13] and applying this algorithm to each point we get $O(n^2 + nk(n - k)^2)$ time.

Theorem 8. *Given a set S of n points in the plane, we can find the smallest area sector of a constrained ring that contains k points $(k \geq \frac{n}{2})$ points of S in $O(n^2 + nk(n - k)^2)$ time using $O(n^2)$ space.*

7 Query rectangle (Problem 6)

Problem: Given a set S of n points in the plane and a number k $(\frac{n}{2} \leq k \leq n)$ we want to preprocess the points in order to answer efficiently whether k or more points are enclosed by a query rectangle. The naive approach to this problem is to build a range tree [3] on the set S. When a query rectangle R is given, we can answer how many points are inside of R in $O(\log n)$ time using the fractional cascading technique of [5]. The preprocessing time and space is $O(n \log n)$. Notice that we did not use the parameter k at all. In order to improve the preprocessing time and space and also the query time we use the following observation.

Observation 9 *In order for the query rectangle to contain at least k points, the vertical strip defined by the vertical sides l_1, l_2 of the query rectangle R must be located between the $n - k$ smallest and $n - k$ largest x values of the points of S and the horizontal strip defined by the horizontal sides l_3, l_4 of the query rectangle R must be located between the $n - k$ smallest and $n - k$ largest y values of the points of S.*

Using this observation we proceed as follows. First we evaluate the smallest and the largest $n - k$ x values of the points of S (denote by S_x) and the smallest and the largest $n - k$ y values of the points of S (denote by S_y). Next, by a binary search, we find how many points are in the left halfplane of l_1, in the right halfplane of l_2, in the upper halpfplane of l_3 and in the lower halpflane of l_4 (See Figure 3 below). Notice that we count twice the points in the regions $R_i, 1 \leq i \leq 4$ in Figure 3. We can compute how many points are in these regions by building, at the beginning of the algorithm, a range search tree but only for the points with either x-coordinate in S_x or y-coordinate in S_y. We have $O(n - k)$ such points. Thus the construction of the tree takes $O((n - k) \log (n - k))$ time with $O((n - k) \log (n - k))$ space. Now we can compute how many points are in the four query rectangles that correspond to the regions $R_i, 1 \leq i \leq 4$ in the Figure 3. Clearly, the query time for such a rectangle is $O(\log (n - k))$. Thus,

Theorem 10. *Given a set S of n points in the plane and a number k $(\frac{n}{2} \leq k \leq n)$, we can preprocess the points of S in $O((n - k) \log (n - k))$ time with $O((n - k) \log (n - k))$ space to answer in $O(\log (n - k))$ time whether k or more points are enclosed by a query rectangle.*

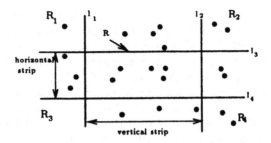

Fig. 3. The strips enclose a query rectangle R.

References

1. M.Aigner "Combinatorical search", Wiley-Teubner Series in CS, John Wiley and Sons, 1988.
2. P.K.Agarwal, M.Sharir "Efficient randomized algorithms for some geometric optimization problems", In *Proc. 11th Annu. ACM Symp. Computational Geometry*, 1995.
3. J.L.Bentley "Decomposable searching problems", *Info. Proc. Lett.* 8, 244-251, 1979.
4. M.Blum, R.Floyd, V. Pratt, R. Rivest, R. Tarjan "Time bounds for selection", *Journal of Computer and System Sciences*, 7(4):448-461, 1973.
5. B.M.Chazelle, L.J.Guibas "Fractional cascading: I. A data structuring technique", *Algorithmica*, 1, 133-162, 1986.
6. M.T.Dickerson, R.L.Scot Drysdale, J-R.Sack "Simple algorithms for enumerating interpoint distances and finding k nearest neighbors", *Internat. J. Comput. Geom. Appl.*, 2(3):221-239, 1992.
7. M.T.Dickerson, J.Shugart "A simple algorithm for enumerating longest distances in the plane", *Inform. Process. Lett.* 45, 269-274, 1993.
8. M.T.Dickerson, D.Eppstein "Algorithms for proximity problems in higher dimensions", *Computational Geometry: Theory and Applications* 5, 277-291, 1996.
9. A. Efrat, M.Sharir, A. Ziv "Computing the smallest k-enclosing circle and related problems", *Computational Geometry* 4, 119-136, 1994.
10. J.Erickson, D.Eppstein "Iterated nearest neighbors and finding minimal polytopes", *Discrete Comput. Geom* 11, 321-350, 1994.
11. R.Y.Flatland, C.H.Stewart "Extending range queries and nearest neighbors", In *Proc. 7th Canad. Conf. Comput. Geom.*, 267-272, 1995.
12. G.Frederickson, D.Johnson, "Generalized selection and ranking: sorted matrices", *SIAM J. Comput.* 13, 14-30, 1984.
13. D.T.Lee, Y.T.Ching "The power of geometric duality revised", *Inf. Proc. Lett.* 21, 117-122, 1985.
14. V.B.Le, D.T.Lee "Out-of-roundness problem revisited", *IEEE trans. Pattern Anal. Mach. Intell* PAMI-13, 217-223, 1991.
15. J. Matoushek "On geometric optimization with few violated constraints", *Discrete Comput. Geom.*,14 (1995), 365-384.
16. H-P.Lenhof, M.Smid "Sequential and parallel algorithms for the k closest pairs problem", *Internat. J. Comput. Geom. Appl.* 5, 273-288, 1995.
17. J.Salowe "Enumerating interdistances in space", *Internat. J. Comput. Geom. Appl.* 2, 49-59, 1992.
18. M.Segal, K.Kedem "Enclosing k points in the smallest axis parallel rectangle", In *Proc. 8th Canad. Conf. Comput. Geom.*, 20-25, 1996.
19. M.Smid, R.Janardan "On the width and roundness of a set of points in the plane", In *Proc. 7th Canad. Conf. Comput. Geom.*, 193-198, 1995.

Constructing Pairwise Disjoint Paths with Few Links

Himanshu Gupta[1] and Rephael Wenger[2]*

[1] Stanford University, Stanford CA 94305 (hgupta@cs.stanford.edu)
[2] The Ohio State University, Columbus, OH 43210 (wenger@cis.ohio-state.edu).

Abstract. Let P be a simple polygon and let $\{(u_i, u_i')\}$ be m pairs of distinct vertices of P where for every distinct $i, j \leq m$, there exist pairwise disjoint paths connecting u_i to u_i' and u_j to u_j'. We wish to construct m pairwise disjoint paths in the interior of P connecting u_i to u_i' for $i = 1, \ldots, m$, with minimal total number of line segments. We give an approximation algorithm which in $O(n \log m + M \log m)$ time constructs such a set of paths using $O(M)$ line segments where M is the number of line segments in the optimal solution.

1 Introduction

Let P be a simple polygon and let u and u' be two distinct vertices of P. The *(interior) link distance* from u to u' is the minimum number of line segments (also called *links*) required to connect u to u' by a polygonal path lying in (the interior of) P. The interior link distance from u to u' may differ greatly from the link distance between the two points. (See Figure 1.) A polygonal path which uses the minimum number of required line segments is called a *minimum link (interior) path*. Suri in [11] gave a linear time algorithm for determining the link distance and a minimal link path between two vertices.

Let u_1, u_1', u_2, u_2' be four vertices lying in the given order around P. By virtue of the relative locations of these four vertices, there are nonintersecting paths, ζ_1 and ζ_2, connecting u_1 to u_1' and u_2 to u_2', respectively. However, it is possible that every minimum interior link path connecting u_1 to u_1' intersects every minimum interior link path connecting u_2 to u_2'. (See Figure 1.) To simultaneously connect u_1 to u_1' and u_2 to u_2' by nonintersecting interior paths requires more line segments. In general, two additional line segments suffice to construct two such nonintersecting interior paths. (See [7].)

A set $\Pi = \{(u_i, u_i')\}$, $i \leq m$, of m pairs of distinct vertices of P is *untangled* if some set of pairwise disjoint paths connects each u_i to u_i'. Let $\Pi = \{(u_i, u_i')\}$, $i \leq m$, be an untangled set of m pairs of distinct vertices of P. Let $l(u_i, u_i')$ be the interior link distance from u to u' and let $L = \sum_{i=1..m} l(u_i, u_i')$ be the sum of those distances. Clearly, L line segments are required to construct a set of pairwise disjoint interior paths connecting u_i to u_i', for $i = 1, \ldots, m$. How many additional line segments are required? In [7] we proved that $O(m \log m)$

---

* Supported by NSA grant MDA904-97-1-10019.

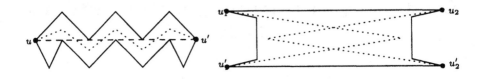

Fig. 1. Minimum link paths, minimum link interior paths and intersecting link paths.

additional line segments suffice and claimed without proof that $\Omega(m \log m)$ additional line segments may be required. A proof of the lower bound is provided in [6].

We define the *pairwise disjoint link paths problem* as: given an untangled set, $\{(u_i, u_i')\}$, of m pairs of distinct vertices of P, find the minimum total number of line segments required by a set of pairwise disjoint interior paths connecting u_i to u_i'. We were unable to give a polynomial time algorithm for this problem or to determine if the problem is NP-complete. Instead we present an algorithm which finds a solution within a constant factor of the optimal solution. Related problems are shown to be NP-complete in [2] and [5], but we do not know if those results can be applied to our problem.

A triangulation T_P of P (possibly with interior vertices) is *isomorphic* to a triangulation T_Q of Q if there is a one-to-one, onto mapping f between the vertices of T_P and the vertices of T_Q such that p, p', p'' are vertices of a triangle in T_P if and only if $f(p), f(p'), f(p'')$ are vertices of a triangle in T_Q. An isomorphic triangulation of P and Q defines a piecewise linear homeomorphism between P and Q. The size of a triangulation is the total number of vertices, edges and triangles in the triangulation.

Algorithms for constructing isomorphic triangulations and piecewise linear homeomorphisms between simple polygons are also given in [1,8,7]. Algorithms for constructing isomorphic triangulations between labelled point sets are described in [9] and [10]. The main result in this paper improves the output size and running time of the approximation algorithm in [7] from $O(M_1 \log n + n \log^2 n)$ to $O(M_1 \log n)$ where n is the input size and M_1 is the size of the optimal solution. The improvement is described in [6].

2 Approximation Algorithm

We first give a an approximation algorithm for connecting a set of vertices $\mathcal{U}$ by pairwise disjoint interior paths to a distinguished edge e^* of P. We start with some definitions.

Point $p \in P$ is *visible* from point $p' \in P$ if P contains the open line segment (p, p'). Point p is *clearly visible* from point $p' \in P$ if the interior of P contains the open line segment (p, p').

Point $p \in P$ is *(clearly) visible* from edge $e \in P$ if there is some point $p' \in e$ such that p is (clearly) visible from p'. (This definition of visibility is sometimes called *weak visibility* as opposed to *strong visibility* where p must be visible from

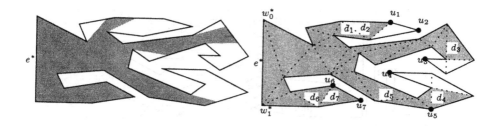

Fig. 2. $\hat{V}$is(e) and Γ_{e^*}.

every point $p' \in e$. Throughout this paper, visibility refers to weak visibility.)
Edge e or triangle t is *(clearly) visible* from edge e' or triangle t' if there are
points $p \in e$ or $p \in t$ and $p' \in e'$ or $p' \in t'$ such that p is (clearly) visible from p'.

We let $\hat{V}$is(p) and $\hat{V}$is(e) denote the points clearly visible from point p and
edge e. (See Figure 2.) Note that $\hat{V}$is(p) and $\hat{V}$is(e) are not necessarily closed
sets.

Let u and e be a vertex and an edge of P, respectively. Edge d of triangulation
T_P *separates* u from e if every interior path from u to the interior of e must
intersect the interior of d. Triangle t of triangulation T_P *separates* u from e if
every interior path from u to the interior of e must intersect the interior of t.

To construct pairwise disjoint paths connecting the vertices $\mathcal{U}$ to edge e^*
of P, we construct a triangulated region Γ_{e^*} which contains and approximates
$\hat{V}$is(e^*), the set of points clearly visible from e^*. For each $u_i \in \mathcal{U}$, let d_i be the
diagonal of Γ_{e^*} farthest from e^* which separates e^* from u_i. Let s_i be the portion
of e^* visible from d_i. Note that s_i is a line segment. (See Figures 2 and 3.)

For each u_i we wish to choose a point p_i on e^* to be the endpoint of the
path from u_i to e^*. Obviously, a point in s_i is a good candidate since it can
reach d_i with a single line segment. However, we also need to choose the p_i
such that their order on e^* is consistent with the order of $\mathcal{U}$ around P. In other
words, u_i, u_j, p_j, p_i should lie clockwise or counter-clockwise around P in the
given order.

We partition the set of line segments $\{s_i\}$ into groups and associate each
such group with a point g_j on e^* which is in the "middle" of the line segments
in the groups. If many of the line segments in the group contain g_j, then the
corresponding diagonals can be connected to g_j by pairwise disjoint line segments. If few line segments contain g_j, then g_j partitions the line segments into
roughly two equals subgroups in e^* with the property that many line segments
connecting d_i to s_i from one subgroup intersect many line segments connecting
d_j to s_j from the other subgroup. In addition, the order that the points g_j lie
on e^* is consistent with the order that the associated vertices of $\mathcal{U}$ lie on the
boundary of P. Partitioning the line segments is conceptually and technically
the most difficult part of the algorithm.

From all the $\hat{V}$is(g_j), we construct another triangulated region $\Gamma \subseteq \Gamma_{e^*}$. We
recursively connect $\mathcal{U}$ by pairwise disjoint paths to the edges on the boundary of

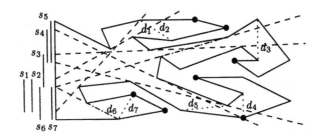

Fig. 3. Line segments $s_i = \text{int}(e^*) \cap \hat{\text{V}}\text{is}(d_i)$.

Γ and then connect those boundary edges by pairwise disjoint line segments to Γ. A careful analysis shows that Γ must contain many line segments in any set of pairwise disjoint paths connecting $\mathcal{U}$ to e^*. Thus the number of line segments in our solution is proportional to the number in the optimal solution.

Lemma 1. *Let P be a simple polygon on n vertices with distinguished edge $e^* = \{w_0^*, w_1^*\}$ and let $\mathcal{U}$ be a subset of $\text{Vert}(P) \setminus \{w_0^*, w_1^*\}$ of size m. A set of m pairwise disjoint interior paths connecting the vertices in $\mathcal{U}$ to the interior of e^* can be constructed in $O(n \log m + M \log m)$ time using a total of at most $240M$ line segments where M is the minimum total number of line segments necessary to connect $\mathcal{U}$ to e^* by m pairwise disjoint paths.*

Proof. Let $u_1, u_2, \ldots, u_m$ be the points in $\mathcal{U}$ labeled in clockwise order around P starting at e^*. Construct a triangulation T_P of P. Let Γ_{e^*} be union of the triangles of T_P which are clearly visible from edge e^*. The region Γ_{e^*} is a simple polygon in P. (See Figure 2.)

For each $u_i \in \mathcal{U}$, let d_i be the diagonal of Γ_{e^*} farthest from e^* which separates e^* from u_i. Let s_i be $\text{int}(e^*) \cap \hat{\text{V}}\text{is}(d_i)$, the interior of e^* which is clearly visible from s_i. Set s_i is an open line segment lying on e^*. Note that s_i may equal s_j (and d_i may equal d_j) for many distinct points $u_i, u_j \in \mathcal{U}$. (See Figure 3.)

Let S be any set of open line segments in $\mathbf{R}^1$, not necessarily distinct. For each point $q \in \mathbf{R}^1$, let $f(q, S)$ be the number of line segments of S which contain the point q. The line segments of S are open and do not contain their endpoints. Let $f^-(q, S)$ and $f^+(q, S)$ be the number of line segments of S contained in the open intervals $(-\infty, q)$ and (q, ∞), respectively. Note that $f(q, S) + f^-(q, S) + f^+(q, S)$ equals $|S|$.

Let $\mathcal{R}$ be the set of midpoints of line segments of S, again not necessarily distinct. The median point of $\mathcal{R}$ is the $\lceil |\mathcal{R}|/2 \rceil$'th point in $\mathcal{R}$, ordered from $-\infty$ to ∞. Let $g(S)$ be this median point of $\mathcal{R}$. At least $|\mathcal{R}|/2 = |S|/2$ points of $\mathcal{R}$ lie in each of the closed intervals $(-\infty, g(S)]$ and $[g(S), \infty)$. If the midpoint of segment $s \in S$ lies in $(-\infty, g(S)]$, then segment s either contains $g(S)$ or lies in the open interval $(-\infty, g(S))$. Thus $f(g(S), S) + f^-(g(S), S)$ is greater than or equal to $\lceil |S|/2 \rceil$. Similarly $f(g(S), S) + f^+(g(S), S)$ is greater than or equal to $\lceil |S|/2 \rceil$. (See Figure 4.)

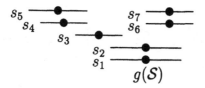

$$f(g(\mathcal{S}), \mathcal{S}) = 2, \ f^-(g(\mathcal{S}), \mathcal{S}) = 3, \ f^+(g(\mathcal{S}), \mathcal{S}) = 2.$$

Fig. 4. $\mathcal{S}$, $g(\mathcal{S})$ and $f(g(\mathcal{S}), \mathcal{S})$.

Now consider two sets of line segments $\mathcal{S}_0$ and $\mathcal{S}_1$ on $\mathbf{R}^1$ and let $\mathcal{S} = \mathcal{S}_0 \cup \mathcal{S}_1$. Define

$$\mathcal{F}(\mathcal{S}_0, \mathcal{S}_1) = f(g(\mathcal{S}), \mathcal{S}) + f^+(g(\mathcal{S}), \mathcal{S}_0) + f^-(g(\mathcal{S}), \mathcal{S}_1).$$

Without loss of generality, assume that w_0^*, e^*, w_1^* appear in counter-clockwise order around P. Let $\mathcal{S}_\mathcal{U}$ be the sequence $(s_1, s_2, \ldots, s_m)$. Embed e^* and the line segments $s_i \in \mathcal{S}$ in the real line $\mathbf{R}^1$, mapping w_0^* to zero and w_1^* to one. In the next section, we describe an algorithm to partition $\mathcal{S}_\mathcal{U}$ into contiguous subsequences $\sigma_1 = (s_1, s_2, \ldots, s_{i_1})$, $\sigma_2 = (s_{i_1+1}, s_{i_1+2}, \ldots, s_{i_2}), \ldots,$ $\sigma_{2h} = (s_{i_{2h-1}+1}, s_{i_{2h-1}+2}, \ldots, s_m)$, such that:

1. $g(\sigma_1 \cup \sigma_2) \le g(\sigma_3 \cup \sigma_4) \le \cdots \cdots \le g(\sigma_{2h-1} \cup \sigma_{2h})$;
2. $|\sigma_{2j-1}| = |\sigma_{2j}|$ $(+1)$ for $1 \le j \le h$;
3. $\sum_{j=1..h} \mathcal{F}(\sigma_{2j-1}, \sigma_{2j}) \ge m/40$.

(One possible partition of the segments in Figure 3 is $\sigma_1 = \{s_1, s_2, s_3\}$, $\sigma_2 = \{s_4, s_5\}$, $\sigma_3 = \{s_6\}$, $\sigma_4 = \{s_7\}$.)

Let g_j equal $g(\sigma_{2j-1} \cup \sigma_{2j})$ for $j = 1, \ldots, h$. Note that $g_1, g_2, \ldots, g_h$ lie in counter-clockwise order around P. Let $\mathcal{U}_j = \{u_i : s_i \in \sigma_j\}$ be the points in $\mathcal{U}$ corresponding to the line segments in σ_j for $j = 1, \ldots, h$. For each g_j, let Γ_j be the union of the triangles of T_P which intersect $\hat{V}\mathrm{is}(g_j)$ and separate some $u \in \mathcal{U}_{2j-1} \cup \mathcal{U}_{2j}$ from e^*. (See Figure 5.) Let Γ be the union of all the Γ_j. Similar to Γ_{e^*}, the region Γ is also a simple polygon in P, its boundary is composed of edges and chords of P, and it has a triangulation T_Γ induced by the triangulation T_P of P.

Let $\mathcal{C}$ be the set of chords of P bounding Γ. Each chord $c \in \mathcal{C}$ separates P into two subpolygons. Let P_c be the subpolygon not containing Γ. Let w_0^c and w_1^c be the endpoints of c. For each chord $c \in \mathcal{C}$, let $\mathcal{U}_c$ be the points of $\mathcal{U} \setminus \{w_0^c, w_1^c\}$ in P_c. Recursively, construct pairwise disjoint paths connecting the points in $\mathcal{U}_c$ to c. (See Figure 6.)

For each $u_i \in \mathcal{U}$, let $\tilde{d}_i$ be the diagonal of Γ farthest from e^* which separates e^* from u_i. Choose the minimum j such that $\tilde{d}_i$ is a diagonal of Γ_j. Connect $\tilde{d}_i$ to g_j by a line segment λ_i in the interior of P. (See Figure 7.) Diagonal $\tilde{d}_i$ may also separate other vertices of $\mathcal{U}$ from e^* and there may be many line segments which intersect $\tilde{d}_i$. The line segments λ_i should be chosen so that their order along $\tilde{d}_i$ corresponds to the order of the vertices around P. The choice of $\tilde{d}_i$

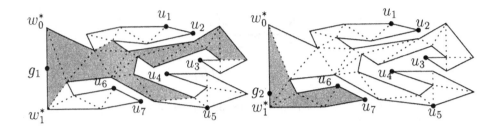

Fig. 5. Γ_1 and Γ_2.

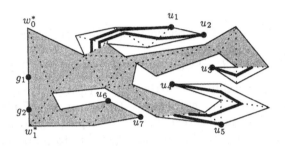

Fig. 6. Γ, triangulation T_Γ and paths to the boundary of Γ.

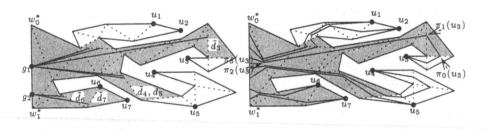

Fig. 7. Diagonals $\bar{d}_i$, line segments λ_i and paths connecting $\mathcal{U}$ to e^*.

and the associated point g_j ensures that line segments λ_i intersect only at their endpoints. (See [6].)

For each point $u_i \in \mathcal{U}(c)$, let $\pi_0(u_i)$ be the endpoint on c of the path connecting u_i to c. For each point $u_i \in \mathcal{U}$ which lies in Γ, let $\pi_0(u_i)$ equal u_i. Let $\pi_1(u_i)$ be the endpoint of λ_i on $\bar{d}_i$. Let $\pi_2(u_i)$ be the first intersection point of λ_i and the triangle containing e^*. Place m points equally spaced on e^*. Let $\pi_3(u_i)$ be the i'th point, ordered counter-clockwise from w_0^*. Connect $\pi_0(u_i)$ to e^* with a polygonal line through $\pi_0(u_i), \pi_1(u_i), \pi_2(u_i), \pi_3(u_i)$. (See Figure 7.)

We claim that this algorithm connects $\mathcal{U}$ to e^* using $O(M)$ links where M is the number of links in some optimal solution. For each $u_i \in \mathcal{U}$, let ζ_i be the path constructed from u_i to e^* by our algorithm while η_i is the path from u_i to e^* in the optimal solution. Path ζ_i has at most three line segments in Γ. Line segment s_i is in $\sigma_{2j-1} \cup \sigma_{2j}$ for some j. If s_i contains g_j, then some point on diagonal d_i is clearly visible from g_j and d_i is a diagonal of $\Gamma_j \subseteq \Gamma$. Since d_i is the farthest diagonal visible from e^* which separates u_i from e^*, any path from u_i to e^* must have at least one line segment contained in $\Gamma_j \subseteq \Gamma$. Thus if s_i contains g_j, then we can charge the three links of ζ_i in Γ to a line segment of η_i in Γ. However, s_i may not contain g_j.

Consider the case where $s_i \in \sigma_{2j-1}$ lies between g_j and w_1^* while $s_{i'} \in \sigma_{2j}$ lies between w_0^* and g_j. Any two paths from u_i to s_i and $u_{i'}$ to $s_{i'}$ must intersect. Since paths η_i and $\eta_{i'}$ are pairwise disjoint, either the endpoint of η_i must lie between w_0^* and g_j or the endpoint of $\eta_{i'}$ must lie between g_j and w_1^*. Without loss of generality, assume that the endpoint ρ of η_i lies between w_0^* and g_j. In that case, g_j lies between ρ and s_i. Let d be the farthest diagonal of P visible from ρ and separating ρ from u_i. By the construction of d_i and s_i, diagonal d separates d_i from e^* and hence is visible to s_i. Since g_j lies between ρ and s_i, diagonal d is also visible to g_j and is contained in $\Gamma_j \subseteq \Gamma$. Thus if ρ lies between w_0^* and g_j, path η_i must have at least one line segment contained in $\Gamma_j \subseteq \Gamma$. Similarly, if the endpoint of $\eta_{i'}$ lies between g_j and w_1^*, path $\eta_{i'}$ must have at least one line segment contained in $\Gamma_j \subseteq \Gamma$. It follows that either η_i or $\eta_{i'}$ must have a line segment contained in $\Gamma_j \subseteq \Gamma$.

Let m_0, m_-, m_+ equal $f(g_j, \sigma_{2j-1})$, $f^-(g_j, \sigma_{2j-1})$ and $f^+(g_j, \sigma_{2j-1})$, respectively, while m_0', m_-', m_+' equal $f(g_j, \sigma_{2j})$, $f^-(g_j, \sigma_{2j})$ and $f^+(g_j, \sigma_{2j})$, respectively. By the arguments above, the paths connecting the points in $\mathcal{U}_{2j-1} \cup \mathcal{U}_{2j}$ to e^* in the optimal solution must have at least $m_0 + m_0' + \min(m_+, m_-')$ line segments contained in Γ. By the choice of point g_j, $m_0 + m_- + m_0' + m_-' \geq |\sigma_{2j-1} \cup \sigma_{2j}|/2$. Since $|\sigma_{2j-1}|$ equals $|\sigma_{2j}|$ or $|\sigma_{2j-1}|+1$, $m_0 + m_- + m_0' + m_-' \geq |\sigma_{2j-1}|$. On the other hand, $m_0 + m_- + m_+ = |\sigma_{2j-1}|$. Subtracting the second equation from the first gives $m_0' + m_-' \geq m_+$. Thus

$$
\begin{aligned}
m_0 + m_0' + \min(m_+, m_-') &= \min(m_0 + m_0' + m_+, m_0 + m_0' + m_-') \\
&\geq \min(m_0 + m_0' + m_+, m_0 + m_+) \\
&= m_0 + m_+
\end{aligned}
$$

Similarly, $m_0 + m_+ \geq m_-'$ and $m_0 + m_0' + \min(m_+, m_-') \geq m_0' + m_-'$. Thus

$$
m_0 + m_0' + \min(m_+, m_-') \geq \max(m_0 + m_+, m_0' + m_-') \geq \mathcal{F}(g_j, \sigma_{2j-1}, \sigma_{2j})/2.
$$

The paths connecting the points in $\mathcal{U}_{2j-1} \cup \mathcal{U}_{2j}$ to e^* in the optimal solution must have at least $\mathcal{F}(g_j, \sigma_{2j-1}, \sigma_{2j})/2$ line segments in Γ. Since $\sum_{j=1..h} \mathcal{F}(\sigma_{2j-1}, \sigma_{2j}) \geq m/40$, any pairwise disjoint paths connecting the points in $\mathcal{U}$ to e^* must have at least $m/80$ line segments contained in Γ. The construction produces at most $3m$ line segments in Γ, so the solution is at most 240 times the optimal.

Finally, we discuss the running time of our algorithm. Constructing the initial triangulation T_P takes $O(n)$ time [3]. As discussed in the next section, partitioning $\mathcal{S}_{\mathcal{U}}$ into the subsequences σ_j takes $O(m \log m)$ time. Constructing Γ_{e^*} takes $O(n^*)$ time where n^* is the number of triangles of T_P intersected by $\hat{\text{Vis}}(e^*)$ [4]. All the other steps in the algorithm can be done in $O(n^* + m)$ time. Thus the

PARTITION($\mathcal{S}$)
/* $\mathcal{S}$ = a sequence of line segments $(s_1, s_2, \ldots, s_m)$ */
/* Returns a linked list of contiguous subsequences of $\mathcal{S}$ */
1. Initialize linked list $\mathcal{A}$ to $\emptyset$;
2. FOR $i = 1$ TO m DO
3.   Create new node a where $a.seq = (s_i)$ and $a.size = 1$;
4.   Add a to the end of linked list $\mathcal{A}$;
5. WHILE $\exists a \in \mathcal{A}$ such that $g(a.seq) > g(a.next.seq)$ DO
6.   Merge a and $a.next$ to form a new node a' in $\mathcal{A}$;
7.   BALANCE-NEXT(a');
8.   BALANCE-PREV(a');
9. Return($\mathcal{A}$).

Fig. 8. Algorithm PARTITION.

non-recursive steps in this algorithm take $O(n^* + m \log m)$ time. A careful accounting for the recursive steps gives the desired $O(n \log m + M \log m)$ bound. Details appear in [6]. $\quad\square$

Using arguments similar to those given in [7], the previous algorithm can be turned into an algorithm for connecting an untangled set of m pairs of vertices of P. The algorithm and its analysis is provided in [6].

Theorem 2. *Let P be a simple polygon on n vertices let $\Pi = \{(u, u')\}$ be an untangled set of m pairs of distinct vertices of P. A set of m pairwise disjoint interior paths connecting u to u' for each $(u, u') \in \Pi$ can be constructed in $O(n \log m + M \log m)$ time using $O(M)$ line segments where M is the minimum total number of line segments necessary to connect all pairs $(s, s') \in \Pi$ by pairwise disjoint paths.*

3 Partition Algorithm

In this section, we describe and analyze the algorithm for partitioning a sequence of line segments. The functions f, f^+, f^- and $\mathcal{F}$ were defined in the previous section.

Lemma 3. *Let $\mathcal{S}$ be a sequence of line segments$(s_1, s_2, \ldots, s_m)$ on the real line $\mathbf{R}^1$. In $O(m \log m)$ time, $\mathcal{S}$ can be partitioned into contiguous subsequences $\sigma_1 = (s_1, s_2, \ldots, s_{i_1}), \sigma_2 = (s_{i_1+1}, s_{i_1+2}, \ldots, s_{i_2}), \ldots, \sigma_{2h} = (s_{i_{2h-1}+1}, s_{i_{2h-1}+2}, \ldots, s_m)$, such that:*

1. *$g(\sigma_1 \cup \sigma_2) \leq g(\sigma_3 \cup \sigma_4) \leq \cdots\cdots \leq g(\sigma_{2h-1} \cup \sigma_{2h})$;*
2. *$|\sigma_{2j-1}| = |\sigma_{2j}|$ (+1) for $1 \leq j \leq h$;*
3. *$\sum_{j=1..h} \mathcal{F}(\sigma_{2j-1}, \sigma_{2j}) \geq m/40$.*

Proof (outline). Split S into m distinct subsequences, (s_i), consisting of one element each. Store the m subsequences in a linked list A in the order they appear in S. Each node $a \in A$ contains a subsequence $a.seq$. Call A *balanced* if the size of each subsequence is at most three times the size of any adjacent subsequence in A.

While A contains two adjacent subsequences, $a.seq$ followed by $a.next.seq$, such that $g(a.seq) > g(a.next.seq)$, merge the subsequences $a.seq$ and $a'.seq$. After each merge of two such subsequences, rebalance list A by merging adjacent subsequences, as necessary. Figure 8 contains the main algorithm. A complete description of the subroutines BALANCE-NEXT and BALANCE-PREV is provided in [6].

Let a_j be the j'th node in A when the algorithm is completed. Partition $a_j.seq = (s_i, \ldots, s_{i'})$ into two approximately equal sized sequences $\sigma_{2j-1} = (s_i, \ldots, s_{\lceil (i+i')/2 \rceil})$ and $\sigma_{2j} = (s_{\lceil (i+i')/2 \rceil + 1}, \ldots, s_{i'})$. We claim that this is a partitioning of S with the desired properties. Initially, the s_j are stored in A in sorted order. The merging and splitting steps in the main algorithm and in the subroutines BALANCE-NEXT and BALANCE-PREV preserve the order of the s_i, so the σ_j properly partition S into contiguous subsequences.

Let g_j be $g(a_j.seq) = g(\sigma_{2j-1} \cup \sigma_{2j})$. The while loop only terminates when $g_1 \le g_2 \le \cdots \le g_h$, so property 1 is clearly satisfied. Sets σ_{2j-1} and σ_{2j} are created by partitioning $a_j.seq$ into two equal sized sequences, so property 2 is satisfied.

To show property 3 holds, note that a_j could be an initial node or it could be created when $g(a.seq) > g(a.next.seq)$ or it could be created in the rebalancing procedure.

If a_j is an initial node, then $a_j.seq = \{s\}$ for some $s \in S$ and $\sigma_{2j-1} = \{s\}$ and $\sigma_{2j} = \emptyset$. Point g_j is the midpoint of s and $\mathcal{F}(\sigma_{2j-1}, \sigma_{2j}) \ge 1 \ge (1/8)|a_j|$.

Assume a_j is created when $g(a.seq)$ is greater than $g(a.next.seq)$ and that $|a.next| \ge |a|$. The sequence $a.seq$ is a subsequence of σ_{2j-1}, so

$$f(g_j, \sigma_{2j-1}) \ge f(g_j, a.seq) \text{ and}$$
$$f^+(g_j, \sigma_{2j-1}) \ge f^+(g_j, a.seq).$$

The point $g_j = g(a_j.seq)$ must lie between $a.g$ and $a.next.g$, so

$$f(g_j, a.seq) + f^+(g_j, a.seq) \ge f(g(a.seq), a.seq) + f^+(g(a.seq), a.seq)$$
$$\ge |a|/2.$$

Since $|a.next| \le 3|a|$, we have $|a_j| \le 4|a|$. Thus,

$$\mathcal{F}(\sigma_{2j-1}, \sigma_{2j}) \ge f(g_j, \sigma_{2j-1}) + f^+(g_j, \sigma_{2j-1}) \ge (1/8)|a_j|.$$

In the case that $|a.next| < |a|$, similar reasoning gives

$$\mathcal{F}(\sigma_{2j-1}, \sigma_{2j}) \ge f(g_j, \sigma_{2j}) + f^-(g_j, \sigma_{2j}) \ge (1/8)|a_j|.$$

Finally, if a_j is created in the rebalancing step, $\mathcal{F}(\sigma_{2j-1}, \sigma_{2j})$ may not have the desired lower bound. However, at most $(4/5)m$ line segments lie in nodes created in the rebalancing step. By counting the $m/5$ line segments which are not in a rebalanced node, we find $\sum_{i=1..k} \mathcal{F}(\sigma_{2j-1}, \sigma_{2j}) \geq m/40$.

A complete description and analysis of the algorithm, its correctness and $O(m \log m)$ running time appears in [6]. □

References

1. ARONOV, B., SEIDEL, R., AND SOUVAINE, D. On compatible triangulations of simple polygons. *Comput. Geom. Theory Appl. 3*, 1 (1993), 27–35.
2. BASTERT, O., AND FECKETE, S. Geometrische verdrahtungsprobleme. Technical Report 96.247, Angewandte Mathematik und Informatik, Universität zu Köln, Köln, Germany, 1996.
3. CHAZELLE, B. Triangulating a simple polygon in linear time. *Discrete Comput. Geom. 6* (1991), 485–524.
4. GUIBAS, L. J., HERSHBERGER, J., LEVEN, D., SHARIR, M., AND TARJAN, R. E. Linear-time algorithms for visibility and shortest path problems inside triangulated simple polygons. *Algorithmica 2* (1987), 209–233.
5. GUIBAS, L. J., HERSHBERGER, J. E., MITCHELL, J. S. B., AND SNOEYINK, J. S. Approximating polygons and subdivisions with minimum link paths. *Internat. J. Comput. Geom. Appl. 3*, 4 (Dec. 1993), 383–415.
6. GUPTA, H., AND WENGER, R. Constructing pairwise disjoint paths with few links. Technical Report OSU-CISRC-2/97-TR16, The Ohio State University, Columbus, Ohio, 1997.
7. GUPTA, H., AND WENGER, R. Constructing piecewise linear homeomorphisms of simple polygons. *J. Algorithms 22* (1997), 142–157.
8. KRANAKIS, E., AND URRUTIA, J. Isomorphic triangulations with small number of Steiner points. In *Proc. 7th Canad. Conf. Comput. Geom.* (1995), pp. 291–296.
9. SAALFELD, A. Joint triangulations and triangulation maps. In *Proc. 3rd Annu. ACM Sympos. Comput. Geom.* (1987), pp. 195–204.
10. SOUVAINE, D., AND WENGER, R. Constructing piecewise linear homeomorphisms. Technical Report 94–52, DIMACS, New Brunswick, New Jersey, 1994.
11. SURI, S. A linear time algorithm for minimum link paths inside a simple polygon. *Comput. Vision Graph. Image Process. 35* (1986), 99–110.

Trans-dichotomous Algorithms Without Multiplication – Some Upper and Lower Bounds

Andrej Brodnik[1]*.   Peter Bro Miltersen[2]**   J. Ian Munro[3]***

[1] Institute of Mathematics, Physics and Mechanics, Ljubljana, Slovenia and Luleå University, Luleå, Sweden. Email Andrej.Brodnik@IMFM.Uni-Lj.SI.
[2] BRICS, Centre of the Danish National Research Foundation, University of Aarhus, Denmark. Email bromille@brics.dk.
[3] Department of Computer Science, University of Waterloo, Canada. Email imunro@uwaterloo.ca.

Abstract. We show that on a RAM with addition, subtraction, bitwise Boolean operations and shifts, but no multiplication, there is a trans-dichotomous solution to the static dictionary problem using linear space and with query time $\sqrt{\log n}(\log \log n)^{1+o(1)}$. On the way, we show that two w-bit words can be multiplied in time $(\log w)^{1+o(1)}$ and that time $\Omega(\log w)$ is necessary, and that $\Theta(\log \log w)$ time is necessary and sufficient for identifying the least significant set bit of a word.

1   Introduction

Consider a problem (like sorting or searching) whose instances consists of collections of members of the universe $U = \{0, 1\}^w$ of w-bit strings (or numbers between 0 and $2^w - 1$). An increasingly popular theoretical model for studying such problems is the *trans-dichotomous* model of computation [13, 14, 1, 7, 8, 3, 2, 20, 18, 9, 4, 21, 6], where one assumes a random access machine where each register is capable of holding exactly one element of the universe, i.e. we assume that the word size of the machine matches the "word size" of the universe. We can operate on the registers using at least a *basic instruction set* consisting of: Direct and indirect addressing, conditional jump, and a number of computational instructions, including addition, subtraction, bitwise Boolean operations and left and right shifts. All operations are unit cost. An important ingredient of many trans-dichotomous algorithms is exploiting the unit cost assumption by doing parallel operations on the word level. It is well known that such "tricks" often speed programs up in practice, and we can view trans-dichotomous algorithms as theory's contribution to the understanding of this phenomenon.

In this paper, we consider trans-dichotomous algorithms for the static dictionary problem: Given a set of keys $S \subseteq \{0, 1\}^w$, construct a data structure using $O(n)$ registers, with $n = |S|$, so that membership queries "Is x in S?" can be answered efficiently, and, if $x \in S$, some information associated with x can be retrived.

---

* Some of the results of this paper appeared in the PhD thesis of this author

** Supported by the ESPRIT Long Term Research Programme of the EU under project number 20244 (ALCOM-IT). Part of this work was done while the author was at the University of Toronto.

*** This research was supported by a grant from the Natural Science and Engineering Council of Canada.

It is well known that if the computational instructions also include multiplication, the static dictionary problem can be solved using only $O(1)$ query time [12, 11] and even in space within an additive low order term of the information theoretic minimum [8]. However, in many architectures encountered in practice, multiplication is more expensive to perform than the basic operations mentioned above, so it seems natural to disallow it as a unit cost operation in our theoretical model. Indeed, this approach is taken in [2, 18, 21], and we shall also take it here.

Andersson et al [4] show that if only the basic instruction set is used, worst case query time $\Omega(\sqrt{\log n / \log \log n})$ is necessary. This lower bound is also valid under extended computational instruction sets, as long as all instructions can be computed by AC^0 circuits (multiplication can not [15]). Andersson et al show that a matching upper bound $O(\sqrt{\log n / \log \log n})$ can be obtained, if various exotic AC^0 instructions (*clustering functions* and *cluster busters*) are allowed.

One of the main results of the present paper is that with only the basic instruction set of addition, subtraction, bitwise Boolean operations, and shifts, worst case query time $\sqrt{\log n}(\log \log n)^{1+o(1)}$ is possible. This leaves only a $(\log \log n)^{1+o(1)}$ gap to the lower bound.

The technique is basically a variation of the technique of Andersson et al [4], combining range reduction with packed B-trees. In order to carry this approach through in the setting of the basic instruction set, we introduce some techniques and subroutines which may be useful for trans-dichotomous computation in general, namely *bit permutation*, *circuit evaluation*, and *locating the least significant bit of a word*. We also show lower bounds for these problems, in some cases establishing optimality of our subroutines. Thus, we note that any fixed permutation of the bits of a word can be realized in time $O(\log w)$ and that reversing a word requires time $\Omega(\log w)$. We show that multiplying two words can be done in time $(\log w)^{1+o(1)}$ and that $\Omega(\log w)$ is a lower bound. We show that finding the least significant 1-bit in a word can be done in time $O(\log \log w)$ and that this is optimal. The lower bounds hold, even if a precomputed table of size $O(2^{w^\epsilon})$ for a small constant $\epsilon > 0$ is allowed. Our lower bound technique is essentially an extension of the *locality*-technique of [17].

The existence of a static dictionary with sublogarithmic query time on a RAM with the basic instruction set disproves a conjecture of the second author [17]. It is, however, *not* a practical solution, and is intended only to demonstrate an asymptotic upper bound; having disallowed unit cost multiplication from the instruction set, we base our upper bound on reintroducing it as a subroutine on subwords: An essential part of the algorithm is the parallel execution of several instances of Schönhage and Strassen's multiplication algorithm on parts of a word using word level parallelism.

The paper is structured as follows: in Section 2, we construct our general subroutines. In Section 3, we show how to use them to solve the static dictionary problem, and in Section 4, we show the lower bounds.

Notation

When x and y are bit strings of equal length, we denote by x AND y, x OR y, x NAND y, NOT x, and x XOR y the bitwise Boolean operations on x and y. $x[i]$ denotes the i'th bit of x from the left. Words are considered bit strings of length w. Note that when a word x is interpreted as an unsigned integer, the least significant bit is $x[w]$ and the most significant bit is $x[1]$. This might be slightly confusing, but most often we view words as strings, rather than integers. If $I = [i, j] = \{i, i+1, \ldots, j\}$ is an interval of bits, $x[I]$ denotes $x[i]x[i+1] \ldots x[j]$. When x is a bit string, and i a positive integer, we denote by $x \uparrow i$ the result of shifting x i positions to the left (padding the rightmost part of the result with zeros, and killing off the i leftmost positions of x). Similarly, $x \downarrow i$ denotes shifting x i positions to the right (padding the leftmost part of the result with zeros, and killing off the i rightmost positions of x). If i is negative, $x \uparrow i = x \downarrow -i$ and vice versa. We denote by w the word size of the machine. We shall assume that w is a power of two. We will assume that bit strings are stored packed, i.e. a bit string of length m is stored in $\lceil m/w \rceil$ machine words. Note that if we want to do the above mentioned operations on strings of length m, we can do so in time $O(\lceil m/w \rceil)$ by using the basic bit manipulation instructions of our machine.

2 Useful subroutines

Permuting and circuit evaluation

Proposition 1. *Consider a fixed permutation π on m symbols. Given a bit vector $x[1]x[2] \ldots x[m]$, we can compute its permutation*

$$\mathrm{perm}_\pi(x) = x[\pi(1)]x[\pi(2)]x[\pi(3)] \ldots x[\pi(m)]$$

in time $O(\lceil m/w \rceil \log m)$.

Proof. Assume without loss of generality that m is a power of two. It is well known that a *butterfly network* with m sources and m sinks is a permutation network for m packages. Given the permutation π, in this graph we can find edge disjoint paths from source $\pi(i)$ to sink i. Given an input $x \in \{0, 1\}^m$, mark all nodes along the path from $\pi(i)$ to i with label $x[\pi(i)]$. Now, each column in the graph is marked with a bit vector in $\{0, 1\}^m$. The first column is marked with the input, the last column with the desired output. It is easily checked that given the mark of one column, we can compute the mark of the next in time $O(\lceil m/w \rceil)$, using only shifts and bitwise Boolean operations (bitwise AND to mask out bits, and bitwise OR to combine masks).

Proposition 2. *Let m, m' be given, and let $i_1 + i_2 + \cdots + i_m = m'$. Consider a map* expand : $\{0, 1\}^m \to \{0, 1\}^{m'}$, *defined by*

$$\mathrm{expand}(x[1]x[2] \ldots x[m]) = x[1]^{i_1}x[2]^{i_2} \ldots x[m]^{i_m}.$$

Then, expand(x) *can be computed in time $O(\lceil (m + m')/w \rceil \log(m + m'))$.*

Proof. For convenience of notation, we assume that $i_j > 0$ for all j and hence $m' \geq m$; the general case is similar. We also assume that m is even. First compute, using Proposition 1, $x' = \text{perm}_\pi(x0^{m'-m}) = 0^{i_1-1}x[1]0^{i_2-1}x[2]\ldots 0^{i_m-1}x[m]$ in time $O(\lceil m'/w\rceil \log m')$. Let $y = 1^{i_1}0^{i_2}1^{i_3}\ldots 1^{i_{m-1}}0^{i_m}$. Now, compute

$z_1 = y$ AND NOT$[(y$ AND $x') + y)]$,

$z_2 = (\text{NOT } y)$ AND NOT$[((\text{NOT } y)$ AND $x') + (\text{NOT } y)]$,

$z = z_1$ OR z_2

Then z is $x[1]^{i_1}x[2]^{i_2}\ldots x[m]^{i_m}$, as desired.

Proposition 3. *Let m', m be given. A monotone projection $f : \{0,1\}^m \to \{0,1\}^{m'}$ is a map of the form $f(x)[i] = \rho(i)$, with $\rho(i) \in \{0, 1, x[1], x[2], \ldots, x[m]\}$. Any monotone projection can be computed in time $O(\lceil (m' + m)/w\rceil \log(m'+m))$.*
Proof. Write the projection as a composition of an expansion, a permutation, and bitwise Boolean operations with masks, and use Propositions 1 and 2.

Lemma 4. *Let C be a Boolean circuit of depth d, fan-in 2, and size $s \geq m, m'$, computing a map $\{0,1\}^m \to \{0,1\}^{m'}$. Given x, $C(x)$ can be computed in time $O(\lceil s/w\rceil d \log s)$.*
Proof. Convert C to C' of the following normal form:

- All negations of C' occur at inputs.
- C' is constructed on levels $0..2d + 1$, with and-gates on even levels and or-gates on odd levels. The inputs to gates at level l are either constants or outputs of gates at level $l - 1$. The inputs and negations to inputs are regarded as level 0 gates, and the outputs as level $2d + 1$ gates. There are exactly s gates on each level, except level 0, with $2m$ gates, and level $2d + 1$, with m' gates.

Note that the connection between two levels can be described as a monotone projection. Given input $x \in \{0,1\}^m$, we can now compute $C(x)$ by first computing the concatenation of x and NOT x and then computing d blocks of operations, each consisting of a monotone projection, a bitwise Boolean OR, a monotone projection, and a bitwise Boolean AND. Finally, we compute a last projection. Total time is $O(\lceil s/w\rceil d \log s)$.

Corollary 5. *Multiplying two w-bit words can be done in time $(\log w)^{3+o(1)}$.*

Proof. Apply Lemma 4 to Schönhage and Strassen's multiplication circuit [19] of size $w(\log w)(\log\log w)$ and depth $(\log w)(\log\log w)$.
Since multiplication is of primary importance, we want to optimize the above corollary a bit. Using the general circuit simulation algorithm is a bit of an overkill. Schönhage and Strassen have two multiplication circuits, one of size $w(\log w)(\log\log w)(\log\log\log w)\ldots$, based on the Fourier transform over the complex numbers, and one of size $w(\log w)(\log\log w)$, based on the Fourier transform over finite rings. By examining the first algorithm in details (which we won't go into here) one finds that it *word parallelizes* perfectly, i.e. it can be implemented using bitwise Boolean operations and shifts (and no additions) with only a constant factor of overhead, i.e. we get the following proposition:

Proposition 6. *Multiplying two w-bit words can be done in time* $O((\log w)(\log \log w)(\log \log \log w)\ldots)$.

The second multiplication circuit does *not* seem to word parallelize perfectly, the obstacle being that different parts of a word must be shifted by different amounts, so we don't know if we can improve this to $O((\log w)(\log \log w))$.

We need Fact 6 to get the $\sqrt{\log n}(\log \log n)^{1+o(1)}$ bound for the static dictionary problem in the next section. However, readers who prefer a presentation without gaps can use Fact 5 instead and still get a $\sqrt{\log n}(\log \log n)^{O(1)}$ solution.

Note that all of the above upper algorithms use various word-sized constants containing masks depending on the word size w. We cannot afford to compute these constants at run-time, so we assume that they have been determined and hard-wired into the algorithm at compile-time. Following the terminology of Ben-Amram and Galil [5], such solutions are called *weakly non-uniform*. Fredman and Willard's fusion tree [13] is another example of a weakly non-uniform algorithm. We do not consider weak non-uniformity as a serious deficiency of the algorithms; computing useful constants at compile-time is hardly an esoteric phenomenon. Note that in our solution to the static dictionary where the subroutines will be used, we can eliminate the non-uniformity by simply making the constants part of the static data structure instead of the query program.

Simulating circuits in trans-dichotomous algorithms is not a new idea; indeed, the word merging algorithm of Albers and Hagerup [1] which is often used in trans-dichotomous algorithms is essentially a simulation of Batcher's bitonic merging network. Thorup, in his paper on sorting with the basic instruction set [21], shows a different simulation result:

Theorem 7 (Thorup). *Given a Boolean circuit C, mapping w bits to w bits. Suppose w instances $x_1, x_2, \ldots, x_w$ are given. The sequence $C(x_1), C(x_2), \ldots, C(x_w)$ can be computed in time $O(s + \log w)$.*

Thus, by applying "mass production" and performing several evaluations simultaneously, Thorup avoids the multiplicative $d \log s$ penalty we have to pay. However, for our purposes, we have to consider the evaluation of a single instance.

The rightmost set bit of a word

Consider the function RightmostOne, giving the position of the rightmost set bit of a word. For example, RightmostOne(0010001010000000) = 9.

Lemma 8. RightmostOne(x) *can be computed in time* $O(\log \log w)$

Proof. 1. Given a word x with least significant 1-bit in position i. Let 1^w be the all-1 string and consider the two words $x + 1^w$ and x XOR 1^w. It is easily seen that they are different on bits $1, 2, \ldots, i$, but the same on bits $i+1, i+2, \ldots, w$. Thus, if we let $y = (x + 1^w)$ XOR $(x$ XOR $1^w)$ and let $z = y$ XOR $(y \uparrow 1)$, z will be the word with a 1 in position i and 0 elsewhere.
   2. Do a binary search for the single 1 bit in z. Stop after $\log \log w$ steps. We now have an integer r, so that we know the 1-bit is in the subword $x' = x[r \ldots r + s - 1]$ with $s = \lceil w/\log w \rceil$.

3. Move x' to the leftmost end of the word, i.e. perform $x := x \uparrow (r-1)$ Note that we now have a word where the least significant bit is between 1 and s. If we find the position and add $r-1$, we are done.

4. Make a bit string x'' containing $j = \lceil \log s \rceil$ consecutive copies of x'. This can be done in time $O(\log s) = O(\log \log w)$ [2]. The bit string x'' can be stored in ≤ 2 words, so we can operate on it with unit cost.

5. To avoid cumbersome notation, the next step is most easily explained by example. Suppose that $w = 64$, then $s = 11$ and $j = 4$. Suppose $x' = 00001000000$. We want to return 5. We have
$$x'' = 00001000000|00001000000|00001000000|00001000000.$$
Let p be the constant bit string
$$p = 00000001111|00011110000|01100110011|10101010101.$$
In general, each block of p contains a sequence of alternating blocks of 0's and 1's, in the i'th block from the right, the blocks have size 2^i. The first 0 of each block is chopped off. Let $g = x''$ AND p. Then
$$g = 00000000000|000010000000|0000000000000|0000100000.$$
Now, the i'th block from the right of g contains a 1, if and only if the i'th least significant bit of the answer is a 1. We want to move these 1's to specified positions.

6. Let $h = g + a$, where
$$a = 01111111111|01111111111|01111111111|01111111111.$$
Now, for $i = 0 \ldots j-1$, position $is+1$ of h is a 1 if and only if the i'th most significant bit of the desired answer is a 1. But using time $O(\log \log w)$ time, we can move position $is+1$ to position $w - (j-1) + i$ for all i, and we are done.

We shall actually use the following slight generalisation, which is a corollary of the proof:

Corollary 9. *Let* RightmostField(x, d) *be the function which looks at bits* $1, d+1, 2d+1, \ldots,$ *in the word* x, *and returns the largest* i, *for which bit* id *is 1.* RightmostField(x, d) *can be computed in time* $O(\log \log(w/d))$.

Remarks: Gerth Brodal (personal communication) has noted that a slight modification of the above algorithm can also be used to find the *most* significant set bit of a word. When multiplication is allowed, both operations can be done in constant time [13, 7, 3].

3 The static dictionary problem

We shall use the subroutine $u \leftarrow$ InterleavedMult(x, y, l, m). Here, l and m are positive integers, and x and y are two bit strings (actually single words) of length at most $lm \leq w$; x and y are interpreted as consisting of the binary notation of m positive integers, each containing at most l bits, stored in *interleaved* fashion (see figure 1). The subroutine returns all of the m pairwise products stored in a bit string of length $2lm$. Using the general circuit simulation lemma, we get

Lemma 10. InterleavedMult(x, y, l, m) *can be computed in time* $O((\log l)^{3+o(1)})$.

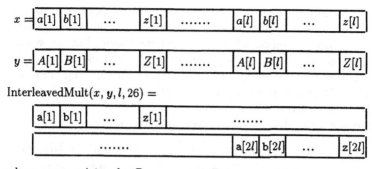

where $a = a * A, b = b * B, \ldots, z = z * Z$

Fig. 1. The InterleavedMult function

Readers preferring a presentation without gaps can use this bound in Theorem 13 and obtain a slightly weaker bound than stated there. For the best bound we know, we need

Lemma 11. InterleavedMult(x, y, l, m) *can be computed in time* $O((\log l)^{1+o(1)})$.
Proof. We use Fact 6. Since the algorithm proving this fact can be implemented without additions, we can execute m instances of the algorithm, with the instances given in interleaved form, without any overhead.

Lemma 12. *Let* $S \subseteq \{0,1\}^b$ *be a set of b-bit keys of size n with $n \geq 4\lceil \log n \rceil^2$. There is an interval $I \subseteq \{0, \ldots, b-1\}$ of size at most $b/\log n$, so that for some $a \in \{0,1\}^b$, $f_a(x) =$ InterleavedMult$(x, a, 4\lceil \log n \rceil^2, \lceil b/4\lceil \log n \rceil^2 \rceil)[I]$ defines a 1-1 function on S. (Recall that $x[I]$ selects the bits marked by I from x).*

Proof. We shall use the following fact, due to Dietzfelbinger *et al* [11]: The class of hash functions $\mathcal{H}_{k,l} = \{h_a : \{0, \ldots, 2^k - 1\} \to \{0, \ldots, 2^l - 1\}| 0 < a < 2^k$ and a odd$\}$ given by $h_a(x) = (ax \bmod 2^k) \operatorname{div} 2^{k-l}$ is *nearly universal*, i.e. for fixed x and y with $0 \leq x, y < 2^k$, if a is chosen at random, then $\Pr[h_a(x) = h_a(y)] \leq 2/2^l$. A value $a \in \{0,1\}^b$ can be interpreted as $\lceil b/4\lceil \log n \rceil^2 \rceil$ $4\lceil \log n \rceil^2$-bit words stored in interleaved fashion. The InterleavedMult function multiplies each of these with the corresponding segment in x. If each of these functions are injective on S (restricted to the corresponding segment), the entire function will be injective. In general, if we are given a set S of size n taken from some universe, and a family of nearly universal hash functions mapping the universe to a set of size n^2, some member of the family will be injective on S [10]. Thus, we have that for each segment, there is a value of the restriction of a to that segment, so that when the multiplication is performed, and a certain sub-segment of length $2\lceil \log n \rceil$ of the result is extracted, this will be an injective function on S, restricted to the corresponding segment. But since we store the segments in interleaved fashion, the union of all the subsegments is an interval of size

$$2\lceil \log n \rceil \lceil b/4\lceil \log n \rceil^2 \rceil \leq 2\lceil \log n \rceil (\frac{b}{4\lceil \log n \rceil^2} + 1) \leq \frac{b}{2\log n} + 2\lceil \log n \rceil \leq b/\log n.$$

This is the interval I, and the proof of the lemma is complete.

Note why we multiply segments stored in interleaved fashion, rather than concatenated fashion: we could also have implemented a BlockMult, multiplying a number of continuous blocks with the same time bound, but we would have no way of collecting the various subsegments from each block into a single interval within a sufficiently small time bound; this can be shown using our lower bounds techniques of Section 4. However, Thorup [21] has found an application for such a BlockMult subroutine.

We are now ready for the main theorem of this section.

Theorem 13. *There is a solution to the static dictionary problem for sets $S \subseteq \{0,1\}^w$ of size n using $n + o(n)$ memory cells, each containing w bits and with worst case query time $\sqrt{\log n}(\log\log n)^{1+o(1)}$ with a query algorithm only using the basic instruction set.*

Proof. Assume first that $w \geq 4\lceil \log n \rceil^2$. Using Lemma 12 with $b = w$, we choose a word a such that f_a is 1-1 on A. We store a in our data structure and use f_a to hash our elements, thereby reducing them to size $b' \leq w/\log n$. By Lemma 11, computing the hash function on a single element requires time $(\log\log n)^{1+o(1)}$. Now, let $b = b'$ and repeat. In general, after t iterations, we have reduced our keys to word size less than $\leq w/(\log n)^t$, *unless* we arrive at a key size which is less than $4\lceil \log n \rceil^2$. If we do arrive at such a word size, we stop. Otherwise we continue for $t = \sqrt{\log n / \log\log n}$ iterations. In the first case we have found a sequence of hash functions whose composition is an injective function mapping S to $\{0,1\}^{4\lceil \log n \rceil^2}$. In the latter case, we have found a sequence of hash functions whose composition is an injective function mapping S to the domain $\{0,1\}^{\lfloor w/(\log n)^t \rfloor}$.

In the first case, we store the reduced elements in a two level hash table, following Fredman, Komlos, and Szemeredi [12], except that we use the hash function of Dietzfelbinger *et al* (the analysis of [12] goes through for any nearly universal family). The multiplicative hash function can be evaluated using lemma 6, using time $(\log\log n)^{1+o(1)}$. The search time is $O((t+1)(\log\log n)^{1+o(1)}) \leq \sqrt{\log n}(\log\log n)^{1+o(1)}$, as desired.

In the second case, we store the reduced elements in a packed B-tree of degree $D \geq \lfloor (\log n)^t \rfloor$, following Andersson [2]. In Andersson's paper, it was shown how to search a packed B-tree in time $O(\log_D n)$ using the basic RAM instruction set. However, he had to use a huge table in order to determine the rank of a given element in a B-tree node and we want linear space. Examining his algorithm, we find that the obstacle is precisely the computation of the RightmostField function, or, to be completely precise, the analogous LeftmostField function. By storing each B-tree node in reverse sorted order, we can use the RightmostField subroutine instead, and search the B-tree in time $O((\log\log D)(\log_D n))$.

We now clearly have a data structure of size $n + o(n)$. The search time is

$$t(\log\log n)^{1+o(1)} + O\left(\frac{\log n}{\log((\log n)^t)}(\log\log(\log n)^t))\right) = \sqrt{\log n}(\log\log n)^{1+o(1)},$$

as desired.

Remark: Combining the above techniques with techniques of the first and third author [8], it is possible to improve the space bound in Theorem 13 to $B + o(B)$ bits, where B is the information theoretic minimum $\log\binom{2^w}{n}$, in the case where we only want to test membership and not retrieve associated information.

4   Lower bounds

In this section, we show lower bounds for some of our subroutines; namely reversing a word, multiplying two words, and locating the least (or most) significant bit of a word, using the basic instruction set. As mentioned in the introduction, a lower bound for the static dictionary problem *itself* is shown in [4]. Note that since the problems only involve $O(1)$ words, they can be solved in constant time if a precomputed array containing a *table* of the function is allowed, but this table will be very large. Our lower bounds hold even in the presence of precomputed tables, as long as the precomputed table has size less than 2^{w^ϵ} for certain constants $\epsilon > 0$.

For the lower bounds, we need to specify and simplify our model slightly: we have a RAM with a certain fixed number of CPU-registers, say registers A, ..., Z, and a random access memory which can be accessed from the CPU using direct and indirect reads and writes. We have conditional jump (if A $= 0$ then branch) and the following computational instructions operating on CPU-registers: $+$, NAND (this is sufficient to simulate all the bitwise Boolean operations) and $\uparrow$.

As an aid in showing our lower bounds, we shall consider non-Boolean *word circuits* computing on words. Each wire of such a circuit C holds a word $x \in W = \{0,1\}^w$. It may contain three kinds of gates: $+$-gates, NAND-gates, and $\uparrow_r$-gates for various constants r, with $\uparrow_r(x) = x \uparrow r$. Arbitrary constants $c \in W$ may also be fed into the circuit. The size of a circuit is the number of gates it contains. A circuit with one source and one sink computes a function $W \to W$ in the natural way.

The following general lemma is the key to all lower bounds.

Lemma 14. *Let C be a word circuit of size s. Let $u \in W$ be a word with a single block of ones in the left end, i.e. $u = 1^k 0^{w-k}$ for some k.*

Then, there is a bit string v with the following properties:

- *v has length $3w$ and will be indexed $v[-w] \ldots v[-1]v[1]v[2] \ldots v[2w]$, so when we do a bitwise AND with v and a word, only the middle part of v will matter.*
- *v has Hamming weight at most $2^s k$*
- *For any i between 0 and $w - k$, if x varies, but the value of x AND $(v \downarrow i)$ is fixed (i.e. we only vary x on the bits where $v \downarrow i$ is 0), $C(x)$ AND $(u \downarrow i)$ takes on at most $2^{2^{2s}}$ values.*

Proof. The proof is an induction in s. For $s = 0$ (i.e. $C(x) = x$ or $C(x) = c$), the claim clearly holds, if we let $v = u$, padded with 0's.

Suppose the top gate of C is a $+$-gate, i.e. $C(x) = C_1(x) + C_2(x)$, where C_1 and C_2 are smaller circuits than C. By induction, let v_1 and v_2 be given, and let $v = v_1$ OR v_2. Now, if the bits of x AND $(v \downarrow i)$ are fixed, $C_1(x)$ AND $(u \downarrow i)$ and

$C_2(x)$ AND $(u \downarrow i)$ each take on at most $2^{2^{2(s-1)}}$ different values. Furthermore, since the word $u \downarrow i$ is a word with only a single interval of set bits, we have that if $C_1(x)$ AND $(u \downarrow i)$ and $C_2(x)$ AND $(u \downarrow i)$ are fixed, $C(x)$ AND $(u \downarrow i)$ take on at most 2 different values, corresponding to whether a carry bit is propagated to the interval or not. Thus, $C(x)$ AND $(u \downarrow i)$ takes on at most $2^{2^{2s-2}} \cdot 2^{2^{2s-2}} \cdot 2 \leq 2^{2^{2s}}$ different values when the bits of x AND $(v \downarrow i)$ are fixed.

The case where the top gate of C is a NAND-gate is handled similarly to the $+$-gate case.

Now suppose the top gate of C is an $\uparrow_r$ gate with input C_1. By induction, let v_1 be given, and let $v = v_1 \downarrow r$. If x AND $(v \downarrow i) = x$ AND $(v_1 \downarrow (i+r))$ is fixed, $C_1(x)$ AND $(u \downarrow (i+r))$ takes on at most $2^{2^{2(s-1)}}$ values, and $C_1(x)$ AND $(u \downarrow (i+r))$ determines $C(x)$ AND $(u \downarrow i)$.

Theorem 15. *Any RAM program computing the rightmost (or leftmost) set bit of a word uses time $\Omega(\log \log w)$. This is true, even if a precomputed table of size $O(2^{w^{1-\epsilon}})$ is allowed, for any constant $\epsilon > 0$.*

Proof. Let w be sufficiently large. Let $e_i = 0^{i-1}10^{w-i}$. Let $U = \{e_1, e_2, \ldots, e_w\}$, i.e., the set of words with Hamming weight one. Given e_i in U, a least significant bit algorithm returns i. Suppose an algorithm exists which gives this answer in less than $(\log \log w)/10$ steps. We shall show that it gives the same answer on $i, j \in U$ with $i \neq j$ and thus cannot be correct.

Consider executing the algorithm on each $x \in U$. After t steps, the random access machine is in some configuration $K(x, t)$.

We shall show that for any $t < \log \log w/10$, we can find a subset $U_t \subseteq U$ and word circuits $A_1, C_1, A_2, C_2, \ldots, A_t, C_t, D_A, \ldots, D_Z$, so that

- $|U_t| \geq |U|/2^{2^{4t}}$,
- the size of each A_i, C_i, and D_i is at most t.
- For all $x \in U_t$, in all configurations $K(x, t)$, the program counter is at the same location, and furthermore, the memory has the same appearance as originally, except that in memory register $A_i(x)$, the new value $C_i(x)$ can be found. If for more than one value $i_1 < i_2 < \cdots < i_r$, we have $A_{i_1}(x) = A_{i_2}(x) = \cdots = A_{i_r}(x)$, the value of the register is $C_{i_r}(x)$, i.e. the largest index "wins". The value of CPU-register i is $D_i(x)$.

Let us first see that if we can show that this invariant can be upheld for any algorithm, we get the theorem. We can assume, without loss of generality, that the value is returned in CPU-register A, i.e. for members of U_T, with $T = (\log \log w)/10$, the returned value is $D_A(x)$. Note that all bits of the result are zero, except the least significant $\log w$. According to Lemma 14, there is a bit string v of Hamming weight at most $(\log w)^2$, so that if the bits of v are all zero, the result is member of a set of size $2^{2^{\log \log w/5}}$. This means that for $x \in U_T - B$, for a certain set B of size $(\log w)^2$, only $2^{2^{\log \log w/5}}$ different answers are obtained. Since $U_T - B \gg 2^{2^{\log \log w/5}}$, two elements return the same answer and the algorithm is incorrect.

We should now check that the invariant can be upheld. The cases of direct and indirect write, conditional jump, and computation using NAND and + are all straightforward. The interesting cases are *read* and *computation using* $\uparrow$. Suppose we do an indirect read, e.g., let CPU register A be equal to memory cell u, where u is the value of CPU register B. The value of u for members $x \in U_t$ is described by a word circuit $D_B(x)$. For each member x of U_t, it is either the case that $D_B(x)$ is the address of one of the undisturbed values in the precomputed table, it is one of the previously written addresses $A_1(x), \ldots, A_t(x)$, or it is neither. Thus, we get the set U_t partitioned into $t + 2$ subset. Take the largest of these, V. If V corresponds to one of the $t+1$ last cases, we can let $U_{t+1} = V$ and easily uphold the invariant. In the first case, we read some precomputed value. Only the $w^{1-\epsilon}$ least significant bits of the address are non-zero. By Lemma 14, there is a bit string v of Hamming weight at most $(\log w)w^{1-\epsilon}$, so that if x AND $v = 0$, the result is member of a set of size $2^{2^{w^t}}$. This means that for $x \in V - B$, for a certain set B of size $(\log w)w^{1-\epsilon}$, only $2^{2^{2^t}}$ different addresses are read. We get $V - B$ partitioned into $2^{2^{2^t}}$ subsets. Let U_{t+1} be the largest of these. The invariant can now easily be upheld.

The case of a computation using $\uparrow$ is very similar. The reason that $\uparrow$ is more difficult to handle than the other computational instructions is that $\uparrow$ is a binary predicate, but in our word circuits, we only allow unary $\uparrow_r$- gates with fixed second argument r. But we can handle $\uparrow$ instructions similarly to reads, by noting that only the $\log w + 1$ least significant bits of the second argument are relevant. By Lemma 14, we find a large subset of U_t for which the second argument is the same and we are done.

The lower bound for reverse follows a slightly different strategy. We consider reversing a subword of length $b \le w$; $\text{reverse}(x[1]x[2] \ldots x[b]) = x[b] \ldots x[2]x[1]$.

Lemma 16. *Let $b \le w$. A word circuit of size $\le \frac{1}{10}\log b$ correctly computes reverse on at most a $2^{-b^{1/4}}$ fraction of $\{0,1\}^b$.*

Proof. Let $k = \lfloor b^{1/3} \rfloor$. Let a word circuit C be given. By Lemma 14, we can find a word v of Hamming weight at most $b^{1/10}b^{1/3}$ so that the statement of the lemma holds for $u = 1^k 0^{w-k}$, i.e. if x AND $(v \downarrow i)$ is fixed, $C(x)$ AND $(u \downarrow i)$ takes on at most $2^{b^{1/5}}$ different values.

Let $T = \{0, 1, 2, \ldots, b - \lfloor b^{1/3} \rfloor - 1\}$. We claim that for some $i \in T$, $v \downarrow i$ and $\text{reverse}(u \downarrow i)$ are *disjoint*, i.e. they do not contain set bits in overlapping positions. Note that $v \downarrow i$ and $\text{reverse}(u \downarrow i)$ are disjoint if $b - u_1 - i \ne v_1 + i$ for all indices u_1, v_1 so that $u[u_1] = v[v_1] = 1$. This is equivalent to $i \ne (b - u_1 - v_1)/2$, i.e. for any particular (u_1, v_1) pair, at most one i goes wrong. Since there are only $|u|_h|v|_h \le b^{1/3}b^{1/10}b^{1/3} < \#T$ pairs, some $i \in T$ is good for all pairs.

Now fix the bits of x which are 1 in $v \downarrow i$ to any value and select all other bits at random. We want to find the probability that $C(x) = \text{reverse}(x)$. We know that $C(x)$ AND $(u \downarrow i)$ takes on at most $2^{b^{1/5}}$ different values. Note that $\text{reverse}(x)$ AND $(u \downarrow i)$ is determined 1-1 by x AND $\text{reverse}(u \downarrow i)$, and since reverse$(u \downarrow i)$ is disjoint from the set of fixed bits of x, all $2^{\lfloor b^{1/3} \rfloor}$ possible values of reverse(x) AND $(u \downarrow i)$ are equally likely. The probability that reverse(x) AND $(u \downarrow i)$ has one of the different possible values of $C(x)$ AND $(u \downarrow i)$ is thus at most $2^{b^{1/5}}/2^{\lfloor b^{1/3} \rfloor} < 2^{-\lfloor b^{1/3} \rfloor/2} < 2^{-b^{1/4}}$. This is also an upper bound on the probability that the circuit is correct for random x (since the bits of x, marked by $v \downarrow i$ were fixed arbitrarily).

Theorem 17. *Computing reverse of a b-bit string, $(\log w)^{10} \leq b \leq w$, requires time $\Omega(\log b)$, even when a precomputed table of size $2^{b^{1/10}}$ is given.*

Proof. The strategy is to find a small set of word circuits, so that for a non-negligible subset of the possible inputs, every value found in the random access memory is the value of one of the circuits on the input. Unlike the lower bound proof for the least significant bit problem, we only have to argue about the set of values in the random access memory, not where they are located.

We will show: For any $t > 0$, there is a subset of U_t of $U = \{0,1\}^b$ and an (unordered) set of word circuits C_t so that

- $\#C_t \leq 2^{b^{1/10}} + t + 1$
- For each $C \in C_t$, the size of C is at most t.
- $\#U_t \geq 2^b/(\#C_t)^{2t}$.
- For any $x \in U_t$, if we run the algorithm on x for t steps, the set of non-zero values in the memory is a subset of $C_t(x)$. Furthermore, the program counter points to the same location in the program for all $x \in U_t$.

We first show why this implies our theorem. After running $T = \frac{1}{10} \log b$ steps, we find slightly more than $2^{b^{1/10}}$ word circuits of size at most T and a set U_T of size at least $2^{b-o(b)}$. Each of these circuits are correct on at most a $2^{-b^{1/4}}$ fraction of U. Therefore on at most a $(\#C_T) \cdot 2^{-b^{1/4}}$ fraction of $x \in U$ will the value of even one of the circuits be reverse(x). Since U_T forms a greater fraction of U than that, we have for some x in U_T, the value reverse(x) is not found in any memory location. Therefore the program is not correct.

The proof that we can maintain the invariant is similar to the proof for the least significant bit, only simpler. The proof is an induction in t. For $t = 0$ the lemma clearly holds, we just let the initial C_i's be trivial circuits containing the constants of the precomputed table, except for one circuit, the identity circuit mapping x to x. Also, $U_0 = U$.

A conditional jump divides U_t in two sets, of which we let U_{t+1} be the larger one. A direct or indirect read or write does not change the set of values. A computation combines two previously computed values. For addition and bitwise NAND, for each $x \in U_t$, the two arguments are described by two circuits from the set C_t. Thus, U_t is partitioned into $(\#C_t)^2$ subsets, of which we let U_{t+1} be the larger. For $\uparrow$, take the most commonly occurring circuit appearing as first argument, and the most commonly occurring integer between 0 and $w - 1$ appearing as second argument, given the first argument. This reduces the set with a factor at most $w(\#C_t)$. Thus, the invariant can be upheld and we are done.

The lower bound for multiplication is most conveniently shown by a reduction. The following lemma, which may be useful in other contexts, shows that if unit cost multiplication is allowed, we can reverse medium sized subwords in constant time.

Lemma 18. *On a RAM with unit cost multiplication in addition to the basic instruction set, reversing a string containing $s \leq \sqrt{w}$ bits can be done in constant time.*

Proof. Assume that the subword is in the right hand side of the word. The algorithm works in two steps. In the first step, it makes copies of the s least significant bits across the word and applies masks to get one copy of each of the bits of s in equidistant positions, but in reverse order. The second step gathers the bits together again. We use a result of Fredman and Willard [13] that guarantees that if y is a word with set bits on equidistant positions only, then for some w constants a and b, the expression $((a * y) \text{ AND } b) \downarrow w$ gathers these bits together on the least significant positions in a word in the same order as they were in y. The entire algorithm for reversing an s-bit string x is:

- $y = ((x \cdot d) \text{ AND } m) \downarrow (s-1)$
- RETURN $(a \cdot y) \downarrow ((s-1)s+1) \text{ AND } b$

where $a = \sum_{i=0}^{s-1} 2^{(s-i-1)s+(i+1)}$, $b = \sum_{i=0}^{s-1} 2^i$, $d = \sum_{i=0}^{s-1} 2^{i(s+1)}$ and $m = \sum_{i=0}^{s-1} 2^{i(s+1)+s-1-i}$.

Theorem 19. *Any RAM program multiplying two w-bit words using the basic instruction set uses time $\Omega(\log w)$. This is true, even if a precomputed table of size $O(2^{w^{1/20}})$ is allowed.*

Proof. Combine Lemma 18 and Theorem 17.

It is interesting to note that the following somewhat weaker lower bound follows more or less directly from Håstad's size-depth tradeoff for unbounded fan-in circuits computing multiplication [16]: time $\Omega(\log w/\log\log w)$ is necessary, even if a precomputed table of size $w^{O(1)}$ is allowed. This is *not* the case for the other two lower bounds, as reverse and least significant bit are AC^0 functions.

References

1. S. Albers and T. Hagerup. Improved parallel integer sorting without concurrent writting. In *3rd ACM-SIAM Symposium on Discrete Algorithms*, pages 463–472, Orlando, Florida, 1992.
2. A. Andersson. Sublogarithmic searching without multiplications. In *36th IEEE Symposium on Foundations of Computer Science*, pages 655–663, 1995.
3. A. Andersson, T. Hagerup, S. Nilsson, and R. Raman. Sorting in linear time? In *27th ACM Symposium on Theory of Computing*, pages 427–436, Las Vegas, Nevada, 1995.
4. A. Andersson, P.B. Miltersen, S. Riis, and M. Thorup. Static dictionaries on AC^0 RAMs: Query time $\Theta(\sqrt{\log n \log\log n})$ is necessary and sufficient. In *37th IEEE Symposium on Foundations of Computer Science*, pages 538–546, Burlington, Vermont, 1996.

5. A.M. Ben-Amram and Z. Galil. When can we sort in $o(n \log n)$ time? In 34th *IEEE Symposium on Foundations of Computer Science*, pages 538–546, Palo Alto, California, 1993.

6. G.S. Brodal. Predecessor queries in dynamic integer sets. In *Proceedings 10th Symposium on Theoretical Aspects of Computer Science*. Springer-Verlag, 1997 (To appear).

7. A. Brodnik. Computation of the least significant set bit. In *Proceedings Electrotechnical and Computer Science Conference*, volume B, pages 7–10, Portorož, Slovenia, 1993.

8. A. Brodnik and J.I. Munro. Membership in a constant time and a minimum space. In *Proceedings 2nd European Symposium on Algorithms*, volume 855 of *Lecture Notes in Computer Science*, pages 72–81. Springer-Verlag, 1994.

9. A. Brodnik and J.I. Munro. Neighbours on a grid. In *Proceedings 5th Scandinavian Workshop on Algorithm Theory*, volume 1097 of *Lecture Notes in Computer Science*, pages 307–320. Springer-Verlag, 1996.

10. J.L. Carter and M.N. Wegman. Universal classes of hash functions. *Journal of Computer and System Sciences*, 18(2):143–154, April 1979.

11. M. Dietzfelbinger, T. Hagerup, J. Katajainen, and M. Penttonen. A reliable randomized algorithm for the closest-pair problem. Technical Report 513, Fachbereich Informatik, Universität Dortmund, Dortmund, Germany, 1993.

12. M.L. Fredman, J. Komlós, and E. Szemerédi. Storing a sparse table with $O(1)$ worst case access time. *Journal of the ACM*, 31(3):538–544, July 1984.

13. M.L. Fredman and D.E. Willard. Surpassing the information theoretic bound with fusion trees. *Journal of Computer and System Sciences*, 47:424–436, 1993.

14. M.L. Fredman and D.E. Willard. Trans-dichotomous algorithms for minimum spanning trees and shortest paths. *Journal of Computer and System Sciences*, 48(3):533–551, June 1994.

15. M. Furst, J.B. Saxe, and M. Sipser. Parity, circuits, and the polynomial-time hierarchy. *Mathematical Systems Theory*, 17(1):13–27, April 1984.

16. J. Håstad. Almost optimal lower bounds for small depth circuits. In *18th ACM Symposium on Theory of Computing*, pages 6–20, Berkeley, California, 1986.

17. P.B. Miltersen. Lower bounds for static dictionaries on RAMs with bit operations but no multiplication. In *Proceedings 23rd International Colloquium on Automata, Languages and Programming*, volume 1099 of *Lecture Notes in Computer Science*, pages 442–451. Springer-Verlag, 1996.

18. R. Raman. Priority queues: Small, monotone, and trans-dichotomus. In *Proceedings 4th European Symposium on Algorithms*, volume 1136 of *Lecture Notes in Computer Science*, pages 121–137. Springer-Verlag, 1996.

19. A. Schönhage and V. Strassen. Schnelle Multiplikation großer Zahlen. *Computing*, 7:281–292, 1971.

20. M. Thorup. On RAM priority queues. In *7th ACM-SIAM Symposium on Discrete Algorithms*, pages 59–67, Atlanta, Georgia, 1996.

21. M. Thorup. Randomized sorting in $O(n \log \log n)$ time and linear space using addition, shift, and bit-wise boolean operations. In *8th ACM-SIAM Symposium on Discrete Algorithms*, pages 352–359, New Orleans, Louisiana, 1997.

An Approximation Algorithm for Stacking Up Bins from a Conveyer onto Pallets

J. Rethmann and E. Wanke

University of Düsseldorf, Department of Computer Science, D-40225 Düsseldorf, Germany

Abstract. Given a sequence of bins $q = (b_1, \ldots, b_n)$ and a positive integer p. Each bin is destined for a pallet. We consider the problem to remove step by step all bins from q such that the positions of the bins removed from q are as less as possible and after each removal there are at most p open pallets. (A pallet t is called open if the first bin for t is already removed from q but the last bin for t is still contained in q. If a bin b is removed from q then all bins to the right of b are shifted one position to the left.)

The maximal position of the removed bins and the maximal number of open pallets are called the storage capacity and the number of stack-up places, respectively. We introduce an $O(n \cdot \log(p))$ time approximation algorithm that processes each sequence q with a storage capacity of at most $s_{\min}(q, p) \cdot \lceil \log_2(p + 1) \rceil$ bins and $p + 1$ stack-up places, where $s_{\min}(q, p)$ is the minimum storage capacity necessary to process q with p stack-up places.

1 Introduction

From a practical point of view, a stack-up system consists of one or more stacker cranes picking up bins from a roller conveyer. It is usually located at the end of a pick-to-belt orderpicking system; see [dK94, LLKS93] for a description of orderpicking systems. A costumer order consists of several bins that arrive the stack-up system on a conveyer. At the end of the conveyer the bins enter a so-called storage conveyer from that they are moved by stacker cranes onto pallets. The pallets are build on so-called stack-up places. Vehicles take full pallets from stack-up places, put them onto trucks, and bring new empty pallets to the stack-up places. A more detailed description of stack-up systems which are sometimes also called pile-up systems is given in [RW97b].

All bins of a costumer order have to be placed onto the same pallet. Unfortunately, the bins arrive the stack-up system not in a succession such that they can be placed one after the other onto pallets. If all available places are occupied by pallets and the storage conveyer is completely filled with bins not destined for the pallets currently stacked up on the places, then the stack-up system is blocked. Controlling a stack-up system means to make the right decision whenever a new pallet has to be opened such that no blocking situation will happen.

From a theoretical point of view, the stack-up problem can be defined as follows. Given a sequence of bins $q = (b_1, \ldots, b_n)$ and two integers s and p. Each bin is destined for some pallet t. A pallet t is called open if the first bin for t is already removed from q but the last bin for t is still contained in q. The stack-up problem is the question of whether all bins can be removed step by step from q such that the positions of the bins removed from q are always not greater than s and after each removal there are at most p open pallets. If a bin b is removed from q then all bins to the right of b are shifted one position to the left. The integers s and p correspond to the capacity of the storage conveyer and the number of available stack-up places, respectively.

In [RW97b], it is shown by a reduction from 3-SAT that the stack-up problem is in general NP-complete [GJ79] but can be solved in polynomial time if the storage capacity or the number of stack-up places is bounded. It is also shown that it can be solved in polynomial time whether or not it is possible to block the stack-up process by making wrong decisions.

In [RW97a], the worst-case behavior of on-line stack-up algorithms is analyzed. Such on-line algorithms make a decision without knowing the complete sequence, but only the next s bins. On-line algorithms are very interesting from a practical point of view. Let $s_{\min}(q,p)$ be the minimum storage capacity necessary to process q with p stack-up places and let $p_{\min}(q,s)$ be the minimum number of stack-up places necessary to process q with storage capacity s. In [RW97a] the performances of on-line algorithms are compared with optimal off-line solutions by competitive analysis [MMS88]. It is shown that there exist on-line algorithms having a logarithmic bound of $O(p_{\min}(q,p) \cdot \log(s))$ on the used number of stack-up places if the storage capacity is bounded by s, but only a linear bound of $(p+1) \cdot s_{\min}(q,p) - p$ on the storage capacity necessary to process q with p stack-up places.

The stack-up problem seems to be not much investigated up to know by other authors, although it really has important practical applications. In this paper, we introduce a polynomial time approximation algorithm for the processing of sequences q with a storage capacity of $\lceil s_{\min}(q,p) \cdot \log_2(p+1) \rceil$ bins and $p+1$ stack-up places. This is the first polynomial time approximation algorithm for the stack-up storage minimization problem with a logarithmic bound on the storage capacity when using only $p+1$ stack-up places. The existence of a polynomial time approximation algorithm that processes each sequence q with a storage capacity of less than $s_{min}(q,p) \cdot \epsilon_1$ bins and $p + \epsilon_2$ stack-up places for some $\epsilon_1, \epsilon_2 > 0$ is still open.

1.1 Preliminaries

We consider *sequences* $q = (b_1, \ldots, b_n)$ of n pairwise distinct *bins*. A sequence $q' = (b_{i_1}, \ldots, b_{i_k})$ is a *subsequence* of q if

$$1 \leq i_1 < i_2 < \cdots < i_{k-1} < i_k \leq n.$$

The *position* of a bin b_{i_j} in a sequence $q' = (b_{i_1}, \ldots, b_{i_k})$ is denoted by $pos(q', b_{i_j}) := j$. If X is the set of bins removed from q to obtain q' then q' is also denoted by $q - X$.

Each bin b of q is associated with a so-called *pallet symbol* $plt(b)$ which is, without loss of generality, a positive integer. We say bin b is *destined* for pallet $plt(b)$. The labeling of the pallets is arbitrary, because we only need to know whether two bins have to be stacked up onto the same pallet or onto different pallets. The set of all pallets of the bins in sequence q is denoted by $plts(q) := \{plt(b) \mid b \in q\}$. The position of the first (last) bin in sequence q destined for pallet $t \in plts(q)$ is denoted by $first(q, t)$ ($last(q, t)$, respectively).

Suppose we remove step by step the bins from sequence $q = (b_1, \ldots, b_n)$. If q has n bins, then the successive removal of all bins yields n subsequences

$$q_1 = q - \{b_{j_1}\}, \ q_2 = q - \{b_{j_1}, b_{j_2}\}, \ \cdots, \ q_n = q - \{b_{j_1}, \cdots, b_{j_n}\}$$

of $q_0 = q$, where q_n is empty. A pallet t is called *open* in q_i with respect to q if some bin for pallet t is already removed from q but sequence q_i still contains some bin for pallet t. The set of all open pallets in q_i with respect to q is denoted by $open(q, q_i)$.

The aim is to remove step by step all bins from q such that the removed bin is always on a position not greater than some integer s, i.e., $pos(q_{i-1}, b_{j_i}) \leq s$ for $1 \leq i \leq n$, and after each removal the number of open pallets is always not greater than some integer p, i.e., $|open(q, q_i)| \leq p$ for $1 \leq i \leq n$. The integers s and p correspond to the used *storage capacity* and the used number of *stack-up places*. Such a step by step transformation of q into some empty sequence is called an (s, p)-*processing* of q.

Let $s_{min}(q, p)$ be the minimum storage capacity necessary to process q with p stack-up places. That is, there is an $(s_{min}(q, p), p)$-processing, but no (s', p)-processing for $s' < s_{min}(q, p)$. Analogously, let $p_{min}(q, s)$ be the minimum number of places necessary to process q with storage capacity s, i.e., there is an $(s, p_{min}(q, s))$-processing, but no (s, p')-processing for $p' < p_{min}(q, s)$.

It will sometimes be convenient to use in examples pallet identifications instead of bin identifications. So we will write $q \hat{=} (plt(b_1), \ldots, plt(b_n))$ if $q = (b_1, \ldots, b_n)$.

Example 1. Table 1 shows a $(3, 2)$-processing of sequence $q = (b_1, \ldots, b_{11})$, where $q \hat{=} (1, 1, 2, 2, 3, 3, 3, 2, 3, 2, 1)$. The entry in the second column specifies the pallet t of the bin removed from sequence q' of the previous row. The removed bin itself (the underlined bin) is always the first bin in sequence q' for pallet t.

Let q' be a subsequence of q. Each such pair (q, q') is called a *configuration*. The pair (q, q) is called the *initial configuration* and the pair $(q, \emptyset)$ is called the *final configuration*. Configuration $(q, q' - \{b\})$ for some bin b of q' is called a *direct successor configuration* of (q, q') with respect to storage capacity s and p stack-up places, if $pos(q', b) \leq s$, $|open(q, q')| \leq p$, and $|open(q, q' - \{b\})| \leq p$. Such a transformation step is denoted by $(q, q') \Rightarrow_{s,p} (q, q' - \{b\})$ and called an (s, p)-*transformation step*. Configuration (q, q') is called (s, p)-*blocking* if q' is not empty and there is no further (s, p)-transformation step $(q, q') \Rightarrow_{s,p} (q, q'')$.

i	$plt(b_{j_i})$	$q' \doteq$	open pallets
		$(1,1,\underline{2},2,3,3,3,2,3,2,1)$	
1	2	$(1,1,\underline{2},3,3,3,2,3,2,1)$	2
2	2	$(1,1,\underline{3},3,3,2,3,2,1)$	2
3	3	$(1,1,\underline{3},3,2,3,2,1)$	2, 3
4	3	$(1,1,\underline{3},2,3,2,1)$	2, 3
5	3	$(1,1,\underline{2},3,2,1)$	2, 3
6	2	$(1,1,\underline{3},2,1)$	2, 3
7	3	$(1,1,\underline{2},1)$	2
8	2	$(\underline{1},1,1)$	
9	1	$(\underline{1},1)$	1
10	1	$(\underline{1})$	1
11	1	$()$	

Table 1. A $(3,2)$-processing of q.

Let $\overset{k}{\Rightarrow}_{s,p}$ be an (s,p)-transformation consisting of k (s,p)-transformation steps one after the other, and let $\overset{*}{\Rightarrow}_{s,p}$ be the reflexive and transitive closure of $\Rightarrow_{s,p}$.

Consider an (s,p)-transformation $(q,q) \overset{k}{\Rightarrow}_{s,p} (q,q')$. If sequence q has n bins then after k transformation steps sequence q' has $n - k$ bins. In each of the k (s,p)-transformation steps, a bin on the first s positions is removed. For some configuration (q,q'), the *active part* of q with respect to storage capacity s consists of the first s bins of q'. Analogously, the last $|q'| - s$ bins of q' represent the so-called *passive part* of q' with respect to s. If configuration (q,q') is obtained by some (s,p)-transformation from (q,q), then the bins in the passive part of q' with respect to s are the last $|q'| - s$ bins of q. Which bins the active part of q' contains depends on the bins removed during the (s,p)-transformation.

2 The approximation algorithm

Consider an (s,p)-processing $(q,q) \overset{n}{\Rightarrow}_{s,p} (q,\emptyset)$. Let $B = (b_{i_1}, \ldots, b_{i_n})$ be the sequence of bins in the order in that they are removed from q and let $P = (t_{j_1}, \ldots, t_{j_m})$ be the sequence of pallets in the order in that they are opened during the (s,p)-processing. B is called an (s,p)-*bin solution* for q and P is called an (s,p)-*pallet solution* for q. The transformation in table 1 defines $(3,2)$-bin solution $B = (3,4,5,6,7,8,9,10,1,2,11)$ and $(3,2)$-pallet solution $P = (2,3,1)$.

Clearly, each (s,p)-bin solution B defines an (s,p)-pallet solution P by the order in which the pallets are opened during the (s,p)-processing given by B.

For an arbitrary sequence $P = (t_{j_1}, \ldots, t_{j_m})$ of the m pallets of $plts(q)$ and a given integer $p \geq 1$, it is easy to find an (s,p)-bin solution B for q that defines P such that s is minimal. This can for example be done with algorithm 1. Algorithm 1 computes for q, P, and p a storage capacity s and an (s,p)-bin solution B that defines pallet solution P. It is easy to see that the storage capacity s is minimal in the sense that there is no (s',p)-bin solution B' for some $s' < s$ that defines pallet solution P. We will apply algorithm 1 for the pallet sequence computed by our stack-up approximation algorithm.

Algorithm 1:

Let $i := 1$, $s := 0$, and $q' := q$;
Repeat the following steps until q' is empty:

1. If $|open(q, q')| = p$ then let b be the first bin of q' destined for some pallet of $open(q, q')$;
2. If $|open(q, q')| < p$ then let b be the first bin of q' destined for some pallet of $open(q, q') \cup \{t_{j_i}\}$;
3. If $pos(q', b) > s$ then let $s := pos(q', b)$;
4. If $plt(b) = t_{j_i}$ then let $i := i + 1$;
5. Remove bin b from q' and output b;

Output s;

Let $h(q, t, i)$ be the number of bins in sequence $q = (b_1, \ldots, b_n)$ destined for pallet t being on a position less than i such that the last bin for t in q is on a position greater than or equal to i. If all bins for pallet t are on a position less than i or on a position greater than or equal to i then let $h(q, t, i)$ be zero. We will write $h(t, i)$ instead of $h(q, t, i)$, because we always apply $h()$ to sequence q.

i	1	2	3	4	5	6	7	8	9	10	11	12
$plt(b_i)$	1	2	2	2	2	3	3	3	3	3	4	4
$h(1,i)$	0	1	1	1	1	1	1	1	1	1	1	1
$h(2,i)$	0	0	1	2	3	4	4	4	4	4	4	4
$h(3,i)$	0	0	0	0	0	0	1	2	3	4	5	5
$h(4,i)$	0	0	0	0	0	0	0	0	0	0	0	1
$h(5,i)$	0	0	0	0	0	0	0	0	0	0	0	0
$h(6,i)$	0	0	0	0	0	0	0	0	0	0	0	0
Q_i	∅	∅	{2}	{2}	{2}	{2}	{2,3}	{2,3}	{2,3}	{2,3}	{2,3}	{2,3}
Q_i	∅	{1}	{1}	{1}	{1}	{1}	{1}	{1}	{1}	{1}	{1}	{1,4}
output	-	2	-	-	-	3	-	-	-	-	-	-

i	13	14	15	16=i	17	18	19	20	21	22	23
$plt(b_i)$	5	5	5	5	1	1	4	3	2	6	1
$h(1,i)$	1	1	1	1	1	2	3	3	3	3	3
$h(2,i)$	4	4	4	4	4	4	4	4	4	0	0
$h(3,i)$	5	5	5	5	5	5	5	5	0	0	0
$h(4,i)$	2	2	2	2	2	2	2	0	0	0	0
$h(5,i)$	0	1	2	3	0	0	0	0	0	0	0
$h(6,i)$	0	0	0	0	0	0	0	0	0	0	0
Q_i	{2,3}	{2,3}	{2,3}	{2,3}	{2,3}	{2,3}	{2,3}	{2,3}	{1,2}	{1}	{1}
Q_i	{1,4}	{1,4}	{1,4,5}	{1,4,5}	{1,4,5}	{1,4}	{1,4}	{1}	∅	∅	∅
output	-	-	-	5	-	-	4	1	-	6	-

Fig. 1. A processing of some q for $p = 2$ by algorithm 2.

The integers $h(t, 1), h(t, 2), \ldots$ are monotone increasing up to the position of the last bin for pallet t. For all $i \leq first(q, t)$ and for all $i > last(q, t)$, we have $h(t, i) = 0$. See figure 1 for an example.

Let $plts(q) = \{t_1, \ldots, t_m\}$ be the set of all pallets of the bins in q and

$$h_\Sigma(q, i) = h(t_1, i) + h(t_2, i) + \cdots + h(t_m, i)$$

be the sum of all bins on a position less than i destined for pallets whose last bin is on a position greater than or equal to i. Let $h_{\Delta_p}(q,i)$ be the sum of the p greatest integers of

$$\{h(t_1,i),\ h(t_2,i),\ \ldots,\ h(t_m,i)\}.$$

Let (q,q') be a configuration obtained by some (s,p)-transformation from the initial configuration (q,q) such that b_i is the first bin of the passive part of q'. If $s \leq h_\Sigma(q,i) - h_{\Delta_p}(q,i)$ then (q,q') is (s,p)-blocking, because at least $h_\Sigma(q,i) - h_{\Delta_p}(q,i)$ bins have to be in the active part of q' if at most p pallets are allowed to be open, i.e., at least $h_\Sigma(q,i) - h_{\Delta_p}(q,i)$ bins are not destined for already open pallets, So non of them can be removed if all p stack-up places are occupied. There has to be at least one additional bin in the active part such that the (s,p)-transformation can continue. That is,

$$1 + \max_{1 \leq i \leq n} \left\{ h_\Sigma(q,i) - h_{\Delta_p}(q,i) \right\}$$

is a lower bound on the storage capacity for each processing of q with p stack-up places.

Our algorithm will determine a processing of q by scanning q from right to left. Variables $Q_1, \ldots, Q_n$ will be used to store pallets. The algorithm tries to keep that pallets in Q_i for which $h(t,i)$ is maximal. It inserts a pallet t into Q_i if the last bin for t is on position i. It removes a pallet t from Q_i and outputs it if either the first bin for t is on position i or if Q_i contains more than p pallets. The pallet removed from Q_i is one of the pallets of Q_i for which $h(t,i)$ is minimal.

Algorithm 2:

For $i := n$ down to 1:

1. If $i = n$ then let $Q_i := \emptyset$ else let $Q_i := Q_{i+1}$;
2. If for some pallet $t \in \{t_1, \ldots, t_m\}$, $last(q,t) = i$ then insert t into Q_i;
3. If Q_i contains some pallet t such that $first(q,t) = i$ then remove t from Q_i, and output t;
4. If $|Q_i| = p + 1$ then remove a pallet t from Q_i for which $h(t,i)$ is minimal, and output t;

Let P_A be the reverse order of the pallet output of algorithm 2. Sequence P_A contains each pallet exactly once. P_A together with integer $p + 1$ can be transformed by algorithm 1 into an $(s, p+1)$-bin solution B_A for q. Pallet solution P_A is designed for a processing of q with $p + 1$ places and not with p places, because we need one place to open the pallets whose bins are all in the active part.

Figure 1 shows an example of the processing of some sequence q from right to left. The pallet sets $Q_1, \ldots, Q_{23}$ shown in the figure are those which remain after the processing of q. The computed pallet sequence for $p = 2$ is $P_A = (2, 3, 5, 4, 1, 6)$. Algorithm 1 computes the following $(4, 3)$-bin solution B_A for P_A.

$$B_A = (b_{i_2}, b_{i_3}, b_{i_4}, b_{i_5}, b_{i_6}, b_{i_7}, b_{i_8}, b_{i_9}, b_{i_{10}}, b_{i_{13}}, b_{i_{14}}, b_{i_{15}}, b_{i_{16}},$$
$$b_{i_{11}}, b_{i_{12}}, b_{i_{19}}, b_{i_1}, b_{i_{17}}, b_{i_{18}}, b_{i_{20}}, b_{i_{21}}, b_{i_{22}}, b_{i_{23}})$$
$$\hat{=} (2, 2, 2, 2, 3, 3, 3, 3, 3, 5, 5, 5, 4, 4, 4, 1, 1, 1, 3, 2, 6, 1)$$

In the remaining part of this section, we estimate the storage capacity s of the $(s, p+1)$-bin solution B_A. Consider again the $(s, p+1)$-processing defined by $(s, p+1)$-bin solution B_A and a configuration (q, q') obtained during the processing such that b_i is the first bin of the passive part of q'. Then there are at most $h_\Sigma(q, i) - \sum_{t \in Q_i} h(t, i)$ bins in the active part of q' that are destined for some pallet t such that $last(q, t) \geq i$ and $t \notin Q_i$. Let $\overline{Q}_i$ be the set of all pallets $t \in \{t_1, \ldots, t_m\} - Q_i$ such that $h(t, i) > 0$. Thus algorithm 2 outputs an $(s, p+1)$-pallet solution P_A for q for some

$$s \leq 1 + \max_{1 < i \leq n} \left\{ h_\Sigma(q, i) - \sum_{t \in Q_i} h(t, i) \right\} = 1 + \max_{1 < i \leq n} \left\{ \sum_{\bar{t} \in \overline{Q}_i} h(\bar{t}, i) \right\}.$$

The sum $\sum_{\bar{t} \in \overline{Q}_i} h(\bar{t}, i)$ is also maximal for some i such that the bin on position i in q is the last bin for some pallet t_{in} with $h(t_{in}, i) > 0$ and $|Q_i| = p$, because the integers $h(t, i)$ are increasing for $i = 1, 2, \ldots, last(q, t)$ and only at these positions i some pallet t_{out} with $h(t_{out}, i) > 0$ will be removed from Q_i and thus becomes a member of $\overline{Q}_i$. Let $\hat{i}$ be such a position i. Then $last(q, t_{in}) = \hat{i}$ and pallet t_{in} with $h(t_{in}, \hat{i}) > 0$ is inserted into $Q_{\hat{i}}$ by step 2 of the algorithm and pallet t_{out} with $h(t_{out}, \hat{i}) > 0$ is removed from $Q_{\hat{i}}$ by step 4 of the algorithm. The inserted pallet t_{in} and the removed pallet t_{out} may be the same. It also follows that for all pallets $t \in Q_{\hat{i}}$, $h(t, \hat{i}) \geq h(t_{out}, \hat{i})$, because the removed pallet t_{out} was one of the pallets t of $Q_{\hat{i}}$ for which integer $h(t, \hat{i})$ was minimal.

In the example from figure 1, we have $\hat{i} = 16$, $Q_{16} = \{2, 3\}$, and $\overline{Q}_{16} = \{1, 4, 5\}$. Bin b_{16} is the last bin for pallet 5. It is inserted into Q_{16} in step 2 and then removed from Q_{16} in step 4, because $h(5, 16) < h(2, 16)$ and $h(5, 16) < h(3, 16)$. That is, $t_{in} = 5$ and $t_{out} = 5$.

Let $\sigma_1, \ldots, \sigma_p$ be the pallets in $Q_{\hat{i}}$. Let $\overline{\sigma}_1, \ldots, \overline{\sigma}_l$ be the pallets in $\overline{Q}_{\hat{i}}$ in the order the algorithm has output them, i.e., ordered by the positions where they are removed from Q_{i_k}. More precisely, let i_k for $1 \leq k \leq l$ be the position where $\overline{\sigma}_k$ becomes a member of $\overline{Q}_{i_k}$, because it is removed from Q_{i_k}, then $\hat{i} = i_l < \cdots < i_2 < i_1$. If the algorithm outputs pallet $\overline{\sigma}_k$, then $\overline{Q}_{i_k}$ contains at least the k pallets $\overline{\sigma}_1, \ldots, \overline{\sigma}_k$. We don't know the pallets in Q_{i_k}, but we know that Q_{i_k} contains p pallets $\varphi_1, \ldots, \varphi_p$ such that $h(\varphi_j, i_k) \geq h(\overline{\sigma}_k, i_k)$ for $1 \leq j \leq p$, because $\overline{\sigma}_k$ is removed from Q_{i_k} by step 4 of the algorithm, i.e., $\overline{\sigma}_k$ was one of the $p+1$ pallets t of Q_{i_k} for which integer $h(t, i_k)$ was minimal.

Consider the following l vectors $M_{\hat{i}, k}$ of integers. For $1 \leq k \leq l$ let

$$p \text{ times}$$
$$M_{\hat{i}, k} := (h(\overline{\sigma}_1, i_k), \ldots, h(\overline{\sigma}_k, i_k), \overbrace{h((\overline{\sigma}_k, i_k), \ldots, h((\overline{\sigma}_k, i_k))}).$$

The first k entries are the integers $h(t, i_k)$ for $t = \overline{\sigma}_1, \ldots, \overline{\sigma}_k$, and each of the last p entries is a lower bound for each $h(\varphi_j, i_k)$. Let $\Sigma M_{\hat{i}, k}$ be the sum of all integers in $M_{\hat{i}, k}$, and let $\Delta_p M_{\hat{i}, k}$ be the sum of the p largest integers of $M_{\hat{i}, k}$. Then a lower bound for $s_{\min}(q, p)$ is given by

$$s_{\min}(q, p) \geq 1 + \max_{1 \leq k \leq l} \left\{ \Sigma M_{\hat{i}, k} - \Delta_p M_{\hat{i}, k} \right\}.$$

This follows immediately by the definition of $M_{\hat{i},k}$ and the following observation. Consider some configuration (q, q') with respect to some (s, p)-transformation such that b_{i_k} is the first bin in the passive part of q'. Since we can open at most p pallets, and all pallets of $\{\bar{\sigma}_1, \ldots, \bar{\sigma}_k\} \cup Q_{i_k}$ have at least one bin in the passive part of q', the sum of all bins in the active part not destined for some open pallet is at least the sum of the k least integers of $M_{\hat{i},k}$.

														$\hat{i}=i_3$		i_2				i_1
1	2	2	2	2	3	3	3	3	4	4	5	5	5	5	1 1	4	3	2	6	1

	$\bar{\sigma}_1 = 1$
	$\sigma_1 = 2$
	$\sigma_2 = 3$
	$\bar{\sigma}_2 = 4$
	$\bar{\sigma}_3 = 5$

Fig. 2. The horizontal lines indicate the range of the pallets. The line for pallet t start in the column of the first bin for t and ends in the column of the last bin for t. At position $\hat{i} = 16$, we have $Q_{16} = \{\sigma_1, \sigma_2\} = \{2, 3\}$ and $\overline{Q}_{16} = \{\bar{\sigma}_1, \bar{\sigma}_2, \bar{\sigma}_3\} = \{1, 4, 5\}$. The other positions are $i_3 = 16$, $i_2 = 19$, $i_1 = 23$.

Since $h(t, \hat{i}) \leq h(t, i_k)$ for each $t \in Q_{i_k} \cup \overline{Q}_{i_k}$, because $\hat{i} = i_l \leq i_k$, and since $h(\bar{\sigma}_k, i_k) \leq h(t, i_k)$ for each $t \in Q_{i_k}$, it follows that $h(\bar{\sigma}_k, \hat{i}) \leq h(t, i_k)$ for each $t \in Q_{i_k}$. Consider the following l vectors $N_{\hat{i},k}$ of integers whose entries are all less than or equal to corresponding entries of $M_{\hat{i},k}$. For $1 \leq k \leq l$ let

$$N_{\hat{i},k} = (h(\bar{\sigma}_1, \hat{i}), \ldots, h(\bar{\sigma}_k, \hat{i}), \overbrace{h(\bar{\sigma}_k, \hat{i}), \ldots, h(\bar{\sigma}_k, \hat{i})}^{p \text{ times}}),$$

let $\Sigma N_{\hat{i},k}$ be the sum of all integers in $N_{\hat{i},k}$, and let $\Delta_p N_{\hat{i},k}$ be the sum of the p largest integers of $N_{\hat{i},k}$. Then a weaker lower bound for $s_{\min}(q, p)$ is given by

$$s_{\min}(q, p) \geq 1 + \max_{1 \leq k \leq l} \left\{ \Sigma N_{\hat{i},k} - \Delta_p N_{\hat{i},k} \right\}.$$

Let $s_A(q, p+1)$ be the storage capacity computed by algorithm 1 applied to P_A. From the discussion above and the choice of $\hat{i}$, we know that $s_A(q, p+1) \leq 1 + \sum_{1 \leq j \leq l} h(\bar{\sigma}_j, \hat{i})$. This together with the lower bound for $s_{\min}(q, p)$ yields an upper bound on the performance of our approximation algorithm by

$$\frac{s_A(q, p+1)}{s_{\min}(q, p)} \leq \frac{1 + \sum_{1 \leq j \leq l} h(\bar{\sigma}_j, \hat{i})}{1 + \max_{1 \leq k \leq l} \{ \Sigma N_{\hat{i},k} - \Delta_p N_{\hat{i},k} \}} =: R.$$

The fraction R is maximal if all $\Sigma N_{\hat{i},k} - \Delta_p N_{\hat{i},k}$ for $1 \leq k \leq l$ are equal, otherwise we can increase $\sum_{1 \leq j \leq l} h(\bar{\sigma}_j, \hat{i})$ by increasing some integers in $N_{\hat{i},k}$ without changing $\max_{1 \leq k \leq l} \{1 + \Sigma N_{\hat{i},k} - \Delta_p N_{\hat{i},k}\}$. Thus, the integers in $N_{\hat{i},l}$

are monotone decreasing. Consider for example the situation shown in table 2, where $l = 6, p = 2, N_{\hat{i},6} = (162, 81, 54, 36, 24, 16, 16, 16), h(\overline{\sigma}_1, \hat{i}) + \ldots + h(\overline{\sigma}_6, \hat{i}) = 162 + 81 + 54 + 36 + 24 + 16 = 373$, and all $\Sigma N_{\hat{i},k} - \Delta_2 N_{\hat{i},k} = 162$ for $1 \leq k \leq 6$.

k	$N_{\hat{i},k}$	$\Sigma N_{\hat{i},k} - \Delta_2 N_{\hat{i},k}$
6	$(162, 81, 54, 36, 24, 16, 16, 16)$	162
5	$(162, 81, 54, 36, 24, 24, 24)$	162
4	$(162, 81, 54, 36, 36, 36)$	162
3	$(162, 81, 54, 54, 54)$	162
2	$(162, 81, 81, 81)$	162
1	$(162, 162, 162)$	162

Table 2. An example of a worst-case vector $N_{\hat{i},6}$ for $p = 2$.

For the first 6 integers $a_1, a_2, a_3, a_4, a_5, a_6$ of $N_{\hat{i},6}$ we have $a_1 = 162$, $a_2 = 1/2 \cdot a_1$, $a_3 = 2/3 \cdot a_2$, $a_4 = 2/3 \cdot a_3$, $a_5 = 2/3 \cdot a_4$, $a_6 = 2/3 \cdot a_5$. That is, $a_1 + \ldots + a_6 = a_1 + \frac{a_1}{2} + \frac{a_1}{3} + \frac{a_1}{9} + \frac{a_1}{27} + \frac{a_1}{81} = a_1 \cdot (1 + \frac{1}{2} + \frac{1}{3} + \frac{1}{9} + \frac{1}{27} + \frac{1}{81}) = 373$ and $R = \frac{1+373}{1+162} < 2.3$.

The largest fractions R can be defined if $p \geq l - 1$. Then we can design $N_{\hat{i},l}$ such that

$$s_A(q, p+1) = h(\overline{\sigma}_1, \hat{i}) + \ldots + h(\overline{\sigma}_l, \hat{i}) = a_1 \cdot \sum_{k=1}^{l} \frac{1}{k}$$

and $\Sigma N_{\hat{i},k} - \Delta_p N_{\hat{i},k} = a_1$ for all $1 \leq k \leq l$ and some positive integer a_1; see figure 3 for an example in which $p = 9$ and $l = 10$.

Thus, our approximation algorithm has an absolute performance

$$\frac{s_A(q, p+1)}{s_{\min}(q, p)} \leq \frac{1 + a_1 (\sum_{k=1}^{p+1} \frac{1}{k})}{1 + a_1}$$

for some positive integer a_1. Since we can assume that $a_1 > 1$ and $p > 1$, because otherwise our approximation algorithm makes no error, we have proved the following theorem.

Theorem 1. *There is a polynomial time approximation algorithm A for the stack-up storage minimization problem that computes for a sequence q and a positive integer $p > 1$ an $(s, p+1)$-pallet solution such that $s < s_{\min}(q, p) \cdot \log_2(p+1)$.*

Algorithm 1 can be implemented such that the processing of sequences with n bins takes time $O(n)$. This implementation is a little bit tricky but possible. Algorithm 2 can be implemented such that the processing of q takes time $O(n \cdot \log(p))$, by using standard data structures like heaps [AHU74, Meh84] for the implementation of Q_i. So the overall running time is $O(n \cdot \log(p))$.

We conjecture that there is no polynomial time algorithm that processes each sequence q using a storage capacity less than $s_{min}(q, p) \cdot \epsilon_1$ and $p + \epsilon_2$ places for some $\epsilon_1, \epsilon_2 > 0$ unless $P = NP$. The performance of our algorithm is in general for practical instances not greater than three. To obtain an error greater than three the pallets have to consists of several thousand of bins, which is quite unrealistic.

k	$N_{i,k}$
10	$(2520, 1260, 840, 630, 504, 420, 360, 315, 280, 252, 252, 252, 252, 252, 252, 252, 252, 252, 252)$
9	$(2520, 1260, 840, 630, 504, 420, 360, 315, 280, 280, 280, 280, 280, 280, 280, 280, 280, 280)$
8	$(2520, 1260, 840, 630, 504, 420, 360, 315, 315, 315, 315, 315, 315, 315, 315, 315, 315)$
7	$(2520, 1260, 840, 630, 504, 420, 360, 360, 360, 360, 360, 360, 360, 360, 360, 360)$
6	$(2520, 1260, 840, 630, 504, 420, 420, 420, 420, 420, 420, 420, 420, 420, 420,)$
5	$(2520, 1260, 840, 630, 504, 504, 504, 504, 504, 504, 504, 504, 504, 504)$
4	$(2520, 1260, 840, 630, 630, 630, 630, 630, 630, 630, 630, 630, 630)$
3	$(2520, 1260, 840, 840, 840, 840, 840, 840, 840, 840, 840)$
2	$(2520, 1260, 1260, 1260, 1260, 1260, 1260, 1260, 1260, 1260)$
1	$(2520, 2520, 2520, 2520, 2520, 2520, 2520, 2520, 2520, 2520)$

Fig. 3. A worst-case vector $N_{i,l}$ for $l = 10$ and $p = 9$, $\Sigma N_{i,k} - \Delta_9 N_{i,k} = 2520$ for $k = 1, \ldots, 10$, and $R = \frac{1+7381}{1+2520} < 3$.

References

[AHU74] A.V. Aho, J.E. Hopcroft, and J.D. Ullman. *The Design and Analysis of Computer Algorithms*. Addison-Wesley Publishing Company, Massachusetts, 1974.

[dK94]   R. de Koster. Performance approximation of pick-to-belt orderpicking systems. *European Journal of Operational Research*, 92:558–573, 1994.

[GJ79]   M.R. Garey and D.S. Johnson. *Computers and Intractability, A Guide to the Theory of NP-Completeness*. W.H. Freeman and Company, San Francisco, 1979.

[LLKS93] E.L. Lawler, J.K. Lenstra, A.H.G. Rinnooy Kan, and D.B. Shmoys. Sequencing and Scheduling: Algorithms and Complexity. In S.C. Graves, A.H.G. Rinnooy kan, and P.H. Zipkin, editors, *Handbooks in Operations Research and Management Science, Vol. 4. Logistics of Production and Inventory*, pages 445–522. North-Holland, Amsterdam, 1993.

[Meh84] K. Mehlhorn. *Data Structures and Algorithms I: Sorting and Searching*. EATCS Monographs on Theoretical Computer Science. Springer-Verlag, 1984.

[MMS88] M.S. Manasse, L.A. McGeoch, and D.D. Sleator. Competitive algorithms for on-line problems. In *Proceedings of the Annual ACM Symposium on Theory of Computing*, pages 322–333. ACM, 1988.

[RW97a] J. Rethmann and E. Wanke. Competitive analysis of on-line stack-up algorithms. In *Proceedings of the Annual European Symposium on Algorithms*, LNCS. Springer-Verlag, 1997. To appear.

[RW97b] J. Rethmann and E. Wanke. Storage controlled pile-up systems, theoretical foundations. *European Journal of Operational Research*, to appear, 1997. A short abstract of this paper will also appear in the Proceedings of SOR '96.

Relaxed Balance through Standard Rotations

Kim S. Larsen[1]*, Eljas Soisalon-Soininen[2], Peter Widmayer[3]

[1] Department of Mathematics and Computer Science, Odense University,
Campusvej 55, DK–5230 Odense M, Denmark. Email: kslarsen@imada.ou.dk.
[2] Laboratory of Information Processing Science, Helsinki University of Technology,
Otakaari 1 A, FIN-02150 Espoo, Finland. E-mail: ess@cs.hut.fi.
[3] Institut für Theoretische Informatik, ETH Zentrum, CH-8092 Zürich, Switzerland.
E-mail: widmayer@inf.ethz.ch.

Abstract. We consider binary search trees, where rebalancing transformations need not be connected with updates but may be delayed. For standard AVL tree rebalancing, we prove that even though the rebalancing operations are uncoupled from updates, their total number is bounded by $O(M \log(M + N))$, where M is the number of updates to an AVL tree of initial size N. Hence, relaxed balancing of AVL trees comes at no extra cost asymptotically. Furthermore, our scheme differs from most other relaxed balancing schemes in an important aspect: No rebalancing transformation can be done in the wrong direction, i.e., no performed rotation can make the tree less balanced. Moreover, each performed rotation indeed corresponds to a real imbalance situation in the tree.

Our results are important in designing efficient concurrency control strategies for main-memory databases. Main-memory search structures have gained new applications in large embedded systems, such as switching systems for mobile telephones.

1 Introduction

Several proposals for solving the problem of relaxed balancing in search tree structures have been presented [3, 5, 6, 7, 8, 9, 10, 11]. By relaxed balancing we mean that the balancing transformations usually connected with insert and delete operations are separated from the actual update and can be arbitrarily delayed. In such an environment, the operations applied to the search tree are search, insert, delete, and rebalance. Moreover, it is required that only a few nodes of the tree are involved with one invocation of the rebalance operation. This is important in a concurrent environment, as all operations applied must be advanced in small steps involving only a small number of nodes at a time.

---

* Some of the work was done while this author was visiting the Department of Computer Sciences, University of Wisconsin at Madison. The work of this author was supported in part by SNF (Denmark), in part by NSF (U.S.) grant CCR-9510244, and in part by the ESPRIT Long Term Research Programme of the EU under project number 20244 (ALCOM-IT).

Relaxed balancing allows an efficient solution to the concurrency control problem for binary search trees. An example of a need of a high degree of concurrency in main-memory search trees are the rapidly extending switching systems of mobile telephones. In Europe, for instance, there are several national networks, each of which contains several local nodes. Each node covers a certain geographic area, and is responsible for phone calls in this area. The node registers automatically all present phones. For example, when a phone with the home location in Helsinki, Finland, is in Zürich, Switzerland, then it will be connected automatically to a currently closest node of a Swiss network system. Such a network system contains highly dynamic search structures, because the location of each phone with respect to the network nodes must be registered, as must the entering and leaving visiting phones.

In such real-time embedded systems as switching systems for mobile telephones, the transactions are typically quite short: update the location information of a phone, determine the location of a phone, insert a visiting phone into the network, delete a visiting phone from the network. The question is then how much parallelism is to be allowed for concurrent transactions. There is a trade-off between the cost of handling the parallelism—implementing various types of locks may be quite costly, for instance—and the speedup obtained by the increased parallelism. In fact, in many current switching systems, incoming transactions are compiled and put into a transaction queue from which they are performed one by one on a first come first served basis. The growing number of transactions has, however, made the implementors of switching systems consider methods to speed up the process.

There are other options than using a search tree for these kind of problems. Various hashing schemes exist, some of which can be used in a concurrent environment. However, search trees can support more operations efficiently, such as nearest neighbor searches and batch updates.

One point to relaxed balancing is that it may improve the efficiency of the concurrent use of search trees considerably. The reason for this is that with relaxed balancing we may perform the insert and delete transactions in two independent parts, the first performing the actual insertion or deletion, and the second the corresponding rebalancing. A detailed account for why rebalancing immediately after insertions or deletions counteracts attempts to increase parallelism can be found in many papers on data structures for shared-memory architectures. In [4, 10], in particular, this is discussed in the context of relaxed balancing. The problem is that the path from the highest "unsafe" node (which might be the root) to the update must be locked, since otherwise a process might loose the path during the rebalancing phase on its way back up the tree. The alternative option of using top-down rebalancing is not attractive either because of all the unnecessary structural changes that top-down methods incur. Locking schemes are also discussed in [4, 10] as well as in many other papers.

A drawback in the previous solutions for the problem of relaxed balancing is that a single balancing transformation need not make the tree more balanced. It is natural to require that each structure changing operation is locally beneficial because in a concurrent environment urgent searches are present all the time. Moreover, it can be expected that the total number of needed balancing transformations will be decreased, at least on the average, when none of them makes the tree less balanced.

Our solution has the distinctive property that it maintains the actual heights of the nodes and will thus base all decisions of whether or not to perform a rotation on the heights of the nodes in the tree. This property is important because then delayed rebalancing tasks (the structural changes) will be forgotten when the ongoing insert and delete operations themselves make the tree balanced. Observe that all previous solutions to relaxed balancing work in such a way that balance conflicts from the history are remembered and gradually resolved. For example, a chromatic tree [5, 11] or a stratified tree [12] may be full of conflicts that must be resolved even though the tree is perfectly balanced.

In the present paper, we propose a solution in which the balancing transformations are the standard ones used in a sequential solution. This means in particular that they are small compared to rebalancing operations in some of the previous proposals. Also this is important in a parallel environment, because fewer locks are necessary at any given time, and they would also be held less long, since less work has to be done in order to carry out the operation. The results of the present paper may also be considered to be of general interest as regards the theory of search tree structures. The question we address is the following: Is it possible to balance a binary search tree globally in such a way that only those portions of the tree which are indeed out of balance are modified? Moreover, changes in the tree should be allowed in the meantime, and the efficiency of balancing from the standard structures should be retained. We have obtained positive answers to these questions in the case of AVL trees.

In [13], a preliminary sketch of our relaxed balancing scheme was presented. In this present paper, that scheme is presented in full. Additionally, this paper contains a proof that each update gives rise to at most a logarithmic number of rebalancing operations. Especially considering how simple the rebalancing scheme is, the proof that rebalancing is efficient is surprisingly difficult.

2   Height-Valued Binary Search Trees

We consider binary search trees as implementing a totally ordered finite set S of *keys* chosen from a given domain. We allow the standard operations *search(k)*, *insert(k)*, and *delete(k)*, that is, search for key k, insert key k into set S, and delete key k from set S.

We assume that the trees are *leaf-oriented* binary search trees, which are full binary trees (each node has either two or no children) with the keys stored in the leaves. The internal nodes contain *routers*, which guide the search from the root to a leaf. The router stored in a node v must be greater than or equal to any key stored in the leaves of v's left subtree and smaller than any key in the leaves of v's right subtree. The routers need not be keys stored in the leaves of the tree.

The height of a node u in a tree, denoted as $height(u)$, is defined as the length of a longest path from u to a leaf in the subtree rooted at u. The *height of a tree* is the height of its root. We will consider search trees where the balance condition is the AVL balance condition, that is, for each internal node the difference of the heights of its two subtrees is at most one.

With each node u of a binary search tree we associate an integer, called a *height value*, denoted $hv(u)$, which is either -1 or $height(u)$. If $hv(u) \neq -1$, that is, $hv(u) = height(u)$, then we require that for the child nodes v_1 and v_2 of u both (i) and (ii) hold:

(i) $hv(v_1) = height(v_1)$, $hv(v_2) = height(v_2)$, and
(ii) the difference of the heights of v_1 and v_2, is at most one, i.e., $|height(v_1) - height(v_2)| \leq 1$.

A binary search tree with the associated height values stored in the nodes is called a *height-valued tree*.

We say that node u exhibits a *balance conflict* if $hv(u) = -1$; otherwise, u is said to be *in balance*. A height-valued tree T is said to be *completely in balance*, if all its nodes are in balance. In this case, T is clearly an AVL tree.

The idea in the use of height-valued trees is that insertions and deletions cause non-negative height values in the search paths to be set to -1. These values of -1 are gradually changed to real heights by balancing transformations.

The insert and delete operations are defined below.

Insert(k): The tree is searched with key k. Whenever along this search path there is a node u with $hv(u) \neq -1$, $hv(u)$ is set to -1. If the key is found, the process terminates.

An unsuccessful search ends up in a leaf, say l. A new internal node u is created in place of l, and l and a new leaf l' containing the key k are made child nodes of u. The children are ordered such that the one containing the smaller key will be the left child of u. The router of u is a copy of the key contained in its left child.

The height value of l' is set to 0, and the height value of u is set to 1.

Delete(k): The tree is searched with key k. Whenever along this search path there is a node u such that $hv(u) \neq -1$, $hv(u)$ is set to -1. If the key is not found, the process terminates. Otherwise the leaf, denoted l, containing the key k is removed. Its parent is replaced by the sibling node of l.

The following observation is immediate.

Observation 1 *When applied to a height-valued tree, the insert and delete operations preserve the height-valued property of the tree.*

3 Rebalancing a Height-Valued Tree

The task of rebalancing a height-valued tree is to remove all balance conflicts from the tree. Moreover, this should be possible by using small local transformations that allow—besides the concurrent searches—new insertions or deletions to occur.

Our strategy in resolving a balance conflict in a height-valued tree is to advance bottom-up so that a conflict in a node u will be resolved only if both subtrees of u contain no conflicts. This can be checked by looking at the child nodes of u only, because the subtrees of u are free from conflicts exactly when the height values of the child nodes of u are their true heights.

Observation 2 *There is always a conflicting node with both child nodes without conflicts, if the tree contains conflicts at all. Moreover, if a node has height value different from* -1, *then no conflicts can exist below this node.*

This observation is an invariant for height-valued trees; also when update operations are interleaved with the rebalancing operations described below.

Let u be a node that exhibits a conflict, and let v_1 and v_2 be the children of u, such that both v_1 and v_2 have height values different from -1. Call u the *root of the operation before rebalancing.* We have two cases to consider.

Case 1: Finishing rebalancing. $|height(v_1) - height(v_2)| \leq 1$. In this case, the AVL balance condition is retained at u, and we simply set

$$hv(u) = \max\{hv(v_1), hv(v_2)\} + 1.$$

Case 2: Rotations. $|height(v_1) - height(v_2)| > 1$. In this case, a single or a double rotation is performed exactly in the same way as is done in the standard AVL tree balancing algorithm [1]. We assume here that v_1 is the left child and that $height(v_1) > height(v_2)$. The left subtree of node v_1 is denoted by A and the right subtree by B. There are two subcases depending on the heights of A and B.

Case 2a: Single rotation. $height(A) \geq height(B)$. In this case, a single rotation to the right at u will be performed, see Fig. 1. After the rotation, the nodes u, v_1, v_2, and the subtrees A and B are denoted by u', v_1', v_2', A', and B', respectively. The height values of v_1' and u' are set to -1. Node v_1' is called the *root of the operation after rebalancing.*

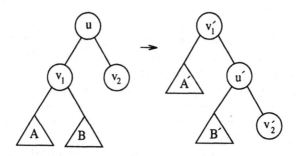

Fig. 1. Single rotation.

Case 2b: Double rotation. $height(A) < height(B)$. The root of B is denoted by w and its subtrees by B_1 and B_2. In this case, a double rotation will be performed, see Fig. 2. The height values of w' and u' are set to -1. Node w' is called the *root of the operation after rebalancing*.

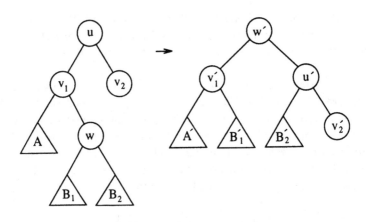

Fig. 2. Double rotation.

Note that due to Observation 2, a tree that is not completely in balance must contain a node to which one of the two cases applies. In the next section, we will analyze the progress towards a balanced tree that such a step is guaranteed to make.

4   The Efficiency of Relaxed Balancing

We assume that at some point we have a height-valued tree which is completely in balance. In this section, we show that rebalancing is logarithmic. More precisely, if M updates are carried out on the tree, then the number of rebalancing operations which can be applied before the tree is again in balance is $O(M \log(N+M))$, where N is the size of the tree when it was last in balance. This result is proven to hold no matter how updates and rebalancing operations are interleaved. The proof technique is amortized analysis [14]. The case where all updates are carried out before rebalancing is initiated was treated in [13].

For the purpose of the analysis, we divide the nodes up into three categories: *passive*, *active*, and *hyperactive*. Initially, all nodes are passive. After that, nodes that are traversed during an update, or that are involved in a rebalancing operation change status as described below.

Leaves are always categorized as passive. For an update, all nodes on the update path become active no matter what their status was before the update. A finishing rebalancing operation changes the status of the root of the operation to passive. Any other rebalancing operation leaves the root of the operation (after it is carried out) as active, and the node u' as hyperactive. Because rebalancing advances from bottom to top, we get the following:

Observation 3

- A passive node can only have passive children.
- The children of a hyperactive node are passive.

Proposition 4. *On a path from the root to a leaf, there is at most one hyperactive node.*

Proof. Only rotations introduce hyperactive nodes. Consider a path from the root through a hyperactive node, newly created by a rotation. The only node on that path which could be hyperactive before the rotation was carried out is the top node of the operation before the rotation is carried out, since, by Observation 3, nodes higher up in the tree do not have passive children. □

The maximal height difference between two subtrees of a node u, both of which are balanced AVL trees, is denoted $mhd(n)$, where n is the number of nodes in the subtree rooted at u. We will abuse notation and frequently write $mhd(u)$ instead of $mhd(n)$. Since the smallest possible subtree is a leaf, $mhd(n)$ must equal the height of the highest AVL tree with $n-2$ nodes. It is well known that the minimum number of nodes in an AVL tree of height h is $F_{h+3} - 1$ [2], where the Fibonacci numbers F_i are defined recursively by $F_1 = F_2 = 1$ and $F_i = F_{i-1} + F_{i-2}$, $i \geq 3$. From the inequality $F_{h+3} - 1 \leq n - 2$, it is easy to show that $h \leq \lfloor \log_\phi(\sqrt{5}n) \rfloor - 3 = mhd(n)$, where ϕ is the golden ratio $\frac{1+\sqrt{5}}{2}$, approximately equal to 1.618.

Proposition 5. *A rebalancing operation op can increase the height difference for at most one node not involved in the operation. If this happens, the increase is one.*

Proof. Clearly, only nodes between the location for the rebalancing operation and the root of the tree can be affected. Let $u_1, u_2, \ldots, u_k$ be these nodes, where u_k is the root of the tree, and u_1 is the root of the operation before rebalancing. Note that the rebalancing operation cannot increase the height of the subtree with root u_1. This height can remain unchanged or decrease by one. This is true for both single and double rotations. Since height is defined recursively, whenever the height of some node u_i remains unchanged, no nodes above u_i will change height. So, assume that the height of u_1 decreases. Let t be the smallest index of a node on the path, the height of which remains unchanged. For any index $j < t$, since the height of u_j decreases, its subtree containing u_1 was the highest before the operation. Thus, the height difference at u_j decreases. This means that only at u_t can the height difference possibly increase; and at most by one. □

Definition 6. We refer to the rebalancing operation op and the node u_t defined in Proposition 5 and the proof of that proposition. We call op a height difference increasing operation targeted at u_t.

Let us now define a potential function that measures the degree of imbalance in a tree, relative to the AVL tree requirements. Intuitively, in the course of events, a node u will become active at some point in time, and then stay active and stay the root of its subtree for a while. During that time, updates may pass through u down into u's subtree, and rotations targeted at u may be carried out in u's subtree. Both intuitively contribute to the imbalance at u. In addition, u starts with an initial imbalance that may be high if its state went from hyperactive to active, and that is zero or one if u was passive before becoming hyperactive. On the other hand, the imbalance at u intuitively cannot exceed the worst imbalance in an AVL tree, $mhd(n)$. The potential function takes these three effects into account, in the following way. For an active node u, let $upd(u)$ be the number of updates passing through u since u was last non-active (hyperactive or passive). Let $hio(u)$ be the number of height difference increasing operations targeted at u since u was last non-active. Let $lna(u)$ be the height difference between the two subtrees of u when u was last non-active.

Definition 7. The potential of a height-valued tree T is defined as the sum of the potentials of its nodes. The potential of a node u with children v and w is defined as follows, depending on the state of the node:

State	Potential
passive	0
active	$2[\min(upd(u) + hio(u) + lna(u), mhd(n)) - 1] + 1$
hyperactive	$2[|height(v) - height(w)| - 1] + 1$

where

$$[x] = \begin{cases} x, & \text{if } x \geq 0 \\ 0, & \text{otherwise} \end{cases}$$

In the following, for a node u, we refer to $2[mhd(u) - 1] + 1$ as u's *maximal potential*.

More intuition concerning the potential function: The -1 reflects that AVL trees allow heights to differ with one. The $+1$ reflects that even if height difference is zero, if the node is still active (because rebalancing is currently taking place beneath it), we must be able to provide a decrease in the potential function when we return to perform a finishing rebalancing operation. The multiplication with 2 ensures that when a rebalancing operation decreases a height difference with one, two units are "released". One of these will give us the potential decrease, whereas the other will become the $+1$ on a node which becomes active due to the rebalancing operation.

Our strategy for obtaining the complexity result is to prove that an update increases the potential with at most $O(\log n)$, where n is the current size of the tree, and that a rebalancing operation decreases the potential with at least one.

Lemma 8. *Fix any path from a leaf to the root, let u_1, u_2, ..., u_k be those nodes on this path for which the potential of u_i, $i = 1, \ldots, k$, is non-zero and less than maximal. Let n_i be the size of the subtree rooted at u_i, and assume that* the nodes are listed in the order they are encountered. Then $k \leq \lfloor \log_\phi(\sqrt{5}n_k) \rfloor - 3 + \log_\phi(2(n_k + 1))$.

Proof. Consider the node u_k. Since its potential is not maximal, fewer than $mhd(n_k) - 1$ updates have gone through u_k since it was last passive or hyperactive. Since, by Observation 3, a node which is passive or hyperactive has balanced subtrees, u_k had a balanced subtree at the time it was last non-active, and the height of that subtree was at least $h = k - 1 - upd(u_k) \geq k - mhd(n_k)$. Such a tree contains at least $F_{h+3} - 1$ nodes. So, $n_k \geq F_{h+3} - 1 - upd(u_k) \geq F_{h+3} - 1 - mhd(n_k) - 1$. This gives an upper bound on k:

Since $F_i \geq \phi^{i-2}$,

$$\phi^{k-mhd(n_k)} - 2 \leq F_{k-1-mhd(n_k)+3} - 2 \leq n_k + mhd(n_k).$$

So,

$$\phi^k \leq \phi^{mhd(n_k)}(n_k + mhd(n_k) + 2).$$

From the value of $mhd(n_k)$ and the (trivial) fact that $mhd(n_k) \leq n_k$,

$$k \leq \lfloor \log_\phi(\sqrt{5}n_k) \rfloor - 3 + \log_\phi(2(n_k + 1)).$$

□

Lemma 9. *For any path from a leaf to the root, let $u_1, u_2, \ldots, u_k$ be those nodes on the path with maximum potential, listed in the order they are encountered. Let n_i denote the size of the subtree rooted at u_i. The potential increase for these k nodes due to an insertion below u_1 is bounded by $2(\text{mhd}(n_k + 1) - \text{mhd}(n_1))$.*

Proof. First note that since u_i is in the subtree of u_{i+1}, $i \in \{1, \ldots, k-1\}$, we have that $n_i < n_{i+1}$, so $n_i + 1 \leq n_{i+1}$ and $mhd(n_i + 1) \leq mhd(n_{i+1})$, since this function is nondecreasing.

Now, the potential increase, I, is

$$I = \Sigma_{i=1}^k(2(mhd(n_i + 1) - 1) + 1) - \Sigma_{i=1}^k(2(mhd(n_i) - 1) + 1)$$
$$= 2\Sigma_{i=1}^k(mhd(n_i + 1) - mhd(n_i))$$
$$\leq 2(mhd(n_k + 1) - mhd(n_1)), \text{ by the observation above}$$

□

Lemma 10. *For any path from a leaf to the root, let $u_1, u_2, \ldots, u_k$ be all those nodes on the path with zero potential, listed in the order they are encountered. Then $k \leq \lfloor \log_\phi(\sqrt{5}(n_k + 1) + 1) \rfloor - 2$.*

Proof. Since the nodes have zero potential, they are all passive. By Observation 3, u_k is the root of a subtree where all nodes are passive, i.e., the subtree is a standard AVL tree. Thus, $F_{k-1+3} - 1 \leq n_k$ from which the result follows. □

Lemma 11. *An update increases the potential by at most $O(\log(N + M))$.*

Proof. By Proposition 4, there is at most one hyperactive node on an update path from the root to a leaf. This node becomes active instead. The potential increase for this node is certainly bounded by $O(mhd(N))$. The rest of the nodes are either active or passive. Active nodes remain active, but since the number of nodes in their subtrees may change due to the update, the potential may increase. The active nodes may either have reached their maximal potential or not. By Lemma 8, the total potential increase for the active nodes with less than maximal potential is logarithmically bounded, and by Lemma 9, the total potential increase for the nodes of maximal potential is logarithmically bounded. Finally, passive nodes become active and their potential increases from 0 to a constant (at most 3). Due to Lemma 10, the total increase for these nodes is also logarithmically bounded. We conclude that an update gives rise to a potential increase of at most $O(\log(N + M))$. □

Lemma 12. *A rebalancing operation decreases the potential by at least 1.*

Proof. There are three types of rebalancing operations: finishing, single rotation, and double rotation. We treat them separately.

Finishing: Only the top node of the operation changes status from active or hyperactive to passive. By definition, active and hyperactive nodes have potential at least 1 and passive nodes have potential 0.

Rotations: For both types of rotations, the root u of the operation *before* the operation is carried out has its status changed from active to hyperactive (if it is not already hyperactive). We analyze this first.

When u was last hyperactive or passive, it had a potential of $2[x-1]+1$, where x was the height difference of its subtrees. Since then, the height difference of u's subtrees can only increase if there is an update, and in that case clearly by at most 1, or if a height difference increasing operation targeted at u is carried out, and in that case, by Proposition 5, by at most one. So, $upd(u) + hio(u) + lna(u)$ is at least the current height difference of its subtrees. On the other hand, u's children are roots of balanced subtrees, so the height difference cannot exceed $mhd(n)$, where n is the size of the subtree rooted at u. In summary, changing the status of u from active to hyperactive cannot increase the potential. In the rest of this proof, we can therefore safely assume that u is hyperactive.

Single rotation: Referring to figure 1, the assumption is that $height(A) \geq height(B)$. Since the subtree rooted at v_1 is balanced, $height(A) = height(B)$ or $height(A) = height(B) + 1$. We treat these two cases separately.

If $height(A) = height(B)+1$, then the height difference between the subtrees of u' is two less than the height difference between the subtrees of u. This gives a potential decrease of 4. The node v_1' becomes active, but the height difference between its subtrees is zero, so it only needs a potential of 1. Even if there is a potential increase of 2 due to this operation increasing the height difference between two subtrees elsewhere in the tree (by Proposition 5, this can happen to at most one node), the total potential change is still negative.

If $height(A) = height(B)$, then the height of the root of the operation before it is carried out is equal to the height of the root after the operation is carried out. Thus, the operation will not increase the height difference between the subtrees of any other nodes in the tree. The height difference at u' is one less than it was at u, decreasing the potential by 2, allowing the potential at v'_1 to be set to 1, while still getting a total potential decrease.

Double rotation: Referring to figure 2, note that v_1 remains passive. The height difference at u' is two less than it was at u, and the height of the subtree decreases by one, so the argument is the same as for the "$height(A) = height(B) + 1$" case above. □

Theorem 13. *Rebalancing is amortized logarithmic.*

Proof. We assume that we have an AVL tree T with N nodes. The potential of such a tree is zero. In Lemma 11 it is shown that an update increases the potential by at most $O(\log(N + M))$, and in Lemma 12 it is shown that a rebalancing operation decreases the potential by at least one. Since the potential cannot be negative, the result follows from these two parts. □

5 Concluding remarks

We summarize the important aspects of the result in this paper with regards to an implementation on a shared-memory system.

First, rebalancing operations are small. Our scheme uses standard rotations as known from the sequential case instead of the larger operations proposed elsewhere. This means that the number of exclusive locks that has to be held at the same time is smaller. Since fewer locks must be obtained and fewer pointers moved, the locks that we have to hold are also released earlier than in other proposals. This is important because other processors are prevented from entering the whole subtree of a node which is exclusively locked.

Second, the number of rebalancing operations which has to be carried out in response to an update is fairly small. We have proven that it is logarithmic, but the constant in front of the logarithmic term is also quite small. From the potential function, it is clear that the constant is at most two, possible smaller.

Third, all the rebalancing operations that are carried out are in some sense necessary. When we rebalance, there really is a problem of imbalance, and the rebalancing operation always improves on this situation. In the other proposals for relaxed balance, trees can be full of conflicts which rebalancing operations must work on, despite of actually being in balance. We explain this in greater detail for the proposals for relaxed red-black trees (chromatic trees). In those proposals, conflicts are sequences of red nodes in a row, or overweighted nodes, i.e., nodes that are not red or black, but double black, triple black, etc. It is possible to have lots of such conflicts in a tree despite of the fact that the nodes of the tree could be colored red and black in such a way that there would be no conflicts at all. Thus, in such a proposal, conflicts do not necessarily correspond to a real problem of imbalance.

References

1. G.M.Adel'son-Vels'kii and E.M.Landis, An algorithm for the organisation of information. *Soviet Math. Dokl.* **3** (1962) 1259–1262.
2. A.V.Aho, J.E.Hopcroft and J.D.Ullman, *Data structures and algorithms*. Addison-Wesley, 1983.
3. J.Boyar, R.Fagerberg, and K.S.Larsen, Amortization results for chromatic search trees, with an application to priority queues, Fourth International Workshop on Algorithms and Data Structures, *Lecture Notes in Computer Science* **955** (1994) 270-281.
4. J.Boyar, R.Fagerberg, and K.S.Larsen, Amortization results for chromatic search trees, with an application to priority queues, *Journal of Computer and System Sciences* (to appear).
5. J.Boyar and K.S.Larsen, Efficient rebalancing of chromatic search trees. *Journal of Computer and System Sciences* **49** (1994) 667–682.
6. S.Hanke, T.Ottmann, and E.Soisalon-Soininen, Relaxed balanced red-black trees, Algorithms and Complexity, Third Italian Conference, *Lecture Notes in Computer Science* **1203** (1997) 193–204.
7. J.L.W.Kessels, On-the-fly optimization of data structures. *Comm. ACM* **26** (1983) 895–901.
8. K.S.Larsen, AVL trees with relaxed balance. In: *Proc. 8th International Parallel Processing Symposium*, IEEE Computer Society Press, 1994, pp. 888–893.
9. K.S.Larsen and R.Fagerberg, Efficient rebalancing of B-trees with relaxed balance. *International Journal of Foundations of Computer Science*, **7**:2 (1996) 196–202.
10. O.Nurmi and E.Soisalon-Soininen, Uncoupling updating and rebalancing in chromatic binary search trees. In: *Proc. 10th ACM Symposium on Principles of Database Systems*, 1991, pp. 192–198.
11. O.Nurmi, E.Soisalon-Soininen and D.Wood, Concurrency control in database structures with relaxed balance. In: *Proc. 6th ACM Symposium on Principles of Database Systems*, 1987, pp. 170–176.
12. T.Ottmann and E.Soisalon-Soininen, Relaxed balancing made simple. Technical Report 71, Institut für Informatik, Universität Freiburg, Germany, 1995.
13. E.Soisalon-Soininen and P.Widmayer, Relaxed balancing in search trees. In: *Advances in Algorithms, Languages, and Complexity* (D.-Z.Du and K.-I.Ko, eds.), Kluwer Academic Publishers, 1997, pp. 267–283.
14. R.E.Tarjan, Amortized computational complexity. *SIAM J. Alg. Disc. Meth.* **6**:2 (1985) 306–318.

Efficient Breakout Routing
in Printed Circuit Boards

(Extended Abstract)

John Hershberger* and Subhash Suri**

Abstract. Breakout routing is a single-layer wire-routing problem in which each of a set of pins must be connected to one of a set of vias, but no matching is prespecified. We propose a network-flow approach to breakout routing in which the wiring grid is modeled by a more compact graph. Our graph is a factor of $\Theta(\kappa^2)$ smaller than the wiring grid, where κ is the ratio of via spacing to pin spacing, which improves both the space and run time efficiency of the flow computation. A flow in the compact graph can be transformed into a wire layout, and vice versa.

1 Introduction

Breakout routing is a crucial problem in routing printed circuit boards (PCBs) and multichip modules (MCMs). The problem arises because a typical PCB or MCM has multiple layers available for routing, but electronic components are often connected to only a single layer of the board. Completing the routing requires that these surface-mounted components be connected to other layers by vias. The most efficient way to do this is to place a set of breakout vias near the component, connect each surface-mounted pin to a unique via, and then finish routing the connections from via to via using all the routing layers.

Breakout routing is often the hardest part of routing a PCB or MCM—once the pins have been connected to all routing layers by breakout vias, finishing the schematic-specified connections is straightforward. Breakout vias are often chosen to lie on a grid that is significantly coarser than the component pin spacing. While in current systems the ratio of via spacing to pin spacing is still under ten, technology trends (such as MCMs and ball grid arrays) are expected to increase the ratio to "tens."

Ball grid arrays (BGAs) also pose problems closely related to breakout routing. When a BGA component is mounted on a PCB, it connects to a coarse rectangular grid of pins on the board surface. Connections from these pins must be routed outward from the component, either on a single layer, or on multiple layers using vias. The outward connections on each layer must be planar, so the problem is similar to breakout routing, with the BGA pins playing the role of the breakout vias and the component boundary playing the role of the pins.

* Mentor Graphics, 1001 Ridder Park Drive, San Jose, CA 94301
** Department of Computer Science, Washington University, St. Louis, MO 63130. Research supported in part by NSF Grant CCR-9501494

A related problem also arises in VLSI processor arrays, in the context of fault-tolerant reconfiguration [5]. The *reconfiguration problem* in processor arrays is to match a set of N nodes (faulty processors) in an $m \times n$ rectangular grid with nodes on the grid boundary using disjoint paths. The fastest known algorithm for this problem is due to Chan and Chin [1]—their algorithm runs in time $O(d\sqrt{nmN})$, where d is the maximum number of disjoint paths found. More interestingly, they show that the size of the grid graph can be compressed from $O(nm)$ to $O(\sqrt{nmN})$ for the reconfiguration problem. While our results are somewhat incomparable, our compressed graph is *smaller by a factor of* κ in the extreme case of full via grid; in particular, the compressed graph of Chan and Chin has size $O\left(\frac{nm}{\kappa}\right)$, whereas our graph has size $O\left(\frac{nm}{\kappa^2}\right)$.

1.1 Previous Work

Due to the importance of breakout routing in design automation, there has been a lot of work on the problem. Not all techniques necessarily appear in the scientific literature, and some are even protected by patents. The two methods most relevant to our approach are:

1. [**Pattern routing in a grid.**] In the special case in which the vias lie in a rectangular grid and pins are located on its periphery, connect pins to vias with a regular pattern of wires. This method is guaranteed to find at least $1/\sqrt{2}$ of the maximum possible number of connections [7], but it is limited to the case of a fully populated via grid.
2. [**Network flow.**] The network flow approach is to solve the routing problem by computing a "flow" on the grid graph, where pins and vias are taken as sources and sinks, and where the grid spacing is the same as the minimum wire spacing. This method clearly finds a feasible routing if one exists, but it appears to be more popular in theory [2, p. 625] than in practice, perhaps because of the asymptotic complexity of network flow algorithms. One big disadvantage of the "vanilla" network flow method is that the size of the grid graph can be quadratic in the number of pins.
   Recent work by Yu, Darnauer, and Dai [8] uses network flow on a graph derived from a constrained Delaunay triangulation to solve the breakout problem approximately. The resulting flow may not correspond to a feasible routing, but the algorithm is applicable to a broader class of problems than the grid-based approach.

1.2 Our Contribution

Our method can be viewed as combining the best features of pattern routing in a grid and network flow. Like pattern routing in a grid, it exploits the fact that vias lie on a grid, but it allows some sites in the via grid to be inactive. It uses a network flow framework and always finds an optimal number of routes, but it does not explicitly build the entire grid graph. In particular, suppose that n breakout vias lie within a rectangular grid of size $p \times q$, and m component pins lie on the boundary of the rectangle, where $m \leq n \leq pq$. Our algorithm computes network flow in a network of size $O(pq)$, whereas the conventional approach will construct a grid graph of size at least $\Omega(pq\kappa^2)$, where $\kappa + 1$ is the ratio of via spacing to pin spacing. (In Figure 1, $p = q = 5$, $m = 6$, and $\kappa = 3$.) We believe that the improvement by a factor of κ^2 in running time and, more significantly, in space can make our approach a viable candidate for breakout routing in practice.

2 Preliminaries and the Combinatorial Wire Flow

Consider a $(\kappa + 1) \times (\kappa + 1)$ square, partitioned into $(\kappa + 1)^2$ unit squares by a square grid, for a nonnegative integer κ. Let us call the big square a *via cell*, its four corners *via sites*, the unit grid the *wire grid*, and vertices of the unit grid *grid points*.[3] For purposes of breakout routing, a printed circuit board can be modeled as a rectangular tiling of via cells. The parameter κ, which we call *via separation*, determines the maximum number of wires that can be routed between two adjacent vias in any row or column. Figure 1 shows an example.

A *wire element* is an edge of a unit square in the wire grid. A *wire path* is a simple (non-self-intersecting) path made up of wire elements. In this paper, all our wire paths will connect a pin location to a via location, and we assume throughout that these paths are *directed* from the pin to the via. Each via can either act as a *sink* node for a wire path, or as a *transit* node, but not both. If a via can act only as a sink, and not a transit node, then our algorithm and method still apply, with only a minor modification in the construction.

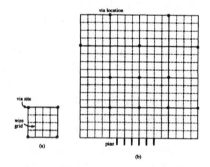

Fig. 1. (a) Via cell; $\kappa = 3$. (b) Tiling of cells.

A *matching* between a set of pins $P = \{p_1, p_2, \ldots, p_m\}$ and a set of vias $V = \{v_1, v_2, \ldots, v_n\}$ is a pairing $\{(p_1, v_1), (p_2, v_2), \ldots, (p_k, v_k)\}$ where each pin or via appears at most once. A *complete matching* is a matching where each pin appears exactly once; that is, $k = m$. A matching is *embeddable* if each matched pair can be joined by a wire path in the wire grid so that all paths are pairwise non-intersecting. In this paper, we will be interested only in embeddable matchings, and so we assume in the following that a matching is always embeddable. As mentioned before, we assume that each wire path is directed from its source pin to its sink via. Finally, we define the notion of a wire element entering or leaving a cell. We say that a wire element x *enters* a cell c if x's head is on the boundary of c but its tail is outside c. Similarly, x *exits* a cell c if x's tail is on the boundary of c but its head is outside c.

---

[3] Thus, a via site is also a grid point.

In order to discuss the wire layout in a via cell, it is convenient to separate the boundary of the cell from its interior. Let us define the *core* of a cell c to be the $\kappa \times \kappa$ subgrid in the center of c. The *boundary* of c is the square defining c. In Figure 1 (a), the central 3×3 grid is the core of the via cell. The main property of a core that we need in the following discussion is that a core has no vias, and therefore no wire terminates inside a core.

Fix a matching $\mathcal{M}$ for the ensuing discussion, and consider the core b of a via cell c. Let $e_i(b)$, for $1 \leq i \leq 4$, denote the edges of the square b in clockwise order, starting from the left edge. Let $f_i(b)$ denote the number of wire elements in $\mathcal{M}$ that enter b through $e_i(b)$, and let $g_i(b)$ denote the number of wire elements in $\mathcal{M}$ that leave b through $e_i(b)$, where $1 \leq i \leq 4$. (We omit b from the notation for f_i, g_i whenever the identity of the square is clear from the context.) The following two constraints are easily established:

1. [Capacity Constraint] $f_i(b) + g_i(b) \leq \kappa$.
2. [Wire Conservation] $\sum_{i=1}^{4} f_i(b) - \sum_{i=1}^{4} g_i(b) = 0$.

The capacity constraint follows from the fact that wires in $\mathcal{M}$ are disjoint and that each side of b has wire capacity κ. Wire conservation follows from the fact each wire of $\mathcal{M}$ terminates at a via and the core b is via-free. The number of wire elements in $\mathcal{M}$ entering or leaving b through each of its four sides describes the *combinatorial wire flow* in b. The capacity constraint and wire conservation are clearly necessary for a combinatorial wire flow to be embeddable. Unfortunately, they are not at all sufficient, as the following example shows: Let $f_1 = f_2 = \kappa$, $f_3 = f_4 = 0$; $g_1 = g_2 = 0$; and $g_3 = g_4 = \kappa$. It can easily be verified that this combinatorial flow satisfies the capacity constraint and wire conservation, but cannot be embedded.

It turns out that the maximum number of wire elements entering b in a legal embedding can vary from κ to 2κ depending on the topology of wires. Our next lemma gives a full characterization of this combinatorial fact.

Let $x_i = f_i - g_i$ denote the net flow of wires into b through the edge e_i. The edge e_i is called a *surplus edge* if $x_i > 0$, a *deficit edge* if $x_i < 0$, and neutral otherwise. We define the *type* of a square b depending on the number of surplus edges; see Figure 2, where (a), (b), (c) illustrate types I, II, and III.

Type I: b has exactly one surplus edge or exactly one deficit edge.
Type II: b has two surplus edges sharing a common vertex.
Type III: b has two surplus edges that are opposite (nonadjacent) edges of b.

Lemma 1. *The maximum wire inflow into a square core b satisfies the following bounds:*

$$\sum_{i=1}^{4} \max(f_i - g_i, 0) \leq \begin{cases} \kappa & \text{if } b \text{ is Type I or II} \\ 2\kappa & \text{if } b \text{ is Type III and } \kappa \text{ is even} \\ 2\kappa - 1 & \text{if } b \text{ is Type III and } \kappa \text{ is odd.} \end{cases}$$

These bounds are tight.

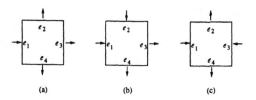

Fig. 2. Three types of cells: (a) Type I; (b) Type II; (c) Type III.

3 A Network Flow Formulation

In Subsection 3.1, we introduce the required network flow concepts; and in Subsection 3.2, we describe the network used to model our problem and show that a matching between pins and vias always corresponds to a valid flow in the network.

3.1 Network Flow Preliminaries

We begin by recalling the definition of a flow in a network. Let $G = (V, E)$ be a directed graph, with a positive capacity $cap(u, v)$ for each edge $(u, v) \in E$. Assume $cap(u, v) = 0$ for $(u, v) \notin E$. Each node $u \in V$ has an integer $\sigma(u)$ representing its supply or demand. The node u is a *supply* node if $\sigma(u) > 0$, a *demand node* if $\sigma(u) < 0$, and *transshipment* node otherwise. We assume that total supply equals total demand, that is, $\sum_u \sigma(u) = 0$. A *flow* on G is a real-valued function ϕ on vertex pairs satisfying the following: (**Anti-symmetry**) $\phi(u, v) = -\phi(v, u)$, (**Capacity Constraint**) $\phi(u, v) \leq cap(u, v)$, and (**Flow Conservation**) $\sum_v \phi(u, v) - \sum_v \phi(v, u) = \sigma(u)$, for all $u \in V$.

The value of the flow is the sum of $\sigma(u)$ over all supply nodes. In some cases, each node u also has a nonnegative *capacity* $cap(u)$, denoting the maximum flow leaving u. In this case, a flow satisfies the following additional constraint:

Node Capacity: $\sum_v \phi(u, v) \leq cap(u)$.

A *cut* in a graph $G = (V, E)$ is a partition of the nodes V into two nonempty sets V_1 and V_2. We denote this cut by $[V_1, V_2]$. Given a cut $[V_1, V_2]$, an edge $(u_1, u_2) \in E$ is called a *forward* edge if $u_1 \in V_1$ and $u_2 \in V_2$. Similarly, (u_1, u_2) is a *backward* edge if $u_1 \in V_2$ and $u_2 \in V_1$. The *capacity* of a cut $[V_1, V_2]$ equals the total capacity of all forward edges. The following celebrated theorem of network flow will be used throughout [2, p. 593]

Theorem 2 (Max-Flow Min-Cut). *Let V_s and V_d, respectively, denote the set of supply and demand nodes in a network $G = (V, E)$. Then a flow satisfying the supplies and demands of V_s and V_d exists if and only if the total supply does not exceed the minimum capacity of a cut $[V_1, V_2]$ where $V_s \subseteq V_1$ and $V_d \subseteq V_2$.*

Let ϕ be a flow in $\mathcal{N}$. We say that ϕ is *canonical* if for each pair of nodes (u, v), the flow value $\phi(u, v)$ is integral. The following lemma is well-known in network flow theory; a constructive proof of this fact follows easily from the correctness of the classical Ford-Fulkerson flow augmentation algorithm.

Lemma 3. *For every integer-valued flow ϕ, there exists a canonical flow of the same value.*

3.2 Network Model for the Via Grid

In our construction, the following *diamond network* will play a crucial role; see Figure 3. The four nodes on the outer cycle are labeled a_1, a_2, a_3, a_4, joined by edges of bidirectional capacity $\kappa_1 = \lfloor \kappa/2 \rfloor$; whenever an edge is drawn undirected, it should be interpreted as having the same capacity in both directions. In addition, there is a central node a_0, joined to the four other nodes by edges of bidirectional capacity $\kappa_2 = \kappa - 2\lfloor \kappa/2 \rfloor$. The central node a_0 also has a node capacity constraint $\sigma(a_0) = 1$; other nodes have no capacity constraints. Notice that $\kappa - 2\lfloor \kappa/2 \rfloor = 0$ if κ is even, and so the middle portion of the network exists only for odd values of κ. In our discussion, a_0 always is a transshipment node, while others can be either supply or demand

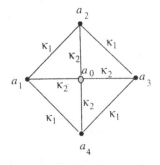

Fig. 3. The diamond network corresponding to the core of a cell.

nodes. The following lemma characterizes the admissible flow based on which nodes on the outer cycle a_1, a_2, a_3, a_4 are supply nodes.

Lemma 4. *The diamond network admits a flow for each of the following cases.*

1. *Exactly one node on the cycle, say a_1, is a supply node with $\sigma(a_1) \leq \kappa$.*
2. *Two adjacent nodes on the cycle, say a_1 and a_2, are supply nodes with $\sigma(a_1) + \sigma(a_2) \leq \kappa$.*
3. *Two opposite nodes on the cycle, say a_1 and a_3, are supply nodes with $\sigma(a_1) + \sigma(a_3) \leq 2\kappa - \kappa_2$.*
4. *Three nodes on the cycle, say a_1, a_2, and a_3, are supply nodes with $\sigma(a_1) + \sigma(a_2) + \sigma(a_3) \leq \kappa$.*

A via cell is modeled using the network shown in Figure 4. We call this a *network element*. The nodes v_1, v_2, v_3, v_4 represent the four via sites; the central diamond represents the core b of the cell; and the nodes u_1, u_2, u_3, u_4 act as connections between the core and the cell boundary. For ease of reference, we will often use the term *connection nodes* for the u_i nodes, *connection edges* for the edges (u_i, a_i), and *boundary edges* for the edges (v_i, u_j).

The via nodes v_i and the middle diamond node a_0 have node capacity 1; the connection nodes u_i have capacity κ each; all other nodes have no capacity constraint. Let $\mathcal{N}$ denote the network corresponding to the entire via grid. (See Figure 5.) We complete the network by adding two artificial nodes s and d, called *supply* and *demand* nodes. We join each via location to d with a unit-capacity edge. We join s to each node on the boundary of the via grid. The capacity of the edge joining s to u is the number of "pins" on the portion of the boundary corresponding to u: if u is a via site, then the capacity is at most 1; otherwise, the capacity is the number of pins lying between the via site neighbors of u. Finally, we set $\sigma(s) = |P|$ and $\sigma(d) = -|P|$. (For example, in Figure 1, there are 9 nodes on the bottom row, of which five are via sites. The capacities of the edges between these nodes and s are, in left to right order, $0, 0, 0, 3, 1, 2, 0, 0, 0$.)

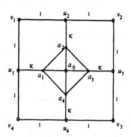

Fig. 4. Network corresponding to a via cell.

Let $\mathcal{M}$ be a matching of the pins and vias, and let ϕ be a flow in $\mathcal{N}$. We say that ϕ is consistent with $\mathcal{M}$ at a via node v_i if there is a unit flow along an edge incident to v_i if and only if there is a wire path in $\mathcal{M}$ incident to v_i along that edge. We also say that ϕ is consistent with a cell c if $\phi(u_i, a_i) = f_i$, and $\phi(a_i, u_i) = g_i$, where f_i and g_i, for $1 \leq i \leq 4$, are the number of wire elements entering and leaving the core of c through the edge e_i. Finally, we say that ϕ is *consistent* with $\mathcal{M}$ if it is consistent with each cell and each via.

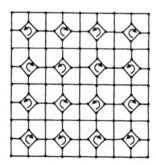

Fig. 5. Illustration for the network $\mathcal{N}$.

Lemma 5. *Given a matching $\mathcal{M}$, there is always a flow in $\mathcal{N}$ consistent with $\mathcal{M}$.*

4   Converting a Flow into a Wire Layout

The preceding section has shown that, given a matching $\mathcal{M}$, we can define a flow consistent with it. We now describe the converse step: to construct a matching consistent with a given flow in $\mathcal{N}$. It turns out that we need to modify the flow values locally, while maintaining the total flow, in order to realize the layout. We call the modified flow *tail-ordered*, and explain it below.

4.1   Cell Orientation

Our method for converting a flow into a matching is a local method: it computes the embedding on a cell-by-cell basis. In order to ensure that the embeddings in different cells are globally compatible in that the wires on common cell boundaries line up properly, we adopt a cyclic ordering for the layout in each cell. Specifically, we orient each cell c either clockwise or counterclockwise, which enforces a direction on each boundary edge of c. If an edge (a, b) is oriented from a to b, then our wiring layout pushes all wires on (a, b) to the b end.

In order to orient the cells consistently, so that the direction enforced on an edge by its two neighboring cells is the same, we assign the orientations in a checkerboard fashion, alternating between clockwise and counterclockwise. The arrows in Figure 5 illustrate this ordering.

Let u be a connection node in a cell c, and let v be the via site at the head of u's edge, as determined by c's orientation. Cell c has a *tail violation* at u if there is positive (or negative) flow from c's core to u and from u to v. The full paper shows that a flow with a tail violation may not be embeddable. A flow is *tail-ordered* if no cell has a tail violation.

Lemma 6. *Given an integer-valued canonical flow ϕ in $\mathcal{N}$, one can always construct an equal-valued flow ϕ' that is tail-ordered.*

4.2   Wire Layout in a Core

Consider the network element corresponding to a cell c. We are interested in ϕ's restriction to the subgraph spanned by c's diamond and the connection nodes—this subgraph corresponds to the core b of the cell c. We will show how to compute a layout consistent with the flow in this subgraph.

First, since ϕ is anti-symmetric, each connection edge (u_i, a_i) has flow in only one direction. Thus, we can use the direction of flow in (u_i, a_i) to label the corresponding core boundary edge. In order to avoid unnecessary new notation, let us use the same symbols for the corresponding edges of b and c. Thus, the boundary of b has e_1, e_2, e_3, e_4 in clockwise order, starting from the left edge. We say that the edge e_i is surplus if $\phi(u_i, a_i) > 0$; it is deficit if $\phi(a_i, u_i) < 0$; and neutral otherwise. The core b can now be classified as Type I, II, or III, depending on which of its edges are surplus, exactly as in Section 2. (It should be clear now that no wiring needs to be done in b if all edges of b are neutral.)

The following three lemmas show how to do the layout in each of these cases, under the further restriction that wiring must be consistent with the cell orientation of c. The core b *inherits* the orientation of its cell c—each of the boundary edges of b has the same direction as the corresponding edge of c. In the layout of b, all wires along an edge are pushed toward the head of the directed edge. We call such a wiring *orientation preserving*. Without loss of generality, we assume in the following that the core is clockwise oriented; the construction is completely symmetric for the other case.

Lemma 7. *Let b be the core of a cell, and suppose that ℓ_1 wires enter b through e_1 and j_i leave through e_i, for $i = 2, 3, 4$, where $\ell_1 = j_2 + j_3 + j_4 \leq \kappa$. Then an orientation preserving wiring exists for b.*

Lemma 8. *Let b be the core of a cell, and suppose that ℓ_1 and ℓ_2 wires enter b through e_1 and e_2, respectively, and j_3 and j_4 wires leave through e_3 and e_4, respectively, where $\ell_1 + \ell_2 = j_3 + j_4 \leq \kappa$. Then an orientation preserving wiring exists for b.*

Lemma 9. *Let b be the core of a cell, and suppose that ℓ_1 and ℓ_3 wires enter b through e_1 and e_3, respectively, and j_2 and j_4 wires leave through e_2 and e_4, respectively, where $\ell_1 + \ell_3 = j_2 + j_4$, and $\ell_1 + \ell_3$ is at most 2κ when κ is even and at most $2\kappa - 1$ otherwise. Then an orientation preserving wiring exists for b.*

4.3 Wiring a Cell

Once the core wiring is complete, the only remaining wires are the ones that correspond to flow along boundary edges of a network element. It suffices to consider a side of a cell, and show how the wires from its two neighboring cores can be augmented with the boundary wires. Due to space limitations, we omit the details.

5 Algorithm and Complexity Analysis

All the pieces of our breakout routing algorithm are in place now, and the algorithm can be described as follows at a high level of modularity:

1. Construct the network $\mathcal{N}$ corresponding to the via grid, with each via location represented by a sink node, and each pin location represented by a source node. Make each sink node a demand node with value one, and make each source node a supply node with value equal to the number of pins it represents. Assign capacities to nodes and edges as described in Section 3.2.
2. Run a flow algorithm to compute a maximum flow ϕ between the source and sink nodes [2].
3. Modify the flow, if necessary, to make it tail-ordered.
4. Use the tail-ordered flow ϕ to compute a wire layout consistent with ϕ in each cell.
5. Post-process to remove unnecessary bends in the wire layout [4, 3].

The time complexity of our method is dominated by the network flow algorithm in Step 2. The value of the maximum flow is the number of pins m, and the size of our network is likely to be linearly related to the number of vias m, and consequently even the simple Ford-Fulkerson flow algorithm will take $O(mn)$ time. In practice, pins are located preferentially on the long sides of the rectangle, and hence the average wire length will tend to be short, Thus, Dinic's algorithm [6, p. 102] will run in roughly $O(m\sqrt{n})$ time in our network. More sophisticated algorithms may perform even better.

6 Future Directions

Power and ground wires are often wider than signal wires; likewise, certain applications require two or more wires to be routed together as if they were a single wire. How can one extend the network model to this case? A second important problem is to model cells that are partially obstructed, and are therefore non-rectangular.

References

1. W.-T. Chan and F. Y. L. Chin. Efficient algorithms for finding disjoint paths in grids. In *Proceedings of ACM-SIAM Symposium on Discrete Algorithms*, pages 454–463, 1997.
2. T. H. Cormen, C. E. Leiserson, and R. L. Rivest. *Introduction to Algorithms*. The MIT Press, Cambridge, Mass., 1990.
3. W. W.-M. Dai, R. Kong, and M. Sato. Routability of a rubber-band sketch. In *Proceedings of 28th IEEE/ACM Design Automation Conf.*, pages 45–48, 1991.
4. F. M. Maley. *Single-Layer Wire Routing and Compaction*. The MIT Press, Cambridge, Mass., 1990.
5. V. P. Roychowdhury, J. Bruck, and T. Kailath. Efficient algorithms for reconfiguration in VLSI/WSI arrays. *IEEE Transactions on Computers*, pages 480–489, 1990.
6. R. E. Tarjan. *Data Structures and Network Algorithms*, volume 44 of *CBMS-NSF Regional Conference Series in Applied Mathematics*. Society for Industrial Applied Mathematics, 1983.
7. M.-F. Yu and W. W.-M. Dai. Single-layer fanout routing and routability analysis for ball grid arrays. In *Proceedings of IEEE/ACM International Conf. on CAD*, pages 581–586, 1995.
8. M.-F. Yu, J. Darnauer, and W. W.-M. Dai. Interchangeable pin routing with application to package layout. In *Proceedings of IEEE/ACM International Conf. on CAD*, pages 668–673, 1996.

Planarity, Revisited (Extended Abstract)

Zhi-Zhong Chen*, Michelangelo Grigni**, Christos H. Papadimitriou***

Planarity, the property of certain graphs to be embeddable in the plane without crossings of edges, is one of the most fundamental, ancient, and well-studied aspects of graph theory. The characterization of planar graphs in terms of the forbidden minors K_5 and $K_{3,3}$ due to Kuratowski is well-known, the four-color conjecture has arguably been the most influential problem in graph theory, and planar graphs play an important role in the recent work by Robertson and Seymour. Furthermore, the development of efficient planarity tests has motivated some fundamental algorithmic techniques such as depth-first search, PQ-trees, and ear decomposition.

One reason why planar graphs are so interesting is because they are the duals of political maps. That is, planar graphs are precisely the intersection graphs of closed planar regions without holes (homeomorphs of discs) with disjoint iteriors, *as long as no four regions intersect at the same point.* But what happens when the emphasized restriction is relaxed? Astonishingly, we could find very little previous work on this important class of graphs, which we call *planar map graphs.* The plausible explanation for this absence of attention is that the theory of this variant of planar graphs is very complex in comparison. For example, at present we do not know of a polynomial algorithm for recognizing planar map graphs —and establishing that the problem is in NP is not totally trivial.

Membership in NP is a rather straightforward consequence of the following characterization of planar map graphs (by *left square* of a bipartite graph we mean the square of the bipartite graph projected to one of its two sets of nodes):

Theorem 1. *A graph is a planar map graph if and only if it is the left square of a planar bipartite graph.*

Notice that the intersection graph of the fourty-eight contiguous states of the United States is an interesting example of a planar map graph which happens to be non-planar! Which brings us to an interesting further variant: A realization of a planar map graph may contain "lakes," that is, the union of the regions may have holes. If we disallow this, an interesting class that may be called *strict planar map graphs* results. This class has a similar characterization, except that in the planar bipartite graph all faces must be of length either four or six.

Perhaps the most striking difference between planar map graphs and traditional planar graphs is that the former may contain arbitrarily large cliques —for

---

* Computer Science Division, University of California at Berkeley.

** Department of Mathematics and Computer Science, Emory University.

*** Computer Science Division, University of California at Berkeley. Supported in part by the National Science Foundation.

example, when many countries meet at the same point, like slices of a pizza. If this were the only way in which large cliques are realizable as planar maps, the recognition problem would be amenable to the following algorithm:

Let G' be a graph with the same nodes as G,
and initially with no edges.
For each maximal clique C of G do
   add to G' a new node v_C, and edges from v_C to each node of C
If there are more than $4n$ maximal cliques of G, then stop and output no
If G' is planar, then output yes else output no

That this algorithm runs in polynomial time relies on the facts that the maximal cliques of a graph can be enumerated in time polynomial in the size of the graph *and* the number of maximal cliques of the graph, and that planar map graphs can have no more than $4n$ maximal cliques.

Unfortunately, the problem with this algorithm is that not all cliques are pizzas! In fact, there are four possible distinct realizations of cliques: *pizza, pizza with crust* (a pizza with an enveloping region that touches all of its slices), *hamantaschen*, ("long" regions divided into three sets, each set with two common endpoints, and joined in a triangle) and *riceballs* (the planar relization of K_4). It turns out that these are all possible realizations of cliques, a fact that can be proved starting from the characterization above and asking, which planar bipartite graphs have a left square that is a clique? Four families result, corresponding to the above mouth-watering realizations.

It is open whether this restriction of the possibilities for realizing a clique leads to a polynomial recognition algorithm for planar map graphs, by some appropriate modification of the above scheme.

However, we have recently discovered a polynomial-time algorithm for recognizing *4-planar map graphs*, that is, intersection graphs of maps in which no more than 4 countries meet at a point —equivalently, left squares of planar bipartite graphs with maximum degree of a right-side node four. Notice that 3-planar map graphs are the ordinary planar graphs, and that the United States is a 4-planar map graph.

Theorem 2. *There is a polynomial-time algorithm for recognizing 4-planar map graphs.*

This proof (forthcoming) is a modification of the above scheme, and relies on a way of classifying a maximal clique depending on its local context (intersecting cliques, and components of the graph with the clique deleted). One of the major advantages of the 4-planar map graph special case is that there are no K_7s in the graph, and all K_6s must be hamantaschens. We conjecture that planar map graphs can also be recognized in polynomial time.

Author Index

Lecture Notes in Computer Science

For information about Vols. 1–1192

please contact your bookseller or Springer-Verlag